Design and Simulation of Two-Stroke Engines

Gordon P. Blair
Professor of Mechanical Engineering
The Queen's University of Belfast

Society of Automotive Engineers, Inc.
Warrendale, Pa.

> **Library of Congress Cataloging-in-Publication Data**
>
> Blair, Gordon P.
>   Design and simulation of two-stroke engines / Gordon P. Blair.
>   p. cm.
>   Includes bibliographical references and index.
>   ISBN 1-56091-685-0
>   1. Two-stroke cycle engines--Design and construction. I. Title
> TJ790.B58 1996
> 621.43--dc20   95-25748
>   CIP

Copyright © 1996 Society of Automotive Engineers, Inc.
            400 Commonwealth Drive
            Warrendale, PA 15096-0001
            Phone: (724)776-4841; Fax: (724)776-5760
            http://www.sae.org

ISBN 1-56091-685-0

All rights reserved. Printed in the United States of America.

Cover photo courtesy of Mercury Marine, a Brunswick Company (3.0-liter V-6 outboard engine).

Permission to photocopy for internal or personal use, or the internal or personal use of specific clients, is granted by SAE for libraries and other users registered with the Copyright Clearance Center (CCC), provided that the base fee of $.50 per page is paid directly to CCC, 222 Rosewood Dr., Danvers, MA 01923. Special requests should be addressed to the SAE Publications Group. 1-56091-685-0/96 $.50.

SAE Order No. R-161

A Second Mulled Toast

When as a student a long time ago
my books gave no theory glimmers,
why two-strokes ended in second place slow,
and four-strokes were always the winners.

Williams and Craig were heroes enough
whose singles thumped to Tornagrough,
such as black 7R or silver Manx,
on open megas they enthused the cranks.

Wallace and Bannister gave me the start
into an unsteady gas dynamic art,
where lambdas and betas meshed in toil
for thirty years consumed midnight oil.

With the parrot on Bush a mental penny
into slot in brain fell quite uncanny.
Lubrication of grey cells finally gave
an alternative way to follow a wave.

That student curiosity is sated today
and many would describe that as winning.
Is this then the end of the way?
No, learning is aye a beginning.

Gordon Blair
July 1994

Foreword

Several years ago I wrote a book, *The Basic Design of Two-Stroke Engines*. It was no sooner published than a veritable revolution took place in several areas of two-stroke design and development. Thus, some time ago, I settled to update that book to a Second Edition. It soon became very obvious that the majority of the material was so different, and the changes so extensive, that to label the book as simply a "second edition" was not only unreasonable but would be both inaccurate and misleading. Nevertheless, the basic premise for providing the book had not changed so the original Foreword is still germane to the issue and is produced below, virtually in its entirety. The fundamental approach is no different the second time around, simply that the material is much more detailed and much more extensive. So here it is, with a postscript added:

This book is intended to be an information source for those who are involved in the design of two-stroke engines. In particular, it is a book for those who are already somewhat knowledgeable on the subject, but who have sometimes found themselves with a narrow perspective on the subject, perhaps due to specialization in one branch of the industry. For example, I am familiar with many who are expert in tuning racing motorcycle engines, but who would freely admit to being quite unable to design for good fuel economy or emission characteristics, should their industry demand it. It is my experience that the literature on the spark-ignition two-stroke engine is rich in descriptive material but is rather sparse in those areas where a designer needs specific guidance. As the two-stroke engine is currently under scrutiny as a future automobile engine, this book will help to reorient the thoughts of those who are more expert in designing camshafts than scavenge ports. Also, this book is intended as a textbook on design for university students in the latter stages of their undergraduate studies, or for those undertaking postgraduate research or course study.

Perhaps more important, the book is a design aid in the areas of gas dynamics, fluid mechanics, thermodynamics and combustion. To stop you from instantly putting the book down in terror at this point, rest assured that the whole purpose of this book is to provide design assistance with the actual mechanical design of the engine, in which the gas dynamics, fluid mechanics, thermodynamics and combustion have been optimized so as to provide the required performance characteristics of power or torque or fuel consumption.

Therefore, the book will attempt to explain, inasmuch as I understand, the intricacies of, for example, scavenging, and then provide you with computer programs written in Basic which will assist with the mechanical design to produce, to use the same example, better scavenging in any engine design. These are the very programs which I have written as my own mechanical design tools, as I spend a large fraction of my time designing engines for test at The Queen's University of Belfast (QUB) or for prototype or production development in industry. Many of the design programs which have been developed at QUB over the last twenty-five years have become so complex, or require such detailed input data, that the operator cannot see the design wood for the data trees. In consequence, these simpler, often empirical, programs have been developed to guide me as to the data set before applying a complex unsteady gas-dynamic or computational fluid dynamic analysis package. On many occasions that complex package merely confirms that the empirical program, containing as it does the distilled experience of several generations, was sufficiently correct in the first place.

At the same time, as understanding unsteady gas dynamics is the first major step to becoming a competent designer of reciprocating IC engines, the book contains a major section dealing with that subject and you are provided with an engine design program of the complete pressure wave motion form, which is clearly not an empirical analytical package.

The majority of the book is devoted to the design of the two-stroke spark-ignition (SI) engine, but there will be remarks passed from time to time regarding the two-stroke diesel or compression-ignition (CI) engine; these remarks will be clearly identified as such. The totality of the book is just as applicable to the design of the diesel as it is to the gasoline engine, for the only real difference is the methodology of the combustion process.

I hope that you derive as much use from the analytic packages as do I. I have always been somewhat lazy of mind and so have found the accurate repetitive nature of the computer solution to be a great saviour of mental perspiration. At the same time, and since my schooldays, I have been fascinated with the two-stroke cycle engine and its development and improvement. In those far-off days in the late 1950s, the racing two-stroke motorcycle was a music-hall joke, whereas a two-stroke-engined car won the Monte Carlo Rally. Today, there are no two-stroke-engined cars and four-stroke engines are no longer competitive in Grand Prix motorcycle racing! Is tomorrow, or the 21st Century, going to produce yet another volte-face?

I have also had the inestimable privilege of being around at precisely that point in history when it became possible to unravel the technology of engine design from the unscientific black art which had surrounded it since the time of Otto and Clerk. That unravelling occurred because the digital computer permitted the programming of the fundamental unsteady gas-dynamic theory which had been in existence since the time of Rayleigh, Kelvin, Stokes and Taylor.

The marriage of these two interests, computers and two-stroke engines, has produced this book and the material within it. For those in this world who are of a like mind, this book should prove to be useful.

Postscript

The above was the original Foreword, but the changes have been so great that, as explained before, this is a new book. The original book had the computer program listings at the back of it, occupying some 270 pages. This book is already larger without them than the original book was with them! However, the computer programs have been extended in number and are available from SAE on diskette as applications for either the IBM® PC or Macintosh® platforms.

Foreword

    The changes to the fundamental chapters on unsteady gas flow, scavenging, combustion and emissions, and noise are very extensive. The new material is synthesized to illustrate engine design by simulation through modeling. These enhanced modeling methods are the product of the last five years of activity. They have been the busiest, and perhaps the most satisfying, years I have known in my career. I find it very hard to come to terms with the irrefutable, namely that I have made more progress in the thermo-fluids design area for reciprocating engines in general, and two-stroke engines in particular, than in the previous twenty-five years put together!

    I hope that you will agree.

Gordon P. Blair
14 April 1995

Acknowledgments

As explained in the Foreword, this is a new book, but the acknowledgments in that first book are just as pertinent today as they were then. So here they are, with an important postscript added:

The first acknowledgment is to those who enthused me during my schooldays on the subject of internal-combustion engines in general, and motorcycles in particular. They set me on the road to a thoroughly satisfying research career which has never seen a hint of boredom. The two individuals were my father, who had enthusiastically owned many motorcycles in his youth, and Mr. Rupert Cameron, who had owned but one and had ridden it everywhere—a 1925 350 cc Rover. Of the two, Rupert Cameron was the greater influence, for he was a walking library of the Grand Prix races of the '20s and '30s and would talk of engine design, and engineering design, in the most knowledgeable manner. He was actually the senior naval architect at Harland and Wolff's shipyard in Belfast and was responsible for the design of some of the grandest liners ever to sail the oceans.

I have to acknowledge that this book would not be written today but for the good fortune that brought Dr. Frank Wallace (Professor at Bath University since 1965) to Belfast in the very year that I wished to do postgraduate research. At that time, Frank Wallace was one of perhaps a dozen people in the world who comprehended unsteady gas dynamics, which was the subject area I already knew I had to understand if I was ever to be a competent engine designer. However, Frank Wallace taught me something else as well by example, and that is academic integrity. Others will judge how well I learned either lesson.

Professor Bernard Crossland deserves a special mention, for he became the Head of the Department of Mechanical Engineering at QUB in the same year I started as a doctoral research student. His drive and initiative set the tone for the <u>engineering</u> research which has continued at QUB until the present day. The word engineering in the previous sentence is underlined because he instilled in me, and a complete generation, that real "know how" comes from using the best theoretical science available, at the same time as conducting related experiments of a product design, manufacture, build and test nature. That he became, in latter years, a Fellow of the Royal Society, a Fellow of the Fellowship of Engineering and a President of the Institution of Mechanical Engineers *(PS...and subsequently knighted..)* seems no more than justice.

I have been very fortunate in my early education to have had teachers of mathematics who taught me the subject not only with enthusiasm but, much more importantly, from the point of view of application. I refer particularly to Mr. T.H. Benson at Larne Grammar School and to Mr. Scott during my undergraduate studies at The Queen's University of Belfast. They gave me a lifelong interest in the application of mathematics to problem solving which has never faded.

The next acknowledgment is to those who conceived and produced the Macintosh® computer. Without that machine, on which I have typed this entire manuscript, drawn every figure which is not from SAE archives, and developed all of the computer programs, there would be no book. In short, the entire book, and the theoretical base for much of it, is there because the Macintosh has such superbly integrated hardware and software so that huge workloads can be tackled rapidly and efficiently.

Postscript

The influence of Frank Wallace (and Professor Bannister) turned out to be even more profound than I had realized, for it was a re-examination of their approach to unsteady gas dynamics which led me to produce the unsteady gas dynamic simulation technique described herein.

I wish to acknowledge the collaboration of: Dr. Sam Kirkpatrick, in the correlation work on the QUB SP rig which ultimately sophisticated the theoretical simulation of unsteady gas flow; Charles McCartan, without whose elegant software for tackling multiple polynomial equations much of the theory would be insoluble; Dr. Brendan Carberry, in the model for the formation of the exhaust emissions of nitric oxide; Dr. John Magee, Dr. Sam Kirkpatrick and Dermot Mackey, in the investigation of unsteady gas flow in tapered pipes.

Proofreading of this text was provided by the post-doctoral and doctoral assistants at QUB and many invaluable comments and criticisms came from that quarter, principally from Brendan Carberry and John Magee, but also from Dermot Mackey and Barry Raghunathan.

David Holland, a QUB engineering technician, requires a special mention for the expert production of many of the photographs which illustrate this book.

Gordon P. Blair
The Queen's University of Belfast
14 April 1995

Table of Contents

Nomenclature xv

Chapter 1 Introduction to the Two-Stroke Engine 1
   1.0   Introduction to the two-stroke cycle engine 1
   1.1   The fundamental method of operation of a simple two-stroke engine 6
   1.2   Methods of scavenging the cylinder 8
       1.2.1   Loop scavenging 8
       1.2.2   Cross scavenging 10
       1.2.3   Uniflow scavenging 11
       1.2.4   Scavenging not employing the crankcase as an air pump 12
   1.3   Valving and porting control of the exhaust, scavenge and inlet processes ... 15
       1.3.1   Poppet valves 16
       1.3.2   Disc valves 16
       1.3.3   Reed valves 17
       1.3.4   Port timing events 18
   1.4   Engine and porting geometry 20
       1.4.1   Swept volume 21
       1.4.2   Compression ratio 22
       1.4.3   Piston position with respect to crankshaft angle 22
       1.4.4   Computer program, Prog.1.1, PISTON POSITION 23
       1.4.5   Computer program, Prog.1.2, LOOP ENGINE DRAW 23
       1.4.6   Computer program, Prog.1.3, QUB CROSS ENGINE DRAW ... 25
   1.5   Definitions of thermodynamic terms used in connection with engine
       design and testing 26
       1.5.1   Scavenge ratio and delivery ratio 26
       1.5.2   Scavenging efficiency and purity 28
       1.5.3   Trapping efficiency 28
       1.5.4   Charging efficiency 29
       1.5.5   Air-to-fuel ratio 29
       1.5.6   Cylinder trapping conditions 30
       1.5.7   Heat released during the burning process 31
       1.5.8   The thermodynamic cycle for the two-stroke engine 31
       1.5.9   The concept of mean effective pressure 34
       1.5.10 Power and torque and fuel consumption 34

1.6 Laboratory testing of two-stroke engines 35
    1.6.1 Laboratory testing for power, torque, mean effective pressure and specific fuel consumption 35
    1.6.2 Laboratory testing for exhaust emissions from two-stroke engines 38
    1.6.3 Trapping efficiency from exhaust gas analysis 41
1.7 Potential power output of two-stroke engines 43
    1.7.1 Influence of piston speed on the engine rate of rotation 44
    1.7.2 Influence of engine type on power output 45
Subscript notation for Chapter 1 46
References for Chapter 1 47

Chapter 2 Gas Flow through Two-Stroke Engines 49
2.0 Introduction 49
2.1 Motion of pressure waves in a pipe 52
    2.1.1 Nomenclature for pressure waves 52
    2.1.2 Propagation velocities of acoustic pressure waves 54
    2.1.3 Propagation and particle velocities of finite amplitude waves 55
    2.1.4 Propagation and particle velocities of finite amplitude waves in air 58
    2.1.5 Distortion of the wave profile 62
    2.1.6 The properties of gases 64
2.2 Motion of oppositely moving pressure waves in a pipe 69
    2.2.1 Superposition of oppositely moving waves 69
    2.2.2 Wave propagation during superposition 72
    2.2.3 Mass flow rate during wave superposition 73
    2.2.4 Supersonic particle velocity during wave superposition 74
2.3 Friction loss and friction heating during pressure wave propagation 77
    2.3.1 Friction factor during pressure wave propagation 81
    2.3.2 Friction loss during pressure wave propagation in bends in pipes 83
2.4 Heat transfer during pressure wave propagation 84
2.5 Wave reflections at discontinuities in gas properties 85
2.6 Reflection of pressure waves 88
    2.6.1 Notation for reflection and transmission of pressure waves in pipes 90
2.7 Reflection of a pressure wave at a closed end in a pipe 91
2.8 Reflection of a pressure wave at an open end in a pipe 92
    2.8.1 Reflection of a compression wave at an open end in a pipe 92
    2.8.2 Reflection of an expansion wave at a bellmouth open end in a pipe 93
    2.8.3 Reflection of an expansion wave at a plain open end in a pipe 95
2.9 An introduction to reflection of pressure waves at a sudden area change 97
2.10 Reflection of pressure waves at an expansion in pipe area 101

	2.10.1	Flow at pipe expansions where sonic particle velocity is encountered 104
2.11	Reflection of pressure waves at a contraction in pipe area 105	
	2.11.1	Flow at pipe contractions where sonic particle velocity is encountered 107
2.12	Reflection of waves at a restriction between differing pipe areas 108	
	2.12.1	Flow at pipe restrictions where sonic particle velocity is encountered 112
	2.12.2	Examples of flow at pipe expansions, contractions and restrictions 113
2.13	An introduction to reflections of pressure waves at branches in a pipe 114	
2.14	The complete solution of reflections of pressure waves at pipe branches 117	
	2.14.1	The accuracy of simple and more complex branched pipe theories 122
2.15	Reflection of pressure waves in tapered pipes 124	
	2.15.1	Separation of the flow from the walls of a diffuser 126
2.16	Reflection of pressure waves in pipes for outflow from a cylinder 127	
	2.16.1	Outflow from a cylinder where sonic particle velocity is encountered 132
	2.16.2	Numerical examples of outflow from a cylinder 133
2.17	Reflection of pressure waves in pipes for inflow to a cylinder 135	
	2.17.1	Inflow to a cylinder where sonic particle velocity is encountered 139
	2.17.2	Numerical examples of inflow into a cylinder 140
2.18	The simulation of engines by the computation of unsteady gas flow 142	
	2.18.1	The basis of the GPB computation model 144
	2.18.2	Selecting the time increment for each step of the calculation 146
	2.18.3	The wave transmission during the time increment, dt 147
	2.18.4	The interpolation procedure for wave transmission through a mesh 147
	2.18.5	Singularities during the interpolation procedure 150
	2.18.6	Changes due to friction and heat transfer during a computation step 151
	2.18.7	Wave reflections at the inter-mesh boundaries after a time step 151
	2.18.8	Wave reflections at the ends of a pipe after a time step 154
	2.18.9	Mass and energy transport along the duct during a time step 156
	2.18.10	The thermodynamics of cylinders and plenums during a time step 162
	2.18.11	Air flow, work, and heat transfer during the modeling process 166
	2.18.12	The modeling of engines using the GPB finite system method 170
2.19	The correlation of the GPB finite system simulation with experiments 170	
	2.19.1	The QUB SP (single pulse) unsteady gas flow experimental apparatus 170
	2.19.2	A straight parallel pipe attached to the QUB SP apparatus 173
	2.19.3	A sudden expansion attached to the QUB SP apparatus 177

	2.19.4	A sudden contraction attached to the QUB SP apparatus 179
	2.19.5	A divergent tapered pipe attached to the QUB SP apparatus 181
	2.19.6	A convergent tapered pipe attached to the QUB SP apparatus 183
	2.19.7	A longer divergent tapered pipe attached to the QUB SP apparatus 185
	2.19.8	A pipe with a gas discontinuity attached to the QUB SP apparatus 187
2.20	Computation time 191
2.21	Concluding remarks 192

References for Chapter 2 193
Appendix A2.1 The derivation of the particle velocity for unsteady gas flow 197
Appendix A2.2 Moving shock waves in unsteady gas flow 201
Appendix A2.3 Coefficients of discharge in unsteady gas flow 205

Chapter 3 Scavenging the Two-Stroke Engine 211

- 3.0 Introduction 211
- 3.1 Fundamental theory 211
  - 3.1.1 Perfect displacement scavenging 213
  - 3.1.2 Perfect mixing scavenging 214
  - 3.1.3 Combinations of perfect mixing and perfect displacement scavenging 215
  - 3.1.4 Inclusion of short-circuiting of scavenge air flow in theoretical models 216
  - 3.1.5 The application of simple theoretical scavenging models 216
- 3.2 Experimentation in scavenging flow 219
  - 3.2.1 The Jante experimental method of scavenge flow assessment 219
  - 3.2.2 Principles for successful experimental simulation of scavenging flow 223
  - 3.2.3 Absolute test methods for the determination of scavenging efficiency 224
  - 3.2.4 Comparison of loop, cross and uniflow scavenging 227
- 3.3 Comparison of experiment and theory of scavenging flow 233
  - 3.3.1 Analysis of experiments on the QUB single-cylinder gas scavenging rig 233
  - 3.3.2 A simple theoretical scavenging model which correlates with experiments 237
  - 3.3.3 Connecting a volumetric scavenging model with engine simulation 241
  - 3.3.4 Determining the exit properties by mass 242
- 3.4 Computational fluid dynamics 244
- 3.5 Scavenge port design 250
  - 3.5.1 Uniflow scavenging 250
  - 3.5.2 Conventional cross scavenging 253
  - 3.5.3 Unconventional cross scavenging 257

		3.5.3.1 The use of Prog.3.3(a) GPB CROSS PORTS	259

- 3.5.4 QUB type cross scavenging 261
  - 3.5.4.1 The use of Prog.3.3(b) QUB CROSS PORTS 261
- 3.5.5 Loop scavenging 263
  - 3.5.5.1 The main transfer port 265
  - 3.5.5.2 Rear ports and radial side ports 266
  - 3.5.5.3 Side ports 266
  - 3.5.5.4 Inner wall of the transfer ports 266
  - 3.5.5.5 Effect of bore-to-stroke ratio on loop scavenging 267
  - 3.5.5.6 Effect of cylinder size on loop scavenging 267
  - 3.5.5.7 The use of Prog.3.4, LOOP SCAVENGE DESIGN 268
- 3.5.6 Loop scavenging design for external scavenging 269
  - 3.5.6.1 The use of Prog.3.5 BLOWN PORTS 270
- 3.6 Scavenging design and development 273
- References for Chapter 3 276

Chapter 4 Combustion in Two-Stroke Engines 281
- 4.0 Introduction 281
- 4.1 The spark-ignition engine 282
  - 4.1.1 Initiation of ignition 282
  - 4.1.2 Air-fuel mixture limits for flammability 284
  - 4.1.3 Effect of scavenging efficiency on flammability 285
  - 4.1.4 Detonation or abnormal combustion 285
  - 4.1.5 Homogeneous and stratified combustion 286
  - 4.1.6 Compression ignition 288
- 4.2 Heat released by combustion 289
  - 4.2.1 The combustion chamber 289
  - 4.2.2 Heat release prediction from cylinder pressure diagram 289
  - 4.2.3 Heat release from a two-stroke loop-scavenged engine 294
  - 4.2.4 Combustion efficiency 294
- 4.3 Heat availability and heat transfer during the closed cycle 296
  - 4.3.1 Properties of fuels 296
  - 4.3.2 Properties of exhaust gas and combustion products 297
    - 4.3.2.1 Stoichiometry and equivalence ratio 298
    - 4.3.2.2 Rich mixture combustion 299
    - 4.3.2.3 Lean mixture combustion 300
    - 4.3.2.4 Effects of dissociation 301
    - 4.3.2.5 The relationship between combustion and exhaust emissions 302
  - 4.3.3 Heat availability during the closed cycle 303
  - 4.3.4 Heat transfer during the closed cycle 305
  - 4.3.5 Internal heat loss by fuel vaporization 308
  - 4.3.6 Heat release data for spark-ignition engines 309
  - 4.3.7 Heat release data for compression-ignition engines 314

		4.3.7.1 The direct injection diesel (DI) engine 314
		4.3.7.2 The indirect injection diesel (IDI) engine 316
4.4	Modeling the closed cycle theoretically 318
	4.4.1	A simple closed cycle model within engine simulations 318
	4.4.2	A closed cycle model within engine simulations 319
	4.4.3	A one-dimensional model of flame propagation in spark-ignition engines 322
4.4.4	Three-dimensional combustion model for spark-ignition engines 323
4.5	Squish behavior in two-stroke engines 325
	4.5.1	A simple theoretical analysis of squish velocity 325
	4.5.2	Evaluation of squish velocity by computer 330
	4.5.3	Design of combustion chambers to include squish effects 331
4.6	Design of combustion chambers with the required clearance volume 334
4.7	Some general views on combustion chambers for particular applications 336
	4.7.1	Stratified charge combustion 337
	4.7.2	Homogeneous charge combustion 338
References for Chapter 4 339
Appendix A4.1 Exhaust emissions 343
Appendix A4.2 A simple two-zone combustion model 347

Chapter 5 Computer Modeling of Engines 357
    5.0   Introduction 357
    5.1   Structure of a computer model 358
    5.2   Physical geometry required for an engine model 359
        5.2.1   The porting of the cylinder controlled by the piston motion 359
        5.2.2   The porting of the cylinder controlled externally 363
        5.2.3   The intake ducting 370
        5.2.4   The exhaust ducting 371
    5.3   Heat transfer within the crankcase 375
    5.4   Mechanical friction losses of two-stroke engines 378
    5.5   The thermodynamic and gas-dynamic engine simulation 379
        5.5.1   The simulation of a chainsaw 380
        5.5.2   The simulation of a racing motorcycle engine 394
        5.5.3   The simulation of a multi-cylinder engine 402
    5.6   Concluding remarks 409
References for Chapter 5 410
Appendix A5.1 The flow areas through poppet valves 412

Chapter 6 Empirical Assistance for the Designer 415
    6.0   Introduction 415
    6.1   Design of engine porting to meet a given performance characteristic 416
        6.1.1   Specific time areas of ports in two-stroke engines 417
        6.1.2   The determination of specific time area of engine porting 424

		6.1.3	The effect of changes of specific time area in a chainsaw 426
	6.2	Some practical considerations in the design process 431
		6.2.1	The acquisition of the basic engine dimensions 431
		6.2.2	The width criteria for the porting 432
		6.2.3	The port timing criteria for the engine 434
		6.2.4	Empiricism in general 434
		6.2.5	The selection of the exhaust system dimensions 435
		6.2.6	Concluding remarks on data selection 445
	6.3	Empirical design of reed valves for two-stroke engines 446
		6.3.1	The empirical design of reed valve induction systems 447
		6.3.2	The use of specific time area information in reed valve design 450
		6.3.3	The design process programmed into a package, Prog.6.4 454
		6.3.4	Concluding remarks on reed valve design 455
	6.4	Empirical design of disc valves for two-stroke engines 456
		6.4.1	Specific time area analysis of disc valve systems 456
		6.4.2	A computer solution for disc valve design, Prog.6.5 459
	6.5	Concluding remarks 460
	References for Chapter 6 461

Chapter 7 Reduction of Fuel Consumption and Exhaust Emissions 463
    7.0 Introduction 463
    7.1 Some fundamentals of combustion and emissions 465
        7.1.1 Homogeneous and stratified combustion and charging 466
    7.2 The simple two-stroke engine 469
        7.2.1 Typical performance characteristics of simple engines 471
            7.2.1.1 Measured performance data from QUB 400 research engine 472
            7.2.1.2 Typical performance maps for simple two-stroke engines 476
    7.3 Optimizing fuel economy and emissions for the simple two-stroke engine 483
        7.3.1 The effect of scavenging on performance and emissions 484
        7.3.2 The effect of air-fuel ratio 486
        7.3.3 The effect of optimization at a reduced delivery ratio 486
        7.3.4 The optimization of combustion 490
        7.3.5 Conclusions regarding the simple two-stroke engine 492
    7.4 The more complex two-stroke engine 494
        7.4.1 Stratified charging with homogeneous combustion 497
        7.4.2 Homogeneous charging with stratified combustion 512
    7.5 Compression-ignition engines 531
    7.6 Concluding comments 531
    References for Chapter 7 532

Appendix A7.1 The effect of compression ratio on performance characteristics and exhaust emissions 536

Chapter 8 Reduction of Noise Emission from Two-Stroke Engines 541
8.0   Introduction 541
8.1   Noise 541
      8.1.1 Transmission of sound 542
      8.1.2 Intensity and loudness of sound 542
      8.1.3 Loudness when there are several sources of sound 544
      8.1.4 Measurement of noise and the noise-frequency spectrum 545
8.2   Noise sources in a simple two-stroke engine 546
8.3   Silencing the exhaust and inlet system 547
8.4   Some fundamentals of silencer design 548
      8.4.1 The theoretical work of Coates 548
      8.4.2 The experimental work of Coates 550
      8.4.3 Future work for the prediction of silencer behavior 554
8.5   Acoustic theory for silencer attenuation characteristics 555
      8.5.1 The diffusing type of exhaust silencer 555
      8.5.2 The side-resonant type of exhaust silencer 560
      8.5.3 The absorption type of exhaust silencer 563
      8.5.4 The laminar flow exhaust silencer 565
      8.5.5 Silencing the intake system 567
      8.5.6 Engine simulation to include the noise characteristics 570
      8.5.7 Shaping the ports to reduce high-frequency noise 577
8.6   Silencing the tuned exhaust system 579
      8.6.1 A design for a silenced expansion chamber exhaust system 580
8.7   Concluding remarks on noise reduction 583
      References for Chapter 8 584

Postscript 587

Appendix Listing of Computer Programs 589

Index 591

Nomenclature

NAME	SYMBOL	UNIT (SI)
Coefficients		
Coefficient of heat transfer, conduction	C_k	W/mK
Coefficient of heat transfer, convection	C_h	W/m²K
Coefficient of heat transfer, radiation	C_r	W/m²K⁴
Coefficient of friction	C_f	
Coefficient of discharge	C_d	
Coefficient of contraction	C_c	
Coefficient of velocity	C_s	
Coefficient of loss of pressure, etc.	C_L	
Squish area ratio	C_{sq}	
Coefficient of combustion equilibrium	K_p	
Area ratio of engine port to engine duct	k	
Dimensions and physical quantities		
area	A	m²
diameter	d	m
length	x	m
length of computation mesh	L	m
mass	m	kg
molecular weight	M	kg/kgmol
radius	r	m
time	t	s
volume	V	m³
force	F	N
pressure	p	Pa
pressure ratio	P	
pressure amplitude ratio	X	
mass flow rate	$\dot{m}$	kg/s
volume flow rate	$\dot{V}$	m³/s
velocity of gas particle	c	m/s
velocity of pressure wave propagation	α	m/s
velocity of acoustic wave (sound)	a	m/s
Young's modulus	Y	N/m²
wall shear stress	τ	N/m²
gravitational acceleration	g	m/s²

Dimensionless numbers

Froude number	**Fr**
Grashof number	**Gr**
Mach number	**M**
Nusselt number	**Nu**
Prandtl number	**Pr**
Reynolds number	**Re**

Energy-, work- and heat-related parameters

system energy	E	J
specific system energy	e	J/kg
internal energy	U	J
specific internal energy	u	J/kg
specific molal internal energy	$\bar{u}$	J/kgmol
potential energy	PE	J
specific potential energy	pe	J/kg
kinetic energy	KE	J
specific kinetic energy	ke	J/kg
heat	Q	J
specific heat	q	J/kg
enthalpy	H	J
specific enthalpy	h	J/kg
specific molal enthalpy	$\bar{h}$	J/kgmol
entropy	S	J/K
specific entropy	s	J/kgK
work	W	J
work, specific	w	J/kg

Engine, physical geometry

number of cylinders	n	
cylinder bore	d_{bo}	mm
cylinder stroke	L_{st}	mm
bore-to-stroke ratio	C_{bs}	
connecting rod length	L_{cr}	mm
crank throw	L_{ct}	mm
swept volume	V_{sv}	m^3
swept volume, trapped	V_{ts}	m^3
clearance volume	V_{cv}	m^3
compression ratio, crankcase	CR_{cc}	
compression ratio, geometric	CR_g	
compression ratio, trapped	CR_t	
speed of rotation	N	rev/min
speed of rotation	rpm	rev/min

speed of rotation	rps	rev/s
speed of rotation	ω	rad/s
mean piston speed	c_p	m/s
crankshaft position at top dead center	tdc	
crankshaft position at bottom dead center	bdc	
crankshaft angle before top dead center	°btdc	degrees
crankshaft angle after top dead center	°atdc	degrees
crankshaft angle before bottom dead center	°bbdc	degrees
crankshaft angle after bottom dead center	°abdc	degrees
crankshaft angle	θ	degrees
combustion period	b°	degrees
throttle area ratio	C_{thr}	
reed tip lift to length ratio	C_{rdt}	

Engine-, performance-related parameters

mean effective pressure, brake	bmep	Pa
mean effective pressure, indicated	imep	Pa
mean effective pressure, friction	fmep	Pa
mean effective pressure, pumping	pmep	Pa
power output	$\dot{W}$	kW
power output, brake	$\dot{W}_b$	kW
power output, indicated	$\dot{W}_i$	kW
torque output	Z	Nm
torque output, brake	Z_b	Nm
torque output, indicated	Z_i	Nm
air-to-fuel ratio	AFR	
air-to-fuel ratio, stoichiometric	AFR_s	
air-to-fuel ratio, trapped	AFR_t	
equivalence ratio	λ	
equivalence ratio, molecular	λ_m	
specific emissions of hydrocarbons	bsHC	g/kWh
specific emissions of oxides of nitrogen	$bsNO_x$	g/kWh
specific emissions of carbon monoxide	bsCO	g/kWh
specific emissions of carbon dioxide	$bsCO_2$	g/kWh
specific fuel consumption, brake	bsfc	kg/kWh
specific fuel consumption, indicated	isfc	kg/kWh
air flow, scavenge ratio	SR	
air flow, delivery ratio	DR	
air flow, volumetric efficiency	η_v	
charging efficiency	CE	
trapping efficiency	TE	
scavenging efficiency	SE	

thermal efficiency	η_t	
thermal efficiency, brake	η_b	
thermal efficiency, indicated	η_i	
mechanical efficiency	η_m	
fuel calorific value (lower)	C_{fl}	MJ/kg
fuel calorific value (higher)	C_{fh}	MJ/kg
fuel latent heat of vaporization	h_{vap}	kJ/kg
mass fraction burned	B	
heat release rate	$\dot{Q}_{R\theta}$	J/deg
combustion efficiency	η_c	
relative combustion efficiency with respect to purity	η_{se}	
relative combustion efficiency with respect to fueling	η_{af}	
index of compression	n_e	
index of expansion	n_c	
flame velocity	c_{fl}	
flame velocity, laminar	c_{lf}	
flame velocity, turbulent	c_{trb}	
squish velocity	c_{sq}	

Gas properties

gas constant	R	J/kgK
universal gas constant	$\overline{R}$	J/kgmolK
density	ρ	kg/m³
specific volume	v	m³/kg
specific heat at constant volume	C_v	J/kgK
specific heat at constant pressure	C_p	J/kgK
molal specific heat at constant volume	$\overline{C}_v$	J/kgmolK
molal specific heat at constant pressure	$\overline{C}_p$	J/kgmolK
ratio of specific heats	γ	
purity	Π	
temperature	T	K
viscosity	μ	kg/ms
kinematic viscosity	ν	m²/s
volumetric ratio of a gas mixture	υ	
mass ratio of a gas mixture	ε	

Noise

sound pressure level	β	dB
sound intensity	I	W/m²
sound frequency	f	Hz

attenuation or transmission loss	β_{tr}	dB
wavelength of sound	Λ	m

General

vectors and coordinates	x, y, z	
differential prefixes, exact, inexact, partial and incremental	$d, \delta, \partial, \Delta$	

Chapter 1

Introduction to the Two-Stroke Engine

1.0 Introduction to the two-stroke cycle engine

It is generally accepted that the two-stroke cycle engine was invented by Sir Dugald Clerk in England at the end of the 19th Century. The form of the engine using crankcase compression for the induction process, including the control of the timing and area of the exhaust, transfer and intake ports by the piston, was patented by Joseph Day in England in 1891. His engine was the original "three-port" engine and is the forerunner of the simple two-stroke engine which has been in common usage since that time.

Some of the early applications were in motorcycle form and are well recorded by Caunter [1.5]. The first engines were produced by Edward Butler in 1887 and by J.D. Roots, in the form of the Day crankcase compression type, in 1892; both of these designs were for powered tricycles. Considerable experimentation and development was conducted by Alfred Scott, and his Flying Squirrel machines competed very successfully in Tourist Trophy races in the first quarter of the 20th Century. They were designed quite beautifully in both the engineering and in the aesthetic sense. After that, two-stroke engines faded somewhat as competitive units in racing for some years until the supercharged DKW machines of the '30s temporarily revived their fortunes. With the banning of supercharging for motorcycle racing after the Second World War, the two-stroke engine lapsed again until 1959 when the MZ machines, with their tuned exhaust expansion chambers and disc valve induction systems, introduced a winning engine design which has basically lasted to the present day. A machine typical of this design approach is shown in Plate 1.1, a 250 cm^3 twin-cylinder engine with a rotary sleeve induction system which was built at QUB about 1969. Today, two-stroke-engined motorcycles, scooters and mopeds are still produced in very large numbers for general transport and for recreational purposes, although the legislative pressure on exhaust emissions in some countries has produced a swing to a four-stroke engine replacement in some cases. Whether the two-stroke engine will return as a mass production motorcycle engine will depend on the result of research and development being conducted by all of the manufacturers at the present time. There are some other applications with engines which are very similar in design terms to those used for motorcycles, and the sports of go-kart and hydroplane racing would fall into this category.

The two-stroke engine is used for lightweight power units which can be employed in various attitudes as handheld power tools. Such tools are chainsaws, brushcutters and concrete saws, to name but a few, and these are manufactured with a view to lightness and high

Plate 1.1 A QUB engined 250 cc racing motorcycle showing the tuned exhaust pipes.

specific power performance. One such device is shown in Plate 1.2. The manufacturing numbers involved are in millions per annum worldwide.

The earliest outboard motors were pioneered by Evinrude in the United States about 1909, with a 1.5 hp unit, and two-stroke engines have dominated this application until the present day. Some of the current machines are very sophisticated designs, such as 300 hp V6- and V8-engined outboards with remarkably efficient engines considering that the basic simplicity of the two-stroke crankcase compression engine has been retained. Although the image of the outboard motor is that it is for sporting and recreational purposes, the facts are that the product is used just as heavily for serious employment in commercial fishing and for everyday water transport in many parts of the world. The racing of outboard motors is a particularly exciting form of automotive sport, as seen in Plate 1.3.

Some of the new recreational products which have appeared in recent times are snowmobiles and water scooters, and the engine type almost always employed for such machines is the two-stroke engine. The use of this engine in a snowmobile is almost an ideal application, as the simple lubrication system of a two-stroke engine is perfectly suited for sub-zero temperature conditions. Although the snowmobile has been described as a recreational vehicle, it is actually a very practical means of everyday transport for many people in an Arctic environment.

The use of the two-stroke engine in automobiles has had an interesting history, and some quite sophisticated machines were produced in the 1960s, such as the Auto-Union vehicle from West Germany and the simpler Wartburg from East Germany. The Saab car from Sweden actually won the Monte Carlo Rally with Eric Carlson driving it. Until recent times, Suzuki built a small two-stroke-engined car in Japan. With increasing ecological emphasis on fuel consumption rate and exhaust emissions, the simple two-stroke-engined car disappeared,

Plate 1.2 A Homelite chainsaw engine illustrating the two-stroke powered tool (courtesy of Homelite Textron).

but interest in the design has seen a resurgence in recent times as the legislative pressure intensifies on exhaust acid emissions. Almost all car manufacturers are experimenting with various forms of two-stroke-engined vehicles equipped with direct fuel injection, or some variation of that concept in terms of stratified charging or combustion.

The two-stroke engine has been used in light aircraft, and today is most frequently employed in the recreational microlite machines. There are numerous other applications for the spark-ignition (SI) engine, such as small electricity generating sets or engines for remotely piloted vehicles, i.e., aircraft for meteorological data gathering or military purposes. These are but two of a long list of multifarious examples.

The use of the two-stroke engine in compression ignition (CI) or diesel form deserves special mention, even though it will not figure hugely in terms of specific design discussion within this book. The engine type has been used for trucks and locomotives, such as the designs from General Motors in America or Rootes-Tilling-Stevens in Britain. Both of these have been very successful engines in mass production. The engine type, producing a high specific power output, has also been a favorite for military installations in tanks and fast naval patrol boats. Some of the most remarkable aircraft engines ever built have been two-stroke diesel units, such as the Junkers Jumo and the turbo-compounded Napier Nomad. There is no doubt that the most successful of all of the applications is that of the marine diesel main propulsion unit, referred to in my student days in Harland and Wolff's shipyard in Belfast as a "cathedral" engine. The complete engine is usually some 12 m tall, so the description is rather apt. Such engines, the principal exponents of which were Burmeister and Wain in

Plate 1.3 A high-performance multi-cylinder outboard motor in racing trim (courtesy of Mercury Marine).

Copenhagen and Sulzer in Winterthur, were typically of 900 mm bore and 1800 mm stroke and ran at 60-100 rpm, producing some 4000 hp per cylinder. They had thermal efficiencies in excess of 50%, making them the most efficient prime movers ever made. These engines are very different from the rest of the two-stroke engine species in terms of scale but not in design concept, as Plate 1.4 illustrates.

The diesel engine, like its spark-ignition counterpart, is also under legislative pressure to conform to ever-tighter emissions standards. For the diesel engine, even though it provides very low emissions of carbon monoxide and of hydrocarbons, does emit visible smoke in the form of carbon particulates and measurable levels of nitrogen oxides. The level of emission of both of these latter components is under increasing environmental scrutiny and the diesel engine must conform to more stringent legislative standards by the year 2000. The combination of very low particulate and NO_x emission is a tough R&D proposition for the designer of CI engines to be able to meet. As the combustion is lean of the stoichiometric mixture by some 50% at its richest setting to avoid excessive exhaust smoke, the exhaust gas is oxygen rich and so only a lean burn catalyst can be used on either a two-stroke or a four-stroke cycle engine. This does little, if anything at all, to reduce the nitrogen oxide emissions. Thus the manufacturers are again turning to the two-stroke cycle diesel engine as a potential alternative powerplant for cars and trucks, as that cycle has inherently a significantly lower NO_x emission characteristic. Much R&D is taking place in the last decade of the 20th Century with a view to eventual manufacture, if the engine meets all relevant criteria on emissions, thermal efficiency and durability.

It is probably true to say that the two-stroke engine has produced the most diverse opinions on the part of both the users and the engineers. These opinions vary from fanatical enthu-

Chapter 1 - Introduction to the Two-Stroke Engine

Plate 1.4 Harland and Wolff uniflow-scavenged two-stroke diesel ship propulsion engine of 21,000 bhp (courtesy of Harland and Wolff plc).

siasm to thinly veiled dislike. Whatever your view, at this early juncture in reading this book, no other engine type has ever fascinated the engineering world to quite the same extent. This is probably because the engine seems so deceptively simple to design, develop and manufacture. That the very opposite is the case may well be the reason that some spend a lifetime investigating this engineering curiosity. The potential rewards are great, for no other engine cycle has produced, in one constructional form or another, such high thermal efficiency or such low specific fuel consumption, such high specific power criteria referred to either swept volume, bulk or weight, nor such low acid exhaust emissions.

Design and Simulation of Two-Stroke Engines

1.1 The fundamental method of operation of a simple two-stroke engine

The simple two-stroke engine is shown in Fig. 1.1, with the various phases of the filling and emptying of the cylinder illustrated in (a)-(d). The simplicity of the engine is obvious, with all of the processes controlled by the upper and lower edges of the piston. A photograph of a simple two-stroke engine is provided in Plate 1.5. It is actually a small chainsaw engine, giving some further explanation of its construction.

In Fig. 1.1(a), above the piston, the trapped air and fuel charge is being ignited by the spark plug, producing a rapid rise in pressure and temperature which will drive the piston

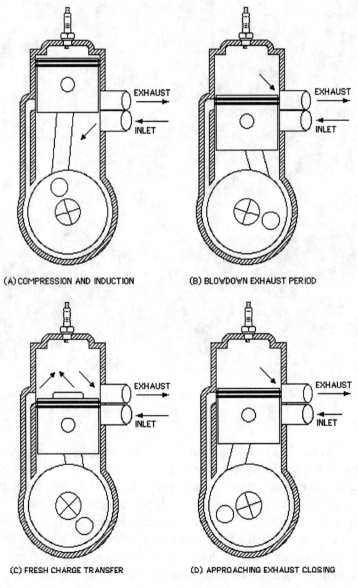

Fig. 1.1 Various stages in the operation of the two-stroke cycle engine.

Chapter 1 - Introduction to the Two-Stroke Engine

Plate 1.5 An exploded view of a simple two-stroke engine.

down on the power stroke. Below the piston, the opened inlet port is inducing air from the atmosphere into the crankcase due to the increasing volume of the crankcase lowering the pressure below the atmospheric value. The crankcase is sealed around the crankshaft to ensure the maximum depression within it. To induce fuel into the engine, the various options exist of either placing a carburetor in the inlet tract, injecting fuel into the inlet tract, injecting fuel into the crankcase or transfer ducts, or injecting fuel directly into the cylinder before or after the closure of the exhaust port. Clearly, if it is desired to operate the engine as a diesel power unit, the latter is the only option, with the spark plug possibly being replaced by a glow plug as an initial starting aid and the fuel injector placed in the cylinder head area.

In Fig. 1.1(b), above the piston, the exhaust port has been opened. It is often called the "release" point in the cycle, and this allows the transmission into the exhaust duct of a pulse of hot, high-pressure exhaust gas from the combustion process. As the area of the port is increasing with crankshaft angle, and the cylinder pressure is falling with time, it is clear that the exhaust duct pressure profile with time is one which increases to a maximum value and then decays. Such a flow process is described as unsteady gas flow and such a pulse can be reflected from all pipe area changes, or at the pipe end termination to the atmosphere. These reflections have a dramatic influence on the engine performance, as later chapters of this book describe. Below the piston, the compression of the fresh charge is taking place. The pressure and temperature achieved will be a function of the proportionate reduction of the crankcase volume, i.e., the crankcase compression ratio.

In Fig. 1.1(c), above the piston, the initial exhaust process, referred to as "blowdown," is nearing completion and, with the piston having uncovered the transfer ports, this connects the cylinder directly to the crankcase through the transfer ducts. If the crankcase pressure exceeds the cylinder pressure then the fresh charge enters the cylinder in what is known as the *scavenge process*. Clearly, if the transfer ports are badly directed then the fresh charge can exit directly out of the exhaust port and be totally lost from the cylinder. Such a process, referred to as "short-circuiting," would result in the cylinder being filled only with exhaust gas at the onset of the next combustion process, and no pressure rise or power output would ensue. Worse, all of the fuel in a carburetted configuration would be lost to the exhaust with a consequential monstrous emission rate of unburned hydrocarbons. Therefore, the directioning of the fresh charge by the orientation of the transfer ports should be conducted in such a manner as to maximize the retention of it within the cylinder. This is just as true for the diesel engine, for the highest trapped air mass can be burned with an appropriate fuel quantity to attain the optimum power output. It is obvious that the scavenge process is one which needs to be optimized to the best of the designer's ability. Later chapters of this book will concentrate heavily on the scavenge process and on the most detailed aspects of the mechanical design to improve it as much as possible. It should be clear that it is not possible to have such a process proceed perfectly, as some fresh charge will always find a way through the exhaust port. Equally, no scavenge process, however extensive or thorough, will ever leach out the last molecule of exhaust gas.

In Fig. 1.1(d), in the cylinder, the piston is approaching what is known as the "trapping" point, or exhaust closure. The scavenge process has been completed and the cylinder is now filled with a mix of air, fuel if a carburetted design, and exhaust gas. As the piston rises, the cylinder pressure should also rise, but the exhaust port is still open and, barring the intervention of some unsteady gas-dynamic effect generated in the exhaust pipe, the piston will spill fresh charge into the exhaust duct to the detriment of the resulting power output and fuel consumption. Should it be feasible to gas-dynamically plug the exhaust port during this trapping phase, then it is possible to greatly increase the performance characteristics of the engine. In a single-cylinder racing engine, it is possible to double the mass of the trapped air charge using a tuned pipe, which means doubling the power output; such effects are discussed in Chapter 2. After the exhaust port is finally closed, the true compression process begins until the combustion process is commenced by ignition. Not surprisingly, therefore, the compression ratio of a two-stroke engine is characterized by the cylinder volume after exhaust port closure and is called the *trapped compression ratio* to distinguish it from the value commonly quoted for the four-stroke engine. That value is termed here as the *geometric compression ratio* and is based on the full swept volume.

In summary, the simple two-stroke engine is a double-acting device. Above the piston, the combustion and power processes take place, whereas below the piston in the crankcase, the fresh charge is induced and prepared for transfer to the upper cylinder. What could be simpler to design than this device?

1.2 Methods of scavenging the cylinder
1.2.1 Loop scavenging

In Chapter 3, there is a comprehensive discussion of the fluid mechanics and the gas dynamics of the scavenging process. However, it is important that this preliminary descriptive section introduces the present technological position, from a historical perspective. The scavenging process depicted in Fig. 1.1 is described as *loop scavenging*, the invention of which is credited to Schnurle in Germany about 1926. The objective was to produce a scav-

Chapter 1 - Introduction to the Two-Stroke Engine

enge process in a ported cylinder with two or more scavenge ports directed toward that side of the cylinder away from the exhaust port, but across a piston with essentially a flat top. The invention was designed to eliminate the hot-running characteristics of the piston crown used in the original method of scavenging devised by Sir Dugald Clerk, namely the deflector piston employed in the *cross scavenging* method discussed in Sec. 1.2.2. Although Schnurle was credited with the invention of loop scavenging, there is no doubt that patents taken out by Schmidt and by Kind fifteen years earlier look uncannily similar, and it is my understanding that considerable litigation regarding patent ownership ensued in Germany in the 1920s and 1930s. In Fig. 1.2, the various layouts observed in loop-scavenged engines are shown. These are plan sections through the scavenge ports and the exhaust port, and the selection shown is far from an exhaustive sample of the infinite variety of designs seen in two-stroke engines. The common element is the sweep back angle for the "main" transfer port, away from the exhaust port; the "main" transfer port is the port next to the exhaust port. Another advantage of the loop scavenge design is the availability of a compact combustion chamber above the flat topped piston which permits a rapid and efficient combustion process. A piston for such an engine is shown in Plate 1.6. The type of scavenging used in the engine in Plate 1.5 is loop scavenging.

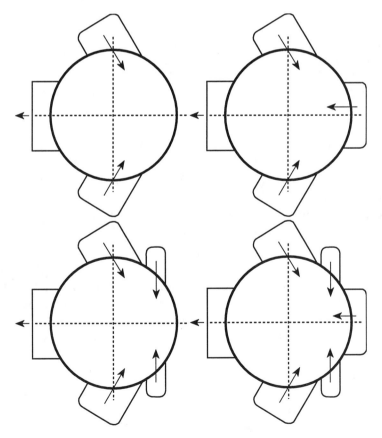

Fig. 1.2 Various scavenge port plan layouts found in loop scavenging.

Design and Simulation of Two-Stroke Engines

Plate 1.6 The pistons, from L to R, for a QUB-type cross-scavenged, a conventional cross-scavenged and a loop-scavenged engine.

1.2.2 Cross scavenging

This is the original method of scavenging proposed by Sir Dugald Clerk and is widely used for outboard motors to this very day. The modern deflector design is illustrated in Fig. 1.3 and emanates from the Scott engines of the early 1900s, whereas the original deflector was a simple wall or barrier on the piston crown. To further illustrate that, a photograph of this type of piston appears in Plate 1.6. In Sec. 3.2.4 it will be shown that this has good scavenging characteristics at low throttle openings and this tends to give good low-speed and low-power characteristics, making it ideal for, for example, small outboard motors employed in sport fishing. At higher throttle openings the scavenging efficiency is not particularly good and, combined with a non-compact combustion chamber filled with an exposed protuberant deflector, the engine has rather unimpressive specific power and fuel economy characteristics (see Plate 4.2). The potential for detonation and for pre-ignition, from the high surface-to-volume ratio combustion chamber and the hot deflector edges, respectively, is rather high and so the compression ratio which can be employed in this engine tends to be somewhat lower than for the equivalent loop-scavenged power unit. The engine type has some considerable packaging and manufacturing advantages over the loop engine. In Fig. 1.3 it can be seen from the port plan layout that the cylinder-to-cylinder spacing in a multi-cylinder configuration could be as close as is practical for inter-cylinder cooling considerations. If one looks at the equivalent situation for the loop-scavenged engine in Fig. 1.2 it can be seen that the transfer ports on the side of the cylinder prohibit such close cylinder spacing; while it is possible to twist the cylinders to alleviate this effect to some extent, the end result has further packaging, gas-dynamic and scavenging disadvantages [1.12]. Further, it is possible to drill the scavenge and the exhaust ports directly, in-situ and in one operation, from the exhaust port side, and

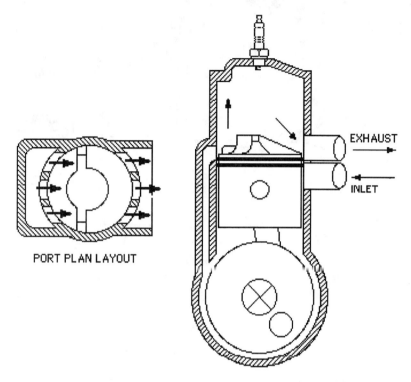

Fig. 1.3 Deflector piston of cross-scavenged engine.

thereby reduce the manufacturing costs of the cross-scavenged engine by comparison with an equivalent loop- or uniflow-scavenged power unit.

One design of cross-scavenged engines, which does not have the disadvantages of poor wide-open throttle scavenging and a non-compact combustion chamber, is the type designed at QUB [1.9] and sketched in Fig. 1.4. A piston for this design is shown in Plate 1.6. However, the cylinder does not have the same manufacturing simplicity as that of the conventional deflector piston engine. I have shown in Ref. [1.10] and in Sec. 3.2.4 that the scavenging is as effective as a loop-scavenged power unit and that the highly squished and turbulent combustion chamber leads to good power and good fuel economy characteristics, allied to cool cylinder head running conditions at high loads, speeds and compression ratios [1.9] (see Plate 4.3). Several models of this QUB type are in series production at the time of writing.

1.2.3 Uniflow scavenging

Uniflow scavenging has long been held to be the most efficient method of scavenging the two-stroke engine. The basic scheme is illustrated in Fig. 1.5 and, fundamentally, the methodology is to start filling the cylinder with fresh charge at one end and remove the exhaust gas from the other. Often the charge is swirled at both the charge entry level and the exhaust exit level by either suitably directing the porting angular directions or by masking a poppet valve. The swirling air motion is particularly effective in promoting good combustion in a diesel configuration. Indeed, the most efficient prime movers ever made are the low-speed marine

Design and Simulation of Two-Stroke Engines

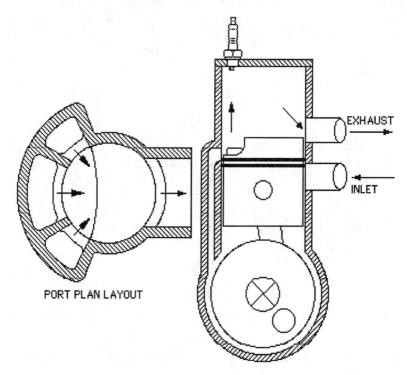

Fig. 1.4 QUB type of deflector piston of cross-scavenged engine.

diesels of the uniflow-scavenged two-stroke variety with thermal efficiencies in excess of 50%. However, these low-speed engines are ideally suited to uniflow scavenging, with cylinder bores about 1000 mm, a cylinder stroke about 2500 mm, and a bore-stroke ratio of 0.4. For most engines used in today's motorcycles and outboards, or tomorrow's automobiles, bore-stroke ratios are typically between 0.9 and 1.3. For such engines, there is some evidence (presented in Sec. 3.2.4) that uniflow scavenging, while still very good, is not significantly better than the best of loop-scavenged designs [1.11]. For spark-ignition engines, as uniflow scavenging usually entails some considerable mechanical complexity over simpler methods and there is not in reality the imagined performance enhancement from uniflow scavenging, this virtually rules out this method of scavenging on the grounds of increased engine bulk and cost for an insignificant power or efficiency advantage.

1.2.4 Scavenging not employing the crankcase as an air pump

The essential element of the original Clerk invention, or perhaps more properly the variation of the Clerk principle by Day, was the use of the crankcase as the air-pumping device of the engine; all simple designs use this concept. The lubrication of such engines has traditionally been conducted on a total-loss basis by whatever means employed. The conventional method has been to mix the lubricant with the petrol (gasoline) and supply it through the carburetor in ratios of lubricant to petrol varying from 25:1 to 100:1, depending on the application, the skill of the designers and/or the choice of bearing type employed as big-ends or as main crankshaft bearings. The British term for this type of lubrication is called "petroil"

Chapter 1 - Introduction to the Two-Stroke Engine

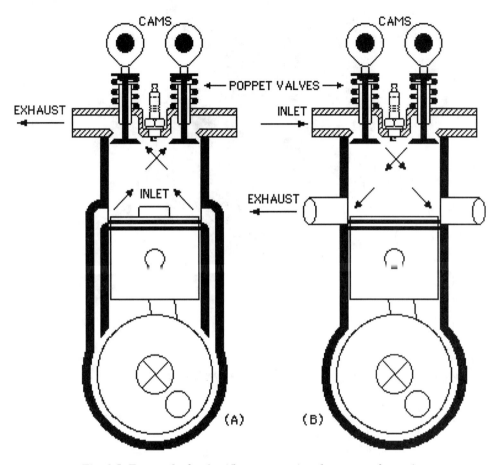

Fig. 1.5 Two methods of uniflow scavenging the two-stroke engine.

lubrication. As the lubrication is of the total-loss type, and some 10-30% of the fuel charge is short-circuited to the exhaust duct along with the air, the resulting exhaust plume is rich in unburned hydrocarbons and lubricant, some partially burned and some totally unburned, and is consequently visible as smoke. This is ecologically unacceptable in the latter part of the 20th Century and so the manufacturers of motorcycles and outboards have introduced separate oil-pumping devices to reduce the oil consumption rate, and hence the oil deposition rate to the atmosphere, be it directly to the air or via water. Such systems can reduce the effective oil-to-petrol ratio to as little as 200 or 300 and approach the oil consumption rate of four-stroke cycle engines. Even so, any visible exhaust smoke is always unacceptable and so, for future designs, as has always been the case for the marine and automotive two-stroke diesel engine, a crankshaft lubrication system based on pressure-fed plain bearings with a wet or dry sump may be employed. One of the successful compression-ignition engine designs of this type is the Detroit Diesel engine shown in Plate 1.7.

By definition, this means that the crankcase can no longer be used as the air-pumping device and so an external air pump will be utilized. This can be either a positive displacement blower of the Roots type, or a centrifugal blower driven from the crankshaft. Clearly, it would

Design and Simulation of Two-Stroke Engines

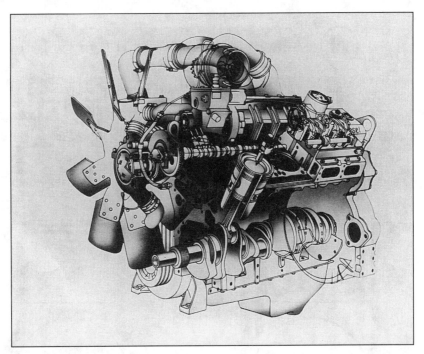

Plate 1.7 The Detroit Diesel Allison Series 92 uniflow-scavenged, supercharged and turbocharged diesel engine for truck applications (courtesy of General Motors).

be more efficient thermodynamically to employ a turbocharger, where the exhaust energy to the exhaust turbine is available to drive the air compressor. Such an arrangement is shown in Fig. 1.6 where the engine has both a blower and a turbocharger. The blower would be used as a starting aid and as an air supplementary device at low loads and speeds, with the turbocharger employed as the main air supply unit at the higher torque and power levels at any engine speed. To prevent short-circuiting fuel to the exhaust, a fuel injector would be used to supply petrol directly to the cylinder, hopefully after the exhaust port is closed and not in the position sketched, at bottom dead center (bdc). Such an engine type has already demonstrated excellent fuel economy behavior, good exhaust emission characteristics of unburned hydrocarbons and carbon monoxide, and superior emission characteristics of oxides of nitrogen, by comparison with an equivalent four-stroke engine. This subject will be elaborated on in Chapter 7. The diesel engine shown in Plate 1.7 is just such a power unit, but employing compression ignition.

Nevertheless, in case the impression is left that the two-stroke engine with a "petroil" lubrication method and a crankcase air pump is an anachronism, it should be pointed out that this provides a simple, lightweight, high-specific-output powerplant for many purposes, for which there is no effective alternative engine. Such applications range from the agricultural for chainsaws and brushcutters, where the engine can easily run in an inverted mode, to small outboards where the alternative would be a four-stroke engine resulting in a considerable weight, bulk, and manufacturing cost increase.

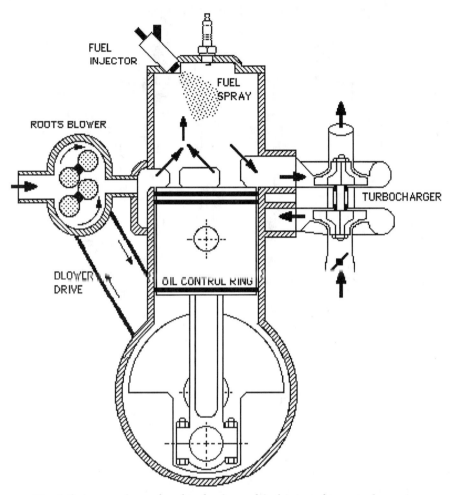

Fig. 1.6 A supercharged and turbocharged fuel-injected two-stroke engine.

1.3 Valving and porting control of the exhaust, scavenge and inlet processes

The simplest method of allowing fresh charge access into, and exhaust gas discharge from, the two-stroke engine is by the movement of the piston exposing ports in the cylinder wall. In the case of the simple engine illustrated in Fig. 1.1 and Plate 1.5, this means that all port timing events are symmetrical with respect to top dead center (tdc) and bdc. It is possible to change this behavior slightly by offsetting the crankshaft centerline to the cylinder centerline, but this is rarely carried out in practice as the resulting improvement is hardly worth the manufacturing complication involved. It is possible to produce asymmetrical inlet and exhaust timing events by the use of disc valves, reed valves and poppet valves. This permits the phasing of the porting to correspond more precisely with the pressure events in the cylinder or the crankcase, and so gives the designer more control over the optimization of the exhaust or intake system. The use of poppet valves for both inlet and exhaust timing control is sketched, in the case of uniflow scavenging, in Fig. 1.5. Fig. 1.7 illustrates the use of disc and reed valves for the asymmetrical timing control of the inlet process into the engine crankcase. It is

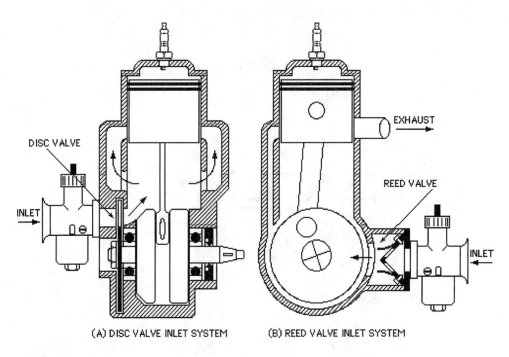

Fig. 1.7 Disc valve and reed valve control of the inlet system.

virtually unknown to attempt to produce asymmetrical timing control of the scavenge process from the crankcase through the transfer ports.

1.3.1 Poppet valves

The use and design of poppet valves is thoroughly covered in texts and papers dealing with four-stroke engines [1.3], so it will not be discussed here, except to say that the flow area-time characteristics of poppet valves are, as a generality, considerably less than are easily attainable for the same geometrical access area posed by a port in a cylinder wall. Put in simpler form, it is difficult to design poppet valves so as to adequately flow sufficient charge into a two-stroke engine. It should be remembered that the actual time available for any given inlet or exhaust process, at the same engine rotational speed, is about one half of that possible in a four-stroke cycle design.

1.3.2 Disc valves

The disc valve design is thought to have emanated from East Germany in the 1950s in connection with the MZ racing motorcycles from Zchopau, the same machines that introduced the expansion chamber exhaust system for high-specific-output racing engines. A twin-cylinder racing motorcycle engine which uses this method of induction control is shown in Plate 1.8. Irving [1.1] attributes the design to Zimmerman. Most disc valves have timing characteristics of the values shown in Fig. 1.8 and are usually fabricated from a spring steel, although discs made from composite materials are also common. To assist with comprehension of disc valve operation and design, you should find useful Figs. 6.28 and 6.29 and the discussion in Sec. 6.4.

Chapter 1 - Introduction to the Two-Stroke Engine

Plate 1.8 A Rotax disc valve racing motorcycle engine with one valve cover removed exposing the disc valve.

1.3.3 Reed valves

Reed valves have always been popular in outboard motors, as they provide an effective automatic valve whose timings vary with both engine load and engine speed. In recent times, they have also been designed for motorcycle racing engines, succeeding the disc valve. In part, this technical argument has been settled by the inherent difficulty of easily designing multi-cylinder racing engines with disc valves, as a disc valve design demands a free crankshaft end for each cylinder. The high-performance outboard racing engines demonstrated that high specific power output was possible with reed valves [1.12] and the racing motorcycle organizations developed the technology further, first for motocross engines and then for Grand Prix power units. Today, most reed valves are designed as V-blocks (see Fig. 1.7 and Plates 1.9 and 6.1) and the materials used for the reed petals are either spring steel or a fiber-reinforced composite material. The composite material is particularly useful in highly stressed racing engines, as any reed petal failure is not mechanically catastrophic as far as the rest of the engine is concerned. Further explanatory figures and detailed design discussions regarding all such valves and ports will be found in Section 6.3.

Fig. 1.7 shows the reed valve being given access directly to the crankcase, and this would be the design most prevalent for outboard motors where the crankcase bottom is accessible (see Plate 5.2). However, for motorcycles or chainsaws, where the crankcase is normally "buried" in a transmission system, this is somewhat impractical and so the reed valve feeds the fresh air charge to the crankcase through the cylinder. An example of this is illustrated in Plate 4.1, showing a 1988 model 250 cm^3 Grand Prix motorcycle racing engine. This can be effected [1.13] by placing the reed valve housing at the cylinder level so that it is connected to the transfer ducts into the crankcase.

Design and Simulation of Two-Stroke Engines

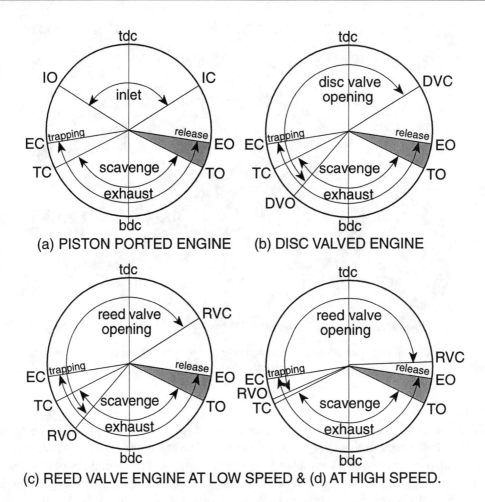

Fig. 1.8 Typical port timing characteristics for piston ported, reed and disc valve engines.

1.3.4 Port timing events

As has already been mentioned in Sec. 1.3.1, the port timing events in a simple two-stroke engine are symmetrical around tdc and bdc. This is defined by the connecting rod-crank relationship. Typical port timing events, for piston port control of the exhaust, transfer or scavenge, and inlet processes, disc valve control of the inlet process, and reed valve control of the inlet process, are illustrated in Fig. 1.8. The symmetrical nature of the exhaust and scavenge processes is evident, where the exhaust port opening and closing, EO and EC, and transfer port opening and closing, TO and TC, are under the control of the top, or timing, edge of the piston.

Where the inlet port is similarly controlled by the piston, in this case the bottom edge of the piston skirt, is sketched in Fig. 1.8(a); this also is observed to be a symmetrical process. The shaded area in each case between EO and TO, exhaust opening and transfer opening, is called the blowdown period and has already been referred to in Sec. 1.1. It is also obvious from various discussions in this chapter that if the crankcase is to be sealed to provide an

Chapter 1 - Introduction to the Two-Stroke Engine

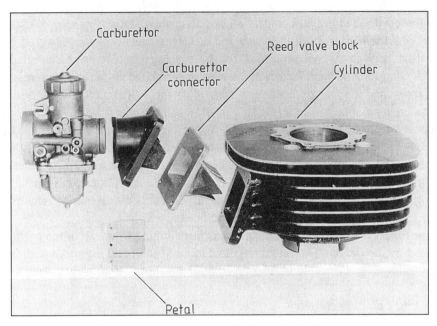

Plate 1.9 An exploded view of a reed valve cylinder for a motorcycle.

effective air-pumping action, there must not be a gas passage from the exhaust to the crankcase. This means that the piston must always totally cover the exhaust port at tdc or, to be specific, the piston length must be sufficiently in excess of the stroke of the engine to prevent gas leakage from the crankcase. In Chapter 6 there will be detailed discussions on porting design. However, to set the scene for that chapter, Fig. 1.9 gives some preliminary facts regarding the typical port timings seen in some two-stroke engines. It can be seen that as the demand rises in terms of specific power output, so too does the porting periods. Should the engine be designed with a disc valve, then the inlet port timing changes are not so dramatic with increasing power output.

Engine Type	Piston Port Control			Disc Valve Control of Inlet Port	
	Exhaust Opens °ATDC	Transfer Opens °ATDC	Inlet Opens °BTDC	Opens °BTDC	Opens °ATDC
Industrial, Moped, Chainsaw, Small Outboard	110	122	65	130	60
Enduro, Snowmobile, RPV, Large Outboard	97	120	75	120	70
Motocross, GP Racer	82	113	100	140	80

Fig. 1.9 Typical port timings for two-stroke engine applications.

Design and Simulation of Two-Stroke Engines

For engines with the inlet port controlled by a disc valve, the asymmetrical nature of the port timing is evident from both Figs. 1.8 and 1.9. However, for engines fitted with reed valves the situation is much more complex, for the opening and closing characteristics of the reed are now controlled by such factors as the reed material, the crankcase compression ratio, the engine speed and the throttle opening. Figs. 1.8(c) and 1.8(d) illustrate the typical situation as recorded in practice by Fleck [1.13]. It is interesting to note that the reed valve opening and closing points, marked as RVO and RVC, respectively, are quite similar to a disc valve engine at low engine speeds and to a piston-controlled port at higher engine speeds. For racing engines, the designer would have wished those characteristics to be reversed! The transition in the RVO and the RVC points is almost, but not quite, linear with speed, with the total opening period remaining somewhat constant. Detailed discussion of matters relating specifically to the design of reed valves is found in Sec. 6.3.

Examine Fig. 6.1, which shows the port areas in an engine where all of the porting events are controlled by the piston. The actual engine data used to create Fig. 6.1 are those for the chainsaw engine design discussed in Chapter 5 and the geometrical data displayed in Fig. 5.3.

1.4 Engine and porting geometry

Some mathematical treatment of design will now be conducted, in a manner which can be followed by anyone with a mathematics education of university entrance level. The fundamental principle of this book is not to confuse, but to illuminate, and to arrive as quickly as is sensible to a working computer program for the design of the particular component under discussion.

(a) Units used throughout the book

Before embarking on this section, a word about units is essential. This book is written in SI units, and all mathematical equations are formulated in those units. Thus, all subsequent equations are intended to be used with the arithmetic values inserted for the symbols of the SI unit listed in the Nomenclature before Chapter 1. If this practice is adhered to, then the value computed from any equation will appear also as the strict SI unit listed for that variable on the left-hand side of the equation. Should you desire to change the unit of the ensuing arithmetic answer to one of the other units listed in the Nomenclature, a simple arithmetic conversion process can be easily accomplished. One of the virtues of the SI system is that strict adherence to those units, in mathematical or computational procedures, greatly reduces the potential for arithmetic errors. I write this with some feeling, as one who was educated with great difficulty, as an American friend once expressed it so well, in the British "furlong, hundredweight, fortnight" system of units!

(b) Computer programs presented throughout the book

The listing of all computer programs connected with this book is contained in the Appendix Listing of Computer Programs. Logically, programs coming from, say, Chapter 3, will appear in the Appendix as Prog.3. In the case of the first programs introduced below, they are to be found as Prog.1.1, Prog.1.2, and Prog.1.3. As is common with computer programs, they also have names, in this case, PISTON POSITION, LOOP ENGINE DRAW, and QUB CROSS ENGINE DRAW, respectively. All of the computer programs have been written in Microsoft® QuickBASIC for the Apple Macintosh® and this is the same language prepared by Microsoft

Corp. for the IBM® PC and its many clones. Only some of the graphics statements are slightly different for various IBM-like machines.

Almost all of the programs are written in, and are intended to be used in, the interpreted QuickBASIC mode. However, the speed advantage in the compiled mode makes for more effective use of the software. In Microsoft QuickBASIC, a "user-friendy" computer language and system, it is merely a flick of a mouse to obtain a compiled version of any program listing. The software is available from SAE in disk form for direct use on either Macintosh or IBM PC (or clone) computers.

1.4.1 Swept volume

If the cylinder of an engine has a bore, d_{bo}, and a stroke, L_{st}, as sketched in Fig. 4.2, then the total *swept volume*, V_{sv}, of an engine with n cylinders having those dimensions, is given by:

$$V_{sv} = n \frac{\pi}{4} d_{bo}^2 L_{st} \qquad (1.4.1)$$

The total swept volume of any one cylinder of the engine is given by placing n as unity in the above equation.

If the exhaust port closes some distance called the *trapped stroke*, L_{ts}, before tdc, then the *trapped swept volume* of any cylinder, V_{ts}, is given by:

$$V_{ts} = n \frac{\pi}{4} d_{bo}^2 L_{ts} \qquad (1.4.2)$$

The piston is connected to the crankshaft by a connecting rod of length, L_{cr}. The throw of the crank (see Fig. 1.10) is one-half of the stroke and is designated as length, L_{ct}. As with four-stroke engines, the connecting rod-crank ratios are typically in the range of 3.5 to 4.

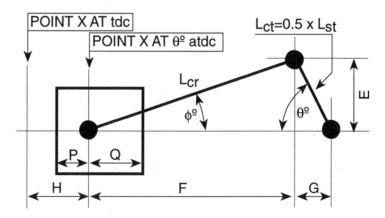

Fig. 1.10 Position of a point on a piston with respect to top dead center.

1.4.2 Compression ratio

All compression ratio values are the ratio of the maximum volume in any chamber of an engine to the minimum volume in that chamber. In the crankcase that ratio is known as the *crankcase compression ratio*, CR_{cc}, and is defined by:

$$CR_{cc} = \frac{V_{cc} + V_{sv}}{V_{cc}} \qquad (1.4.3)$$

where V_{cc} is the crankcase clearance volume, or the crankcase volume at bdc.

While it is true that the higher this value becomes, the stronger is the crankcase pumping action, the actual numerical value is greatly fixed by the engine geometry of bore, stroke, con-rod length and the interconnected value of flywheel diameter. In practical terms, it is rather difficult to organize the CR_{cc} value for a 50 cm^3 engine cylinder above 1.4 and almost physically impossible to design a 500 cm^3 engine cylinder to have a value less than 1.55. Therefore, for any given engine design the CR_{cc} characteristic is more heavily influenced by the choice of cylinder swept volume than by the designer. It then behooves the designer to tailor the engine air-flow behavior around the crankcase pumping action, defined by the inherent CR_{cc} value emanating from the cylinder size in question. There is some freedom of design action, and it is necessary for it to be taken in the correct direction.

In the cylinder shown in Fig. 4.2, if the clearance volume, V_{cv}, above the piston at tdc is known, then the geometric *compression ratio*, CR_g, is given by:

$$CR_g = \frac{V_{sv} + V_{cv}}{V_{cv}} \qquad (1.4.4)$$

Theoretically, the actual compression process occurs after the exhaust port is closed, and the compression ratio after that point becomes the most important one in design terms. This is called the *trapped compression ratio*. Because this is the case, in the literature for two-stroke engines the words "compression ratio" are sometimes carelessly applied when the precise term "trapped compression ratio" should be used. This is even more confusing because the literature for four-stroke engines refers to the geometric compression ratio, but describes it simply as the "compression ratio." The trapped compression ratio, CR_t, is then calculated from:

$$CR_t = \frac{V_{ts} + V_{cv}}{V_{cv}} \qquad (1.4.5)$$

1.4.3 Piston position with respect to crankshaft angle

At any given crankshaft angle, θ, after tdc, the connecting rod centerline assumes an angle, ϕ, to the cylinder centerline. This angle is often referred to in the literature as the "angle of obliquity" of the connecting rod. This is illustrated in Fig. 1.10 and the piston position of any point, X, on the piston from the tdc point is given by length H. The controlling trigonometric equations are:

Chapter 1 - Introduction to the Two-Stroke Engine

as
$$H + F + G = L_{cr} + L_{ct} \tag{1.4.6}$$

and
$$E = L_{ct} \sin \theta = L_{cr} \sin \phi \tag{1.4.7}$$

and
$$F = L_{cr} \cos \phi \tag{1.4.8}$$

and
$$G = L_{ct} \cos \theta \tag{1.4.9}$$

by Pythagoras
$$L_{ct}^2 = E^2 + G^2 \tag{1.4.10}$$

by Pythagoras
$$L_{cr}^2 = E^2 + F^2 \tag{1.4.11}$$

then
$$H = L_{cr} + L_{ct}(1 - \cos \theta) - \sqrt{L_{cr}^2 - (L_{ct} \sin \theta)^2} \tag{1.4.12}$$

Clearly, it is essential for the designer to know the position of the piston at salient points such as exhaust, transfer, and inlet port opening, closing and the fully open points as well, should the latter not coincide with either tdc or bdc. These piston positions define the port heights, and the mechanical drafting of any design requires these facts as precise numbers. In later chapters, design advice will be presented for the detailed porting design, but often connected with general, rather than detailed, piston and rod geometry. Consequently, the equations shown in Fig. 1.10 are programmed into three computer programs, Prog.1.1, Prog.1.2 and Prog.1.3, for use in specific circumstances.

1.4.4 Computer program, Prog.1.1, PISTON POSITION

This is a quite unsophisticated program without graphics. When RUN, the prompts for the input data are self-evident in nature. An example of the simplistic nature of the input and output data is shown in Fig. 1.11. If you are new to the ways of the Macintosh or IBM system of operation then this straightforward program will provide a useful introduction. The printout which appears on the line printer contains further calculated values of use to the designer. The program allows for the calculation of piston position from bdc or tdc for any engine geometry between any two crankshaft angles at any interval of step between them. The output shown in Fig. 1.11 is exactly what you would see on the Macintosh computer screen.

1.4.5 Computer program, Prog.1.2, LOOP ENGINE DRAW

This program is written in a much more sophisticated manner, using the facilities of the Macintosh software to speed the process of program operation, data handling and decision making by the user. Fig. 1.12 illustrates a completed calculation which you can print in that form on demand.

Design and Simulation of Two-Stroke Engines

```
RUNNING THE PROGRAM(Y OR N?)?  Y
enter BORE in mm 60
enter STROKE in mm 60
enter CON-ROD LENGTH 1n mm 110
enter  CRANK-ANGLE  after tdc at  start  of calculation 100
enter  CRANK-ANGLE  after tdc at end of calculation 120
enter  CRANK-ANGLE interval for the calculation step from start to finish  5

Swept volume,CM3,= 169.6
    crank-angle after tdc      height from tdc       height from bdc
           100.00                  39.25                 20.75
           105.00                  41.65                 18.35
           110.00                  43.93                 16.07
           115.00                  46.09                 13.91
           120.00                  48.11                 11.89
WANT A PRINT-OUT(Y OR N?)?  N
```

Fig. 1.11 *Example of a calculation from Prog.1.1, PISTON POSITION.*

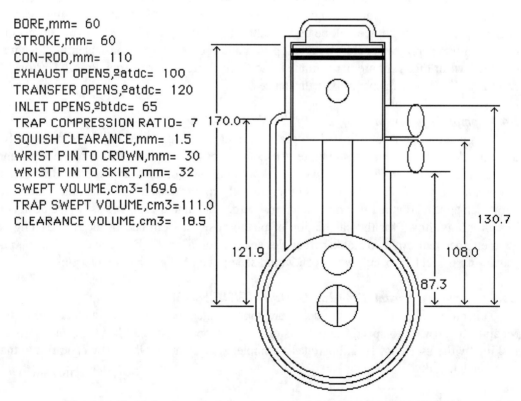

Fig. 1.12 *Example of a calculation from Prog.1.2, LOOP ENGINE DRAW.*

The data input values are written on the left-hand side of the figure from "BORE" down to "WRIST PIN TO SKIRT." All other values on the picture are output values. "Wrist pin" is the American term for a gudgeon pin and the latter two data input values correspond to the dimensions P and Q on Fig. 1.10. It is also assumed in the calculation that all ports are opened by their respective control edges to the top or bottom dead center positions. The basic geometry of the sketch is precisely as Fig. 1.1; indeed that sketch was created using this particular program halted at specific crankshaft angular positions.

When you run this program you will discover that the engine on the screen rotates for one complete cycle, from tdc to tdc. When the piston comes to rest at tdc, the linear dimensions of all porting positions from the crankshaft centerline are drawn, as illustrated. By this means, as the drawing on the screen is exactly to scale, you can be visually assured that the engine has no unusual problems in geometrical terms. You can observe, for example, that the piston is sufficiently long to seal the exhaust port and preserve an effective crankcase pumping action!

The inlet port is shown in Figs. 1.1 and 1.12 as being underneath the exhaust port for reasons of diagrammatic simplicity. There have been engines produced this way, but they are not as common as those with the inlet port at the rear of the cylinder, opposite to the cylinder wall holding the exhaust port. However, the simple two-stroke engine shown in Plate 1.5 is just such an engine. It is a Canadian-built chainsaw engine with the engine cylinder and crankcase components produced as high-pressure aluminum die castings with the cylinder bore surface being hard chromium plated. The open-sided transfer ports are known as "finger ports."

1.4.6 Computer program, Prog.1.3, QUB CROSS ENGINE DRAW

This program is exactly similar in data input and operational terms to Prog.1.2. However, it designs the basic geometry of the QUB type of cross-scavenged engine, as shown in Fig. 1.4 and discussed in Sec. 1.2.2.

You might well inquire as to the design of the conventional cross-scavenged unit as sketched in Fig. 1.3 and also discussed in the same section. The timing edges for the control of both the exhaust and transfer ports are at the same height in the conventional deflector piston engine. This means that the same geometry of design applies to it as for the loop-scavenged engine. Consequently, Prog.1.2 applies equally well.

Because the exhaust and transfer port timing edges are at different heights in the QUB type of engine, separated by the height of the deflector, a different program is required and is listed as Prog.1.3 in the Appendix. An example of the calculation is presented in Fig. 1.13. As with Prog.1.2, the engine rotates for one complete cycle. One data input value deserves an explanation: the "wrist pin to crown" value, shown in the output Fig. 1.13 as 25 mm, is that value from the gudgeon pin to the crown on the scavenge side of the piston, and is not the value to the top of the deflector. As the engine rotates, one of the interesting features of this type of engine appears: the piston rings are below the bottom edge of the exhaust ports as the exhaust flow is released by the deflector top edge. Consequently, the exhaust flame does not partially burn the oil on the piston rings, as it does on a loop-scavenged design. As mentioned earlier, the burning of oil on the piston rings and within the ring grooves eventually causes the rings to stick in their grooves and deteriorates the sealing effect of the rings during the com-

Design and Simulation of Two-Stroke Engines

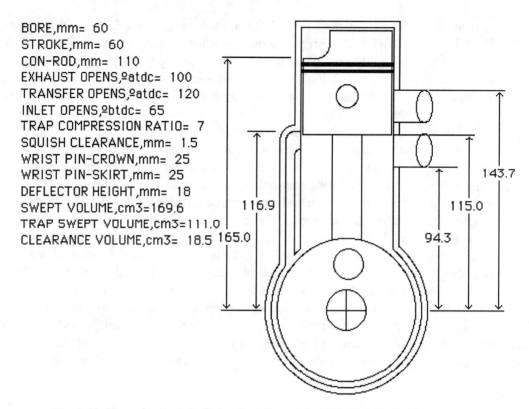

Fig. 1.13 Example of a calculation from Prog.1.3, QUB CROSS ENGINE DRAW.

pression and expansion strokes, reducing both power output and fuel economy. Therefore, the QUB-type engine has enhanced engine reliability and efficiency in this regard over the life span of the power unit.

The data values used for the basic engine geometry in Figs. 1.10-1.13 are common for all three program examples, so it is useful to compare the actual data output values for similarities and differences. For example, it can be seen that the QUB cross-scavenged engine is taller to the top of the deflector, yet is the same height to the top of the combustion chamber as the loop-scavenged power unit. In a later chapter, Fig. 4.13 shows a series of engines which are drawn to scale and the view expressed above can be seen to be accurate from that comparative sketch. Indeed, it could be argued that a QUB deflector engine can be designed to be a shorter engine overall than an equivalent loop-scavenged unit with the same bore, stroke and rod lengths.

1.5 Definitions of thermodynamic terms used in connection with engine design and testing

1.5.1 Scavenge ratio and delivery ratio

In Fig. 1.1(c), the cylinder has just experienced a scavenge process in which a mass of fresh charge, m_{as}, has been supplied through the crankcase from the atmosphere. By measuring the atmospheric, i.e., the ambient pressure and temperature, p_{at} and T_{at}, the air density will be given by ρ_{at} from the thermodynamic equation of state, where R_a is the gas constant for air:

$$\rho_{at} = \frac{p_{at}}{R_a T_{at}} \quad (1.5.1)$$

The *delivery ratio*, DR, of the engine defines the mass of air supplied during the scavenge period as a function of a reference mass, m_{dref}, which is that mass required to fill the swept volume under the prevailing atmospheric conditions, i.e.:

$$m_{dref} = \rho_{at} V_{sv} \quad (1.5.2)$$

$$DR = \frac{m_{as}}{m_{dref}} \quad (1.5.3)$$

The *scavenge ratio*, SR, of a naturally aspirated engine defines the mass of air supplied during the scavenge period as a function of a reference mass, m_{sref}, which is the mass that could fill the entire cylinder volume under the prevailing atmospheric conditions, i.e.:

$$m_{sref} = \rho_{at}(V_{sv} + V_{cv}) \quad (1.5.4)$$

$$SR = \frac{m_{as}}{m_{sref}} \quad (1.5.5)$$

The SAE Standard J604d [1.24] refers to and defines delivery ratio. For two-stroke engines the more common nomenclature in the literature is "scavenge ratio," but it should be remembered that the definitions of these air-flow ratios are mathematically different.

Should the engine be supercharged or turbocharged, then the new reference mass, m_{sref}, for the estimation of scavenge ratio is calculated from the state conditions of pressure and temperature of the scavenge air supply, p_s and T_s.

$$\rho_s = \frac{p_s}{R_a T_s} \quad (1.5.6)$$

$$SR = \frac{m_{as}}{\rho_s(V_{sv} + V_{cv})} \quad (1.5.7)$$

The above theory has been discussed in terms of the air flow referred to the swept volume of a cylinder as if the engine is a single-cylinder unit. However, if the engine is a multi-cylinder device, it is the total swept volume of the engine that is under consideration.

1.5.2 Scavenging efficiency and purity

In Chapter 3 it will be shown that for a perfect scavenge process, the very best which could be hoped for is that the *scavenging efficiency*, SE, would be equal to the scavenge ratio, SR. The scavenging efficiency is defined as the mass of delivered air that has been trapped, m_{tas}, by comparison with the total mass of charge, m_{tr}, that is retained at exhaust closure. The trapped charge is composed only of fresh charge trapped, m_{tas}, and exhaust gas, m_{ex}, and any air remaining unburned from the previous cycle, m_{ar}, where:

$$m_{tr} = m_{tas} + m_{ex} + m_{ar} \tag{1.5.8}$$

Hence, scavenging efficiency, SE, defines the effectiveness of the scavenging process, as can be seen from the following statement:

$$SE = \frac{m_{tas}}{m_{tr}} = \frac{m_{tas}}{m_{tas} + m_{ex} + m_{ar}} \tag{1.5.9}$$

However, the ensuing combustion process will take place between all of the air in the cylinder with all of the fuel supplied to that cylinder, and it is important to define the purity of the trapped charge in its entirety. The purity of the trapped charge, Π, is defined as the ratio of air trapped in the cylinder before combustion, m_{ta}, to the total mass of cylinder charge, where:

$$m_{ta} = m_{tas} + m_{ar} \tag{1.5.10}$$

$$\Pi = \frac{m_{ta}}{m_{tr}} \tag{1.5.11}$$

In many technical papers and textbooks on two-stroke engines, the words "scavenging efficiency" and "purity" are somewhat carelessly interchanged by the authors, assuming prior knowledge by the readers. They assume that the value of m_{ar} is zero, which is generally true for most spark-ignition engines and particularly when the combustion process is rich of stoichiometric, but it would not be true for two-stroke diesel engines where the air is never totally consumed in the combustion process, and it would not be true for similar reasons for a stratified combustion process in a gasoline-fueled spark-ignition engine. More is written on this subject in Secs. 1.5.5 and 1.6.3 where stoichiometry and trapping efficiency measurements are debated, respectively.

1.5.3 Trapping efficiency

Definitions are also to be found in the literature [1.24] for *trapping efficiency*, TE. Trapping efficiency is the capture ratio of mass of delivered air that has been trapped, m_{tas}, to that supplied, m_{as}, or:

$$TE = \frac{m_{tas}}{m_{as}} \tag{1.5.12}$$

It will be seen that expansion of Eq. 1.5.12 gives:

$$TE = \frac{m_{tr}SE}{m_{sref}SR} \quad (1.5.13)$$

It will also be seen under ideal conditions, in Chapter 3, that m_{tr} can be considered to be equal to m_{sref} and that Eq. 1.5.13 can be simplified in an interesting manner, i.e.,

$$TE = \frac{SE}{SR}$$

In Sec. 1.6.3, a means of measuring trapping efficiency in a firing engine from exhaust gas analysis will be described.

1.5.4 Charging efficiency

Charging efficiency, CE, expresses the ratio of the filling of the cylinder with air, by comparison with filling that same cylinder perfectly with air at the onset of the compression stroke. After all, the object of the design exercise is to fill the cylinder with the maximum quantity of air in order to burn a maximum quantity of fuel with that same air. Hence, charging efficiency, CE, is given by:

$$CE = \frac{m_{tas}}{m_{sref}} \quad (1.5.14)$$

It is also the product of trapping efficiency and scavenge ratio, as shown here:

$$CE = \frac{m_{tas}}{m_{as}} \times \frac{m_{as}}{m_{sref}} = TE \times SR \quad (1.5.15)$$

It should be made quite clear that this definition is not precisely as defined in SAE J604d [1.24]. In that SAE nomenclature Standard, the reference mass is declared to be m_{dref} from Eq. 1.5.2, and not m_{sref} as used from Eq. 1.5.4. My defense for this is "custom and practice in two-stroke engines," the fact that it is all of the cylinder space that is being filled and not just the swept volume, the convenience of charging efficiency assessment by the relatively straightforward experimental acquisition of trapping efficiency and scavenge ratio, and the opinion of Benson [1.4, Vol. 2].

1.5.5 Air-to-fuel ratio

It is important to realize that there are narrow limits of acceptability for the combustion of air and fuel, such as gasoline or diesel. In the case of gasoline, the ideal fuel is octane, C_8H_{18}, which burns "perfectly" with air in a balanced equation called the stoichiometric equation. Most students will recall that air is composed, volumetrically and molecularly, of 21 parts oxygen and 79 parts nitrogen. Hence, the chemical equation for complete combustion becomes:

$$2C_8H_{18} + 25\left[O_2 + \frac{79}{21}N_2\right] = 16CO_2 + 18H_2O + 25\frac{79}{21}N_2 \tag{1.5.16}$$

This produces the information that the ideal stoichiometric *air-to-fuel ratio*, AFR, is such that for every two molecules of octane, we need 25 molecules of air. As we normally need the information in mass terms, then as the molecular weights of O_2, H_2, N_2 are simplistically 32, 2 and 28, respectively, and the atomic weight of carbon C is 12, then:

$$\text{AFR} = \frac{25 \times 32 + 25 \times 28 \times \frac{79}{21}}{2(8 \times 12 + 18 \times 1)} = 15.06 \tag{1.5.17}$$

As the equation is balanced, with the exact amount of oxygen being supplied to burn all of the carbon to carbon dioxide and all of the hydrogen to steam, such a burning process yields the minimum values of carbon monoxide emission, CO, and unburned hydrocarbons, HC. Mathematically speaking they are zero, and in practice they are also at a minimum level. As this equation would also produce the maximum temperature at the conclusion of combustion, this gives the highest value of emissions of NO_x, the various oxides of nitrogen. Nitrogen and oxygen combine at high temperatures to give such gases as N_2O, NO, etc. Such statements, although based in theory, are almost exactly true in practice as illustrated by the expanded discussion in Chapters 4 and 7.

As far as combustion limits are concerned, although Chapter 4 will delve into this area more thoroughly, it may be helpful to point out at this stage that the rich misfire limit of gasoline-air combustion probably occurs at an air-fuel ratio of about 9, peak power output at an air-fuel ratio of about 13, peak thermal efficiency (or minimum specific fuel consumption) at an air-fuel ratio of about 14, and the lean misfire limit at an air-fuel ratio of about 18. The air-fuel ratios quoted are those in the combustion chamber at the time of combustion of a homogeneous charge, and are referred to as the *trapped air-fuel ratio*, AFR_t. The air-fuel ratio derived in Eq. 1.5.17 is, more properly, the trapped air-fuel ratio, AFR_t, needed for stoichiometric combustion.

To briefly illustrate that point, in the engine shown in Fig. 1.6 it would be quite possible to scavenge the engine thoroughly with fresh air and then supply the appropriate quantity of fuel by direct injection into the cylinder to provide a AFR_t of, say, 13. Due to a generous oversupply of scavenge air the overall AFR_o could be in excess of, say, 20.

1.5.6 Cylinder trapping conditions

The point of the foregoing discussion is to make you aware that the net effect of the cylinder scavenge process is to fill the cylinder with a mass of air, m_{ta}, within a total mass of charge, m_{tr}, at the trapping point. This total mass is highly dependent on the trapping pressure, as the equation of state shows:

Chapter 1 - Introduction to the Two-Stroke Engine

$$m_{tr} = \frac{p_{tr} V_{tr}}{R_{tr} T_{tr}} \tag{1.5.18}$$

where
$$V_{tr} = V_{ts} + V_{cv} \tag{1.5.19}$$

In any given case, the trapping volume, V_{tr}, is a constant. This is also true of the gas constant, R_{tr}, for gas at the prevailing gas composition at the trapping point. The gas constant for exhaust gas, R_{ex}, is almost identical to the value for air, R_a. Because the cylinder gas composition is usually mostly air, the treatment of R_{tr} as being equal to R_a invokes little error. For any one trapping process, over a wide variety of scavenging behavior, the value of trapping temperature, T_{tr}, would rarely change by 5%. Therefore, it is the value of trapping pressure, P_{tr}, that is the significant variable. As stated earlier, the value of trapping pressure is directly controlled by the pressure wave dynamics of the exhaust system, be it a single-cylinder engine with or without a tuned exhaust system, or a multi-cylinder power unit with a branched exhaust manifold. The methods of design and analysis for such complex systems are discussed in Chapters 2 and 5. The value of the trapped fuel quantity, m_{tf}, can be determined from:

$$m_{tf} = \frac{m_{ta}}{AFR_t} \tag{1.5.20}$$

1.5.7 Heat released during the burning process

The total value of the heat that will be released from the combustion of this quantity of fuel will be Q_R:

$$Q_R = \eta_c m_{tf} C_{fl} \tag{1.5.21}$$

where η_c is the *combustion efficiency* and C_{fl} is the (lower) *calorific value* of the fuel in question.

A further discussion of this analysis, in terms of an actual experimental example, is given in Sec. 4.2.

1.5.8 The thermodynamic cycle for the two-stroke engine

This is often referred to as a derivative of the Otto Cycle, and a full discussion can be found in many undergraduate textbooks on internal combustion engines or thermodynamics, e.g., Taylor [1.3]. The result of the calculation of a theoretical cycle can be observed in Figs. 1.14 and 1.15, by comparison with measured pressure-volume data from an engine of the same compression ratios, both trapped and geometric. In the measured case, the cylinder pressure data are taken from a 400 cm^3 single-cylinder two-stroke engine running at 3000 rpm at wide open throttle. In the theoretical case, and this is clearly visible on the log p-log V plot in Fig. 1.15, the following assumptions are made: (a) compression begins at trapping, (b) all heat release (combustion) takes place at tdc at constant volume, (c) the exhaust process is considered as a heat rejection process at release, (d) the compression and expansion processes occur under ideal, or isentropic, conditions with air as the working fluid, and so those processes are calculated as:

$$pV^\gamma = \text{constant}$$

where γ is a constant. For air, the ratio of specific heats, γ, has a value of 1.4. A fundamental theoretical analysis would show [1.3] that the *thermal efficiency*, η_t, of the cycle is given by:

$$\eta_t = 1 - \frac{1}{CR_t^{\gamma-1}} \qquad (1.5.22)$$

Thermal efficiency is defined as:

$$\eta_t = \frac{\text{work produced per cycle}}{\text{heat available as input per cycle}} \qquad (1.5.23)$$

As the actual 400 cm³ engine has a trapped compression value of 7, and from Eq. 1.5.22 the theoretical value of thermal efficiency, η_t, is readily calculated as 0.541, the considerable disparity between fundamental theory and experimentation becomes apparent, for the measured value is about one-half of that calculated, at 27%.

Upon closer examination of Figs. 1.14 and 1.15, the theoretical and measured pressure traces look somewhat similar and the experimental facts do approach the theoretical presumptions. However, the measured expansion and compression indices are at 1.33 and 1.17, respectively, which is rather different from the ideal value of 1.4 for air. On the other hand,

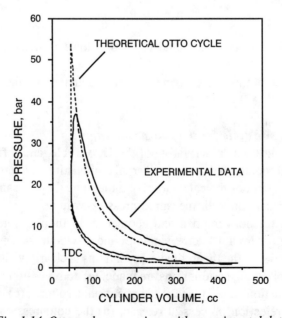

Fig. 1.14 Otto cycle comparison with experimental data.

Chapter 1 - Introduction to the Two-Stroke Engine

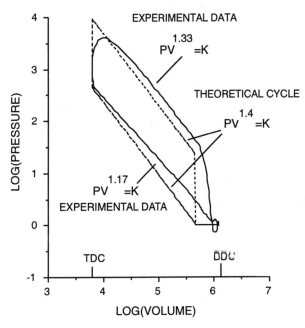

Fig. 1.15 Logarithmic plot of pressure and volume.

the actual compression process clearly begins before the official trapping point at exhaust port closure, and this in an engine with no tuned exhaust pipe. The theoretical assumption of a constant volume process for the combustion and exhaust processes is clearly in error when the experimental pressure trace is examined. The peak cycle pressures of 54 bar calculated and 36 bar measured are demonstrably different. In Chapters 4 and 5 a more advanced theoretical analysis will be seen to approach the measurements more exactly.

The work on the piston during the cycle is ultimately and ideally the work delivered to the crankshaft by the connecting rod. The word "ideal" in thermodynamic terms means that the friction or other losses, like leakage past the piston, are not taken into consideration in the statement made above. Therefore, the ideal work produced per cycle (see Eq. 1.5.24) is that work carried out on the piston by the force, F, created from the gas pressure, p. Work is always the product of force and distance, x, moved by that force, so, where A is the piston area:

$$\text{Work produced per cycle} = \int F dx = \int pA dx = \int p dV \qquad (1.5.24)$$

Therefore, the work produced for any given engine cycle, in the case of a two-stroke engine for one crankshaft revolution from tdc to tdc, is the cyclic integral of the pressure-volume diagram in the cylinder above the piston. By the same logic, the pumping work required in the crankcase is the cyclic integral of the pressure-volume diagram in the crankcase. In both cases, this work value is the enclosed area on the pressure-volume diagram, be it a theoretical cycle or the actual cycle as illustrated in Fig. 1.14. The above statements are illustrated in Fig. 1.16 for the actual data shown previously in Figs. 1.14 and 1.15.

Design and Simulation of Two-Stroke Engines

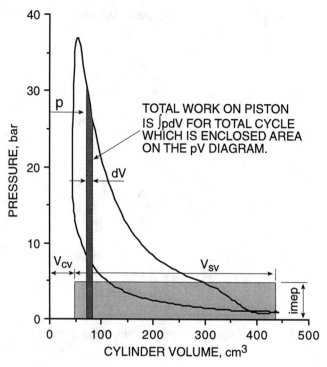

Fig. 1.16 Determination of imep from the cylinder p-V diagram.

1.5.9 The concept of mean effective pressure

As stated above, the enclosed p-V diagram area is the work produced on the piston, in either the real or the ideal cycle. Fig. 1.16 shows a second rectangular shaded area, equal in area to the enclosed cylinder p-V diagram. This rectangle is of height imep and of length V_{sv}, where imep is known as the *indicated mean effective pressure* and V_{sv} is the swept volume. The word "indicated" stems from the historical fact that pressure transducers for engines used to be called "indicators" and the p-V diagram, of a steam engine traditionally, was recorded on an "indicator card." The concept of mean effective pressure is extremely useful in relating one engine development to another for, while the units of imep are obviously that of pressure, the value is almost dimensionless. That remark is sufficiently illogical as to require careful explanation. The point is, any two engines of equal development or performance status will have identical values of mean effective pressure, even though they may be of totally dissimilar swept volume. In other words, Figs. 1.14, 1.15 and 1.16 could have equally well been plotted as pressure-compression (or volume) ratio plots and the values of imep would be identical for two engines of differing swept volume, if the diagrammatic profiles in the pressure direction were also identical.

1.5.10 Power and torque and fuel consumption

Power is defined as the rate of doing work. If the engine rotation rate is rps, revolutions per second, and the two-stroke engine has a working cycle per crankshaft revolution, then the power delivered to the piston crown by the gas force is called the *indicated power output*, $\dot{W}_i$,

Chapter 1 - Introduction to the Two-Stroke Engine

where:

$$\dot{W}_i = \text{imep} \times V_{sv} \times (\text{work cycles per second})$$
$$= \text{imep} \times V_{sv} \times \text{rps} \quad \text{- for a two-stroke engine}$$
$$= \text{imep} \times V_{sv} \times \frac{\text{rps}}{2} \quad \text{- for a four-stroke engine} \quad (1.5.25)$$

For a four-stroke cycle engine, which has a working cycle lasting two crankshaft revolutions, the working cycle rate is 50% of the rps value, and this should be inserted into Eq. 1.5.25 rather than rps. In other words, a four-stroke cycle engine of equal power output and equal swept volume has an imep value which is double that of the two-stroke engine. Such is the actual, if somewhat illogical, convention used in everyday engineering practice.

The *indicated torque*, Z_i, is the turning moment on the crankshaft and is related to power output by the following equation:

$$\dot{W}_i = 2\pi Z_i \text{rps} \quad (1.5.26)$$

Should the engine actually consume fuel of calorific value C_{fl} at the measured (or at a theoretically calculated) mass flow rate of $\dot{m}_f$, then the *indicated thermal efficiency*, η_i, of the engine can be predicted from an extension of Eq. 1.5.23:

$$\eta_i = \frac{\text{power output}}{\text{rate of heat input}} = \frac{\dot{W}_i}{\dot{m}_f C_{fl}} \quad (1.5.27)$$

Of great interest and in common usage in engineering practice is the concept of specific fuel consumption, the fuel consumption rate per unit power output. Hence, to continue the discussion on indicated values, *indicated specific fuel consumption*, isfc, is given by:

$$\text{isfc} = \frac{\text{fuel consumption rate}}{\text{power output}} = \frac{\dot{m}_f}{\dot{W}_i} \quad (1.5.28)$$

It will be observed from a comparison of Eqs. 1.5.27 and 1.5.28 that thermal efficiency and specific fuel consumption are reciprocally related to each other, without the employment of the calorific value of the fuel. As most petroleum-based fuels have virtually identical values of calorific value, then the use of specific fuel consumption as a comparator from one engine to another, rather than thermal efficiency, is quite logical and is more immediately useful to the designer and the developer.

1.6 Laboratory testing of two-stroke engines
1.6.1 Laboratory testing for power, torque, mean effective pressure and specific fuel consumption

Most of the testing of engines for their performance characteristics takes place under laboratory conditions. The engine is connected to a power-absorbing device, called a dynamometer, and the performance characteristics of power, torque, fuel consumption rate, and air consumption rate, at various engine speeds, are recorded. Many texts and papers describe this process and the Society of Automotive Engineers provides a Test Code J1349 for this very purpose [1.14]. There is an equivalent test code from the International Organization for Standardization in ISO 3046 [1.16]. For the measurement of exhaust emissions there is a SAE Code J1088 [1.15] which deals with exhaust emission measurements for small utility engines, and many two-stroke engines fall into this category. The measurement of air flow rate into the engine is often best conducted using meters designed to British Standard BS 1042 [1.17].

Several interesting technical papers have been published in recent times questioning some of the correction factors used within such test codes for the prevailing atmospheric conditions. One of these by Sher [1.18] deserves further study.

There is little point in writing at length on the subject of engine testing and of the correction of the measured performance characteristics to standard reference pressure and temperature conditions, for these are covered in the many standards and codes already referenced. However, some basic facts are relevant to the further discussion and, as the testing of exhaust emissions is a relatively new subject, a simple analytical computer program on that subject, presented in Sec. 1.6.2, should prove to be useful to quite a few readers.

A laboratory engine testing facility is diagrammatically presented in Fig. 1.17. The engine power output is absorbed in the dynamometer, for which the slang word is a "dyno" or a "brake." The latter word is particularly apt as the original dynamometers were, literally, friction brakes. The principle of any dynamometer operation is to allow the casing to swing freely. The reaction torque on the casing, which is exactly equal to the engine torque, is measured on a lever of length, L, from the centerline of the dynamometer as a force, F. This restrains the outside casing from revolving, or the torque and power would not be absorbed. Consequently, the reaction torque measured is the *brake torque*, Z_b, and is calculated by:

$$Z_b = F \times L \qquad (1.6.1)$$

Therefore, the work output from the engine per engine revolution is the distance "traveled" by the force, F, on a circle of radius, L:

$$\text{Work per revolution} = 2\pi FL = 2\pi Z_b \qquad (1.6.2)$$

The measured power output, the *brake power*, $\dot{W}_b$, is the work rate, and at rps rotational speed, is clearly:

$$\dot{W}_b = (\text{Work per rev}) \times (\text{rev/s})$$
$$= 2\pi Z_b \text{rps} = \pi Z_b \frac{\text{rpm}}{30} \qquad (1.6.3)$$

To some, this equation may clear up the apparent mystery of the use of the operator π in the similar theoretical equation, Eq. 1.5.26, when considering the indicated power output, $\dot{W}_i$.

Chapter 1 - Introduction to the Two-Stroke Engine

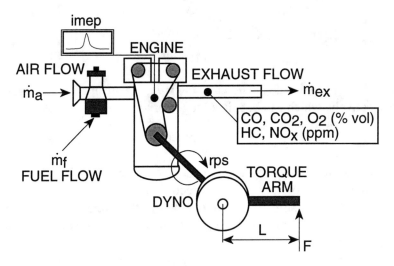

Fig. 1.17 Dynamometer test stand recording of performance parameters.

The *brake thermal efficiency*, η_b, is then given by the corresponding equation to Eq. 1.5.23:

$$\eta_b = \frac{\text{power output}}{\text{rate of heat input}} = \frac{\dot{W}_b}{\dot{m}_f C_{fl}} \qquad (1.6.4)$$

A similar situation holds for *brake specific fuel consumption*, bsfc, and Eq. 1.5.28:

$$\text{bsfc} = \frac{\text{fuel consumption rate}}{\text{power output}} = \frac{\dot{m}_f}{\dot{W}_b} \qquad (1.6.5)$$

However, it is also possible to compute a mean effective pressure corresponding to the measured power output. This is called the *brake mean effective pressure*, bmep, and is calculated from a manipulation of Eq. 1.5.25 in terms of measured values:

$$\text{bmep} = \frac{\dot{W}_b}{V_{sv} \times \text{rps}} \qquad (1.6.6)$$

It is obvious that the brake power output and the brake mean effective pressure are the residue of the indicated power output and the indicated mean effective pressure, after the engine has lost power to internal friction and air pumping effects. These friction and pumping losses deteriorate the indicated performance characteristics by what is known as the engine's *mechanical efficiency*, η_m. Friction and pumping losses are related simply by:

$$\dot{W}_i = \dot{W}_b + \text{friction and pumping power} \qquad (1.6.7)$$

$$\eta_m = \frac{\dot{W}_b}{\dot{W}_i} = \frac{\text{bmep}}{\text{imep}} \qquad (1.6.8)$$

This raises the concept of the *friction* and the *pumping mean effective pressures*, fmep and pmep, respectively, which can be related together as:

$$\text{imep} = \text{bmep} + \text{pmep} + \text{fmep} \qquad (1.6.9)$$

It is often very difficult to segregate the separate contributions of friction and pumping in measurements taken in a laboratory except by recording crankcase pressure diagrams and by measuring friction power using a motoring methodology that eliminates all pumping action at the same time. It is very easy to write the last fifteen words but it is much more difficult to accomplish them in practice.

Finally, the recording of the overall air-fuel ratio, AFR_o, is relatively straightforward as:

$$AFR_o = \frac{\dot{m}_{as}}{\dot{m}_f} \qquad (1.6.10)$$

Continuing the discussion begun in Sec. 1.5.5, this overall air-fuel ratio, AFR_o, is also the trapped air-fuel ratio, AFR_t, if the engine is charged with a homogeneous supply of air and fuel, i.e., as in a carburetted design for a simple two-stroke engine. If the total fuel supply to the engine is, in any sense, stratified from the total air supply, this will not be the case.

1.6.2 Laboratory testing for exhaust emissions from two-stroke engines

There have been quite a few technical contributions in this area [1.7] [1.15] [1.19] [1.25] as the situation for the two-stroke engine is subtly different from the four-stroke engine case. Much of the instrumentation available has been developed for, and specifically oriented toward, four-stroke cycle engine measurement and analysis. As has been pointed out in Secs. 1.1 and 1.2, a significant portion of the scavenge air ends up in the exhaust pipe without enduring a combustion process. Whereas a change from a lean to a rich combustion process, in relation to the stoichiometric air-fuel ratio, for a four-stroke engine might change the exhaust oxygen concentration from 2% to almost 0% by volume, in a two-stroke engine that might produce an equivalent shift from 10% to 8%. Thus, in a two-stroke engine the exhaust oxygen concentration is always high. Equally, if the engine has a simple carburetted fueling device, then the bypassed fuel along with that short-circuited air produces a very large count of unburned hydrocarbon emission. Often, this count is so high that instruments designed for use with four-stroke cycle engines will not record it!

In today's legislative-conscious world, it is important that exhaust emissions are recorded on a mass basis. Most exhaust gas analytical devices measure on a volumetric or molecular basis. It is necessary to convert such numbers to permit comparison of engines on their effectiveness in reducing exhaust emissions at equal power levels. From this logic appears the concept of deriving measured, or brake specific, emission values for such pollutants as carbon monoxide, unburned hydrocarbons, and oxides of nitrogen. The first of these pollutants is toxic, the second is blamed for "smog" formation, and the last is regarded as a major contributor to "acid rain." These and other facets of pollution are discussed more extensively in Chapter 7.

As an example, consider a pollutant gas, PG, with molecular weight, M_g, and a volumetric concentration in the exhaust gas of proportion, V_{cg}. In consequence, the numerical value in ppm, V_{ppmg}, would be $10^6 V_{cg}$ and as % by volume, $V_{\%g}$, it would be $100 V_{cg}$. The average molecular weight of the exhaust gas is M_{ex}. The power output is $\dot{W}_b$ and the fuel consumption rate is $\dot{m}_f$. The total mass flow rate of exhaust gas is $\dot{m}_{ex}$:

$$\dot{m}_{ex} = (1 + AFR_o)\dot{m}_f \quad kg/s$$

$$= \frac{(1 + AFR_o)\dot{m}_f}{M_{ex}} \quad kgmol/s \qquad (1.6.11)$$

$$\text{Pollutant gas flow rate} = (M_g V_{cg})\frac{(1 + AFR_o)\dot{m}_f}{M_{ex}} \quad kg/s \qquad (1.6.12)$$

Brake specific pollutant gas flow rate, bsPG, is then:

$$bsPG = \frac{M_g V_{cg}}{\dot{W}_b} \times \frac{(1 + AFR_o)\dot{m}_f}{M_{ex}} \quad kg/Ws \qquad (1.6.13)$$

$$bsPG = \frac{(M_g V_{cg}) \times (1 + AFR_o) bsfc}{M_{ex}} \quad kg/Ws \qquad (1.6.14)$$

By quoting an actual example, this last equation is readily transferred into the usual units for the reference of any exhaust pollutant. If bsfc is employed in the conventional units of kg/kWh and the pollutant measurement of, say, carbon monoxide is in % by volume, the brake specific carbon monoxide emission rate, bsCO, in g/kWh units is given by:

$$bsCO = 10(1 + AFR_o) \, bsfc \, V_{\%CO} \frac{28}{29} \quad g/kWh \qquad (1.6.15)$$

where the average molecular weights of exhaust gas and carbon monoxide are assumed simplistically and respectively to be 29 and 28.

The actual mass flow rate of carbon monoxide in the exhaust pipe is

$$\dot{m}_{CO} = bsCO \times \dot{W}_b \quad g/h \qquad (1.6.16)$$

These equations are programmed into Prog.1.4, EXHAUST GAS ANALYSIS, and should be useful to those who are involved in this form of measurement in connection with engine research and development. The pollutants covered by Prog.1.4 are carbon monoxide, carbon dioxide, hydrocarbons and oxides of nitrogen. Brake specific values for air and oxygen are also produced. An example of the use of this calculation, an encapsulation of the computer screen during a running of the program, is illustrated in Fig. 1.18.

```
enter air to fuel ratio, AF? 20
enter brake specific fuel consumption, BSFC, as kg/kWh? .315
enter oxygen concentration in the exhaust gas as % vol, O2VOL? 7.1
enter carbon monoxide emission as % vol, COVOL? .12
enter carbon dioxide emission as % vol, CO2VOL? 6.9
enter oxides of nitrogen emission as ppm, NOX? 236
the unburned hydrocarbon emission values will have been measured as-
either HC ppm by a NDIR system as hexane equivalent, C6H14,
or as HC ppm by a FID system as methane equivalent, CH4
type in the name of the type of measurement system, either 'NDIR' or 'FID'? NDIR
enter the HC as ppm measured by NDIR system? 356
OUTPUT DATA
The brake specific air consumption, BSAC, in kg/kWh units, is      6.3
The brake specific emission values printed are in g/kWh units
The brake specific carbon monoxide value, BSCO, is    7.7
The brake specific nitrogen oxide value, BSNOX, is    1.6
The brake specific carbon dioxide value, BSCO2, is  692.5
The brake specific oxygen value, BSO2, is  518.3
The brake specific hydrocarbon value by NDIR system, BSHC, is      7.0
The Trapping Efficiency, TE , as %, is   64.5
WANT A PRINT-OUT(Y OR N?)? N
```

Fig. 1.18 Example of the use of Prog.1.4, EXHAUST GAS ANALYSIS.

It should be pointed out that there are various ways of recording exhaust emissions as values "equivalent to a reference gas." In the measurement of hydrocarbons, either by a NDIR (non-dispersive infrared) device or by a FID (flame ionization detector), the readings are quoted as ppm hexane, or ppm methane, respectively. Therefore Prog.1.4 contains, in the appropriate equations, the molecular weights for hexane, C_6H_{14}, or methane, CH_4. Some FID meters use a $CH_{1.85}$ equivalent [1.15] and in that case, in the relevant equation in Prog.1.4, the molecular weight for HC equivalent would have to be replaced by 13.85. The same holds true for nitrogen oxide emission, and in Prog.1.4 it is assumed that it is to be NO and so the molecular weight for NO, 30, is employed. Should the meter used in a particular laboratory be different, then, as for the HC example quoted above, the correct molecular weight of the reference gas should be employed.

A further complication arises for the comparison of HC values recorded by NDIR and FID instrumentation. This is discussed by Tsuchiya and Hirano [1.19] and they point out that the NDIR reading should be multiplied by a sensitivity factor "K" to obtain equality with that recorded by a FID system. They illustrate this in graphical form (Fig. 1.19). In Prog.1.4, this sensitivity factor "K" is taken as unity.

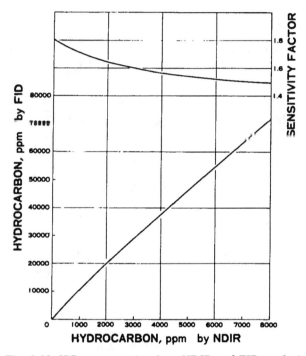

Fig. 1.19 HC concentration from NDIR and FID analysis (from Ref. [1.19]).

1.6.3 Trapping efficiency from exhaust gas analysis

In an engine where the combustion is sufficiently rich, or is balanced as in the stoichiometric equation, Eq. 1.5.16, it is a logical presumption that any oxygen in the exhaust gas must come from scavenge air which has been lost to the exhaust system. Actually, a stoichiometric air-fuel ratio in practice would still have some residual oxygen in the exhaust gas. So, such an experimental test would be conducted with a sufficiently rich mixture during the combustion process as to ensure that no free oxygen remained within the cylinder after the combustion period. Measurements and calculations of exhaust oxygen content as a function of air-to-fuel ratio are shown in Figs. 7.3-7.5 and Figs. 7.13-7.15, respectively, which support this presumption. If the actual combustion process were deliberately stratified, as in a diesel engine, then that would be a very difficult condition to satisfy. However, on the assumption that this condition can be met, and it is possible as most simple two-stroke engines are homogeneously charged, the trapping efficiency, TE, can be calculated from the exhaust gas analysis as follows:

$$\text{Exhaust gas mass flow rate} = \frac{(1 + AFR_o)\dot{m}_f}{M_{ex}} \text{ kgmol/s} \quad (1.6.17)$$

$$\text{Exhaust } O_2 \text{ mass flow rate} = \frac{(1 + AFR_o)\dot{m}_f V_{\%O_2}}{100 M_{ex}} \text{ kgmol/s} \quad (1.6.18)$$

$$\text{Engine } O_2 \text{ mass inflow rate} = 0.2314 \frac{\dot{m}_f AFR_o}{M_{O_2}} \text{ kgmol/s} \quad (1.6.19)$$

The numerical value of 0.2314 is the mass fraction of oxygen in air and 32 is the molecular weight of oxygen. The value noted as $V_{\%O_2}$ is the percentage volumetric concentration of oxygen in the exhaust gas.

Trapping efficiency, TE, is given by:

$$TE = \frac{\text{air trapped in cylinder}}{\text{air supplied}}$$

Hence:

$$TE = 1 - \frac{\text{air lost to exhaust}}{\text{air supplied}} = 1 - \frac{\text{Eq. 1.6.18}}{\text{Eq. 1.6.19}}$$

or:

$$TE = 1 - \frac{(1 + AFR_o)V_{\%O_2}M_{O_2}}{23.14 \times AFR_o M_{ex}} \quad (1.6.20)$$

Assuming simplistically that the average molecular weight of exhaust gas is 29 and that oxygen is 32, and that atmospheric air contains 21% oxygen by volume this becomes:

$$TE = 1 - \frac{(1 + AFR_o)V_{\%O_2}}{21 \times AFR_o} \quad (1.6.21)$$

This equation, produced by Kee [1.20], is programmed together with the other parameters in Prog.1.4.

The methodology emanates from history in a paper by Watson [1.21] in 1908. Huber [1.22] basically uses the Watson approach, but provides an analytical solution for trapping efficiency, particularly for conditions where the combustion process yields some free oxygen. The use of this analytical technique, to measurements taken in a two-stroke engine under firing conditions, is described by Blair and Kenny [1.23]; they provide further data on the in-cylinder conditions at the same time using the experimental device shown in Plate 3.3.

1.7 Potential power output of two-stroke engines

At this stage of the book, it will be useful to be able to assess the potential power output of two-stroke engines. From Eq. 1.6.6, the power output of an engine delivered at the crankshaft is seen to be:

$$\dot{W}_b = bmep_b V_{sv} \text{ rps} \qquad (1.7.1)$$

From experimental work for various types of two-stroke engines, the potential levels of attainment of brake mean effective pressure are well known within quite narrow limits. The literature is full of experimental data for this parameter, and the succeeding chapters of this book provide further direct information on the matter, often predicted directly by engine modeling computer programs. Some typical levels of bmep for a brief selection of engine types are given in Fig. 1.20.

Engine Type	bmep, bar	Piston Speed, m/s	Bore/Stroke Ratio
Single-cylinder spark-ignition engines			
A untuned silenced exhaust	4.5 - 6.0	12 - 14	1.0 - 1.3
B tuned silenced exhaust	8.0 - 9.0	12 - 16	1.0 - 1.3
C tuned unsilenced exhaust	10.0 - 11.0	16 - 22	1.0 - 1.2
Multi-cylinder spark-ignition engines			
D two-cylinder exhaust tuned	6.0 - 7.0	12 - 14	1.0 - 1.2
E 3+ cylinders exhaust tuned	7.0 - 9.0	12 - 20	1.0 - 1.3
Compression-ignition engines			
F naturally aspirated engine	3.5 - 4.5	10 - 13	0.85 - 1.0
G supercharged engine	6.5 - 10.5	10 - 13	0.85 - 1.0
H turbocharged marine unit	8.0 - 14.0	10 - 13	0.5 - 0.9

Fig.1.20 Potential performance criteria for some two-stroke engines.

The engines, listed as A-H, can be related to types which are familiar as production devices. For example, type A could be a chainsaw engine or a small outboard motor of less than 5 hp. The type B engine would appear in a motorcycle for both on- or off-road applications. The type C engine would be used for competition purposes, such as motocross or road-racing. The type D engine could also be a motorcycle, but is more likely to be an outboard motor or a snowmobile engine. The type E engine is almost certain to be an outboard motor. The type F engine is possibly an electricity-generating set engine, whereas type G is a truck power unit, and type H could be either a truck engine or a marine propulsion unit. Naturally, this table contains only the broadest of classifications and could be expanded into many sub-sets, each with a known band of attainment of brake mean effective pressure.

Therefore, it is possible to insert this data into Eq. 1.7.1, and for a given engine total swept volume, V_{sv}, at a rotation rate, rps, determine the power output, $\dot{W}_b$. It is quite clear that this might produce some optimistic predictions of engine performance, say, by assuming a bmep of 10 bar for a single-cylinder spark-ignition engine of 500 cm³ capacity running at an improbable speed of 20,000 rpm. However, if that engine had ten cylinders, each of 50 cm³ capacity, it might be mechanically possible to rotate it safely at that speed! Thus, for any prediction of power output to be realistic, it becomes necessary to accurately assess the possible speed of rotation of an engine, based on criteria related to its physical dimensions.

1.7.1 Influence of piston speed on the engine rate of rotation

The maximum speed of rotation of an engine depends on several factors, but the principal one, as demonstrated by any statistical analysis of known engine behavior, is the mean piston speed, c_p. This is not surprising as a major limiting factor in the operation of any engine is the lubrication of the main cylinder components, the piston and the piston rings. In any given design the oil film between those components and the cylinder liner will deteriorate at some particular rubbing velocity, and failure by piston seizure will result. The mean piston speed, c_p, is given by:

$$c_p = 2 \times L_{st} \times \text{rps} \quad (1.7.2)$$

As one can vary the bore and stroke for any design within a number of cylinders, n, to produce a given total swept volume, the bore-stroke ratio, C_{bs}, is determined as follows:

$$C_{bs} = \frac{d_{bo}}{L_{st}} \quad (1.7.3)$$

The total swept volume of the engine can now be written as:

$$V_{sv} = n \frac{\pi}{4} d_{bo}^2 L_{st} = n \frac{\pi}{4} C_{bs}^2 L_{st}^3 \quad (1.7.4)$$

Substitution of Eqs. 1.7.2 and 1.7.4 into Eq. 1.7.1 reveals:

$$\dot{W}_b = \frac{c_p \text{bmep}}{2} \times (C_{bs} \times V_{sv})^{0.666} \times \left(\frac{\pi n}{4}\right)^{0.333} \quad (1.7.5)$$

This equation is strictly in SI units. Perhaps a more immediately useful equation in familiar working units, where the measured or brake power output is in kW, $\dot{W}_{kW}$, the bmep is in bar, bmep_{bar}, and the total swept volume is in cm³ units, V_{svcc}, is:

$$\begin{aligned}\dot{W}_{kW} &= \frac{c_p \text{bmep}_{bar}}{200} \times (C_{bs} \times V_{svcc})^{0.666} \times \left(\frac{\pi n}{4}\right)^{0.333} \\ &= \frac{c_p \text{bmep}_{bar}}{216.78} \times (C_{bs} \times V_{svcc})^{0.666} \times n^{0.333}\end{aligned} \quad (1.7.6)$$

Chapter 1 - Introduction to the Two-Stroke Engine

The values for bore-stroke ratio and piston speed, which are typical of the engines listed as types A-H, are shown in Fig. 1.20. It will be observed that the values of piston speed are normally in a common band from 12 to 14 m/s for most spark-ignition engines, and those with values about 20 m/s are for engines for racing or competition purposes which would have a relatively short lifespan. The values typical of diesel engines are slightly lower, reflecting not only the heavier cylinder components required to withstand the greater cylinder pressures but also the reducing combustion efficiency of the diesel cycle at higher engine speeds and the longer lifespan expected of this type of power unit. It will be observed that the bore-stroke ratios for petrol engines vary from "square" at 1.0 to "oversquare" at 1.3. The diesel engine, on the other hand, has bore-stroke ratios which range in the opposite direction to "undersquare," reflecting the necessity for suitable proportioning of the smaller combustion chamber of that higher compression ratio power unit.

1.7.2 Influence of engine type on power output

With the theory developed in Eqs. 1.7.5 or 1.7.6, it becomes possible by the application of the bmep, bore-stroke ratio and piston speed criteria to predict the potential power output of various types of engines. This type of calculation would be the opening gambit of theoretical consideration by a designer attempting to meet a required target. Naturally, the statistical information available would be of a more extensive nature than the broad bands indicated in Fig. 1.20, and would form what would be termed today as an "expert system." As an example of the use of such a calculation, three engines are examined by the application of this theory and the results shown in Fig. 1.21.

Engine Type	Type A	Type C	Type G
Input Data	*Chainsaw*	*Racing Motor*	*Truck Diesel*
power, kW	5.2	46.2	186.0
piston speed, m/s	12	20	10
bore/stroke ratio	1.3	1	0.9
bmep, bar	4.5	10	7
number of cylinders	1	2	6
Output Data			
bore, mm	49.5	54.0	106.0
stroke, mm	38.0	54.0	118.0
swept volume, cm^3	73.8	250.9	6300
engine speed, rpm	9440	11,080	2545

Fig. 1.21 Calculation output predicting potential engine performance.

The engines are very diverse in character such as a small chainsaw, a racing motorcycle engine, and a truck diesel powerplant. The input and output data for the calculation are declared in Fig. 1.21 and are culled from those applicable to the type of engine postulated in Fig. 1.20. The target power output in the data table are in kW units, but in horsepower values are 7 bhp for the chainsaw, 62 bhp for the racing motorcycle engine, and 250 bhp for the truck diesel engine.

The physical dimensions predicted for the three engines are seen to be very realistic, and from a later discussion in Chapters 5 and 6, the reported behavior of engines such as the chainsaw and the racing motorcycle engine will confirm that statement. The varied nature of the specific power performance from these very different engines is observed from the 95 bhp/liter for the chainsaw, 247 bhp/liter for the racing motorcycle engine, and 39.7 bhp/liter for the truck diesel power unit. The most useful part of this method of initial prediction of the potential power performance of an engine is that some necessary pragmatism is injected into the selection of the data for the speed of rotation of the engine.

Subscript notation for Chapter 1

a	air
ar	air retained
as	air supplied
at	atmosphere
b	brake
c	combustion
cg	gas concentration by proportion
CO	carbon monoxide
CO_2	carbon dioxide
dref	reference for delivery ratio
ex	exhaust gas
f	fuel
fl	fuel calorific value
g	gas
i	indicated
o	overall
O_2	oxygen
%g	gas concentration by % volume
ppmg	gas concentration by ppm
sref	reference for scavenge ratio
t	trapped
ta	trapped air
tas	trapped air supplied
tf	trapped fuel
tr	trapping

References for Chapter 1

1.1 P.E. Irving, Two-Stroke Power Units, their Design and Application, Temple Press Books for Newnes, London, 1967.

1.2 K.G. Draper, The Two-Stroke Engine, Design and Tuning, Foulis, Oxford, 1968.

1.3 C.F. Taylor, E.S. Taylor, The Internal Combustion Engine, International Textbook Company, Scranton, Pa., 1962.

1.4 R.S. Benson, N.D. Whitehouse, Internal Combustion Engines, Vols. 1 and 2, Pergamon, Oxford, 1979.

1.5 C.F. Caunter, Motor Cycles, a Technical History, Science Museum, London, HMSO, 1970.

1.6 P.H. Schweitzer, Scavenging of Two-Stroke Cycle Diesel Engines, Macmillan, New York, 1949.

1.7 G.P. Blair, R.R. Booy, B.L. Sheaffer, (Eds.), Technology Pertaining to Two-Stroke Cycle Spark-Ignition Engines, SAE PT-26, Society of Automotive Engineers, Warrendale, Pa., 1982.

1.8 W.J.D. Annand, G.E. Roe, Gas Flow in the Internal Combustion Engine, Foulis, Yeovil, Somerset, 1974.

1.9 G.P. Blair, R.A.R. Houston, R.K. McMullan, N. Steele, S.J. Williamson, "A New Piston Design for a Cross-Scavenged Two-Stroke Cycle Engine with Improved Scavenging and Combustion Characteristics," SAE Paper No. 841096, Society of Automotive Engineers, Warrendale, Pa., 1984.

1.10 G.P. Blair, R.G. Kenny, J.G. Smyth, M.E.G. Sweeney, G.B. Swann, "An Experimental Comparison of Loop and Cross Scavenging of the Two-Stroke Cycle Engine," SAE Paper No. 861240, Society of Automotive Engineers, Warrendale, Pa., 1986.

1.11 G.P. Blair, "Correlation of Theory and Experiment for Scavenging Flow in Two-Stroke Cycle Engines," SAE Paper No. 881265, Society of Automotive Engineers, Warrendale, Pa., 1988.

1.12 J.D. Flaig, G.L. Broughton, "The Design and Development of the OMC V-8 Outboard Motor," SAE Paper No. 851517, Society of Automotive Engineers, Warrendale, Pa., 1985.

1.13 R. Fleck, G.P. Blair, R.A.R. Houston, "An Improved Model for Predicting Reed Valve Behavior in Two-Stroke Cycle Engines," SAE Paper No. 871654, Society of Automotive Engineers, Warrendale, Pa., 1987.

1.14 SAE J1349, "Engine Power Test Code, Spark Ignition and Diesel," Society of Automotive Engineers, Warrendale, Pa., June 1985.

1.15 SAE J1088, "Test Procedure for the Measurement of Exhaust Emissions from Small Utility Engines," Society of Automotive Engineers, Warrendale, Pa., June 1983.

1.16 ISO 3046, "Reciprocating Internal Combustion Engines: Performance-Parts 1, 2, and 3," International Organization for Standardization, 1981.

1.17 BS 1042, "Fluid Flow in Closed Conduits," British Standards Institution, 1981.

1.18 E. Sher, "The Effect of Atmospheric Conditions on the Performance of an Air-Borne Two-Stroke Spark-Ignition Engine," Proc. I. Mech. E., Vol 198D, No 15, pp239-251 and as SAE Paper No. 844962.

1.19 K. Tsuchiya, S. Hirano, "Characteristics of 2-Stroke Motorcycle Exhaust Emission and Effects of Air-Fuel Ratio and Ignition Timing," SAE No. 750908, Society of Automotive Engineers, Warrendale, Pa., 1975.

1.20 R.J. Kee, "Stratified Charging of a Cross-Scavenged Two-Stroke Cycle Engine," Doctoral Thesis, The Queen's University of Belfast, October, 1988.

1.21 W. Watson, "On the Thermal and Combustion Efficiency of a Four-Cylinder Petrol Motor," *Proc. I. Auto. E.*, Vol 2, p387, 1908-1909.

1.22 E.W. Huber, "Measuring the Trapping Efficiency of Internal Combustion Engines Through Continuous Exhaust Gas Analysis," SAE Paper No. 710144, Society of Automotive Engineers, Warrendale, Pa., 1971.

1.23 G.P. Blair, R.G. Kenny, "Further Developments in Scavenging Analysis for Two-Cycle Engines," SAE Paper No. 800038, Society of Automotive Engineers, Warrendale, Pa., 1980.

1.24 SAE J604d, "Engine Terminology and Nomenclature," Society of Automotive Engineers, Warrendale, Pa., June 1979.

1.25 G.P. Blair, B.L. Sheaffer, G.G. Lassanske, (Eds.), <u>Advances in Two-Stroke Cycle Engine Technology</u>, SAE PT-33, Society of Automotive Engineers, Warrendale, Pa., 1989.

Chapter 2

Gas Flow through Two-Stroke Engines

2.0 Introduction

The gas flow processes into, through, and out of an engine are all conducted in an unsteady manner. The definition of unsteady gas flow is where the pressure, temperature and gas particle velocity in a duct are variable with time. In the case of exhaust flow, the unsteady gas flow behavior is produced because the cylinder pressure falls with the rapid opening of the exhaust port by the piston. This gives an exhaust pipe pressure that changes with time. In the case of induction flow into the crankcase through an intake port whose area changes with time, the intake pipe pressure alters because the cylinder pressure, or crankcase pressure in a simple two-stroke engine, is affected by the piston motion, causing volumetric change within that space.

To illustrate the dramatic variations of pressure wave and particle motion caused by unsteady flow in comparison to steady flow, the series of photographs taken by Coates [8.2] are shown in Plates 2.1-2.4. These photographs were obtained using the Schlieren method [3.14], which is an optical means of using the variation of the refractive index of a gas with its density. Each photograph was taken with an electronic flash duration of 1.5 μs and the view observed is around the termination of a 28-mm-diameter exhaust pipe to the atmosphere. The exhaust pulsations occurred at a frequency of 1000 per minute. The first photograph, Plate 2.1, shows the front of an exhaust pulse about to enter the atmosphere. Note that it is a plane front, and its propagation within the pipe up to the pipe termination is clearly one-dimensional. The next picture, Plate 2.2, shows the propagation of the pressure wave into the atmosphere in a three-dimensional fashion with a spherical front being formed. The beginning of rotational movement of the gas particles at the pipe edges is now evident. The third picture, Plate 2.3, shows the spherical wave front fully formed and the particles being impelled into the atmosphere in the form of a toroidal vortex, or a spinning donut of gas particles. Some of you may be more familiar with the term "smoke ring." That propagating pressure wave front arrives at the human eardrum, deflects it, and the nervous system reports it as "noise" to the brain. The final picture of the series, Plate 2.4, shows that the propagation of the pressure wave front has now passed beyond the frame of the photograph, but the toroidal vortex of gas particles is proceeding downstream with considerable turbulence. Indeed, the flow through the eye of the vortex is so violent that a new acoustic pressure wave front is forming in front of that vortex. The noise that emanates from these pressure pulsations is composed of the basic pressure front propagation and also from the turbulence of the fluid motion in the

Design and Simulation of Two-Stroke Engines

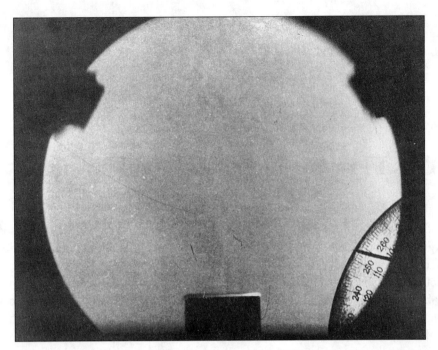

Plate 2.1 Schlieren picture of an exhaust pulse at the termination of a pipe.

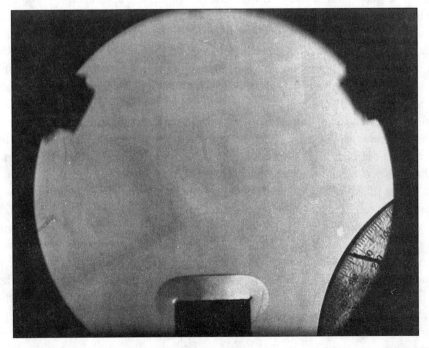

Plate 2.2 The exhaust pulse front propagates into the atmosphere.

Chapter 2 - Gas Flow through Two-Stroke Engines

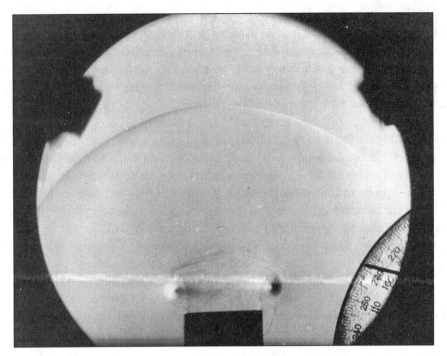

Plate 2.3 Further pulse propagation followed by the toroidal vortex of gas particles.

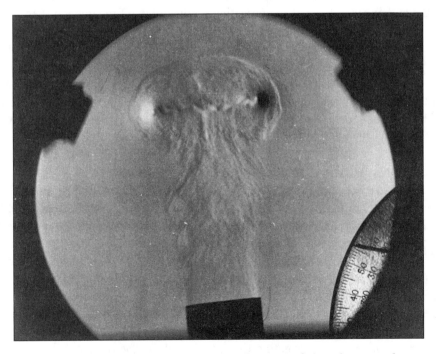

Plate 2.4 The toroidal vortex of gas particles proceeds into the atmosphere.

vortex. Further discussion on the noise aspects of this flow is given in Chapter 8. I have always found this series of photographs, obtained during research at QUB, to be particularly illuminating. When I was a schoolboy on a farm in Co. Antrim too many years ago, the milking machines were driven by a single-cylinder diesel engine with a long straight exhaust pipe and, on a frosty winter's morning, it would blow a "smoke ring" from the exhaust on start-up. That schoolboy used to wonder how it was possible; I now know.

As the resulting performance characteristics of an engine are significantly controlled by this unsteady gas motion, it behooves the designer of engines to understand this flow mechanism thoroughly. This is true for all engines, whether they are destined to be a 2 hp outboard motor or a 150 hp Grand Prix power unit. A simple example will suffice to illustrate the point. If one were to remove the tuned exhaust pipe from a single-cylinder racing engine while it is running at peak power output, the pipe being an "empty" piece of fabricated sheet metal, the engine power output would fall by at least 50% at that engine speed. The tuned exhaust pipe harnesses the pressure wave motion of the exhaust process to retain a greater mass of fresh charge within the cylinder. Without it, the engine will trap only about half as much fresh air and fuel in the cylinder. To design such exhaust systems, and the engine that will take advantage of them, it is necessary to have a good understanding of the mechanism of unsteady gas flow. For the more serious student interested in the subject of unsteady gas dynamics in depth, the series of lectures [2.2] given by the late Prof. F.K. Bannister of the University of Birmingham is an excellent introduction to the topic; so too is the book by Annand and Roe [1.8], and the books by Rudinger [2.3] and Benson [2.4]. The references cited in this chapter will give even greater depth to that study.

Therefore, this chapter will explain the fundamental characteristics of unsteady gas flow in the intake and exhaust ducts of reciprocating engines. Such fundamental theory is just as applicable to four-stroke engines as it is to two-stroke engines, although the bias of the discussion will naturally be for the latter type of power unit. As with other chapters of this book, computer programs are available for you to use for your own education and experience. Indeed, these programs are a means of explaining the behavior of pressure waves and their effect on the filling and emptying of engine cylinders. In the later sections of the chapter, there will be explanations of the operation of tuned exhaust and intake systems for two-stroke engines.

2.1 Motion of pressure waves in a pipe
2.1.1 Nomenclature for pressure waves

You are already familiar with the motion of pressure waves of small amplitude, for they are acoustic waves, or sound. Some of our personal experience with sound waves is helpful in understanding the fundamental nature of the flow of the much larger-amplitude waves to be found in engine ducts. Pressure waves, and sound waves, are of two types: (1) compression waves or (2) expansion waves. This is illustrated in Fig. 2.1. In both Fig. 2.1(a) and (b), the undisturbed pressure and temperature in the pipe ahead of the pressure wave are p_0 and T_0, respectively.

The compression wave in the pipe is shown in Fig. 2.1(a) and the expansion wave in Fig. 2.1(b), and both waves are propagating toward the right in the diagram. At a point on the compression wave the pressure is p_e, where $p_e > p_0$, and it is being propagated at a velocity,

Chapter 2 - Gas Flow through Two-Stroke Engines

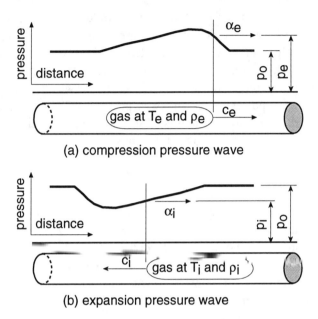

(a) compression pressure wave

(b) expansion pressure wave

Fig. 2.1 Pressure wave nomenclature.

α_e. It is also moving gas particles at a gas particle velocity of c_e and in the same direction as the wave is being propagated. At a point on the expansion wave in the pipe the pressure is p_i, where $p_i < p_0$, and it is being propagated at a velocity, α_i. It is also moving gas particles at a gas particle velocity of c_i, but in the opposite direction to that which the wave is being propagated.

At this point you can draw on your personal experience of sound waves to help understand the physical nature of the statements made in the preceding paragraph. Imagine standing several meters away from another person, Fred. Fred produces a sharp exhalation of breath, for example, he says "boo" somewhat loudly. He does this by raising his lung pressure above the atmospheric pressure due to a muscular reduction of his lung volume. The compression pressure wave produced, albeit of small amplitude, leaves his mouth and is propagated at the local acoustic velocity, or speed of sound, to your ear. The speed of sound involved is on the order of 350 m/s. The gas particles involved with the "boo" leaving Fred's mouth have a much lower velocity, probably on the order of 1 m/s. However, that gas particle velocity is in the same direction as the propagation of the compression pressure wave, i.e., toward your ear. Contrast this simple experiment with a second test. Imagine that Fred now produces a sharp inhalation of breath. This he accomplishes by expanding his lung volume so that his lung pressure falls sharply below the atmospheric pressure. The resulting "u....uh" you hear is caused by the expansion pressure wave leaving Fred's mouth and propagating toward your ear at the local acoustic velocity. In short, the direction of propagation is the same as before with the compression wave "boo," and the propagation velocity is, to all intents and purposes, identical. However, as the gas particles manifestly entered Fred's mouth with the creation of this expansion wave, the gas particle velocity is clearly opposite to the direction of the expansion wave propagation.

It is obvious that exhaust pulses resulting from cylinder blowdown come under the heading of compression waves, and expansion waves are generated by the rapidly falling crankcase pressure during induction in the case of the two-stroke engine, or cylinder pressure for a four-stroke cycle unit. However, as in most technologies, other expressions are used in the literature as jargon to describe compression and expansion waves. Compression waves are variously called "exhaust pulses," "compression pulses" or "ramming waves." Expansion waves are often described as "suction pulses," "sub-atmospheric pulses" or "intake pulses." However, as will be seen from the following sections, expansion and compression waves do appear in both inlet and exhaust systems.

2.1.2 Propagation velocities of acoustic pressure waves

As already pointed out, acoustic pressure waves are pressure waves where the pressure amplitudes are small. Let dp be the pressure difference from atmospheric pressure, i.e., $(p_e - p_0)$ or $(p_0 - p_i)$, for the compression or expansion wave, respectively. The value of dp for Fred's "boo" would be on the order of 0.2 Pa. The pressure ratio, P, for any pressure wave is defined as the pressure, p, at any point on the wave under consideration divided by the undisturbed pressure, p_0. The undisturbed pressure is more commonly called the reference pressure. Here, the value for Fred's "boo" would be:

$$P = \frac{p}{p_0} = \frac{101,325.2}{101,325} = 1.000002$$

For the loudest of acoustic sounds, say a rifle shot at about 0.2 m away from the human ear, dp could be 2000 Pa and the pressure ratio would be 1.02. That such very loud sounds are still small in pressure wave terms can be gauged from the fact that a typical exhaust pulse in an engine exhaust pipe has a pressure ratio of about 1.5.

According to Earnshaw [2.1], the velocity of a sound wave in air is given by a_0, where:

$$a_0 = \sqrt{\gamma R T_0} \qquad (2.1.1)$$

or

$$a_0 = \sqrt{\frac{\gamma p_0}{\rho_0}} \qquad (2.1.2)$$

The value denoted by γ is the ratio of specific heats for air. In the above equations, T_0 is the reference temperature and ρ_0 is the reference density, which are related to the reference pressure, p_0, by the state equation:

$$p_0 = \rho_0 R T_0 \qquad (2.1.3)$$

For sound waves in air, p_0, T_0 and ρ_0 are the values of the atmospheric pressure, temperature and density, respectively, and R is the gas constant for the particular gas involved.

2.1.3 Propagation and particle velocities of finite amplitude waves

Particle velocity

Any pressure wave with a pressure ratio greater than an acoustic wave is called a wave of finite amplitude. Earnshaw [2.1] showed that the gas particle velocity associated with a wave of finite amplitude was given by c, where:

$$c = \frac{2}{\gamma - 1} a_0 \left[\left(\frac{p}{p_0}\right)^{\frac{\gamma-1}{2\gamma}} - 1 \right] \qquad (2.1.4)$$

Bannister's [2.2] derivation of this equation is explained with great clarity in Appendix A2.1. Within the equation, shorthand parameters can be employed that simplify the understanding of much of the further analysis. The symbol P is referred to as the pressure ratio of a point on the wave of absolute pressure, p. The notation of X is known as the pressure amplitude ratio and G represents various functions of γ which is the ratio of specific heats for the particular gas involved. These are set down as:

pressure ratio
$$P = \frac{p}{p_0}$$

pressure amplitude ratio
$$X = \left(\frac{p}{p_0}\right)^{\frac{\gamma-1}{2\gamma}} = P^{\frac{\gamma-1}{2\gamma}} \qquad (2.1.5)$$

Incorporation of the above shorthand notation within the complete equation for Eq. 2.1.4 gives:

$$c = \frac{2}{\gamma - 1} a_0 (X - 1) \qquad (2.1.6)$$

If the gas in which this pressure wave is propagating has the properties of air, then these properties are:

Gas constant $\qquad R = 287$ J/kgK

Specific heats ratio $\qquad \gamma = 1.4$

Specific heat at constant pressure $\qquad C_P = \dfrac{\gamma R}{\gamma - 1} = 1005$ J/kgK

Specific heat at constant volume $\qquad C_V = \dfrac{R}{\gamma - 1} = 718$ J/kgK

Various functions of the ratio of specific heats, G_5, G_7, etc., which are useful as shorthand notation in many gas-dynamic equations in this theoretical area, are given below. The logic of the notation for G is that the value of G_5 for air is 5, G_7 for air is 7, etc.

$$G_3 = \frac{4 - 2\gamma}{\gamma - 1} \quad \text{for air where } \gamma = 1.4 \text{ then } G_3 = 3$$

$$G_4 = \frac{3 - \gamma}{\gamma - 1} \quad \text{for air where } \gamma = 1.4 \text{ then } G_4 = 4$$

$$G_5 = \frac{2}{\gamma - 1} \quad \text{for air where } \gamma = 1.4 \text{ then } G_5 = 5$$

$$G_6 = \frac{\gamma + 1}{\gamma - 1} \quad \text{for air where } \gamma = 1.4 \text{ then } G_6 = 6$$

$$G_7 = \frac{2\gamma}{\gamma - 1} \quad \text{for air where } \gamma = 1.4 \text{ then } G_7 = 7$$

$$G_{17} = \frac{\gamma - 1}{2\gamma} \quad \text{for air where } \gamma = 1.4 \text{ then } G_{17} = \frac{1}{7}$$

$$G_{35} = \frac{\gamma}{\gamma - 1} \quad \text{for air where } \gamma = 1.4 \text{ then } G_{35} = 3.5$$

$$G_{67} = \frac{\gamma + 1}{2\gamma} \quad \text{for air where } \gamma = 1.4 \text{ then } G_{67} = \frac{6}{7}$$

This useful notation simplifies analysis in gas dynamics, particularly as it should be noted that the equations are additive or subtractive by numbers, thus:

$$G_4 = G_5 - 1 = \frac{2}{\gamma - 1} - 1 = \frac{2 - \gamma + 1}{\gamma - 1} = \frac{1 - \gamma}{\gamma - 1}$$

or $G_7 = G_5 + 2$ or $G_3 = G_5 - 2$ or $G_6 = G_3 + 3$.

However, it should be noted in applications of such functions that they are generally neither additive nor operable, thus:

$$G_7 \neq G_4 + G_3 \quad \text{and} \quad G_3 \neq \frac{G_6}{2}$$

Chapter 2 - Gas Flow through Two-Stroke Engines

Gas mixtures are commonplace within engines. Air itself is a mixture, fundamentally of oxygen and nitrogen. Exhaust gas is principally composed of carbon monoxide, carbon dioxide, steam and nitrogen. Furthermore, the properties of gases are complex functions of temperature. Thus a more detailed discussion on this topic is given in Sec. 2.1.6.

If the gas properties are assumed to be as for air given above, then Eq. 2.1.4 for the gas particle velocity reduces to the following:

$$c = \frac{2}{\gamma - 1} a_0 (X - 1) = G_5 a_0 (X - 1) = 5 a_0 (X - 1) \tag{2.1.7}$$

where

$$X = \left(\frac{p}{p_0}\right)^{\frac{\gamma-1}{2\gamma}} = \left(\frac{p}{p_0}\right)^{G_{17}} = \left(\frac{p}{p_0}\right)^{\frac{1}{7}} \tag{2.1.8}$$

Propagation velocity

The propagation velocity at any point on a wave where the pressure is p and the temperature is T is like a small acoustic wave moving at the local acoustic velocity at those conditions, but on top of gas particles which are already moving. Therefore, the absolute propagation velocity of any wave point is the sum of the local acoustic velocity and the local gas particle velocity. The propagation velocity of any point on a finite amplitude wave is given by α, where:

$$\alpha = a + c \tag{2.1.9}$$

and a is the local acoustic velocity at the elevated pressure and temperature of the wave point, p and T.

However, acoustic velocity, a, is given by Earnshaw [2.1] from Eq. 2.1.1 as:

$$a = \sqrt{\gamma RT} \tag{2.1.10}$$

Assuming a change of state conditions from p_0 and T_0 to p and T to be isentropic, then for such a change:

$$\frac{T}{T_0} = \left(\frac{p}{p_0}\right)^{\frac{\gamma-1}{\gamma}} \tag{2.1.11}$$

$$\frac{a}{a_0} = \sqrt{\frac{T}{T_0}} = \left(\frac{p}{p_0}\right)^{\frac{\gamma-1}{2\gamma}} = P^{G_{17}} = X \tag{2.1.12}$$

Hence, the absolute propagation velocity, α, defined by Eq. 2.1.9, is given by the addition of information within Eqs. 2.1.6 and 2.1.12:

$$\alpha = a_0 X + \frac{2}{\gamma - 1} a_0 (X - 1) = a_0 \left[\frac{\gamma + 1}{\gamma - 1} \left(\frac{p}{p_0} \right)^{\frac{\gamma-1}{2\gamma}} - \frac{2}{\gamma - 1} \right] \quad (2.1.13)$$

In terms of the G functions already defined,

$$\alpha = a_0 [G_6 X - G_5] \quad (2.1.14)$$

If the properties of air are assumed for the gas, then this reduces to:

$$\alpha = a_0 [6X - 5] \quad (2.1.15)$$

The density, ρ, at any point on a wave of pressure, p, is found from an extension of the isentropic relationships in Eqs. 2.1.11 and 2.1.14 as follows:

$$\frac{\rho}{\rho_0} = \left(\frac{p}{p_0} \right)^{\frac{1}{\gamma}} = X^{\frac{2}{\gamma-1}} = X^{G5} \quad (2.1.16)$$

For air, where γ is 1.4, the density ρ at a pressure p on the wave translates to:

$$\rho = \rho_0 X^5 \quad (2.1.17)$$

2.1.4 Propagation and particle velocities of finite amplitude waves in air

From Eqs. 2.1.4 and 2.1.15, the propagation velocities of finite amplitude waves in air in a pipe are calculated by the following equations:

Propagation velocity $\quad\quad \alpha = a_0 [6X - 5] \quad\quad\quad\quad\quad (2.1.18)$

Particle velocity $\quad\quad\quad\ c = 5 a_0 (X - 1) \quad\quad\quad\quad\quad (2.1.19)$

Pressure amplitude ratio $\quad X = \left(\dfrac{p}{p_0} \right)^{\frac{1}{7}} = P^{\frac{1}{7}} \quad\quad\quad (2.1.20)$

The reference conditions of acoustic velocity and density are found as follows:

Reference acoustic velocity $\quad a_0 = \sqrt{1.4 \times 287 \times T_0} \ \ \text{m/s} \quad (2.1.21)$

Reference density $\quad\quad \rho_0 = \dfrac{p_0}{287 \times T_0} \quad kg/m^3 \quad\quad\quad\quad (2.1.22)$

It is interesting that these equations corroborate the experiment which you conducted with your imagination regarding Fred's lung-generated compression and expansion waves.

Fig. 2.1 shows compression and expansion waves. Let us assume that the undisturbed pressure and temperature in both cases are at standard atmospheric conditions. In other words, p_0 and T_0 are 101,325 Pa and 20°C, or 293 K, respectively. The reference acoustic velocity, a_0, and reference density, ρ_0, are, from Eqs. 2.1.1 and 2.1.3 or Eqs. 2.1.21 and 2.1.22:

$$a_0 = \sqrt{1.4 \times 287 \times 293} = 343.11 \quad m/s$$

$$\rho_0 = \dfrac{101,325}{287 \times 293} = 1.2049 \quad kg/m^3$$

Let us assume that the pressure ratio, P_e, of a point on the compression wave is 1.2 and that of a point on the expansion wave is P_i with a value of 0.8. In other words, the compression wave has a pressure differential as much above the reference pressure as the expansion wave is below it. Let us also assume that the pipe has a diameter, d, of 25 mm.

(a) *The compression wave*

First, consider the compression wave of pressure, p_e. This means that p_e is:

$$p_e = P_e \times p_0 = 1.2 \times 101,325 = 121,590 \quad Pa$$

The pressure amplitude ratio, X_e, is calculated as:

$$X_e = 1.2^{\frac{1}{7}} = 1.02639$$

Therefore, the propagation and particle velocities, α_e and c_e, are found from:

$$\alpha_e = 343.11 \times (6 \times 1.02639 - 5) = 397.44 \quad m/s$$

$$c_e = 5 \times 343.11 \times (1.02639 - 1) = 45.27 \quad m/s$$

From this it is clear that the propagation of the compression wave is faster than the reference acoustic velocity, a_0, and that the air particles move at a considerably slower rate. The compression wave is moving rightward along the pipe at 397.4 m/s and, as it passes from particle to particle, it propels each particle in turn in a rightward direction at 45.27 m/s. This is deduced from the fact that the sign of the numerical values of α_e and c_e are identical.

The local particle Mach number, M_e, is defined as the ratio of the particle velocity to the local acoustic velocity, a_e, where:

$$M_e = \frac{c_e}{a_e} = \frac{G_5(X_e - 1)}{X_e} \qquad (2.1.23)$$

From Eq. 2.1.12:

$$a_e = a_0 X_e$$

Hence, $\quad a_e = 343.11 \times 1.02639 = 352.16 \text{ m/s}$

and the local particle Mach number,

$$M_e = \frac{45.27}{352.16} = 0.1285$$

The mass rate of gas flow, $\dot{m}_e$, caused by the passage of this point of the compression wave in a pipe of area, A_e, is calculated from the thermodynamic equation of continuity with the multiplication of density, area and particle velocity:

$$\dot{m}_e = \rho_e A_e c_e$$

From Eq. 2.1.17:

$$\rho_e = \rho_e X_e^5 = 1.2049 \times 1.02639^5 = 1.2049 \times 1.1391 = 1.3725 \text{ kg/m}^3$$

The pipe area is given by:

$$A_e = \frac{\pi d^2}{4} = \frac{3.14159 \times 0.025^2}{4} = 0.000491 \text{ m}^2$$

Therefore, as the arithmetic signs of the propagation and particle velocities are both positive, the mass rate of flow is in the same direction as the wave propagation:

$$\dot{m}_e = \rho_e A_e c_e = 1.3725 \times 0.000491 \times 45.27 = 0.0305 \text{ kg/s}$$

(b) *The expansion wave*
Second, consider the expansion wave of pressure, p_i. This means that p_i is given by:

$$p_i = P_i \times p_0 = 0.8 \times 101{,}325 = 81{,}060 \text{ Pa}$$

The pressure amplitude ratio, X_i, is calculated as:

$$X_i = 0.8^{\frac{1}{7}} = 0.9686$$

Therefore, the propagation and particle velocities, α_i and c_i, are found from:

$$\alpha_i = 343.11 \times (6 \times 0.9686 - 5) = 278.47 \text{ m/s}$$

$$c_i = 5 \times 343.11 \times (0.9686 - 1) = 53.87 \text{ m/s}$$

From this it is clear that the propagation of the expansion wave is slower than the reference acoustic velocity but the air particles move somewhat faster. The expansion wave is moving rightward along the pipe at 278.47 m/s and, as it passes from particle to particle, it propels each particle in turn in a leftward direction at 53.87 m/s. This is deduced from the fact that the numerical values of α_i and c_i are of <u>opposite</u> sign.

The local particle Mach number, M_i, is defined as the ratio of the particle velocity to the local acoustic velocity, a_i, where:

$$M_i = \frac{c_i}{a_i}$$

From Eq. 2.1.12,

$$a_i = a_0 X_i$$

Hence, $\qquad a_i = 343.11 \times 0.9686 = 332.3 \text{ m/s}$

and the local particle Mach number is,

$$M_i = \frac{-53.87}{332.3} = -0.1621$$

The mass rate of gas flow, $\dot{m}_i$, caused by the passage of this point of the expansion wave in a pipe of area, A_i, is calculated by:

$$\dot{m}_i = \rho_i A_i c_i$$

From Eq. 2.1.17:

$$\rho_i = \rho_i X_i^5 = 1.2049 \times 0.9686^5 = 1.2049 \times 0.8525 = 1.0272 \text{ kg/m}^3$$

It will be seen that the density is reduced in the more rarified expansion wave. The pipe area is identical as the diameter is unchanged, i.e., $A_i = 0.000491$ m². The mass rate of flow is in the opposite direction to the wave propagation as:

$$\dot{m}_i = \rho_i A_i c_i = 1.0272 \times 0.000491 \times (-53.87) = -0.0272 \quad \text{kg/s}$$

2.1.5 Distortion of the wave profile

It is clear from the foregoing that the value of propagation velocity is a function of the wave pressure and wave temperature at any point on that pressure wave. It should also be evident that, as all of the points on a wave are propagating at different velocities, the wave must change its shape in its passage along any duct. To illustrate this the calculations conducted in the previous section are displayed in Fig. 2.2. In Fig. 2.2(a) it can be seen that both the front and tail of the wave travel at the reference acoustic velocity, a_0, which is 53 m/s slower than the peak wave velocity. In their travel along the pipe, the front and the tail will keep station with each other in both time and distance. However, at the front of the wave, all of the pressure points between it and the peak are traveling faster and will inevitably catch up with it. Whether that will actually happen before some other event intrudes (for instance, the wave front could reach the end of the pipe) will depend on the length of the pipe and the time interval between the peak and the wave front. Nevertheless, there will always be the tendency for the wave peak to get closer to the wave front and further away from the wave tail. This is known as "steep-fronting." The wave peak could, in theory, try to pass the wave front, which is what happens to a water wave in the ocean when it "crests." In gas flow, "cresting" is impossible and the reality is that a shock wave would be formed. This can be analyzed theoretically and Bannister [2.2] gives an excellent account of the mathematical solution for the particle velocity and the propagation velocity, α_{sh}, of a shock wave of pressure ratio, P_{sh}, propagating into an undisturbed gas medium at reference pressure, p_0, and acoustic velocity, a_0. The derivation of the equations set out below is presented in Appendix A2.2. The theoretically derived expressions for propagation velocity, α_{sh}, and particle velocity, c_{sh}, of a compression shock front are:

$$P_{sh} = \frac{p_e}{p_0}$$

$$\alpha_{sh} = a_0 \sqrt{\frac{\gamma+1}{2\gamma} P_{sh} + \frac{\gamma-1}{2\gamma}}$$
$$= a_0 \sqrt{G_{67} P_{sh} + G_{17}} \tag{2.1.24}$$

$$c_{sh} = \frac{2}{\gamma+1}\left(\alpha_{sh} - \frac{a_0^2}{\alpha_{sh}}\right)$$
$$= \frac{a_0(P_{sh} - 1)}{\gamma\sqrt{G_{67} P_{sh} + G_{17}}} \tag{2.1.25}$$

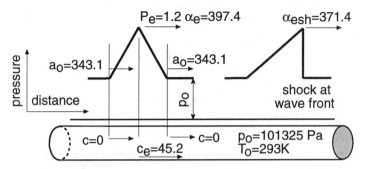

(a) distortion of compression pressure wave profile

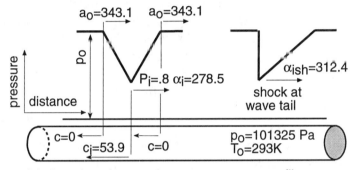

(b) distortion of expansion pressure wave profile

Fig. 2.2 Distortion of wave profile and possible shock formation.

The situation for the expansion wave, shown in Fig. 2.2(b), is the reverse, in that the peak is traveling 64.6 m/s slower than either the wave front or the wave tail. Thus, any shock formation taking place will be at the tail of an expansion wave, and any wave distortion will be where the tail of the wave attempts to overrun the peak.

In this case of shock at the tail of the expansion wave, the above equations also apply, but the "compression" shock is now at the tail of the wave and running into gas which is at acoustic state a_i and moving at particle velocity c_i. Thus the propagation velocity and particle velocity of the shock front at the tail of the expansion wave, which has an undisturbed state at p_0 and a_0 behind it, are given by:

$$P_{sh} = \frac{p_0}{p_i}$$

$$\begin{aligned}
\alpha_{sh} &= \alpha_{sh \text{ relative to gas i}} + c_i \\
&= a_i\sqrt{G_{67}P_{sh} + G_{17}} + c_i \\
&= a_0 X_i \sqrt{G_{67}P_{sh} + G_{17}} + G_5 a_0(X_i - 1)
\end{aligned} \quad (2.1.26)$$

$$c_{sh} = c_{sh \text{ relative to gas i}} + c_i$$

$$= \frac{a_i(P_{sh} - 1)}{\gamma\sqrt{G_{67}P_{sh} + G_{17}}} + c_i$$

$$= \frac{a_0 X_i(P_{sh} - 1)}{\gamma\sqrt{G_{67}P_{sh} + G_{17}}} + G_5 a_0 (X_i - 1) \qquad (2.1.27)$$

For the compression wave illustrated in Fig. 2.2, the use of Eqs. 2.1.24 and 2.1.25 yields finite amplitude propagation and particle velocities of 397.4 and 45.27 m/s, and for the shock wave of the same amplitude, 371.4 and 45.28 m/s, respectively. The difference in propagation velocity is some 7% less but that for particle velocity is negligible.

For the expansion wave illustrated in Fig. 2.2, the use of Eqs. 2.1.26 and 2.1.27 yields finite amplitude propagation and particle velocities of 278.5 and –53.8 m/s, and for the shock wave of the same amplitude, 312.4 and 0.003 m/s, respectively. The difference in propagation velocity is some 12% greater but that for particle velocity is considerable in that the particle velocity at, or immediately behind, the shock is effectively zero.

2.1.6 The properties of gases

It will be observed that the propagation of pressure waves and the mass flow rate which they induce in gases is dependent on the gas properties, particularly that of the gas constant, R, and the ratio of specific heats, γ. The value of the gas constant, R, is dependent on the composition of the gas, and the ratio of specific heats, γ, is dependent on both gas composition and temperature. It is essential to be able to index these properties at every stage of a simulation of gas flow in engines. Much of this information can be found in many standard texts on thermodynamics, but it is essential for reasons of clarity to repeat it here briefly.

The gas constant, R, of any gas can be found from the relationship relating the universal gas constant, $\overline{R}$, and the molecular weight, M, of the gas:

$$R = \frac{\overline{R}}{M} \qquad (2.1.28)$$

The universal gas constant, $\overline{R}$, has a value of 8314.4 J/kgmolK. The specific heats at constant pressure and temperature, C_P and C_V, are determined from their defined relationship with respect to enthalpy, h, and internal energy, u:

$$C_P = \frac{dh}{dT} \qquad C_V = \frac{du}{dT} \qquad (2.1.29)$$

The ratio of specific heats, γ, is found simply as:

$$\gamma = \frac{C_P}{C_V} \qquad (2.1.30)$$

It can be seen that if the gases have internal energies and enthalpies which are nonlinear functions of temperature then neither C_P, C_V nor γ is a constant. If the gas is a mixture of gases then the properties of the individual gases must be assessed separately and then combined to produce the behavior of the mixture.

To illustrate the procedure to determine the properties of gas mixtures, let air be examined as a simple example of a gas mixture with an assumed volumetric composition, υ, of 21% oxygen and 79% nitrogen while ignoring the small but important trace concentration of argon. The molecular weights of oxygen and nitrogen are 31.999 and 28.013, respectively.

The average molecular weight of air is then given by:

$$M_{air} = \sum (\upsilon_{gas} M_{gas}) = 0.21 \times 31.999 + 0.79 \times 28.013 = 28.85$$

The mass ratios, ε, of oxygen and nitrogen in air are given by:

$$\varepsilon_{O_2} = \frac{\upsilon_{O_2} M_{O_2}}{M_{air}} = \frac{0.21 \times 31.999}{28.85} = 0.233$$

$$\varepsilon_{N_2} = \frac{\upsilon_{N_2} M_{N_2}}{M_{air}} = \frac{0.79 \times 28.013}{28.85} = 0.767$$

The molal enthalpies, $\overline{h}$, for gases are given as functions of temperature with respect to molecular weight, where the κ values are constants:

$$\overline{h} = \kappa_0 + \kappa_1 T + \kappa_2 T^2 + \kappa_3 T^3 \quad \text{J/kgmol} \tag{2.1.31}$$

In which case the molal internal energy of the gas is related thermodynamically to the enthalpy by:

$$\overline{u} = \overline{h} - \overline{R}T \tag{2.1.32}$$

Consequently, from Eq. 2.1.29, the molal specific heats are found by appropriate differentiation of Eqs. 2.1.31 and 32:

$$\overline{C}_P = \kappa_1 + 2\kappa_2 T + 3\kappa_3 T^2 \tag{2.1.33}$$

$$\overline{C}_V = \overline{C}_P - \overline{R} \tag{2.1.34}$$

The molecular weights and the constants, κ, for many common gases are found in Table 2.1.1 and are reasonably accurate for a temperature range of 300 to 3000 K. The values of the molal specific heats, internal energies and enthalpies of the individual gases can be found at a particular temperature by using the values in the table.

Considering air as the example gas at a temperature of 20°C, or 293 K, the molal specific heats of oxygen and nitrogen are found using Eqs. 2.1.33 and 34 as:

Oxygen, O_2: $\overline{C}_P = 31{,}192$ J/kgmol $\overline{C}_V = 22{,}877$ J/kgmol

Nitrogen, N_2: $\overline{C}_P = 29{,}043$ J/kgmol $\overline{C}_V = 20{,}729$ J/kgmol

Table 2.1.1 Properties of some common gases found in engines

Gas	M	κ_0	κ_1	κ_2	κ_3
O_2	31.999	−9.3039E6	2.9672E4	2.6865	−2.1194E−4
N_2	28.013	−8.503.3E6	2.7280E4	3.1543	−3.3052E−4
CO	28.011	−8.3141E6	2.7460E4	3.1722	−3.3416E−4
CO_2	44.01	−1.3624E7	4.1018E4	7.2782	−8.0848E−4
H_2O	18.015	−8.9503E6	2.0781E4	7.9577	−7.2719E−4
H_2	2.016	−7.8613E6	2.6210E4	2.3541	−1.2113E−4

From a mass standpoint, these values are determined as follows:

$$C_P = \frac{\overline{C}_P}{M} \qquad C_V = \frac{\overline{C}_V}{M} \qquad (2.1.35)$$

Hence the mass related values are:

Oxygen, O_2: $C_P = 975$ J/kgK $C_V = 715$ J/kgK

Nitrogen, N_2: $C_P = 1037$ J/kgK $C_V = 740$ J/kgK

For the mixture of oxygen and nitrogen which is air, the properties of air are given generally as:

$$R_{air} = \Sigma(\varepsilon_{gas} R_{gas}) \qquad C_{Pair} = \Sigma(\varepsilon_{gas} C_{P_{gas}})$$

$$C_{Vair} = \Sigma(\varepsilon_{gas} C_{V_{gas}}) \qquad \gamma_{air} = \frac{C_{Pair}}{C_{Vair}} \qquad (2.1.36)$$

Taking just one as a numeric example, the gas constant, R, which it will be noted is not temperature dependent, is found by:

$$R_{air} = \Sigma(\varepsilon_{gas} R_{gas}) = 0.233 \left(\frac{8314.4}{31.999}\right) + 0.767 \left(\frac{8314.4}{28.011}\right) = 288 \text{ J/kgK}$$

The other equations reveal for air at 293 K:

$$C_P = 1022 \text{ J/kgK} \quad C_V = 734 \text{ J/kgK} \quad \gamma = 1.393$$

It will be seen that the value of the ratio of specific heats, γ, is not precisely 1.4 at standard atmospheric conditions as stated earlier in Sec. 2.1.3. The reason is mostly due to the fact that air contains argon, which is not included in the above analysis and, as argon has a value of γ of 1.667, the value deduced above is weighted downward arithmetically.

The most important point to make is that these properties of air are a function of temperature, so if the above analysis is repeated at 500 and 1000 K the following answers are found:

for air:
$$T = 500\text{K} \quad C_P = 1061 \text{ J/kgK} \quad C_V = 773 \text{ J/kgK} \quad \gamma = 1.373$$
$$T = 1000\text{K} \quad C_P = 1113 \text{ J/kgK} \quad C_V = 855 \text{ J/kgK} \quad \gamma = 1.337$$

As air can be found within an engine at these state conditions it is vital that any simulation takes these changes of property into account as they have a profound influence on the characteristics of unsteady gas flow.

Exhaust gas

Clearly exhaust gas has a quite different composition as a mixture of gases by comparison with air. Although this matter is discussed in much greater detail in Chapter 4, consider the simple and ideal case of stoichiometric combustion of octane with air. The chemical equation, which has a mass-based air-fuel ratio, AFR, of 15, is as follows:

$$2C_8H_{18} + 25\left[O_2 + \frac{79}{21}N_2\right] = 16CO_2 + 18H_2O + 94.05N_2$$

The volumetric concentrations of the exhaust gas can be found by noting that if the total moles are 128.05, then:

$$\upsilon_{CO_2} = \frac{16}{128.05} = 0.125 \quad \upsilon_{H_2O} = \frac{9}{128.05} = 0.141 \quad \upsilon_{N_2} = \frac{94.05}{128.05} = 0.734$$

This is precisely the same starting point as for the above analysis for air so the procedure is the same for the determination of all of the properties of exhaust gas which ensue from an ideal stoichiometric combustion. A full discussion of the composition of exhaust gas as a function of air-to-fuel ratio is in Chapter 4, Sec. 4.3.2, and an even more detailed debate is in the Appendices A4.1 and A4.2, on the changes to that composition, at any fueling level, as a function of temperature and pressure.

In reality, even at stoichiometric combustion there would be some carbon monoxide in existence and minor traces of oxygen and hydrogen. If the mixture were progressively richer than stoichiometric, the exhaust gas would contain greater amounts of CO and a trace of H_2

but would show little free oxygen. If the mixture were progressively leaner than stoichiometric, the exhaust gas would contain lesser amounts of CO and no H_2 but would show higher concentrations of oxygen. The most important, perhaps obvious, issue is that the properties of exhaust gas depend not only on temperature but also on the combustion process that created them. Tables 2.1.2 and 2.1.3 show the ratio of specific heats, γ, and gas constant, R, of exhaust gas at various temperatures emanating from the combustion of octane at various air-fuel ratios. The air-fuel ratio of 13 represents rich combustion, 15 is stoichiometric and an AFR of 17 is approaching the normal lean limit of gasoline burning. The composition of the exhaust gas is shown in Table 2.1.2 at a low temperature of 293 K and its influence on the value of gas constant and the ratio of specific heats is quite evident. While the tabular values are quite typical of combustion products at these air-fuel ratios, naturally they are approximate as they are affected by more than the air-fuel ratio, for the local chemistry of the burning process and the chamber geometry, among many factors, will also have a profound influence on the final composition of any exhaust gas. At higher temperatures, to compare with the data for air and exhaust gas at 293 K in Table 2.1.2, this same gaseous composition shows markedly different properties in Table 2.1.3, when analyzed by the same theoretical approach.

Table 2.1.2 Properties of exhaust gas at low temperature

T=293 K		% by Volume					
AFR	%CO	%CO_2	%H_2O	%O_2	%N_2	R	γ
13	5.85	8.02	15.6	0.00	70.52	299.8	1.388
15	0.00	12.50	14.1	0.00	73.45	290.7	1.375
17	0.00	11.14	12.53	2.28	74.05	290.4	1.376

Table 2.1.3 Properties of exhaust gas at elevated temperatures

T=500 K			T=1000 K		
AFR	R	γ	AFR	R	γ
13	299.8	1.362	13	299.8	1.317
15	290.7	1.350	15	290.8	1.307
17	290.4	1.352	17	290.4	1.310

From this it is evident that the properties of exhaust gas are quite different from air, and while they are as temperature dependent as air, they are not influenced by air-fuel ratio, particularly with respect to the ratio of specific heats, as greatly as might be imagined. The gas constant for rich mixture combustion of gasoline is some 3% higher than that at stoichiometric and at lean mixture burning.

Chapter 2 - Gas Flow through Two-Stroke Engines

What is evident, however, is that during any simulation of unsteady gas flow or of the thermodynamic processes within engines, it is imperative for its accuracy to use the correct value of the gas properties at all locations within the engine.

2.2 Motion of oppositely moving pressure waves in a pipe

In the previous section, you were asked to conduct an imaginary experiment with Fred, who produced compression and expansion waves by exhaling or inhaling sharply, producing a "boo" or a "u...uh," respectively. Once again, you are asked to conduct another experiment so as to draw on your experience of sound waves to illustrate a principle, in this case the behavior of oppositely moving pressure waves. In this second experiment, you and your friend Fred are going to say "boo" at each other from some distance apart, and at the same time. Each person's ears, being rather accurate pressure transducers, will record his own "boo" first, followed a fraction of time later by the "boo" from the other party. Obviously, the "boo" from each passed through the "boo" from the other and arrived at both Fred's ear and your ear with no distortion caused by their passage through each other. If distortion had taken place, then the sensitive human ear would have detected it. At the point of meeting, when the waves were passing through each other, the process is described as "superposition." The theoretical treatment below is for air, as this simplifies the presentation and enhances your understanding of the theory; the extension of the theory to the generality of gas properties is straightforward.

2.2.1 Superposition of oppositely moving waves

Fig. 2.3 illustrates two oppositely moving pressure waves in air in a pipe. They are shown as compression waves, ABCD and EFGH, and are sketched as being square in profile, which is physically impossible but it makes the task of mathematical explanation somewhat easier. In Fig. 2.3(a) they are about to meet. In Fig. 2.3(b) the process of superposition is taking place for the front EF on wave top BC, and for the front CD on wave top FG. The result is the creation of a superposition pressure, p_s, from the separate wave pressures, p_1 and p_2. Assume that the reference acoustic velocity is a_0. Assuming also that the rightward direction is mathematically positive, the particle and the propagation velocity of any point on the wave top, BC, will be c_1 and α_1. From Eqs. 2.1.18-20:

$$c_1 = 5a_0(X_1 - 1) \qquad \alpha_1 = a_0(6X_1 - 5)$$

Similarly, the values for the wave top FG will be (with rightward regarded as the positive direction):

$$c_2 = -5a_0(X_2 - 1) \qquad \alpha_2 = -a_0(6X_2 - 5)$$

From Eq. 2.1.14, the local acoustic velocities in the gas columns BE and DG during superposition will be:

$$a_1 = a_0 X_1 \qquad a_2 = a_0 X_2$$

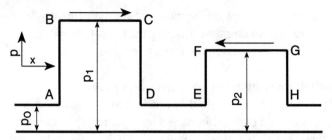

(a) two pressure waves approach each other in a duct

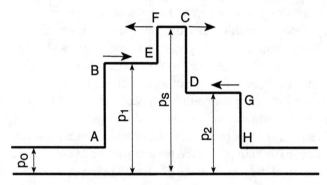

(b) two pressure waves partially superposed in a duct

Fig. 2.3 Superposition of pressure waves in a pipe.

During superposition, the wave top F is now moving into a gas with a new reference pressure level at p_1. The particle velocity of F relative to the gas in BE will be:

$$c_{FrelBE} = -5a_1\left[\left(\frac{p_s}{p_1}\right)^{\frac{1}{7}} - 1\right] = 5a_0X_1\left[\left(\frac{p_s}{p_1}\right)^{\frac{1}{7}} - 1\right] = -5a_0(X_s - X_1)$$

The absolute particle velocity of F, c_s, will be given by the sum of c_{FrelBE} and c_1, as follows:

$$c_s = c_{FrelBE} + c_1 = -5a_0(X_s - X_1) + 5a_0(X_1 - 1) = 5a_0(2X_1 - X_s - 1)$$

Applying the same logic to wave top C proceeding into wave top DG gives another expression for c_s, as F and C are at precisely the same state conditions:

$$c_s = c_{CrelDG} + c_2 = 5a_0(X_s - X_2) - 5a_0(X_2 - 1) = -5a_0(2X_2 - X_s - 1)$$

Equating the two expressions for c_s for the same wave top FC gives two important equations as a conclusion, one for the pressure of superposition, p_s, and the other for the particle velocity of superposition, c_s:

$$X_s = X_1 + X_2 - 1 \qquad (2.2.1)$$

or

$$\left(\frac{p_s}{p_0}\right)^{\frac{1}{7}} = \left(\frac{p_1}{p_0}\right)^{\frac{1}{7}} + \left(\frac{p_2}{p_0}\right)^{\frac{1}{7}} - 1 \qquad (2.2.2)$$

$$c_s = 5a_0(X_1 - 1) - 5a_0(X_2 - 1) = 5a_0(X_1 - X_2) \qquad (2.2.3)$$

or

$$c_s = c_1 + c_2 \qquad (2.2.4)$$

Note that the expressions for superposition particle velocity reserves the need for a sign convention, i.e., a declaration of a positive direction, whereas Eqs. 2.2.1 and 2.2.2 are independent of direction. The more general expression for gas properties other than air is easily seen from the above equations as:

$$X_s = X_1 + X_2 - 1 \qquad (2.2.5)$$

then

$$\left(\frac{p_s}{p_0}\right)^{G17} = \left(\frac{p_1}{p_0}\right)^{G17} + \left(\frac{p_2}{p_0}\right)^{G17} - 1 \qquad (2.2.6)$$

as

$$c_s = c_1 + c_2 \qquad (2.2.7)$$

then

$$c_s = G_5 a_0(X_1 - 1) - G_5 a_0(X_2 - 1) = G_5 a_0(X_1 - X_2) \qquad (2.2.8)$$

At any location within the pipes of an engine, the superposition process is the norm as pressure waves continually pass to and fro. Further, if we place a pressure transducer in the wall of a pipe, it is the superposition pressure-time history that is recorded, not the individual pressures of the rightward and the leftward moving pressure waves. This makes it very difficult to interpret recorded pressure-time data in engine ducting. A simple example will make the point. In Sec. 2.1.4 and in Fig. 2.2, there is an example of two pressure waves, p_e and p_i, with pressure ratio values of 1.2 and 0.8, respectively. Suppose that they are in a pipe but are the oppositely moving waves just discussed and are in the position of precise superposition. There is a pressure transducer at the point where the wave peaks coincide and it records a superposition pressure, p_s. What will it be and what is the value of the superposition particle velocity, c_s?

The values of X_e, X_i, a_0, c_e, and c_i were 1.0264, 0.9686, 343.1, 45.3, and 53.9, respectively, in terms of a positive direction for the transmission of each wave. In other words, the properties of waves p_e and p_i are to be assigned to waves 1 and 2, respectively, merely to reduce the arithmetic clutter both within the text and in your mind.

If p_1 is regarded as moving rightward and that is defined as the positive direction, and consequently, p_2 is moving leftward as in a negative direction, then Eqs. 2.2.1 and 2.2.2 show:

Hence,
$$X_s = X_1 + X_2 - 1 = 1.0264 - 0.9868 - 1 = 0.995$$

$$P_s = X_s^7 = 0.965 \quad \text{and} \quad p_s = P_s p_0 = 0.965 \times 101{,}325 = 97{,}831 \text{ Pa}$$

$$c_s = 45.3 + [-(-53.9)] = 99.2 \text{ m/s}$$

Thus, the pressure transducer in the wall of the pipe would show little of this process, as the summation of the two waves would reveal a trace virtually indistinguishable from the atmospheric line, and exhibit nothing of the virtual doubling of the particle velocity.

The opposite effect takes place when two waves of the same type undergo a superposition process. For example, if wave p_1 met another wave p_1 going in the other direction and in the plane of the pressure transducer, then:

$$X_s = 1.0264 + 1.0264 - 1 = 1.0528 \quad \text{and} \quad P_s = X_s^7 = 1.0528^7 = 1.434$$

$$c_s = 45.3 + [-(+45.3)] = 0 \text{ m/s}$$

The pressure transducer would show a large compression wave with a pressure ratio of 1.434 and tell nothing of the zero particle velocity at the same spot and time.

This makes the interpretation of exhaust and intake pressure records a most difficult business if it is based on experimentation alone. Not unnaturally, the engineering observer will interpret a measured pressure trace which exhibits a large number of pressure oscillations as being evidence of lots of wave activity. This may well be so, but some of these fluctuations will almost certainly be periods approaching zero particle velocity also while yet other periods of this superposition trace exhibiting "calm" could very well be operating at particle velocities approaching the sonic value! As this is clearly a most important topic, it will be returned to in later sections of this chapter.

2.2.2 Wave propagation during superposition

The propagation velocity of the two waves during the superposition process must also take into account direction. The statement below is accurate for the superposition condition where the rightward direction is considered positive.

where,
$$a_s = a_0 X_s$$

As propagation velocity is given by the sum of the local acoustic and particle velocities, as in Eq. 2.1.9

$$\alpha_{s\ rightward} = a_s + c_s$$
$$= a_0(X_1 + X_2 - 1) + G_5 a_0(X_1 - X_2) \quad (2.2.9)$$
$$= a_0(G_6 X_1 - G_4 X_2 - 1)$$

$$\alpha_{s\ leftward} = -a_s + c_s$$
$$= -a_0(X_1 + X_2 - 1) + G_5 a_0(X_1 - X_2) \quad (2.2.10)$$
$$= -a_0(G_6 X_2 - G_4 X_1 - 1)$$

Regarding the propagation velocities during superposition in air of the two waves, p_e and p_i, as presented above in Sec. 2.2.1, where the values of c_s, a_0, X_1, X_2 and X_s were 99.2, 343.1, 1.0264, 0.9686, and 0.995, respectively:

$$a_s = a_0 X_s = 343.1 \times 0.995 = 341.38 \text{ m/s}$$

$$\alpha_s \text{ rightward} = a_s + c_s = 341.38 + 99.2 = 440.58 \text{ m/s}$$

$$\alpha_s \text{ leftward} = -a_s + c_s = -341.38 + 99.2 = -242.18 \text{ m/s}$$

This could equally have been determined more formally using Eqs. 2.2.9 and 2.2.10 as,

$$\alpha_{s\ rightward} = a_0(G_6 X_1 - G_4 X_2 - 1) = 343.38(6 \times 1.0264 - 4 \times 0.9686 - 1) = 440.58 \text{ m/s}$$

$$\alpha_{s\ leftward} = a_0(G_6 X_2 - G_4 X_1 - 1) = 343.38(6 \times 0.9686 - 4 \times 1.0264 - 1) = -242.18 \text{ m/s}$$

As the original propagation velocities of waves 1 and 2, when they were traveling "undisturbed" in the duct into a gas at the reference acoustic state, were 397.44 m/s and 278.47 m/s, it is clear from the above calculations that wave 1 has accelerated, and wave 2 has slowed down, by some 10% during this particular example of a superposition process. This effect is often referred to as "wave interference during superposition."

During computer calculations it is imperative to rely on formal equations such as Eqs. 2.2.9 and 2.2.10 to provide computed values.

2.2.3 Mass flow rate during wave superposition

As the superposition process accelerates some waves and decelerates others, so too must the mass flow rate be affected. This also must be capable of computation at any position within a duct. The continuity equation provides the necessary information.

Design and Simulation of Two-Stroke Engines

$$\text{Mass flow rate} = \text{density} \times \text{area} \times \text{velocity} = \rho_s A c_s$$

where,
$$\rho_s = \rho_0 X_s^{G5} \qquad (2.2.11)$$

Hence,
$$\dot{m} = G_5 a_0 \rho_0 A (X_1 + X_2 - 1)^{G5}(X_1 - X_2) \qquad (2.2.12)$$

In terms of the numerical example used in Sec. 2.2.2, the values of a_0, ρ_0, X_1, and X_2 were 343.1, 1.2049, 1.0264, and 0.9686, respectively. The pipe area is that of the 25-mm-diameter duct, or 0.000491 m². The gas in the pipe is air.

We can solve for the mass flow rate by using the previously known value of X_s, which was 0.995, or that for particle velocity c_s which was 99.2 m/s, and determine the superposition density ρ_s thus:

$$\rho_s = \rho_0 X_s^{G5} = 1.2049 \times 0.995^5 = 1.1751 \ \text{kg}/\text{m}^3$$

Hence, the mass flow rate during superposition is given by:

$$\dot{m}_{\text{right}} = 1.1751 \times 0.000491 \times 99.2 = 0.0572 \ \text{kg/s}$$

The sign was known by having available the information that the superposition particle movement was rightward and inserting c_s as +99.2 and not −99.2. Alternatively, the formal Eq. 2.2.12 gives a numeric answer indicating direction of mass or particle flow. This is obtained by solving Eq. 2.2.12 with the lead term in any bracket, i.e., X_1, as that value where wave motion is considered to be in a positive direction.

Hence, mass flow rate <u>rightward</u>, as direction of wave 1 is called positive, is:

$$\begin{aligned}\dot{m}_{\text{right}} &= G_5 a_0 \rho_0 A (X_1 + X_2 - 1)^{G5}(X_1 - X_2) \\ &= 5 \times 343.11 \times 1.2049 \times (1.0264 + 0.9686 - 1) \times (1.0264 - 0.9686)^5 \\ &= +0.0572 \ \text{kg/s}\end{aligned}$$

It will be observed, indeed it is imperative to satisfy the equation of continuity, that the superposition mass flow rate is the sum of the mass flow rate induced by the individual waves. The mass flow rates in a rightward direction of waves 1 and 2, computed earlier in Sec. 2.1.4, were 0.0305 and 0.0272 kg/s, respectively.

2.2.4 Supersonic particle velocity during wave superposition

In typical engine configurations it is rare for the magnitude of finite amplitude waves which occur to provide a particle velocity that approaches the sonic value. As the Mach number **M** is defined in Eq. 2.1.23 for a pressure amplitude ratio of X as:

$$M = \frac{c}{a} = \frac{G_5 a_0 (X-1)}{a_0 X} = \frac{G_5(X-1)}{X} \qquad (2.2.13)$$

For this to approach unity then:

$$X = \frac{G_5}{G_4} = \frac{2}{3-\gamma} \quad \text{and for air} \quad X = 1.25$$

In air, as seen above, this would require a compression wave of pressure ratio, P, where:

$$P = \frac{p}{p_0} = X^{G_7} \quad \text{and in air} \quad P = 1.25^7 = 4.768$$

Even in high-performance racing engines a pressure ratio of an exhaust pulse greater than 2.2 atmospheres is very unusual, thus sonic particle velocity emanating from that source is not likely. What is a more realistic possibility is that a large exhaust pulse may encounter a strong oppositely moving expansion wave in the exhaust system and the superposition particle velocity may approach or attempt to exceed unity.

Unsteady gas flow does not permit supersonic particle velocity. It is self-evident that the gas particles cannot move faster than the pressure wave disturbance which is giving them the signal to move. As this is not possible gas dynamically, the theoretical treatment supposes that a "weak shock" occurs and the particle velocity reverts to a subsonic value. The basic theory is to be found in any standard text [2.4] and the resulting relationships are referred to as the Rankine-Hugoniot equations. The theoretical treatment is almost identical to that given here, as Appendix A2.2, for moving shocks where the particle velocity behind the moving shock is also subsonic.

Consider two oppositely moving pressure waves in a superposition situation. The individual pressure waves are p_1 and p_2 and the gas properties are γ and R with a reference temperature and pressure denoted by p_0 and T_0. From Eqs. 2.2.5 to 2.2.8 the particle Mach number **M** is found from:

pressure amplitude ratio $\qquad X_s = X_1 + X_2 - 1 \qquad (2.2.14)$

acoustic velocity $\qquad a_s = a_0 X_s \qquad (2.2.15)$

particle velocity $\qquad c_s = G_5 a_0 (X_1 - X_2) \qquad (2.2.16)$

Mach number (+ only) $\qquad M_s = \dfrac{c_s}{a_s} = \left| \dfrac{G_5 a_0 (X_1 - X_2)}{a_0 X_s} \right| \qquad (2.2.17)$

Note that the modulus of the Mach number is acquired to eliminate directionality in any inquiry as to the magnitude of it in absolute terms. If that inquiry reveals that $\mathbf{M}_s$ is greater than unity then the individual waves p_1 and p_2 are modified by internal reflections to provide a shock to a gas flow at subsonic particle velocity. The superposition pressure after the shock transition to subsonic particle flow at Mach number $\mathbf{M}_{s\ new}$ is labeled as $p_{s\ new}$. The "new" pressure waves that travel onward after superposition is completed become labeled as $p_{1\ new}$ and $p_{2\ new}$. The Rankine-Hugoniot equations that describe this combined shock and reflection process are:

pressure
$$\frac{p_{s\ new}}{p_s} = \frac{2\gamma}{\gamma+1}\mathbf{M}_s^2 - \frac{\gamma-1}{\gamma+1} \qquad (2.2.18)$$

Mach number
$$\mathbf{M}_{s\ new}^2 = \frac{\mathbf{M}_s^2 + \dfrac{2}{\gamma-1}}{\dfrac{2\gamma}{\gamma-1}\mathbf{M}_s^2 - 1} \qquad (2.2.19)$$

After the shock the "new" pressure waves are related by:

pressure
$$X_{s\ new} = X_{1\ new} + X_{2\ new} - 1 \qquad (2.2.20)$$

pressure
$$p_{s\ new} = p_0 X_{s\ new}^{G7} \qquad (2.2.21)$$

Mach number
$$\mathbf{M}_{s\ new} = \frac{c_{s\ new}}{a_{s\ new}} = \frac{G_5 a_0 (X_{1\ new} - X_{2\ new})}{a_0 X_{s\ new}} \qquad (2.2.22)$$

pressure wave 1
$$p_{1\ new} = p_0 X_{1\ new}^{G7} \qquad (2.2.23)$$

pressure wave 2
$$p_{2\ new} = p_0 X_{2\ new}^{G7} \qquad (2.2.24)$$

From the knowledge that the Mach number in Eq. 2.2.17 has exceeded unity, the two Eqs. 2.2.18 and 2.2.19 of the Rankine-Hugoniot set provide the basis of the simultaneous equations needed to solve for the two unknown pressure waves $p_{1\ new}$ and $p_{2\ new}$ through the connecting information in Eqs. 2.2.20 to 2.2.22. For simplicity of presentation of this theory, it is predicated that $p_1 > p_2$, i.e., that the sign of any particle velocity is positive. In any application of this theory this point must be borne in mind and the direction of the analysis adjusted accordingly. The solution of the two simultaneous equations reveals, in terms of complex functions Γ_1 to Γ_4 composed of known pre-shock quantities:

$$\Gamma_1 = \frac{M_s^2 + \frac{2}{\gamma-1}}{\frac{2\gamma}{\gamma-1}M_s^2 - 1} \quad \Gamma_2 = \frac{2\gamma}{\gamma+1}M_s^2 - \frac{\gamma-1}{\gamma+1} \quad \Gamma_3 = \frac{\gamma-1}{2}\sqrt{\Gamma_1} \quad \Gamma_4 = X_s\Gamma_2^{\frac{\gamma-1}{2\gamma}}$$

then
$$X_{1\,new} = \frac{1+\Gamma_4+\Gamma_3\Gamma_4}{2} \quad \text{and} \quad X_{2\,new} = \frac{1+\Gamma_4-\Gamma_3\Gamma_4}{2} \quad (2.2.25)$$

The new values of particle velocity, Mach number, wave pressure or other such parameters can be found by substitution into Eqs. 2.2.20 to 2.2.24.

Consider a simple numeric example of oppositely moving waves. The individual pressure waves are p_1 and p_2 with strong pressure ratios of 2.3 and 0.5, and the gas properties are air where the specific heats ratio, γ, is 1.4 and the gas constant, R, is 287 J/kgK. The reference temperature and pressure are denoted by p_0 and T_0 and are 101,325 Pa and 293 K, respectively. The conventional superposition computation as carried out previously in this section would show that the superposition pressure ratio, P_s, is 1.2474, the superposition temperature, T_s, is 39.1°C, and the particle velocity is 378.51 m/s. This translates into a Mach number, M_s, during superposition of 1.0689, clearly just sonic. The application of the above theory reveals that the Mach number, $M_{s\,new}$, after the weak shock is 0.937 and the ongoing pressure waves, $p_{1\,new}$ and $p_{2\,new}$, have modified pressure ratios of 2.2998 and 0.5956, respectively.

From this example it is obvious that it takes waves of uncommonly large amplitude to produce even a weak shock and that the resulting modifications to the amplitude of the waves are quite small. Nevertheless, it must be included in any computational modeling of unsteady gas flow that has pretensions of accuracy.

In this section we have implicitly introduced the concept that the amplitude of pressure waves can be modified by encountering some "opposition" to their perfect, i.e., isentropic, progress along a duct. This also implicitly introduces the concept of reflections of pressure waves, i.e., the taking of some of the energy away from a pressure wave and sending it in the opposite direction. This theme is one which will appear in almost every facet of the discussions below.

2.3 Friction loss and friction heating during pressure wave propagation

Particle flow in a pipe induces forces acting against the flow due to the viscous shear forces generated in the boundary layer close to the pipe wall. Virtually any text on fluid mechanics or gas dynamics will discuss the fundamental nature of this behavior in a comprehensive fashion [2.4]. The frictional effect produces a dual outcome: (a) the frictional force results in a pressure loss to the wave opposite to the direction of particle motion and, (b) the viscous shearing forces acting over the distance traveled by the particles with time means that the work expended appears as internal heating of the local gas particles. The typical situation is illustrated in Fig. 2.4, where two pressure waves, p_1 and p_2, meet in a superposition process. This makes the subsequent analysis more generally applicable. However, the following analysis applies equally well to a pressure wave, p_1, traveling into undisturbed conditions, as

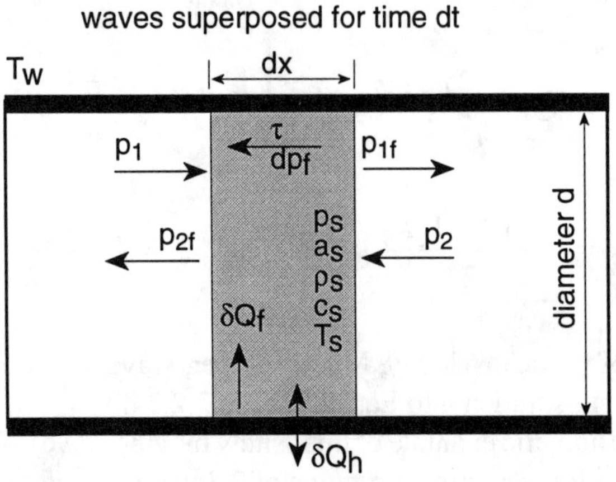

Fig. 2.4 Friction loss and heat transfer in a duct.

it remains only to nominate that the value of p_2 has the same pressure as the undisturbed state p_0.

In the general analysis, pressure waves p_1 and p_2 meet in a superposition process and due to the distance, dx, traveled by the particles during a time dt, engender a friction loss which gives rise to internal heating, dQ_f, and a pressure loss, dp_f. By definition both these effects constitute a gain of entropy, so the friction process is non-isentropic as far as the wave propagation is concerned.

The superposition process produces all of the velocity, density, temperature, and mass flow charactaristics described in Sec. 2.2. However, what is required from any theoretical analysis regarding friction pressure loss and heating is not only the data regarding pressure loss and heat generated, but more importantly the altered amplitudes of pressure waves p_1 and p_2 after the friction process is completed.

The shear stress, τ, at the wall as a result of this process is given by:

Shear stress $$\tau = C_f \frac{\rho_s c_s^2}{2} \tag{2.3.1}$$

The friction factor, C_f, is usually in the range 0.003 to 0.008, depending on factors such as fluid viscosity or pipe wall roughness. The direct assessment of the value of the friction factor is discussed later in this section.

The force, F, exerted at the wall on the pressure wave by the wall shear stress in a pipe of diameter, d, during the distance, dx, traveled by a gas particle during a time interval, dt, is expressed as:

Distance traveled $$dx = c_s dt$$

Force $$F = \pi d \tau dx = \pi d \tau c_s dt \tag{2.3.2}$$

Chapter 2 - Gas Flow through Two-Stroke Engines

This force acts over the entire pipe flow area, A, and provides a loss of pressure, dp_f, for the plane fronted wave that is inducing the particle motion. The pressure loss due to friction is found by incorporating Eq. 2.3.1 into Eq. 2.3.2:

$$\text{Pressure loss} \qquad dp_f = \frac{F}{A} = \frac{\pi d \tau c_s dt}{\frac{\pi d^2}{4}} = \frac{4\tau c_s dt}{d} = \frac{2 C_f \rho_s c_s^3 dt}{d} \qquad (2.3.3)$$

Notice that this equation contains a cubed term for the velocity, and as there is a sign convention for direction, this results in a loss of pressure for compression waves and a pressure rise for expansion waves, i.e., a loss of wave strength and a reduction of particle velocity in either case.

As this friction loss process is occurring during the superposition of waves of pressure p_1 and p_2 as in Fig. 2.4, values such as superposition pressure amplitude ratio X_s, density ρ_s, and particle velocity c_s can be deduced from the equations given in Sec. 2.2. They are repeated here:

$$X_s = X_1 + X_2 - 1 \qquad c_s = G_5 a_0 (X_1 - X_2) \qquad \rho_s = \rho_0 X_s^{G5}$$

The absolute superposition pressure, p_s, is given by:

$$p_s = p_0 X_s^{G7}$$

After the loss of friction pressure the new superposition pressure, p_{sf}, and its associated pressure amplitude ratio, X_{sf}, will be, depending on whether it is a compression or expansion wave,

$$p_{sf} = p_s \pm dp_f \qquad X_{sf} = \left(\frac{p_{sf}}{p_0}\right)^{G17} \qquad (2.3.4)$$

The solution for the transmitted pressure waves, p_{1f} and p_{2f}, after the friction loss is applied to both, is determined using the momentum and continuity equations for the flow regime before and after the event, thus:

Continuity $\qquad\qquad\qquad \dot{m}_s = \dot{m}_{sf}$

Momentum $\qquad\qquad\qquad p_s A - p_{sf} A = \dot{m}_s c_s - \dot{m}_{sf} c_{sf}$

which becomes $\qquad\qquad A(p_s - p_{sf}) = \dot{m}_s c_s - \dot{m}_{sf} c_{sf}$

As the mass flow is found from:

$$\dot{m}_s = \rho_s A c_s = G_5 a_s (X_1 - X_2) A \rho_0 X_s^{G5}$$

the transmitted pressure amplitude ratios, X_{1f} and X_{2f}, and superposition particle velocity, c_{sf}, are related by:

$$X_{sf} = (X_{1f} + X_{2f} - 1) \quad \text{and} \quad c_{sf} = G_5 a_0 (X_{1f} - X_{2f})$$

The momentum and continuity equations become two simultaneous equations for the two unknown quantities, X_{1f} and X_{2f}, which are found by determining c_{sf},

$$c_{sf} = c_s + \frac{p_{sf} - p_s}{\rho_s c_s} \qquad (2.3.5)$$

whence

$$X_{1f} = \frac{1 + X_{sf} + \dfrac{c_{sf}}{G_5 a_0}}{2} \qquad (2.3.6)$$

and

$$X_{2f} = 1 + X_{sf} - X_{1f} \qquad (2.3.7)$$

Consequently the pressures of the ongoing pressure waves p_{1f} and p_{2f} after friction has been taken into account, are determined by:

$$p_{1f} = p_0 X_{1f}^{G7} \quad \text{and} \quad p_{2f} = p_0 X_{2f}^{G7} \qquad (2.3.8)$$

Taking the data for the two pressure waves of amplitude 1.2 and 0.8 which have been used in previous sections of this chapter, consider them to be superposed in a pipe of 25 mm diameter with the compression wave p_1 moving rightward. The reference conditions are also as used before at T_0 is 20°C or 293K and p_0 as 101,325 Pa. However, pressure drop occurs only as a result of particle movement so it is necessary to define a time interval for the superposition process to occur. Consider that the waves are superposed at the pressure levels indicated for a period of 2° crankshaft in the duct of an engine running at 1000 rpm. This represents a time interval, dt, of :

$$dt = \frac{\theta}{360} \times \frac{60}{N} = \frac{\theta}{6N} \text{ s} \qquad (2.3.9)$$

or

$$dt = \frac{2}{6 \times 1000} = 0.333 \times 10^{-3} \text{ s}$$

whence the particle movement, dx, is given by:

$$dx = c_s dt = 99.1 \times 0.333 \times 10^{-3} = 33 \times 10^{-3} \text{ m}$$

From the above superposition time element it is seen that this computes to a numerical value of 0.333 ms. Assuming a friction factor C_f of 0.004, the above equations in the section show that the loss of superposition pressure occurs over a distance dx of 33 mm within the duct and has a magnitude of 122 Pa. The pressure ratio of the rightward wave drops from 1.2 to 1.198 and that of the leftward wave rises from 0.8 to 0.8023. The rightward propagation velocity of the compression wave during superposition drops from 440.5 to 439.5 m/s while that of the leftward expansion wave rises from 242.3 to 243.4 m/s. The superposition particle velocity drops from 99.1 m/s to 98.05 m/s.

It is evident that the pressure loss due to friction reduces the amplitude of compression waves and slows them down. The opposite effect applies to an expansion wave: it raises its absolute pressure, i.e., weakens the wave, and thereby moves its propagation velocity from a subsonic value toward sonic velocity.

While the likelihood of a single traverse of a pressure wave in a duct of an engine is remote, nevertheless it should be considered theoretically. By definition, friction opposes the motion of a pressure wave and does so continuously. This means that a train of pressure waves is sent off in the opposite direction to the propagation of the wave train and with a magnitude which can be calculated from the above equations. Use the data above, but with the exception that the single wave, p_1, traveling rightward, i.e., in the positive direction as far as the sign convention is concerned, has a pressure ratio of 1.2. All other data remain the same and a fixed friction factor of 0.004 is employed. All of the above equations can be used with the value of p_2 inserted as being identical to p_0, i.e., with a pressure ratio of 1.0. The results show that the ongoing wave pressure ratio, P_{1f}, is reduced to 1.1994 and the reflected wave, P_{2f}, is 1.0004. If the calculation is repeated to find the effect of friction on a single traverse of an expansion wave, i.e., by inserting P_1 as 0.8 and P_2 as 1.0, then P_{1f} and P_{2f} become 0.8006 and 0.9995, respectively.

In Sec. 2.19 the traverse of a single pressure wave in a duct is described as both theory and experiment.

2.3.1 Friction factor during pressure wave propagation

It is possible to predict the value of friction factor more closely by considering further information available in the literature of experimental and theoretical fluid mechanics. The properties of air for thermal conductivity, C_k, and viscosity, μ, are a function of absolute temperature, T, and are required for the calculation of the shearing forces in air from which friction factor can be assessed. The interconnection between friction factor and shear stress has been set out in Eq. 2.3.1. The thermal conductivity and viscosity of air can be found from data tables and curve fitted to provide high accuracy from values of T from 300 to 2000 K as:

$$C_k = 6.1944 \times 10^{-3} + 7.3814 \times 10^{-5} T - 1.2491 \times 10^{-8} T^2 \text{ W/mK} \qquad (2.3.10)$$

$$\mu = 7.457 \times 10^{-6} + 4.1547 \times 10^{-8}\,T - 7.4793 \times 10^{-12}\,T^2 \text{ kg/ms} \qquad (2.3.11)$$

While the data above for air are not the same as that for exhaust gas, the differences are sufficiently small as to warrant describing exhaust gas by the same relationships for the purposes of determining Reynolds, Nusselt and other dimensionless parameters common in fluid mechanics.

The Reynolds number at any local point in space and time in a duct of diameter, d, at superposition temperature, T_s, density, ρ_s, and particle velocity, c_s, is given by:

$$\mathbf{Re} = \frac{\rho_s d c_s}{\mu_{Ts}} \qquad (2.3.12)$$

Much experimental work in fluid mechanics has related friction factor to Reynolds number and that ascribed to Blasius to describe fully turbulent flow is typical:

$$C_f = \frac{0.0791}{\mathbf{Re}^{0.25}} \quad \text{for } \mathbf{Re} \geq 4000 \qquad (2.3.13)$$

Almost all unsteady gas flow in engine ducting is turbulent, but where it is not and is close to laminar, or is laminar, then it is simple and accurate to assign the friction factor as 0.01. This threshold number can be easily deduced from the Blasius formula by inserting the Reynolds number as 4000.

Using all the data from the example cited in the previous section above, it was stated that the calculated superposition particle velocity is 99.1 m/s. The superposition density is 1.1752 kg/m³ and can be calculated using Eq. 2.2.11. The relationship for superposition temperature is given by Eq. 2.1.11, therefore:

$$\frac{T_s}{T_0} = \left(\frac{p_s}{p_0}\right)^{\frac{\gamma-1}{\gamma}} = X_s^2 \qquad (2.3.14)$$

Using the numerical data provided, the superposition temperature is 290.1 K or 17.1°C. The viscosity of air at 290.1 K is determined from Eq. 2.3.11 as 1.888×10^{-5} kg/ms. Consequently, the Reynolds number **Re** in the 25-mm-diameter pipe, from Eq. 2.3.12, is 154,210; it is clearly turbulent flow. Thence, from Eq. 2.3.13, the friction factor is calculated as 0.00399, which is not far removed from the assumption used not only in the previous section but which also has appeared frequently in the literature!

The use of this theoretical approach for the determination of friction factor, allied to the general theory regarding friction loss in the previous section, permits the accurate assessment of the friction loss on pressure waves in unsteady flow.

It is important to stress that the action of friction on a pressure wave passing through the pipe is a non-isentropic process. The manifestation of this is the heating of the local gas as friction occurs and the continual decay of the pressure wave due to its application. The heating effect can be calculated as the friction force, F, is available from the combination of Eqs. 2.3.1 and 2.3.2 and the work done by this force acting through distance, dx, appears as heat in the gas element involved.

$$F = \pi d \left[C_f \frac{\rho_s c_s^2}{2} \right] c_s dt \tag{2.3.15}$$

Thus the work, dW_f, resulting in the heat generated, dQ_f, can be calculated by:

$$\delta W_f = Fdx = Fc_s dt = \frac{\pi d C_f \rho_s c_s^4 dt^2}{2} = \delta Q_f \tag{2.3.16}$$

All of the relevant data for the numerical example used in this section are available to insert into Eq. 2.3.16, from which it is calculated that the internal heating due to friction is 1.974 mJ. While this value may appear miniscule, remember that this is a continuous process occurring for a pressure wave during its excursion throughout a pipe, and that this heating effect of 1.974 mJ takes place in a time frame of 0.333 ms. This represents a heating rate of 5.93 W which puts the heating effect due to friction into a physical context which can be more readily comprehended.

One issue which must be taken into account by those concerned with computation of wave motion is that all of the above equations use a length term within the calculation for friction force with respect to the work done or heat generated by opposition to it. This length term is quite correctly computed from the particle velocity, c_s, and the time period, dt, for the motion of those particles. However, should the computation method purport to represent a group of gas particles within a pipe by the behavior of those particles at the wave point under calculation, then the length term in the ensuing calculation for force, F, must be replaced by that length occupied by the said group of particles. The subsequent calculation to compute the work, dW_f, i.e., the heat quantity dQ_f generated by friction, is the force due to friction for all of the group of particles multiplied by the distance moved by any one of the particles in this group, which distance remains as the "$c_s dt$" term. This is discussed at greater length in Sec. 2.18.6.

2.3.2 Friction loss during pressure wave propagation in bends in pipes

This factor is seldom considered of pressing significance in the simulation of engine ducting because pressure waves travel around quite sharp kinks, bends and radiused corners in ducting with very little greater pressure loss than is normally associated with friction, as has been discussed above. It is not a subject that has been researched to any great extent as can be seen from the referenced literature of those who have, namely Blair *et al.* [2.20]. The basic mechanism of analysis is to compute the pressure loss, dp_b, in the segment of pipe of length which is under analysis. The procedure is similar to that for friction:

Pressure loss $\quad\quad\quad dp_b = C_b \rho_s c_s^2 \tag{2.3.17}$

where the pressure loss coefficient, C_b, is principally a function of Reynolds number and the deflection angle per unit length around the bend:

$$C_b = f\left(\mathbf{Re}, \frac{d\theta}{dx}\right) \tag{2.3.18}$$

The extra pressure loss as a function of being deflected by an angle appropriate to the pipe segment length, dx, can then be added to the friction loss term in Eq. 2.3.3 and the analysis continued for the segment of pipe length, dx, under scrutiny. There is discussion in Sec. 2.14 for pressure losses at branches in pipes and Eq. 2.14.1 gives an almost identical relationship for the pressure loss in deflecting flows around the corners of a pipe branch. All such pressure loss equations relate that loss to a gain of entropy through a decrease in the kinetic energy of the gas particles during superposition.

It is a subject deserving of painstaking measurement through research using the sophisticated experimental QUB SP apparatus described in Sec. 2.19. The aim would be to derive accurately the values of the pressure loss coefficient, C_b, and ultimately report them in the literature; I am not aware of such information having been published already.

2.4 Heat transfer during pressure wave propagation

Heat can be transferred to or from the wall of the duct and the gas as the unsteady flow process is conducted. While all three processes of conduction, convection and radiation are potentially involved, it is much more likely that convection heat transfer will be the predominant phenomenon in most cases. This is certainly true of induction systems but some of you will remember exhaust manifolds glowing red and ponder the potential errors of considering convection heat transfer as the sole mechanism. There is no doubt that in such circumstances radiation heat transfer should be seriously considered for inclusion in any theoretical treatment. It is not an easy topic and the potential error of its inclusion could actually be more serious than its exclusion. As a consequence, only convection heat transfer will be discussed.

The information to calculate the normal and relevant parameters for convection is available from within most analysis of unsteady gas flow. The physical situation is illustrated in Fig. 2.4. A superposition process is underway. The gas is at temperature, T_s, particle velocity, c_s, and density, ρ_s.

In Sec. 2.3.1, the computation of the friction factor, C_f, and the Reynolds number, $\mathbf{Re}$, was described. From the Reynolds analogy of heat transfer with friction it is possible to calculate the Nusselt number, $\mathbf{Nu}$, thus;

$$\mathbf{Nu} = \frac{C_f \mathbf{Re}}{2} \tag{2.4.1}$$

The Nusselt number contains a direct relationship between the convection heat transfer coefficient, C_h, the thermal conductivity of the gas, C_k, and the effective duct diameter, d. The standard definitions for these parameters can be found in any conventional text in fluid mechanics or heat transfer [2.4]. The definition for the Nusselt number is:

$$\text{Nu} = \frac{C_h d}{C_k} \tag{2.4.2}$$

From Eqs. 2.4.1 and 2.4.2, the convection heat transfer coefficient can be determined:

$$C_h = \frac{C_k \text{Nu}}{d} = \frac{C_k C_f \text{Re}}{2d} \tag{2.4.3}$$

The relationship for the thermal conductivity for air is given in Eq. 2.3.10 as a function of the gas temperature, T. In any unsteady gas flow process, the time element, dt, and the distance of exposure of the gas element to the wall, dx, is available from the computation or from the input data. In which case, knowing the pipe wall temperature, the heat transfer, dQ_h, from the gas to the pipe wall can be assessed. The direction of this heat transfer is clear from the ensuing theory:

$$\delta Q_h = \pi d C_h dx (T_w - T_s) dt \tag{2.4.4}$$

If we continue the numeric example with the same input data as given previously in Sec. 2.3, but with the added information that the pipe wall temperature is 100°C, then the output data using the theory shown in this section give the magnitude and direction of the heat transfer. As the wall at 100°C is hotter than the gas at superposition temperature, T_s, at 17.1°C, the heat transfer direction is positive as it is added to the gas in the pipe. The value of heat transferred is 23.44 mJ which is considerably greater than the 1.974 mJ attributable to friction. During the course of this computation the Nusselt number, **Nu**, is calculated as 307.8 and the convection heat transfer coefficient, C_h, as 326.9 W/m²K.

The total heating or cooling of a gas element undergoing an unsteady flow process is a combination of that which emanates externally from the convection heat transfer and internally from the friction. This total heat transfer is defined as dQ_{fh} and is obtained thus:

$$dQ_{fh} = dQ_f + dQ_h \tag{2.4.5}$$

For the same numeric data the value of dQ_{fh} is the sum of +1.974 mJ and +23.44 mJ which would result in dQ_{fh} = +25.44 mJ.

2.5 Wave reflections at discontinuities in gas properties

As a pressure wave propagates within a duct, it is highly improbable that it will always encounter ahead of it gas which is at precisely the same state conditions or has the same gas properties as that through which it is currently traveling. In physical terms, many people have experienced an echo when they have shouted in foggy surroundings on a cold damp morning with the sun warming some sections of the local atmosphere more strongly than others. In acoustic terms this is precisely the situation that commonly exists in engine ducting. It is particularly pronounced in the inlet ducting of a four-stroke engine which has experienced a strong backflow of hot exhaust gas at the onset of the intake process. Two-stroke engines have an exhaust process which sends hot exhaust gas into the pipe during the blowdown phase and then short-circuits a large quantity of much colder air into this same duct during the scavenge process. Pressure waves propagating through such pipes encounter gas at varying temperatures, and reflections ensue. While this is referred to in the literature as a "temperature" discontinuity, this is quite misleading. Pressure wave reflection will take place at such boundaries, even at constant temperature, if other gas properties such as gas constant or density are variables across it.

The physical and thermodynamic situation is set out in sketch format in Fig. 2.5, where a pressure wave p_1 is meeting pressure wave p_2. Clearly, a superposition process takes place. In Sec. 2.2.1, where the theory of superposition is set down, once the superposition was completed the pressure waves, p_1 and p_2, would have proceeded onward with unaltered amplitudes. However, this superposition process is taking place with the waves arriving having traversed through gas at differing thermodynamic states. Wave p_1 is coming from side "a" where the gas has properties of γ_a, R_a and reference conditions of density ρ_{0a} and temperature T_{0a}. Wave p_2 is coming from side "b" where the gas has properties of γ_b, R_b and reference conditions of density ρ_{0b} and temperature T_{0b}. As with most reflection processes, the momentum equation is that which best describes a "bounce" behavior, and this proves to be so in this case. As the superposition process occurs, and the reflection is taking place at the interface, the transmitted pressure waves have amplitudes of p_{1d} and p_{2d}. The theory to describe this states that the laws of conservation of mass and momentum must be upheld.

Continuity $\quad\quad\quad\quad\quad\quad\quad\quad \dot{m}_{\text{side a}} = \dot{m}_{\text{side b}}$ \quad\quad\quad (2.5.1)

Momentum $\quad A(p_{s\ \text{side a}} - p_{s\ \text{side b}}) = \dot{m}_{\text{side a}} c_{s\ \text{side a}} - \dot{m}_{\text{side b}} c_{s\ \text{side b}}$ \quad\quad (2.5.2)

It can be seen that this produces a relatively simple solution where the superposition pressure and the superposition particle velocity are identical on either side of the thermodynamic discontinuity.

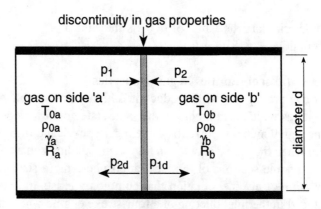

Fig. 2.5 Wave reflection at a temperature discontinuity.

$$p_s \text{ side a} = p_s \text{ side b} \tag{2.5.3}$$

$$c_s \text{ side a} = c_s \text{ side b} \tag{2.5.4}$$

The solution divides into two different cases, one simple and the other more complex, depending on whether or not the gas composition is identical on both sides of the boundary.

(i) The simple case of common gas composition

The following is the solution of the simple case where the gas is identical in composition on both sides of the boundary, i.e., γ_a and R_a are identical to γ_b and R_b. Eq. 2.5.4 reduces to:

$$G_{5a} a_{0a} (X_1 - X_{2d}) = G_{5b} a_{0b} (X_{1d} - X_2) \tag{2.5.5}$$

As the gas composition is common (and although the G_5 terms are actually equal they are retained for completeness), this reduces to:

$$\left(\frac{a_{0a} G_{5a}}{a_{0b} G_{5b}} \right) (X_1 - X_{2d}) = (X_{1d} - X_2) \tag{2.5.6}$$

Eq. 2.5.3 reduces to:

$$(X_1 + X_{2d} - 1)^{G_{7a}} = (X_{1d} + X_2 - 1)^{G_{7b}} \tag{2.5.7}$$

In this simple case, the values of G_{7a} and G_{7b} are identical and are simply G_7:

$$X_1 + X_{2d} - 1 = X_{1d} + X_2 - 1 \tag{2.5.8}$$

The solution becomes straightforward:

$$X_{2d} = \frac{2X_2 - X_1 \left(1 - \frac{a_{0a} G_{5a}}{a_{0b} G_{5b}} \right)}{1 + \frac{a_{0a} G_{5a}}{a_{0b} G_{5b}}} \quad \text{hence} \quad p_{2d} = p_0 X_{2d}^{G_7} \tag{2.5.9}$$

$$X_{1d} = X_1 + X_{2d} - X_2 \quad \text{hence} \quad p_{1d} = p_0 X_{1d}^{G_7} \tag{2.5.10}$$

(ii) The more complex case of variable gas composition

The simplicity of reduction of Eqs. 2.5.6 to 2.5.7, and also Eqs. 2.5.7 to 2.5.8, is no longer possible as Eq. 2.5.7 remains as a polynomial function. The method of solution is to eliminate one of the unknowns from Eqs. 2.5.6 and 2.5.7, either X_{1d} or X_{2d}, and solve for the remaining unknown by the Newton-Raphson method. The remaining step is as in Eqs. 2.5.9 and 2.5.10, but with the gas composition inserted appropriate to the side of the discontinuity:

$$p_{2d} = p_0 X_{2d}^{G_{7a}} \tag{2.5.11}$$

$$p_{1d} = p_0 X_{1d}^{G_{7b}} \tag{2.5.12}$$

A numerical example will assist understanding of the above theory. Consider, as in Fig. 2.5, a pressure wave p_1 arriving at a boundary where the reference temperature T_{01} on side "a" is 200°C and T_{02} is 100°C on side "b." The gas is colder on side "b" and is more dense. An echo should ensue. The gas is air on both sides, i.e., γ is 1.4 and R is 287 J/kgK. The situation is undisturbed on side "b," i.e., the pressure p_2 is the same as the reference pressure p_0. The pressure wave on side "a" has a pressure ratio P_1 of 1.3. Both the simple solution and the more complex solution will give precisely the same answer. The transmitted pressure wave into side "b" is stronger with a pressure ratio, P_{1d}, of 1.32 while the reflected pressure wave, the echo, has a pressure ratio, P_{2d}, of 1.016.

If the above calculation is repeated with just one exception, the gas on side "b" is exhaust gas with the appropriate properties, i.e., γ is 1.36 and R is 300 J/kgK, then the simple solution can be shown to be inaccurate. In this new case, the simple solution will predict that the transmitted pressure ratio, P_{1d}, is 1.328 while the reflected pressure wave, the echo, has a pressure ratio, P_{2d}, of 1.00077. The accurate solution, solving the equations and taking into account the variable gas properties, will predict that the transmitted pressure ratio, P_{1d}, is 1.314 while the reflected pressure wave, the echo, has a pressure ratio, P_{2d}, of 1.011. The difference in answer is considerable, making the unilateral employment of a simple solution, based on the assumption of a common gas composition throughout the ducting, inappropriate for use in engine calculations. In real engines, and therefore also in calculations with pretensions of accuracy, the variation of gas properties and composition throughout the ducting is the normality rather than the exception, and must be simulated as such.

2.6 Reflection of pressure waves

Oppositely moving pressure waves arise from many sources, but principally from reflections of pressure waves at the boundaries of the inlet or the exhaust duct. Such boundaries include junctions at the open or closed end of a pipe and also from any change in area, gradual or sudden, within a pipe. You are already aware of sound reflections, as an echo is a classic example of a closed-ended reflection. All reflections are, by definition, a superposition process, as the reflection proceeds to move oppositely to the incident pressure wave causing that reflection. Fig. 2.6 shows some of the possibilities for reflections in a three-cylinder engine. Whether the engine is a two-stroke or a four-stroke engine is of no consequence, but the air inflow and the exhaust outflow at the extremities are clearly marked. Expansion pressure waves are produced into the intake system by the induction process and exhaust pressure waves are sent into the exhaust system in the sequence of the engine firing order.

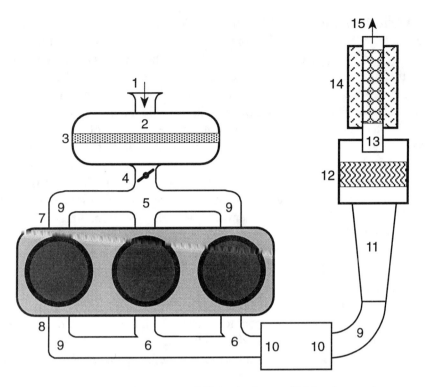

Fig. 2.6 Wave reflection possibilities in the manifolds of an engine.

The potential for wave reflection in the ducting, and its location, is marked by a number. It is obvious that the location accompanies a change of section of the ducting or a change in its direction.

At point 1 is the intake bellmouth where the induction pressure wave is reflected at the atmosphere. At 2 is a plenum chamber where pressure wave oscillations are damped out in amplitude. At 3 is an air filter which provides a restriction to the flow and the possibility of "echoes" taking place. At 4 is a throttle in the duct and, depending on the throttle opening value, can pose either a major restriction or even none at all. At 5 in the intake manifold is a four-way branch where the pressure wave must divide and send reflections back from the change of section.

At 6 in the exhaust manifold are three-way branches where the pressure wave must also divide and send reflections back from the change of section. When a pressure wave arrives at a three-way junction, the values transmitted into each of the other branches will be a function of the areas of the pipes and all of the reflections present. When the induction pressure wave arrives at the open pipe end at 1, a reflection will take place which will be immediately superposed upon the incident pressure pulse. In short, all reflection processes are superposition processes, which is the fundamental reason why the theoretical text up to this point has set down the wave motion within constant area ducting and the basics of the superposition process in some detail.

At 7 and 8 are the valves or ports into the cylinder where, depending on the valve or port opening schedule, acts as everything from a partially closed or open end to a cylinder at varying pressures, to a complete "echo" situation when these valves or ports are closed. At 9 are bends in the ducting where the pressure wave is reflected from the deflection process in major or minor part depending on the severity of the radius of the bend. At 10 in the exhaust ducting are sudden expansion and contractions in the pipe. At 11 is a tapered exhaust pipe which will act as a diffuser or a nozzle depending on the direction of the particle flow; wave reflections ensue in both cases. At 12 is a restriction in the form of a catalyst, little different from an air filter other than that chemical reactions are taking place at the same time; this sentence is easily written but the theoretical calculations to predict the wave motion and the thermodynamics of the reaction are somewhat more complex. At 13 is a re-entrant pipe to a chamber which is a very common element in any silencer design. At 14 is an absorption silencer element, a length of perforated pipe surrounded by packing, which acts as both diffuser and the trimmer of sharp peaks on exhaust pulses; by definition it provides wave reflections. At 15 is a plain-ended exhaust pipe entering the atmosphere and here too pressure wave reflections take place.

The above tour of the pressure wave routes in and out of an engine is far from a complete description of the processes but it is meant to illustrate both the complexity of the events and to postulate that, without a complete understanding of every and all possibilities for wave reflection in and through an internal-combustion engine, none can seriously declaim to be a designer of engines.

The resulting pressure-time history is complicated and beyond the memory-tracking capability of the human mind. Computers are the type of methodical calculation tool ideal for this pedantic exercise, and you will be introduced to their use for this purpose. Before that juncture, it is essential to comprehend the basic effect of each of these reflection mechanisms, as the mathematics of their behavior must be programmed in order to track the progress of all incident and reflected waves.

The next sections of this text analyze virtually all of the above possibilities for wave reflection due to changes in pipe or duct geometry and analyze the reflection and transmission process which takes place at each juncture.

2.6.1 Notation for reflection and transmission of pressure waves in pipes

A wave arriving at a position where it can be reflected is called the *incident wave*. In the paragraphs that follow, all incident pressure waves, whether they be compression or expansion waves, will be designated by the subscript "i," i.e., pressure p_i, pressure ratio P_i, pressure amplitude ratio X_i, particle velocity c_i, density ρ_i, acoustic velocity a_i, and propagation velocity α_i. All reflections will be designated by the subscript "r," i.e., pressure p_r, pressure ratio P_r, pressure amplitude ratio X_r, particle velocity c_r, density ρ_r, acoustic velocity a_r, and propagation velocity α_r. All superposition characteristics will be designated by the subscript "s," i.e., pressure p_s, pressure ratio P_s, pressure amplitude ratio X_s, particle velocity c_s, density ρ_s, acoustic velocity a_s, and propagation velocity α_s.

Where a gas particle flow regime is taking place, it is always considered to flow from gas in regime subscripted with a "1" and flowing to gas in a regime subscripted by a "2." Thus the gas properties of specific heats ratio and gas constant in the upstream regime are γ_1 and R_1 while the downstream regime contains gas at γ_2 and R_2.

The reference conditions are pressure p_0, temperature T_0, acoustic velocity a_0 and density ρ_0.

2.7 Reflection of a pressure wave at a closed end in a pipe

When a pressure wave arrives at the plane of a closed end in a pipe, a reflection takes place. This is the classic echo situation, so it is no surprise to discover that the mathematics dictate that the reflected pressure wave is an exact image of the incident wave, but traveling in the opposite direction. The one certain fact available, physically speaking, is that the superposition particle velocity is zero in the plane of the closed end, as shown in Fig. 2.7(a).

From Eq. 2.2.3:

$$c_s = c_i + c_r = 0 \qquad (2.7.1)$$

or

$$c_r = -c_i \qquad (2.7.2)$$

From Eq. 2.2.2, as c_s is zero:

$$X_r = X_i \qquad (2.7.3)$$

Hence

$$P_r = P_i \qquad (2.7.4)$$

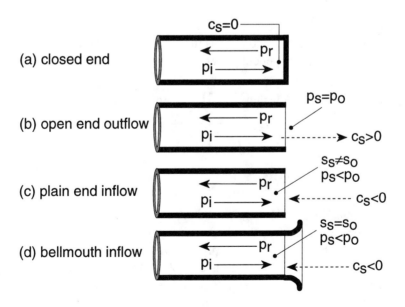

Fig. 2.7 Wave reflection criteria at some typical pipe.

From the combination of Eqs. 2.7.1 and 2.7.3:

$$X_s = 2X_i - 1 \tag{2.7.5}$$

Therefore
$$p_s = p_0(2X_i - 1)^{G7} \tag{2.7.6}$$

2.8 Reflection of a pressure wave at an open end in a pipe

Here the situation is slightly more complex in that the fluid flow behavior is different for compression and expansion waves. The first one to be dealt with is that for compression waves, which, by definition, must produce gas particle outflow from the pipe into the atmosphere.

2.8.1 Reflection of a compression wave at an open end in a pipe

In this case, illustrated in Fig. 2.7(b), the first logical assumption which can be made is that the superposition pressure in the plane of the open end is the atmospheric pressure. The reference pressure, p_0, in such a case is the atmospheric pressure. From Eq. 2.2.1 this gives:

$$X_s = X_i + X_r - 1 = 1$$

Hence
$$X_r = 2 - X_i \tag{2.8.1}$$

Therefore, the amplitude of reflected pressure wave p_r is:

$$p_r = p_0(2 - X_i)^{G7} \tag{2.8.2}$$

For compression waves, as $X_i > 1$ and $X_r < 1$, the reflection is an expansion wave.

From Eq. 2.2.6 for the more general case:

$$\begin{aligned} c_s &= G_5 a_0 (X_i - 1) - G_5 a_0 (X_r - 1) \\ &= G_5 a_0 (X_i - 1 - 2 + X_i + 1) \\ &= 2 G_5 a_0 (X_i - 1) \\ &= 2 c_i \end{aligned} \tag{2.8.3}$$

This equation is applicable as long as the superposition particle velocity, c_s, does not reach the local sonic velocity, when the flow regime must be analyzed by further equations. The sonic regime at an open pipe end is an unlikely event in most engines, but the student will find the analysis described clearly by Bannister [2.2] and in Sec. 2.16 of this text.

Equally it can be shown that, as c_r equals c_i, and as that equation is sign dependent, so the particle flow is in the same direction. This conclusion can also be reached from the earlier conclusion that the reflection is an expansion wave which will impel particles opposite to its

Chapter 2 - Gas Flow through Two-Stroke Engines

direction of propagation. This means that an exhaust pulse arriving at an open end is sending suction reflections back toward the engine which will help to extract exhaust gas particles further down the pipe and away from the engine cylinder. Clearly this is a reflection to be used by the designer so as to aid the scavenging, i.e., emptying, of the cylinder of its exhaust gas. The numerical data below emphasize this point.

To get a basic understanding of the results of employing this theory for the calculation of reflection of compression waves at the atmospheric end of a pipe, consider an example using the compression pressure wave, p_e, previously used in Sec. 2.1.4.

Recall that the wave p_e is a compression wave of pressure ratio 1.2. In the nomenclature of this section it becomes the incident pressure wave p_i at the open end. This pressure ratio is shown to give a pressure amplitude ratio, X_i, of 1.02639. Using Eq. 2.8.1, the reflected pressure amplitude ratio, X_r, is given by:

$$X_r = 2 - X_i = 2 - 1.02639 = 0.9736$$

or
$$P_r = X_r^{G7} = 0.9736^7 = 0.8293$$

That the reflection of a compression wave at the open end is a rarefaction wave is now evident numerically.

2.8.2 Reflection of an expansion wave at a bellmouth open end in a pipe

This reflection process is connected with inflow and therefore it is necessary to consider the fluid mechanics of the flow into a pipe. Inflow of air in an intake system, which is the normal place to find expansion waves, is usually conducted through a bellmouth-ended pipe of the type illustrated in Fig. 2.7. This form of pipe end will be discussed in the first instance.

The analysis of gas flow to and from a thermodynamic system, which may also be experiencing heat transfer and work transfer processes, is analyzed by the First Law of Thermodynamics. The theoretical approach is to be found in any standard textbook on thermodynamics [5.11]. In general this is expressed as:

$$\Delta(\text{heat transfer}) + \Delta(\text{energy entering}) = \Delta(\text{system energy}) + \Delta(\text{energy leaving}) + \Delta(\text{work transfer})$$

The First Law of Thermodynamics for an open system flow from the atmosphere to the superposition station at the full pipe area in Fig. 2.7(d) is as follows:

$$\delta Q_{system} + \Delta m_0 \left(h_0 + \frac{c_0^2}{2} \right) = \delta E_{system} + \Delta m_s \left(h_s + \frac{c_s^2}{2} \right) + \delta W_{system} \qquad (2.8.4)$$

If the flow at the instant in question can be presumed to be quasi-steady and steady-state flow without heat transfer, and also to be isentropic, then ΔQ, ΔW, and ΔE are all zero. The

Design and Simulation of Two-Stroke Engines

mass flow increments must be equal to satisfy the continuity equation. The difference in the enthalpy terms can be evaluated as:

$$h_s - h_0 = C_p(T_s - T_0) = \frac{\gamma R}{\gamma - 1}(T_s - T_0) = \frac{a_s^2 - a_0^2}{\gamma - 1}$$

Although the particle velocity in the atmosphere, c_0, is virtually zero, Eq. 2.8.4 reduces to:

$$c_0^2 + G_5 a_0^2 = c_s^2 + G_5 a_s^2 \tag{2.8.5}$$

As the flow is isentropic, from Eq. 2.1.12:

$$a_s = a_0 X_s \tag{2.8.6}$$

Substituting Eq. 2.8.6 into Eq. 2.8.5 and now regarding c_0 as zero:

$$c_s^2 = G_5 a_0^2 \left(1 - X_s^2\right) \tag{2.8.7}$$

From Eq. 2.2.6:

$$c_s = G_5 a_0 (X_i - X_r) \tag{2.8.8}$$

From Eq. 2.2.1:

$$X_s = X_i + X_r - 1 \tag{2.8.9}$$

Bringing these latter three equations together:

$$G_5 (X_i - X_r)^2 = 1 - (X_i + X_r - 1)^2$$

This becomes:

$$G_6 X_r^2 - (2G_4 X_i + 2) X_r + \left(G_6 X_i^2 - 2X_i\right) = 0 \tag{2.8.10}$$

This is a quadratic equation solution for X_r. Ultimately neglecting the negative sign in the general solution to a quadratic equation, this yields:

Chapter 2 - Gas Flow through Two-Stroke Engines

$$X_r = \frac{(2G_4X_i + 2) \pm \sqrt{(2G_4X_i + 2)^2 - 4G_6(G_6X_i^2 - 2X_i)}}{2G_6}$$

$$= \frac{(G_4X_i + 1) \pm \sqrt{1 + X_i(2G_4 + 2G_6) + X_i^2(G_4^2 - G_6^2)}}{G_6} \quad (2.8.11)$$

For inflow at a bellmouth end when the incoming gas is air where γ equals 1.4, this becomes:

$$X_r = \frac{1 + 4X_i + \sqrt{1 + 20X_i - 20X_i^2}}{6} \quad (2.8.12)$$

This equation shows that the reflection of an expansion wave at a bellmouth end of an intake pipe will be a compression wave. Take as an example the induction pressure wave p_i of Sec. 2.1.4(b), where the expansion wave has a pressure ratio, P_i, of 0.8 and a pressure amplitude ratio, X_i, of 0.9686. Substituting these numbers into Eq. 2.3.17 to determine the magnitude of its reflection at a bellmouth open end gives:

$$X_r = \frac{4 \times 0.9686 + 1 + \sqrt{1 + 20 \times 0.9686 - 20 \times 0.9686^2}}{6} = 1.0238$$

Thus, $$P_r = X_r^7 = 1.0238^7 = 1.178$$

As predicted, the reflection is a compression wave and, if allowed by the designer to arrive at the intake port while the intake process is still in progress, will push further air charge into the cylinder or crankcase. In the jargon of engine design this effect is called "ramming" and will be discussed thoroughly later in this chapter.

2.8.3 Reflection of an expansion wave at a plain open end in a pipe

The reflection of expansion waves at the end of plain pipes can be dealt with in a similar theoretical manner. Bannister [2.2] describes the theory in some detail. It is presented here for completeness.

The First Law of Thermodynamics still applies, as seen in Eq. 2.8.5, and c_0 is regarded as effectively zero.

$$G_5a_0^2 = c_s^2 + G_5a_s^2 \quad (2.8.5)$$

However, Eq. 2.8.6 cannot apply as the particle flow has a distinct vena contracta within the pipe end with an associated turbulent vortex ring, and cannot be regarded as an isentropic process. Thus:

$$a_s \neq a_0 X_s$$

but an entropy gain leaves the value of a_0 as a_0', thus:

$$a_s = a_0' X_s \qquad (2.8.13)$$

The most accurate statement is that the superposition acoustic velocity is given by:

$$a_s = \sqrt{\frac{\gamma p_s}{\rho_s}} \qquad (2.8.14)$$

Another gas-dynamic relationship is required and as is normal in a thermodynamic analysis for non-isentropic flow, the momentum equation is employed. Consider the flow from the atmosphere to the superposition plane of fully developed flow downstream of the vena contracta. Newton's Second Law of Motion can be expressed as:

$$\text{Force} = \text{rate of change of momentum}$$

whence
$$p_0 A - p_s A = \dot{m}_s c_s - \dot{m}_0 c_0$$

or
$$p_0 - p_s = \rho_s c_s^2 \qquad (2.8.15)$$

From the wave summation equations, and ignoring the entropy gain encapsulated in the a_0' term so as to effect an approximate solution, the superposition pressure and particle velocity are related by:

$$X_s = X_i + X_r - 1$$

$$c_s = G_5 a_0 (X_i - X_r)$$

$$P_s = X_s^{G7}$$

Elimination of unwanted terms does not provide a simple solution, but one which requires an iterative approach by a Newton Raphson method to arrive at the unknown value of reflected pressure, X_r, from a known value of the incident pressure, X_i, at the plain open end. The solution is expressed as a function of the unknown quantity, X_r, as follows:

$$f(X_r) = \frac{1}{2}\left[X_i + X_r - \sqrt{\frac{1}{G_5} \times \frac{1-(X_i + X_r - 1)^{G7}}{1+G_6(X_i + X_r - 1)^{G7}}}\right] - X_i = 0 \qquad (2.8.16)$$

This is solved by the Newton Raphson method, i.e., by differentiation of the above equation with respect to the unknown quantity, X_r, and iterating for the answer in the classic mathematical manner. Thence the value of particle velocity, c_r, can be derived "approximately" from substitution into the equation below once the unknown quantity, X_r, is determined.

$$c_s = c_i + c_r - G_5 a_0(X_1 - X_r) \qquad (2.8.17)$$

This is a far from satisfactory solution to an apparently simple problem. In actual fact, the inflow process at a plain open-ended pipe is a singular manifestation of what is generally called "cylinder to pipe outflow from an engine." In this case the atmosphere is the very large "cylinder" flowing gas into a pipe! Thus, as this complex problem is treated in great detail later in Sec. 2.17, and this present boundary condition of inflow at a plain-ended pipe can be calculated by the same theoretical approach as for pipe to cylinder flow, further wasteful simplistic explanation is curtailed until that point in the text has been reached.

2.9 An introduction to reflection of pressure waves at a sudden area change

It is quite common to find a sudden area change within a pipe or duct attached to an engine. In Fig. 2.6 at position 10 there are sudden enlargements and contractions in pipe area. The basic difference, gas dynamically speaking, between a sudden enlargement and contraction in pipe area, such as at position 10, and a plenum or volume, such as at position 2, is that the flow in the duct is considered to be one-dimensional whereas in the plenum or volume it is considered to be three-dimensional. A subsidiary definition is one where the particle velocity in a plenum or volume is so low as to be always considered as zero in any thermodynamic analysis. This will produce a change in amplitude of the transmitted pulse beyond the area change and also cause a wave reflection from it. Such sudden area changes are sketched in Fig. 2.8, and it can be seen that the pipe area can contract or expand at the junction. In each case, the incident wave at the sudden area change is depicted as propagating rightward, with the pipe nomenclature as 1 for the wave arrival pipe, with "i" signifying the incident pulse, "r" the reflected pulse and "s" the superposition condition. The X value shown is the conventional symbolism for pressure amplitude ratio and p is that for absolute pressure. For example, at any instant, the incident pressure pulses at the junction are p_{i1} and p_{i2} which, depending on the areas A_1 and A_2, will give rise to reflected pulses, p_{r1} and p_{r2}.

In either expansion or contraction of the pipe area the particle flow is considered to be proceeding from the upstream superposition station 1 to the downstream superposition station 2. Therefore the properties and composition of the gas particles which are considered to

Design and Simulation of Two-Stroke Engines

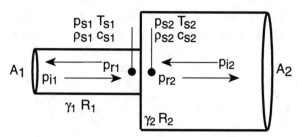

(a) sudden expansion in area in a pipe where $c_s > 0$

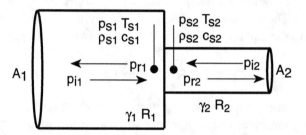

(b) sudden contraction in area in a pipe where $c_s > 0$

Fig. 2.8 Sudden contractions and expansions in area in a pipe.

be flowing in any analysis based on quasi-steady flow are those of the gas at the upstream point. In all of the analyses presented here that nomenclature is maintained. Therefore the various functions of the gas properties are:

$$\gamma = \gamma_1 \quad R = R_1 \quad G_5 = G_{5_1} \quad G_7 = G_{7_1}, \text{ etc.}$$

It was Benson [2.4] who suggested a simple theoretical solution for such junctions. He assumed that the superposition pressure at the plane of the junction was the same in both pipes at the instant of superposition. The assumption is inherently one of an isentropic process. Such a simple junction model will clearly have its limitations, but it is my experience that it is remarkably effective in practice, particularly if the area ratio changes, A_r, are in the band,

$$\frac{1}{6} < A_r < 6$$

The area ratio is defined as:

$$A_r = \frac{A_2}{A_1} \tag{2.9.1}$$

From Benson,

$$p_{s1} = p_{s2} \tag{2.9.2}$$

Consequently, from Eq. 2.2.1:

$$X_{i1} + X_{r1} - 1 = X_{i2} + X_{r2} - 1 \tag{2.9.3}$$

From the continuity equation, equating the mass flow rate in an isentropic process on either side of the junction where,

mass flow rate = (density) × (area) × (particle velocity)

$$\rho_{s1}A_1c_{s1} = \rho_{s2}A_2c_{s2} \tag{2.9.4}$$

Using the theory of Eqs. 2.1.17 and 2.2.2, where the reference conditions are p_0, T_0 and ρ_0, Eq. 2.9.4 becomes, where rightward is decreed as positive particle flow:

$$\rho_0 X_{s1}^{G5} A_1 G_5 a_0 (X_{i1} - X_{r1}) = -\rho_0 X_{s2}^{G5} A_2 G_5 a_0 (X_{i2} - X_{r2}) \tag{2.9.5}$$

As X_{s1} equals X_{s2}, this reduces to:

$$A_1(X_{i1} - X_{r1}) = -A_2(X_{i2} - X_{r2}) \tag{2.9.6}$$

Joining Eqs. 2.9.1, 2.9.3 and 2.9.6, and eliminating each of the unknowns in turn, i.e., X_{r1} or X_{r2}:

$$X_{r1} = \frac{(1 - A_r)X_{i1} + 2X_{i2}A_r}{1 + A_r} \tag{2.9.7}$$

$$X_{r2} = \frac{2X_{i1} - X_{i2}(1 - A_r)}{(1 + A_r)} \tag{2.9.8}$$

To get a basic understanding of the results of employing Benson's simple "constant pressure" criterion for the calculation of reflections of compression and expansion waves at sudden enlargements and contractions in pipe area, consider an example using the two pressure waves, p_e and p_i, previously used in Sec. 2.1.4.

The wave, p_e, is a compression wave of pressure ratio 1.2 and p_i is an expansion wave of pressure ratio 0.8. Such pressure ratios are shown to give pressure amplitude ratios X of 1.02639 and 0.9686, respectively. Each of these waves in turn will be used as data for X_{i1} arriving in pipe 1 at a junction with pipe 2, where the area ratio will be either halved for a contraction or doubled for an enlargement to the pipe area. In each case the incident pressure amplitude ratio in pipe 2, X_{i2}, will be taken as unity, which means that the incident pressure wave in pipe 1 is facing undisturbed conditions in pipe 2.

(a) An enlargement, $A_r = 2$, for an incident compression wave where $P_{i1} = 1.2$ and $X_{i1} = 1.02639$

From Eqs. 2.9.7 and 2.9.8, $X_{r1} = 0.9912$ and $X_{r2} = 1.01759$. Hence, the pressure ratios, P_{r1} and P_{r2}, of the reflected waves are:

$$P_{r1} = 0.940 \text{ and } P_{r2} = 1.130$$

The sudden enlargement behaves like a slightly less-effective "open end," as a completely open-ended pipe from Sec. 2.8.1 would have given a reflected pressure ratio of 0.8293 instead of 0.940. The onward transmitted pressure wave into pipe 2 is also one of compression, but with a reduced pressure ratio of 1.13.

(b) An enlargement, $A_r = 2$, for an incident expansion wave where $P_{i1} = 0.8$ and $X_{i1} = 0.9686$

From Eqs. 2.9.7 and 2.9.8, $X_{r1} = 1.0105$ and $X_{r2} = 0.97908$. Hence, the pressure ratios, P_{r1} and P_{r2}, of the reflected waves are:

$$P_{r1} = 1.076 \text{ and } P_{r2} = 0.862$$

As above, the sudden enlargement behaves as a slightly less-effective "open end" because a "perfect" bellmouth open end to a pipe in Sec. 2.8.2 was shown to produce a stronger reflected pressure ratio of 1.178, instead of the weaker value of 1.076 determined here. The onward transmitted pressure wave in pipe 2 is one of expansion, but with a diminished pressure ratio of 0.862.

(c) A contraction, $A_r = 0.5$, for an incident compression wave where $P_{i1} = 1.2$ and $X_{i1} = 1.02639$

From Eqs. 2.9.7 and 2.9.8, $X_{r1} = 1.0088$ and $X_{r2} = 1.0352$. Hence, the pressure ratios, P_{r1} and P_{r2}, of the reflected waves are:

$$P_{r1} = 1.063 \text{ and } P_{r2} = 1.274$$

The sudden contraction behaves like a partially closed end, sending back a partial "echo" of the incident pulse. The onward transmitted pressure wave is also one of compression, but of increased pressure ratio 1.274.

(d) A contraction, $A_r = 0.5$, for an incident expansion wave where $P_{i1} = 0.8$ and $X_{i1} = 0.9686$

From Eqs. 2.9.7 and 2.9.8, $X_{r1} = 0.9895$ and $X_{r2} = 0.9582$. Hence, the pressure ratios, P_{r1} and P_{r2}, of the reflected waves are:

$$P_{r1} = 0.929 \text{ and } P_{r2} = 0.741$$

Chapter 2 - Gas Flow through Two-Stroke Engines

As in (c), the sudden contraction behaves like a partially closed end, sending back a partial "echo" of the incident pulse. The onward transmitted pressure wave is also one of expansion, but it should be noted that it has an increased expansion pressure ratio of 0.741.

The theoretical presentation here, due to Benson [2.4], is clearly too simple to be completely accurate in all circumstances. It is, however, a very good guide as to the magnitude of pressure wave reflection and transmission. The major objections to its use where accuracy is required are that the assumption of "constant pressure" at the discontinuity in pipe area cannot possibly be tenable over all flow situations and that the thermodynamic assumption is of isentropic flow in all circumstances. A more complete theoretical approach is examined in more detail in the following sections. A full discussion of the accuracy of such a simple assumption is illustrated by numeric examples in Sec. 2.12.2.

2.10 Reflection of pressure waves at an expansion in pipe area

This section contains the non-isentropic analysis of unsteady gas flow at an expansion in pipe area. The sketch in Fig. 2.8(a) details the nomenclature for the flow regime, in precisely the same manner as in Sec. 2.9. However, to analyze the flow completely, the further information contained in sketch format in Figs. 2.9(a) and 2.10(a) must also be considered.

In Fig. 2.10(a) the expanding flow is seen to leave turbulent vortices in the corners of the larger section. That the streamlines of the flow give rise to particle flow separation implies a gain of entropy from area section 1 to area section 2. This is summarized on the temperature-entropy diagram in Fig. 2.9(a) where the gain of entropy for the flow falling from pressure p_{s1} to pressure p_{s2} is clearly visible.

As usual, the analysis of flow in this quasi-steady and non-isentropic context uses, where appropriate, the equations of continuity, the First Law of Thermodynamics and the momen-

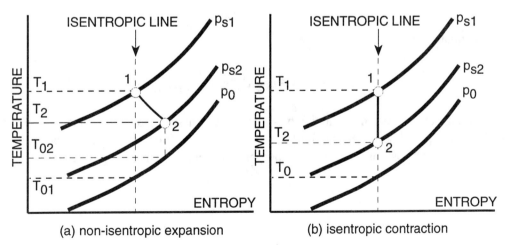

Fig. 2.9 Temperature entropy characteristics for simple expansions and contractions.

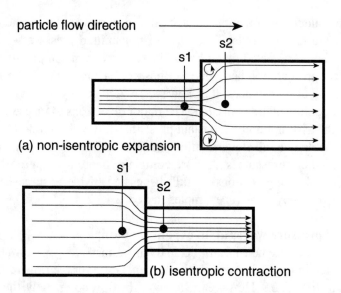

Fig. 2.10 *Particle flow in simple expansions and contractions.*

tum equation. The properties and composition of the gas particles are those of the gas at the upstream point. Therefore, the various functions of the gas properties are:

$$\gamma = \gamma_1 \quad R = R_1 \quad G_5 = G_{5_1} \quad G_7 = G_{7_1}, \text{ etc.}$$

The continuity equation for mass flow in Eq. 2.9.5 is still generally applicable and repeated here, although the entropy gain is reflected in the reference acoustic velocity and density at position 2:

$$\dot{m}_1 - \dot{m}_2 = 0 \qquad (2.10.1)$$

This equation becomes:

$$\rho_{01} X_{s1}^{G5} A_1 G_5 a_{01}(X_{i1} - X_{r1}) + \rho_{02} X_{s2}^{G5} A_2 G_5 a_{02}(X_{i2} - X_{r2}) = 0 \qquad (2.10.2)$$

The First Law of Thermodynamics was introduced for such flow situations in Sec. 2.8. The analysis required here follows similar logical lines. The First Law of Thermodynamics for flow from superposition station 1 to superposition station 2 can be expressed as:

$$h_{s1} + \frac{c_{s1}^2}{2} = h_{s2} + \frac{c_{s2}^2}{2}$$

or,

$$\left(c_{s1}^2 + G_5 a_{s1}^2\right) - \left(c_{s2}^2 + G_5 a_{s2}^2\right) = 0 \qquad (2.10.3)$$

The momentum equation for flow from superposition station 1 to superposition station 2 is expressed as:

$$A_1 p_{s1} + (A_2 - A_1) p_{s1} - A_2 p_{s2} + (\dot{m}_{s1} c_{s1} - \dot{m}_{s2} c_{s2}) = 0$$

The logic for the middle term in the above equation is that the pressure, p_{s1}, is conventionally presumed to act over the annulus area between the two ducts. The momentum equation, also taking into account the information regarding mass flow equality from the continuity equation, reduces to:

$$A_2(p_{s1} - p_{s2}) + \dot{m}_{s1}(c_{s1} - c_{s2}) = 0 \qquad (2.10.4)$$

As with the simplified "constant pressure" solution according to Benson presented in Sec. 2.9, the unknown values will be the reflected pressure waves at the boundary, p_{r1} and p_{r2}, and also the reference temperature at position 2, namely T_{02}. There are three unknowns, necessitating three equations, namely Eqs. 2.10.2, 2.10.3 and 2.10.4. All other "unknown" quantities can be computed from these values and from the "known" values. The known values are the upstream and downstream pipe areas, A_1 and A_2, the reference state conditions at the upstream point, the gas properties at superposition stations 1 and 2, and the incident pressure waves, p_{i1} and p_{i2}.

Recalling that,

$$X_{i1} = \left(\frac{p_{i1}}{p_0}\right)^{G17} \quad \text{and} \quad X_{i2} = \left(\frac{p_{i2}}{p_0}\right)^{G17}$$

The reference state conditions are:

density $\qquad \rho_{01} = \dfrac{p_0}{RT_{01}} \qquad \rho_{02} = \dfrac{p_0}{RT_{02}} \qquad (2.10.5)$

acoustic velocity $\qquad a_{01} = \sqrt{\gamma RT_{01}} \qquad a_{02} = \sqrt{\gamma RT_{02}} \qquad (2.10.6)$

The continuity equation, Eq. 2.10.2, reduces to:

$$\rho_{01}(X_{i1} + X_{r1} - 1)^{G5} A_1 G_5 a_{01}(X_{i1} - X_{r1}) \\ + \rho_{02}(X_{i2} + X_{r2} - 1)^{G5} A_2 G_5 a_{02}(X_{i2} - X_{r2}) = 0 \qquad (2.10.7)$$

The First Law of Thermodynamics, Eq. 2.10.3, reduces to:

$$\left[(G_5 a_{01}(X_{i1} - X_{r1}))^2 + G_5 a_{01}^2(X_{i1} + X_{r1} - 1)^2\right] -$$
$$\left[(G_5 a_{02}(X_{i2} - X_{r2}))^2 + G_5 a_{02}^2(X_{i2} + X_{r2} - 1)^2\right] = 0 \qquad (2.10.8)$$

The momentum equation, Eq. 2.10.4, reduces to:

$$p_0 A_2\left[(X_{i1} + X_{r1} - 1)^{G7} - (X_{i2} + X_{r2} - 1)^{G7}\right]$$
$$+ \left[\rho_{01}(X_{i1} + X_{r1} - 1)^{G5} A_1 G_5 a_{01}(X_{i1} - X_{r1})\right]$$
$$\times \left[G_5 a_{01}(X_{i1} - X_{r1}) + G_5 a_{02}(X_{i2} - X_{r2})\right] = 0 \qquad (2.10.9)$$

The three equations cannot be reduced any further as they are polynomial functions of all three variables. These functions can be solved by a standard iterative method for such problems. I have determined that the Newton-Raphson method for the solution of multiple polynomial equations is stable, accurate and rapid in execution. The arithmetic solution on a computer is conducted by a Gaussian Elimination method.

As with all numerical methods, the computer time required is heavily dependent on the number of iterations needed to acquire a solution of the requisite accuracy, in this case for an error no greater than 0.01% for the solution of any of the variables. The use of the Benson "constant pressure" criterion, presented in Sec. 2.9, is invaluable in this regard by considerably reducing the number of iterations required. Numerical methods of this type are also arithmetically "frail," if the user makes ill-advised initial guesses at the value of any of the unknowns. It is in this context that the use of the Benson "constant pressure" criterion is indispensable. Numeric examples are given in Sec. 2.12.2.

2.10.1 Flow at pipe expansions where sonic particle velocity is encountered

In the above analysis of unsteady gas flow at expansions in pipe area the particle velocity at section 1 will occasionally be found to reach, or even attempt to exceed, the local acoustic velocity. This is not possible in thermodynamic or gas-dynamic terms as the particles in unsteady gas flow cannot move faster than the pressure wave signal that is impelling them. The highest particle velocity permissible is the local acoustic velocity at station 1, i.e., the flow is permitted to become choked. Therefore, during the mathematical solution of Eqs. 2.10.7, 2.10.8 and 2.10.9, the local Mach number at station 1 is monitored and retained at unity if it is found to exceed it.

$$M_{s1} = \frac{c_{s1}}{a_{s1}} = \frac{G_5 a_{01}(X_{i1} - X_{r1})}{a_{01} X_{s1}} = \frac{G_5(X_{i1} - X_{r1})}{X_{i1} + X_{r1} - 1} \qquad (2.10.10)$$

This immediately simplifies the entire procedure as it gives a direct solution for one of the unknowns:

if, $\quad\quad\quad\quad\quad\quad\quad\quad M_{s1} = 1$

then, $\quad\quad\quad X_{r1} = \dfrac{M_{s1} + X_{i1}(G_5 - M_{s1})}{M_{s1} + G_5} = \dfrac{1 + G_4 X_{i1}}{G_6}$ $\quad\quad\quad$ (2.10.11)

The acquisition of all related data for pressure, density, particle velocity and mass flow rate at both superposition stations follows directly from the solution of the three polynomials for X_{r1}, X_{r2} and a_{02}, in the manner indicated in Sec. 2.9.

In many classic analyses of choked flow a "critical pressure ratio" is determined for flow from the upstream point to the throat where sonic flow is occurring. That method assumes zero particle velocity at the upstream point; such is clearly not the case here. Therefore, that concept cannot be employed in this geometry for unsteady gas flow.

2.11 Reflection of pressure waves at a contraction in pipe area

This section contains the isentropic analysis of unsteady gas flow at a contraction in pipe area. The sketch in Fig. 2.8(b) details the nomenclature for the flow regime, in precisely the same manner as in Sec. 2.9. However, to analyze the flow completely, the further information contained in sketch format in Figs. 2.9(b) and 2.10(b) must also be considered.

In Fig. 2.10(b) the contracting flow is seen to flow smoothly from the larger section to the smaller area section. The streamlines of the flow do not give rise to particle flow separation and so it is considered to be isentropic flow. This is in line with conventional nozzle theory as observed in many standard texts in thermodynamics. It is summarized on the temperature-entropy diagram in Fig. 2.9(b) where there is no entropy gain for the flow falling from pressure p_{s1} to pressure p_{s2}.

As usual, the analysis of quasi-steady flow in this context uses, where appropriate, the equations of continuity, the First Law of Thermodynamics and the momentum equation. However, one less equation is required by comparison with the analysis for expanding or diffusing flow in Sec. 2.10. This is because the value of the reference state is known at superposition station 2, for the flow is isentropic:

$$T_{01} = T_{02} \quad \text{or} \quad a_{01} = a_{02} \quad\quad\quad (2.11.1)$$

As there is no entropy gain, that equation normally reserved for the analysis of non-isentropic flow, the momentum equation, can be neglected in the ensuing analytic method.

The properties and composition of the gas particles are those of the gas at the upstream point. Therefore the various functions of the gas properties are:

$$\gamma = \gamma_1 \quad R = R_1 \quad G_5 = G_{5_1} \quad G_7 = G_{7_1}, \text{ etc.}$$

The continuity equation for mass flow in Eq. 2.9.5 is still generally applicable and repeated here:

$$\dot{m}_1 - \dot{m}_2 = 0 \qquad (2.11.2)$$

This equation becomes:

$$\rho_{01} X_{s1}^{G5} A_1 G_5 a_{01}(X_{i1} - X_{r1}) + \rho_{02} X_{s2}^{G5} A_2 G_5 a_{02}(X_{i2} - X_{r2}) = 0 \qquad (2.11.3)$$

or,

$$X_{s1}^{G5} A_1 (X_{i1} - X_{r1}) + X_{s2}^{G5} A_2 (X_{i2} - X_{r2}) = 0$$

The First Law of Thermodynamics was introduced for such flow situations in Sec. 2.8. The analysis required here follows similar logical lines. The First Law of Thermodynamics for flow from superposition station 1 to superposition station 2 can be expressed as:

$$h_{s1} + \frac{c_{s1}^2}{2} = h_{s2} + \frac{c_{s2}^2}{2}$$

or,

$$(c_{s1}^2 + G_5 a_{s1}^2) - (c_{s2}^2 + G_5 a_{s2}^2) = 0 \qquad (2.11.4)$$

As with the simplified "constant pressure" solution according to Benson, presented in Sec. 2.9, the unknown values will be the reflected pressure waves at the boundary, p_{r1} and p_{r2}. There are two unknowns, necessitating two equations, namely Eqs. 2.11.3 and 2.11.4. All other "unknown" quantities can be computed from these values and from the "known" values. The known values are the upstream and downstream pipe areas, A_1 and A_2, the reference state conditions at the upstream and downstream points, the gas properties at superposition stations 1 and 2, and the incident pressure waves, p_{i1} and p_{i2}.

Recalling that,

$$X_{i1} = \left(\frac{p_{i1}}{p_0}\right)^{G17} \quad \text{and} \quad X_{i2} = \left(\frac{p_{i2}}{p_0}\right)^{G17}$$

The reference state conditions are:

density $\qquad \rho_{01} = \rho_{02} = \dfrac{p_0}{RT_{01}} \qquad (2.11.5)$

acoustic velocity $\qquad a_{01} = a_{02} = \sqrt{\gamma R T_{01}} \qquad (2.11.6)$

Chapter 2 - Gas Flow through Two-Stroke Engines

The continuity equation, Eq. 2.11.3, reduces to:

$$\rho_{01}(X_{i1} + X_{r1} - 1)^{G5} A_1 G_5 a_{01}(X_{i1} - X_{r1})$$
$$+\rho_{02}(X_{i2} + X_{r2} - 1)^{G5} A_2 G_5 a_{02}(X_{i2} - X_{r2}) = 0 \quad (2.11.7)$$

or $\quad (X_{i1} + X_{r1} - 1)^{G5} A_1(X_{i1} - X_{r1}) + (X_{i2} + X_{r2} - 1)^{G5} A_2(X_{i2} - X_{r2}) = 0$

The First Law of Thermodynamics, Eq. 2.11.4, reduces to:

$$\left[(G_5 a_{01}(X_{i1} - X_{r1}))^2 + G_5 a_{01}^2(X_{i1} + X_{r1} - 1)^2\right] -$$
$$\left[(G_5 a_{02}(X_{i2} - X_{r2}))^2 + G_5 a_{02}^2(X_{i2} + X_{r2} - 1)^2\right] = 0 \quad (2.11.8)$$

or
$$\left[G_5(X_{i1} - X_{r1})^2 + (X_{i1} + X_{r1} - 1)^2\right] -$$
$$\left[G_5(X_{i2} - X_{r2})^2 + G_5(X_{i2} + X_{r2} - 1)^2\right] = 0$$

The two equations cannot be reduced any further as they are polynomial functions of the two variables. These functions can be solved by a standard iterative method for such problems. I have determined that the Newton-Raphson method for the solution of multiple polynomial equations is stable, accurate and rapid in execution. The arithmetic solution on a computer is conducted by a Gaussian Elimination method. Actually, this is not strictly necessary as a simpler solution can be effected as it devolves to two simultaneous equations for the two unknowns and the corrector values for each of the unknowns.

In Sec. 2.9 there are comments regarding the use of the Benson "constant pressure" criterion, for the initial guesses for the unknowns to the solution, as being indispensable; they are still appropriate. Numerical examples are given in Sec. 2.12.2.

The acquisition of all related data for pressure, density, particle velocity and mass flow rate at both superposition stations follows directly from the solution of the two polynomials for X_{r1} and X_{r2}.

2.11.1 Flow at pipe contractions where sonic particle velocity is encountered

In the above analysis of unsteady gas flow at contractions in pipe area the particle velocity at section 2 will occasionally be found to reach, or even attempt to exceed, the local acoustic velocity. This is not possible in thermodynamic or gas-dynamic terms. The highest particle velocity permissible is the local acoustic velocity at station 2, i.e., the flow is permitted to become choked. Therefore, during the mathematical solution of Eqs. 2.11.7 and 2.11.8, the local Mach number at station 2 is monitored and retained at unity if it is found to exceed it.

As,
$$M_{s2} = \frac{c_{s2}}{a_{s2}} = \frac{G_5 a_{02}|X_{i2} - X_{r2}|}{a_{02} X_{s2}} = \frac{G_5 |X_{i2} - X_{r2}|}{X_{i2} + X_{r2} - 1} \quad (2.11.9)$$

This immediately simplifies the entire procedure for this gives a direct solution for one of the unknowns:

Then if, $\qquad M_{s2} = 1$

$$X_{r2} = \frac{M_{s2} + X_{i2}(G_5 - M_{s2})}{M_{s2} + G_5} = \frac{1 + G_4 X_{i2}}{G_6} \quad (2.11.10)$$

In this instance of sonic particle flow at station 2, the entire solution can now be obtained directly by substituting the value of X_{r2} determined above into either Eq. 2.11.7 or 2.11.8 and solving it by the standard Newton-Raphson method for the one remaining unknown, X_{r1}.

2.12 Reflection of waves at a restriction between differing pipe areas

This section contains the non-isentropic analysis of unsteady gas flow at restrictions between differing pipe areas. The sketch in Fig. 2.8 details much of the nomenclature for the flow regime, but essential subsidiary information is contained in a more detailed sketch of the geometry in Fig. 2.12. However, to analyze the flow completely, the further information contained in sketch format in Figs. 2.11 and 2.12 must be considered completely. The geometry is of two pipes of differing area, A_1 and A_2, which are butted together with an orifice of area, A_t, sandwiched between them. This geometry is very common in engine ducting. For example, it could be the throttle body of a carburetor with a venturi and a throttle plate. It could also be simply a sharp-edged, sudden contraction in pipe diameter where A_2 is less than A_1 and there is no actual orifice of area A_t at all. In the latter case the flow naturally forms a vena contracta with an effective area of value A_t which is less than A_2. In short, the theoretical analysis to be presented here is a more accurate and extended, and inherently more complex, version of that already presented for sudden expansions and contractions in pipe area in Sec. 2.11.

In Fig. 2.12 the expanding flow from the throat to the downstream superposition point 2 is seen to leave turbulent vortices in the corners of that section. That the streamlines of the flow give rise to particle flow separation implies a gain of entropy from the throat to area section 2. On the other hand, the flow from the superposition point 1 to the throat is contracting and can be considered to be isentropic in the same fashion as the contractions debated in Sec. 2.11. This is summarized on the temperature-entropy diagram in Fig. 2.11 where the gain of entropy for the flow rising from pressure p_t to pressure p_{s2} is clearly visible. The isentropic nature of the flow from p_{s1} to p_t can also be observed as a vertical line on Fig. 2.11.

Chapter 2 - Gas Flow through Two-Stroke Engines

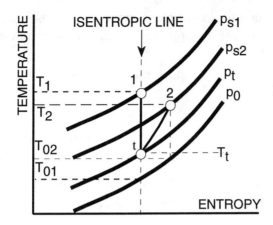

Fig. 2.11 Temperature entropy characteristics for a restricted area change.

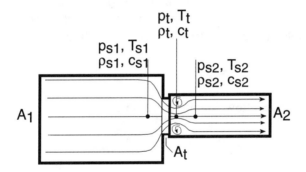

Fig. 2.12 Particle flow regimes at a restricted area change.

The properties and composition of the gas particles are those of the gas at the upstream point. Therefore the various functions of the gas properties are:

$$\gamma = \gamma_1 \quad R = R_1 \quad G_5 = G_{5_1} \quad G_7 = G_{7_1}, \text{ etc.}$$

As usual, the analysis of flow in this context uses, where appropriate, the equations of continuity, the First Law of Thermodynamics and the momentum equation.

The reference state conditions are:

density $\quad\quad\quad \rho_{01} = \rho_{0t} = \dfrac{p_0}{RT_{01}} \quad \rho_{02} = \dfrac{p_0}{RT_{02}}$ $\quad\quad\quad$ (2.12.1)

acoustic velocity $\quad a_{01} = a_{0t} = \sqrt{\gamma RT_{01}} \quad a_{02} = \sqrt{\gamma RT_{02}}$ $\quad\quad\quad$ (2.12.2)

109

The continuity equation for mass flow in previous sections is still generally applicable and repeated here, although the entropy gain is reflected in the reference acoustic velocity and density at position 2:

$$m_1 - m_2 = 0$$
$$m_1 - m_t = 0 \qquad (2.12.3)$$

These equations become, with rightward retained as the positive direction:

$$\rho_{01}X_{s1}^{G5}A_1G_5a_{01}(X_{i1} - X_{r1}) + \rho_{02}X_{s2}^{G5}A_2G_5a_{02}(X_{i2} - X_{r2}) = 0$$
$$\rho_{01}X_{s1}^{G5}A_1G_5a_{01}(X_{i1} - X_{r1}) - \rho_t[C_cA_t][C_sc_t] = 0 \qquad (2.12.4)$$

The above equation for the mass flow continuity for flow from the upstream station 1 to the throat, contains the coefficient of contraction on the flow area, C_c, and the coefficient of velocity, C_s. These are conventionally connected in fluid mechanics theory to a coefficient of discharge, C_d, to give an effective throat area, A_{teff}, as follows:

$$C_d = C_cC_s \quad \text{and} \quad A_{teff} = C_dA_t$$

This latter equation of mass flow continuity becomes:

$$\rho_{01}X_{s1}^{G5}A_1G_5a_{01}(X_{i1} - X_{r1}) - C_d\rho_tA_tc_t = 0$$

The First Law of Thermodynamics was introduced for such flow situations in Sec. 2.8. The analysis required here follows similar logical lines. The First Law of Thermodynamics for flow from superposition station 1 to superposition station 2 can be expressed as:

$$h_{s1} + \frac{c_{s1}^2}{2} = h_{s2} + \frac{c_{s2}^2}{2}$$

or

$$\left(c_{s1}^2 + G_5a_{s1}^2\right) - \left(c_{s2}^2 + G_5a_{s2}^2\right) = 0 \qquad (2.12.5)$$

The First Law of Thermodynamics for flow from superposition station 1 to the throat can be expressed as:

$$h_{s1} + \frac{c_{s1}^2}{2} = h_t + \frac{c_t^2}{2}$$

or

$$C_P(T_{s1} - T_t) + \frac{c_{s1}^2 - c_t^2}{2} = 0 \qquad (2.12.6)$$

Chapter 2 - Gas Flow through Two-Stroke Engines

The momentum equation for flow from the throat to superposition station 2 is expressed as:

$$A_2(p_t - p_{s2}) + m_{sl}(c_t - c_{s2}) = 0 \qquad (2.12.7)$$

The unknown values will be the reflected pressure waves at the boundaries, p_{r1} and p_{r2}, the reference temperature at position 2, namely T_{02}, and the pressure p_t and the velocity c_t at the throat. There are five unknowns, necessitating five equations, namely the two mass flow equations in Eq. 2.12.4, the two First Law equations, Eqs. 2.12.5 and 2.12.6, and the momentum equation, Eq. 2.12.7. All other "unknown" quantities can be computed from these values and from the "known" values. The known values are the upstream and downstream pipe areas, A_1 and A_2, the throat area, A_t, the reference state conditions at the upstream point, the gas properties at superposition stations 1 and 2, and the incident pressure waves, p_{i1} and p_{i2}. A numerical value for the coefficient of discharge, C_d, is also required, but further information on this difficult and often controversial subject is supplied in Sec. 2.19.

Recalling that,

$$X_{i1} = \left(\frac{p_{i1}}{p_0}\right)^{G17} \text{ and } X_{i2} = \left(\frac{p_{i2}}{p_0}\right)^{G17} \text{ and setting } X_t = \left(\frac{p_t}{p_0}\right)^{G17}$$

then due to isentropic flow from station 1 to the throat, the throat density and temperature are given by:

$$\rho_t = \rho_{01} X_t^{G5} \text{ and } T_t = \frac{(a_{01}X_t)^2}{\gamma R}$$

The continuity equations set in Eq. 2.12.4 reduce to:

$$\rho_{01}(X_{i1} + X_{r1} - 1)^{G5} A_1 a_{01}(X_{i1} - X_{r1})$$
$$+ \rho_{02}(X_{i2} + X_{r2} - 1)^{G5} A_2 a_{02}(X_{i2} - X_{r2}) = 0 \qquad (2.12.8)$$

$$(X_{i1} + X_{r1} - 1)^{G5} A_1 G_5 a_{01}(X_{i1} - X_{r1}) - X_t^{G5} C_d A_t c_t = 0 \qquad (2.12.9)$$

The First Law of Thermodynamics in Eq. 2.12.5 reduces to:

$$\left[(G_5 a_{01}(X_{i1} - X_{r1}))^2 + G_5 a_{01}^2 (X_{i1} + X_{r1} - 1)^2\right] -$$
$$\left[(G_5 a_{02}(X_{i2} - X_{r2}))^2 + G_5 a_{02}^2 (X_{i2} + X_{r2} - 1)^2\right] = 0 \qquad (2.12.10)$$

The First Law of Thermodynamics in Eq. 2.12.6 reduces to:

$$G_5\left\{(a_{01}(X_{i1}+X_{r1}-1))^2 - (a_{01}X_t)^2\right\} + (G_5 a_{01}(X_{i1}-X_{r1}))^2 - c_t^2 = 0 \qquad (2.12.11)$$

The momentum equation, Eq. 2.12.7, reduces to:

$$p_0 A_2\left[X_t^{G7} - (X_{i2}+X_{r2}-1)^{G7}\right]$$
$$+ \left[\rho_{01}(X_{i1}+X_{r1}-1)^{G5} A_1 G_5 a_{01}(X_{i1}-X_{r1})\right]\left[c_t + G_5 a_{02}(X_{i2}-X_{r2})\right] = 0 \qquad (2.12.12)$$

The five equations, Eqs. 2.12.8 to 2.12.12, cannot be reduced any further as they are polynomial functions of all five variables, X_{r1}, X_{r2}, X_t, a_{02}, and c_t. These functions can be solved by a standard iterative method for such problems. I have determined that the Newton-Raphson method for the solution of multiple polynomial equations is stable, accurate and rapid in execution. The arithmetic solution on a computer is conducted by a Gaussian Elimination method. Numerical examples are given in Sec. 2.12.2.

It is not easy to supply initial guesses which are close to the final answers for the iteration method on a computer. The Benson "constant pressure" assumption is of great assistance in this matter. Even with this assistance to the numerical solution, this theory must be programmed with great care to avoid arithmetic instability during its execution.

2.12.1 Flow at pipe restrictions where sonic particle velocity is encountered

In the above analysis of unsteady gas flow at restrictions in pipe area the particle velocity at the throat will quite commonly be found to reach, or even attempt to exceed, the local acoustic velocity. This is not possible in thermodynamic or gas-dynamic terms. The highest particle velocity permissible is the local acoustic velocity at the throat, i.e., the flow is permitted to become choked. Therefore, during the mathematical solution of Eqs. 2.12.8 to 2.12.12, the local Mach number at the throat is monitored and retained at unity if it is found to exceed it.

$$\text{As,} \qquad M_t = \frac{c_t}{a_{01}X_t} = 1 \quad \text{then} \quad c_t = a_{01}X_t \qquad (2.12.13)$$

This simplifies the entire procedure for this gives a direct relationship between two of the unknowns and replaces one of the equations employed above. It is probably easier from an arithmetic standpoint to eliminate the momentum equation, but probably more accurate thermodynamically to retain it!

The acquisition of all related data for pressure, density, particle velocity and mass flow rate at both superposition stations and at the throat follows directly from the solution of the four polynomials for X_{r1}, X_{r2}, X_t, a_{02}, and c_t.

2.12.2 Examples of flow at pipe expansions, contractions and restrictions

In Secs. 2.9-2.12 the theory of unsteady flow at these discontinuities in area have been presented. In Sec. 2.9 the simple theory of a constant pressure assumption by Benson was given and the more complete theory in subsequent sections. In various sections the point was made that the simple theory is reasonably accurate. Some numerical examples are given here which will illustrate that point and the extent of that inaccuracy. The input and output data all use the nomenclature of the theory sections and Figs. 2.8, 2.10 and 2.12. The input data for the diameters, d, are in mm units.

Table 2.12.1 Input data to the calculations

No.	d_1	d_2	d_t	C_d	A_r	P_{i1}	P_{i2}
1	25	60	25	1.000	4.0	1.2	1.0
2	25	50	25	0.85	4.0	1.2	1.0
3	50	25	25	1.000	0.25	1.2	1.0
4	50	25	25	0.7	0.25	1.2	1.0
5	50	25	15	0.85	0.25	1.2	1.0
6	25	50	15	0.85	4.0	1.2	1.0

Table 2.12.2 Output data using the constant pressure theory of Sec. 2.9

No.	P_{r1}	P_{r2}	$\dot{m}$ g/s	$\dot{m}$ error %	E or C
1	0.8943	1.0763	45.15	2.65	E
2	0.8943	1.0763	45.15	2.82	E
3	1.1162	1.3357	52.68	−8.18	C
4	1.1162	1.3357	52.68	−17.89	C
5	1.1162	1.3357	52.68	−32.48	C
6	0.8943	1.0763	45.15	−61.37	E

The input data show six different sets of numerical data. In Table 2.12.1 the test data sets 1 and 2 illustrate an expansion of the flow, and test data sets 3 and 4 are for contractions. Test data set 5 is for a restricted pipe at a contraction whereas test data set 6 is for a restricted pipe at an expansion in pipe area. The incident pulse for all tests has a pressure ratio of 1.2, the gas is air and is undisturbed on side 2, and the reference temperature at side 1 is always 20°C or 293 K. There are two output data sets: in Table 2.12.2 where the theory employed is the Benson constant pressure theory, and the second set in Table 2.12.3 using the more complex theory of Secs. 2.10, 2.11 and 2.12 as appropriate. The more complex theory gives information on the entropy gain exhibited by the reference temperature, T_{02}. In the output is shown the mass flow rate and the "error" when comparing that mass flow rate computed by the constant pressure theory with respect to that calculated by the more complex theory. The column labeled "E or C" describes whether the flow encountered was at an expansion (E) or a contraction (C) in pipe area.

Table 2.12.3 Output data from the calculations using the more complex theory

No.	P_{r1}	P_{r2}	$\dot{m}$ g/s	T_{02} K	Theory
1	0.8850	1.0785	46.38	294.5	Sec. 2.10
2	0.9036	1.0746	43.91	296.2	Sec. 2.12
3	1.1227	1.3118	48.70	293.0	Sec. 2.11
4	1.1292	1.2886	44.69	295.2	Sec. 2.12
5	1.1371	1.2595	39.77	298.1	Sec. 2.12
6	1.0174	1.0487	27.98	306.7	Sec. 2.12

For simple expansions, in test data sets 1 and 2, the constant pressure theory works remarkably well with an almost negligible error, i.e., less than 3% in mass flow rate terms. If the expansion has a realistic coefficient of discharge of 0.85 applied to it then that error on mass flow rate rises slightly from 2.65 to 2.8%. The magnitudes of the reflected pressure waves compare quite favorably in most circumstances.

At a simple sudden contraction in test data sets 3 and 4, even with a coefficient discharge of 1.0 the mass flow rate error is significant at 8.2%, rising to 17.9% error when the quite logical C_d value of 0.7 is applied to the sudden contraction. The magnitudes of the reflected pressure waves are significantly different when emanating from the simple and the complex theories.

When any form of restriction is placed in the pipe, i.e., in test data sets 5 and 6 for an expansion and a contraction, respectively, the constant pressure theory is simply not capable of providing any relevant information either for the magnitude of the reflected wave or for the mass flow rates. The error is greater for expansions than contractions, which is the opposite of the situation when restrictions in the pipe are not present. Note the significant entropy gains in tests 4-6.

The constant pressure theory is seen to be reasonably accurate only for flow which encounters sudden expansions in the ducting.

2.13 An introduction to reflections of pressure waves at branches in a pipe

The simple theoretical treatment for this situation was also suggested by Benson [2.4] and in precisely the same form as for the sudden area changes found in the previous section. A sketch of a typical branch is shown in Fig. 2.13. The sign convention for all of the branch theory presented here is that inward propagation of a pressure wave toward the branch will be regarded as "positive." Benson [2.4] postulates that the superposition pressure at the junction, at any instant of wave incidence and reflection, can be regarded as a constant. This is a straightforward extension of his thinking for the expansion and contractions given in Sec. 2.9.

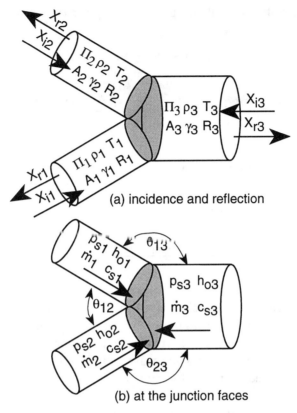

Fig. 2.13 Unsteady flow at a three-way branch.

The incident pressure waves are p_{i1}, p_{i2}, and p_{i3} and the ensuing reflections are of pressures p_{r1}, p_{r2}, and p_{r3}. The superposition states are p_{s1}, p_{s2}, and p_{s3}.

Therefore, the theoretical solution involves expansion of Eqs. 2.9.3 and 2.9.4 to deal with the superposition state and mass flow rate of the extra pipe 3 at the junction. Benson's criterion inherently assumes isentropic flow.

$$p_{s1} = p_{s2} = p_{s3}$$

or
$$X_{i1} + X_{r1} - 1 = X_{i2} + X_{r2} - 1 = X_{i3} + X_{r3} - 1 \tag{2.13.1}$$

The net mass flow rate at the junction is zero:

$$A_1(X_{i1} - X_{r1}) + A_2(X_{i2} - X_{r2}) + A_3(X_{i3} - X_{r3}) = 0 \tag{2.13.2}$$

There are three equations to solve for the three unknowns, X_{r1}, X_{r2} and X_{r3}. It is presumed that in the course of any computation we know the values of all incident pressure waves from one calculation time step to another.

The solution of the above simultaneous equations gives (where the total area, A_t, is defined below):

$$A_t = A_1 + A_2 + A_3$$

$$X_{r1} = \frac{2A_2 X_{i2} + 2A_3 X_{i3} + X_{i1}(A_1 - A_2 - A_3)}{A_t}$$

$$X_{r2} = \frac{2A_1 X_{i1} + 2A_3 X_{i3} + X_{i2}(A_2 - A_3 - A_1)}{A_t}$$

$$X_{r3} = \frac{2A_1 X_{i1} + 2A_2 X_{i2} + X_{i3}(A_3 - A_2 - A_1)}{A_t} \tag{2.13.3}$$

Perhaps not surprisingly, the branched pipe can act as either a contraction of area to the flow or an enlargement of area to the gas flow. In short, two pipes may be supplying one pipe, or one pipe supplying the other two, respectively. Consider these two cases where all of the pipes are of equal area, where the pressure waves employed as examples are the familiar pulses which have been used so frequently throughout this text.

(a) A compression wave is coming down to the branch in pipe 1 through air and all other conditions in the other branches are "undisturbed"

The compression wave has a pressure ratio of $P_{i1} = 1.2$, or $X_{i1} = 1.02639$. The results of the calculation using Eqs. 2.13.3 are:

$$X_{r1} = 0.9911 \quad X_{r2} = X_{r3} = 1.01759 \quad P_{r1} = 0.940 \quad P_{r2} = P_{r3} = 1.13$$

As far as pipe 1 is concerned the result is exactly the same as that for the 2:1 enlargement in area in the previous section. In the branch, the incident wave divides evenly between the other two pipes, transmitting a compression wave onward and reflecting a rarefaction pulse. Pipe 1 is supplying the other two pipes, hence the effect is an expansion.

(b) Compression waves of pressure ratio 1.2 are arriving as incident pulses in pipes 1 and 2 leading up to the branch with pipe 3

Pipe 3 has undisturbed conditions as $P_{i3} = 1.0$. Now the branch behaves as a 2:1 contraction to this general flow, for the solutions of Eqs. 2.13.3 show:

$$X_{r1} = X_{r2} = 1.0088 \quad X_{r3} = 1.0352 \quad P_{r1} = P_{r2} = 1.0632 \quad P_{r3} = 1.274$$

The contracting effect is evidenced by the reflection and transmission of compression waves. These numbers are already familiar as computed data for pressure waves of identical amplitude at the 2:1 contraction discussed in the previous section. Pipes 1 and 2 are supplying the third pipe, hence the effect is a contraction.

When we have dissimilar areas of pipes and a mixture of compression and expansion waves incident upon the branch, the situation becomes much more difficult to comprehend by the human mind. At that point the programming of the mathematics into a computer will leave the designer's mind free to concentrate more upon the relevance of the information calculated and less on the arithmetic tedium of acquiring that data.

It is also obvious that the angle between the several branches must play some role in determining the transmitted and reflected wave amplitudes. This subject was studied most recently by Bingham [2.19] and Blair [2.20] at QUB. While the branch angles do have an influence on wave amplitudes, it is not as great in some circumstances as might be imagined. For those who wish to achieve greater accuracy for all such calculations, the following section is presented.

2.14 The complete solution of reflections of pressure waves at pipe branches

The next step forward historically and theoretically was to attempt to solve the momentum equation to cope with the non-isentropic realism that there are pressure losses for real flows changing direction by moving around the sharp corners of the branches. Much has been written on this subject, and many of the references provide a sustained commentary on the subject over many years. Suffice it to say that the practical approach adopted by Bingham [2.19], incorporating the use of a modified form of the momentum equation to account for the pressure losses around the branch, is the basis of the method used here. Bingham's solution was isentropic.

This same approach was also employed by McGinnity [2.39] in a non-isentropic analysis, but for a single composition fluid only; his solution was further complicated by using a non-homentropic Riemann variable method and it meant that, as gas properties were tied to path lines, they were not as clearly defined as in the method used here. It would appear from the literature [2.39] that the solution was reduced to the search for a single unknown quantity whereas I deduce below that there are actually five such for a complete non-isentropic analysis of a three-way branch.

The merit of the Bingham [2.19] method is that it uses experimentally determined pressure loss coefficients at the branches, an approach capable of being enhanced further by information emanating from data banks of which that published by Miller [2.38] is typical.

An alternative method, perhaps one which is more complete theoretically, is to resolve the momentum of the flow at any branch into its horizontal and vertical components and equate them both to zero. The demerit of that approach is that it does not include the fluid mechanic loss component which the Bingham method incorporates so pragmatically and realistically.

While the discussion here is devoted exclusively to three-way branches, the theoretical process for four-way branches, or n-pipe collectors, is almost identical to that reported below. It will be seen that the basic approach is to identify those pipes that are the suppliers, and

Design and Simulation of Two-Stroke Engines

those that are the supplied pipes, at any junction. For an n-pipe junction, that is basically the only addition to the theory below other than that the number of equations increases by the number of extra junctions. With the mathematical technique of solution by the Newton-Raphson method for multiple polynomial equations, and the matrix arithmetic handled by the Gaussian Elimination method, the additional computational complexity is negligible.

The sign convention for particle flow is declared as "positive" toward the branch and Fig. 2.13 inherently stipulates that convention. The pressure loss criterion for flow from one branch to another is set out by Bingham as,

$$\Delta p = C_L \rho_s c_s^2 \quad (2.14.1)$$

where the loss coefficient C_L is given by the inter-branch angle θ,

$$C_L = 1.6 - \frac{1.6\,\theta}{167} \text{ if } \theta > 167 \text{ then } C_L = 0 \quad (2.14.2)$$

In any branch there are supplier pipes and supplied pipes. There are two possibilities in this regard and these lead to two assumptions for their solution. The more fundamental assumption in much of the theory is that the gas within the pipes is a mixture of two gases and in an engine context this is logically exhaust gas and air. Obviously, the theory is capable of being extended to a mixture of many gases, as indeed air and exhaust gas actually are. Equally, the theory is capable of being extended relatively easily to branches with any number of pipes at the junction. The theory set out below details a three-way branch for greater ease of understanding.

(a) One supplier pipe

Here there is one supplier pipe and two are being supplied, in which case the solution required is for the reflected wave amplitudes, X_r, in all three pipes and for the reference acoustic velocities, a_0, for the gas going toward the two supplied pipes. The word "toward" is used here precisely. This means there are five unknown values needing five equations. It is possible to reduce this number of unknowns by one, if we assume that the reference acoustic state toward the two supplied pipes is common. A negligible loss of accuracy accompanies this assumption. Using the notation of Fig. 2.13, it is implied that pipe 1 is supplying pipes 2 and 3; that notation will be used here only to "particularize" the solution so as to aid understanding of the analysis. As the pressure in the face of pipes 2 and 3 will be normally very close, the difference between T_{02} and T_{03} should be small and, as the reference acoustic velocity is related to the square root of these numbers, the error is potentially even smaller. Irrespective of that assumption, it follows absolutely that the basic properties of the gas entering the supplied pipes is that of the supplier pipe.

(b) Two supplier pipes

Here there is one supplied pipe and two are suppliers, in which case the solution required is for the reflected wave amplitudes, X_r, in all three pipes and for the reference acoustic velocities, a_0, in the supplied pipe. This means that there are four unknown values needing four equations. This is possible only by making the assumption that there is equality of superposition pressure at the faces of the two supplier pipes; this is the same assumption used by Bingham [2.19] and McGinnity [2.39]. It then follows that the properties of the gas entering the supplied pipe are a mass-flow-related mixture of those in the supplier pipes. Using the notation of Fig. 2.13, it is implied that pipes 1 and 2 are supplying pipe 3.

continuity $\quad\quad\quad\quad\quad\quad\quad \dot{m}_{e3} = \dot{m}_1 + \dot{m}_2 \quad\quad\quad\quad\quad\quad\quad$ (2.14.3)

purity $\quad\quad\quad\quad\quad\quad\quad \Pi_{e3} = \dfrac{\dot{m}_1 \Pi_1 + \dot{m}_2 \Pi_2}{\dot{m}_{e3}} \quad\quad\quad\quad\quad\quad\quad$ (2.14.4)

gas constant $\quad\quad\quad\quad\quad R_{e3} = \dfrac{\dot{m}_1 R_1 + \dot{m}_2 R_2}{\dot{m}_{e3}} \quad\quad\quad\quad\quad\quad\quad$ (2.14.5)

specific heats ratio $\quad\quad \gamma_{e3} = \dfrac{C_{pe3}}{C_{ve3}} = \dfrac{\dot{m}_1 C_{p1} + \dot{m}_2 C_{p2}}{\dot{m}_1 C_{v1} + \dot{m}_2 C_{v2}} \quad\quad\quad$ (2.14.6)

The subscript notation of "e3" should be noted carefully, for this details the quantity and quality of the gas entering, i.e., going "toward," the mesh space beyond the pipe 3 entrance, whereas the resulting change of all of the gas properties within that mesh space is handled by the unsteady gas-dynamic method which has already been presented [2.31].

The final analysis then relies on incorporating the equations emanating from all of these previous considerations regarding pressure losses, together with the continuity equation and the First Law of Thermodynamics. The notation of Fig. 2.13 applies together with either Fig. 2.14(a) for two supplier pipes, or Fig. 2.14(b) for one supplier pipe.

(a) for one supplier pipe the following are the relationships for the density, particle velocity, and mass flow rate which apply to the supplier pipe and the two pipes that are being supplied:

$$\rho_{s1} = \rho_{01}(X_{i1} + X_{r1} - 1)^{G51} \quad c_{s1} = G_{51} a_{01}(X_{i1} - X_{r1}) \quad \dot{m}_1 = \rho_{s1} A_1 c_{s1}$$
(2.14.7)

$$\rho_{s2} = \rho_{0e2}(X_{i2} + X_{r2} - 1)^{G51} \quad c_{s2} = G_{51} a_{0e2}(X_{i2} - X_{r2}) \quad \dot{m}_2 = \rho_{s2} A_2 c_{s2}$$
(2.14.8)

$$\rho_{s3} = \rho_{0e3}(X_{i3} + X_{r3} - 1)^{G51} \quad c_{s3} = G_{51} a_{0e3}(X_{i3} - X_{r3}) \quad \dot{m}_3 = \rho_{s3} A_3 c_{s3}$$
(2.14.9)

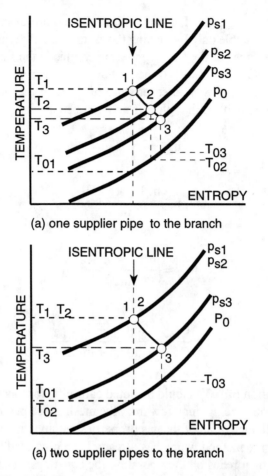

Fig. 2.14 *The temperature-entropy characteristics.*

(b) for two supplier pipes, where the assumption is that the superposition pressure in the faces of pipes 1 and 2 are identical, the equations for pipes 2 and 3 become:

$$\rho_{s2} = \rho_{02}(X_{i2} + X_{r2} - 1)^{G52} \quad c_{s2} = G_{52}a_{02}(X_{i2} - X_{r2}) \quad \dot{m}_2 = \rho_{s2}A_2c_{s2}$$
(2.14.10)

$$\rho_{s3} = \rho_{0e3}(X_{i3} + X_{r3} - 1)^{G51} \quad c_{s3} = G_{51}a_{0e3}(X_{i3} - X_{r3}) \quad \dot{m}_3 = \rho_{s3}A_3c_{s3}$$
(2.14.11)

The First Law of Thermodynamics is, where the local work, heat transfer and system state changes are logically ignored, and the h_0 term is the stagnation specific enthalpy:

$$\dot{m}_1h_{01} + \dot{m}_2h_{02} + \dot{m}_3h_{03} = 0$$
(2.14.12)

This single line expression is the one to be used if the approximation is made that T_{02} equals T_{03} for a single supplier pipe situation; it is also strictly correct for a two supplier pipes model. To solve without that assumption, the equation must be split into two and analyzed specifically for the separate flows from pipe 1 to 2 and from pipe 1 to 3. The reference densities in question emanate from this step, as shown below. The normal isentropic expression for the reference densities is expressed as:

$$\rho_{01} = \frac{p_0}{R_1 T_{01}} \quad \rho_{02} = \frac{p_0}{R_2 T_{02}} \quad \rho_{03} = \frac{p_0}{R_3 T_{03}} \quad (2.14.13)$$

but should gas be entering pipes 1 or 2 as a result of a non-isentropic process then these become for pipes 2 and 3,

$$\rho_{0e2} = \frac{p_0}{R_{e2} T_{02}} \quad \rho_{0e3} = \frac{p_0}{R_{e3} T_{03}} \quad (2.14.14)$$

The stagnation enthalpies appear as:

(a) for one supplier pipe the stagnation enthalpies are,

$$h_{01} = \frac{G_{51} a_{01}^2 X_{s1}^2 + c_{s1}^2}{2} \quad h_{02} = \frac{G_{5e2} a_{e2}^2 X_{s2}^2 + c_{s2}^2}{2} \quad h_{03} = \frac{G_{5e3} a_{e3}^2 X_{s3}^2 + c_{s3}^2}{2} \quad (2.14.15)$$

(b) for two supplier pipes, as with continuity, only the statement for stagnation enthalpy for pipe 2 is altered,

$$h_{02} = \frac{G_{52} a_{02}^2 X_{s2}^2 + c_{s2}^2}{2} \quad (2.14.16)$$

The pressure loss equations first presented in Eq. 2.14.2 devolve to:

(a) for one supplier pipe for flow from pipe 1 to pipes 2 and 3,

$$C_{L12} = 1.6 - \frac{1.6\,\theta_{12}}{167} \quad C_{L13} = 1.6 - \frac{1.6\,\theta_{13}}{167} \quad (2.14.17)$$

$$p_0\left(X_{s1}^{G71} - X_{s2}^{G7e2}\right) = C_{L12} \rho_{s2} c_{s2}^2 \quad p_0\left(X_{s1}^{G71} - X_{s3}^{G7e3}\right) = C_{L13} \rho_{s3} c_{s3}^2 \quad (2.14.18)$$

(b) two supplier pipes, i.e., pipes 1 and 2 supplying pipe 3,

$$C_{L12} = 0 \quad C_{L13} = 1.6 - \frac{1.6\,\theta_{13}}{167} \quad (2.14.19)$$

$$X_{s1}^{G71} - X_{s2}^{G72} = 0 \quad p_0\left(X_{s1}^{G71} - X_{s3}^{G7e3}\right) = C_{L13}\rho_{s3}c_{s3}^2 \quad (2.14.20)$$

In the analysis for the First Law of Thermodynamics and the pressure loss terms it should be noted that the relationships for X_s and c_s for each pipe are written in full in the continuity equation.

These functions can be solved by a standard iterative method for such problems. I have determined that the Newton-Raphson method for the solution of multiple polynomial equations is stable, accurate and rapid in execution. The arithmetic solution on a computer is conducted by a Gaussian Elimination method.

Note that these are the three reflected wave pressures and T_{02} and/or T_{03}, depending on the thermodynamic assumptions debated earlier. In practice it has been found that the "constant pressure" criterion provides excellent initial guesses for the unknown variables. This permits the numerical solution to arrive at the final answers for them in two or three iterations only to a maximum error for the worst case of just 0.05% of its value.

This more sophisticated branched pipe boundary condition can be incorporated into an unsteady gas-dynamic code for implementation on a digital computer, papers on which have already been presented [2.40, 2.59].

2.14.1 The accuracy of simple and more complex branched pipe theories

The assertion is made above that the "constant pressure" theory of Benson is reasonably accurate. The following computations put numbers into that statement. A branch consists of three pipes numbered 1 to 3 where the initial reference temperature of the gas is 20°C and the gas in each pipe is air. The angle θ_{12} between pipe 1 and pipe 2 is 30° and the angle θ_{13} between pipe 1 and pipe 3 is 180°, i.e., it is lying straight through from pipe 1. The input data are shown in Table 2.14.1. The pipes are all of equal diameter, d, in tests numbered 1, 3 and 4 at 25 mm diameter. In test number 2 the pipe 3 diameter, d_3, is 35 mm. In tests numbered 1 and 2 the incident pulse in pipe 1 has a pressure ratio, P_{i1}, of 1.4 and the other pipes have undisturbed wave conditions. In tests numbered 3 and 4 the incident pulse in pipe 1 has a pressure ratio, P_{i1}, of 1.4 and the incident pulse in pipe 3 has a pressure ratio, P_{i3}, of 0.8 and 1.1, respectively.

The output data for the calculations are shown in Table 2.14.2 where the "constant pressure" theory is used. The symbols are P_{r1}, P_{r2}, and P_{r3} for the pressure ratios of the three reflected pressure waves at the branch, for superposition particle velocities c_1, c_2, and c_3. The output data when the more complex theory is employed are shown in Table 2.14.3. The computed mass flow rates (in g/s units) $\dot{m}_1$, $\dot{m}_2$, and $\dot{m}_3$ are shown in Table 2.14.4. In Table 2.14.5 are the "errors" on the computed mass flow between the "constant pressure" theory and the more complex theory. The number of iterations is also shown on the final table; the fact that it requires only two iterations to close a worst-case error of 0.05% on any variable reveals the worth of the Benson "constant pressure" theory for the provision of a very high-quality first guess at the "unknowns."

Table 2.14.1 Input data for the calculations

No.	P_{i1}	P_{i1}	P_{i1}	d_1	d_2	d_3	θ_{12}	θ_{13}	T_{01}
1	1.4	1.0	1.0	25	25	25	30	180	20
2	1.4	1.0	1.0	25	25	35	30	180	20
3	1.4	1.0	0.8	25	25	25	30	180	20
4	1.4	1.0	1.1	25	25	25	30	180	20

Table 2.14.2 Output data from the calculations using "constant pressure" theory

No.	P_{r1}	P_{r1}	P_{r1}	c_1	c_2	c_3
1	0.891	1.254	1.254	112.6	56.3	56.3
2	0.841	1.188	1.188	126.3	42.7	42.7
3	0.767	1.086	1.345	148.5	20.4	128.1
4	0.950	1.333	1.215	97.0	72.0	25.0

Table 2.14.3 Output data from the calculations using complex theory

No.	P_{r1}	P_{r1}	P_{r1}	c_1	c_2	c_3
1	0.905	1.228	1.273	108.7	51.4	60.8
2	0.848	1.171	1.197	124.3	39.5	45.2
3	0.769	1.084	1.350	147.7	19.9	129.7
4	0.974	1.291	1.245	90.9	64.4	31.3

Table 2.14.4 Output on mass flow from constant pressure and complex theory

No.	Constant Pressure Theory			Complex Theory		
	$\dot{m}_1$	$\dot{m}_2$	$\dot{m}_3$	$\dot{m}_1$	$\dot{m}_2$	$\dot{m}_3$
1	78.3	39.1	39.1	76.4	34.6	42.0
2	84.5	28.5	55.9	83.6	25.5	58.1
3	93.2	12.8	80.4	92.9	12.3	80.6
4	70.4	52.3	18.1	67.2	44.8	22.7

Table 2.14.5 Further output data regarding errors on mass flow

Test no.	$\dot{m}_1$ err%	$\dot{m}_2$ err%	$\dot{m}_3$ err%	iterations
1	2.49	13.29	6.73	2
2	1.01	11.70	3.79	2
3	0.32	4.13	0.27	2
4	4.85	16.67	20.0	2

It can be seen in test number 1 that the constant pressure theory takes no account of the branch angle, nor of the non-isentropic nature of the flow, and this induces mass flow differences of up to 13.3% by comparison with the more complex theory. The actual values of the reflected pressure waves are quite close for both theories, but the ensuing mass flow error is significant and is an important argument for the inclusion of the more complex theory in any engine simulation method requiring accuracy. In test number 2 the results are closer, i.e., the mass flow errors are smaller, an effect induced by virtue of the fact that the larger diameter of pipe 3 at 35 mm reduces the particle velocity into pipe 2. As the pressure loss around the intersection into pipe 2, which is angled back from pipe 1 at 30°, is seen from Eq. 2.14.1 to be a function of the square of the superposition velocity c_{s2}^2, then that decreases the pressure loss error within the computation and in reality. This effect is exaggerated in test number 3 where, even though the pipe diameters are equal, the suction wave incident at pipe 3 also reduces the gas particle velocity entering pipe 2; the errors on mass flow are here reduced to a maximum of only 4.1%. The opposite effect is shown in test number 4 where an opposing compression wave incident at the branch in pipe 3 forces more gas into pipe 2; the mass flow errors now rise to a maximum value of 20%.

In all of the tests the amplitudes of the reflected pressure waves are quite close from the application of the two theories but the compounding effect of the pressure error on the density, and the non-isentropic nature of the flow derived by the more complex theory, gives rise to the more serious errors in the computation of the mass flow rate by the "constant pressure" theory.

2.15 Reflection of pressure waves in tapered pipes

The presence of tapered pipes in the ducts of an engine is commonplace. The action of the tapered pipe in providing pressure wave reflections is often used as a tuning element to significantly enhance the performance of engines. The fundamental reason for this effect is that the tapered pipe acts as either a nozzle or as a diffuser, in other words as a more gradual process for the reflection of pressure waves at sudden expansions and contractions previously debated in Secs. 2.10 and 2.11. Almost by definition the process is not only more gradual but more efficient as a reflector of wave energy in that the process is more efficient and spread out in terms of both length and time. As a consequence, any ensuing tuning effect on the engine is not only more pronounced but is effective over a wider speed range.

As a tapered pipe acts to produce a gradual and continual process of reflection, where the pipe area is increasing or decreasing, it must be analyzed in a similar fashion. The ideal would be to conduct the analysis in very small distance steps over the tapered length, but that would be impractical as it would be a very time-consuming process.

A practical method of analyzing the geometry of tapered pipes is shown in Fig. 2.15. The length, L, for the section or sections to be analyzed is usually selected to be compatible with the rest of any computation process for the ducts of the engine [2.31]. The tapered section of the pipe has a taper angle of θ which is the included angle of that taper. Having selected a length, L, over which the unsteady gas-dynamic analysis is to be conducted, it is a matter of simple geometry to determine the diameters at the various locations on the tapered pipe. Consider the sections 1 and 2 in Fig. 2.15. They are of equal length, L. At the commencement of section 1 the diameter is d_a, at its conclusion it is d_b; at the start of section 2 the diameter is d_b and it is d_c at its conclusion.

Any reflection process for sections 1 and 2 will be considered to take place at the interface as a "sudden" expansion or contraction, depending on whether the particle flow is acting in a diffusing manner as in Fig. 2.15(b) or in a nozzle fashion as in Fig. 2.15(c). In short, the flow proceeds in an unsteady gas-dynamic process along section 1 in a parallel pipe of representative diameter d_1 and is then reflected at the interface to section 2 where the representa-

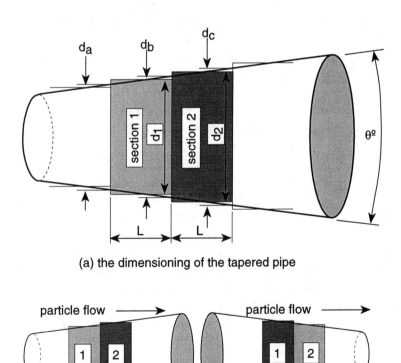

(a) the dimensioning of the tapered pipe

(b) flow in a diffuser   (c) flow in a nozzle

Fig. 2.15 Treatment of tapered pipes for unsteady gas-dynamic analysis.

tive diameter is d_2. This is the analytical case irrespective of whether the flow is acting in a diffusing manner as in Fig. 2.15(b) or in a nozzle fashion as in Fig. 2.15(c). The logical diameter for each of the sections is that area which represents the mean area between the start and the conclusion of each section. This is shown below:

$$A_1 = \frac{A_a + A_b}{2} \quad \text{and} \quad A_2 = \frac{A_b + A_c}{2} \tag{2.15.1}$$

The diameters for each section are related to the above areas by:

$$d_1 = \sqrt{\frac{d_a^2 + d_b^2}{2}} \quad \text{and} \quad d_2 = \sqrt{\frac{d_b^2 + d_c^2}{2}} \tag{2.15.2}$$

The analysis of the flow commences by determining the direction of the particle flow at the interface between section 1 and section 2 and the area change which is occurring at that position. If the flow is behaving as in a diffuser then the ensuing unsteady gas-dynamic analysis is conducted using the theory precisely as presented in Sec. 2.10 for sudden expansions. If the flow is behaving as in a nozzle then the ensuing unsteady gas-dynamic analysis is conducted using the theory precisely as presented in Sec. 2.11 for sudden contractions.

2.15.1 Separation of the flow from the walls of a diffuser

One of the issues always debated in the literature is flow separation from the walls of a diffuser, the physical situation being as in Fig. 2.15(b). In such circumstances the flow detaches from the walls in a central highly turbulent core. As a consequence the entropy gain is much greater in the thermodynamic situation shown in Fig. 2.9(a), for the pressure drop is not as large and the temperature drop is also reduced due to energy dissipation in turbulence. It is postulated in such circumstances of flow separation that the flow process becomes almost isobaric and can be represented as such in the analysis set forth in Sec. 2.10. Therefore, if flow separation in a diffuser is estimated to be possible, the analytical process set forth in Sec. 2.9 should be amended to replace the equation that tracks the non-isentropic flow in the normal attached mode, namely the momentum equation, with another equation that simulates the greater entropy gain of separated flow, namely a constant pressure equation.

Hence, in Sec. 2.9, the set of equations to be analyzed should delete Eq. 2.10.4 (or as Eq. 2.10.9 in its final format) and replace it with Eq. 2.15.3 (or the equivalent Eq. 2.15.4) below.

The assumption is that the particle flow is moving, and diffusing, from section 1 to section 2 as in Fig. 2.15(b) and that separation has been detected. Constant superposition pressure at the interface between sections 1 and 2 produces the following function, using the same variable nomenclature as in Sec. 2.9.

$$p_{s1} - p_{s2} = 0 \tag{2.15.3}$$

This "constant pressure" equation is used to replace the final form of the momentum equation in Eq. 2.10.9. The "constant pressure" equation can be restated in the form below as that most likely to be used in any computational process:

$$X_{s1}^{G7} - X_{s2}^{G7} = 0 \tag{2.15.4}$$

You may well inquire at what point in a computation should this change of tack analytically be conducted? In many texts in gas dynamics, where <u>steady flow</u> is being described, either theoretically or experimentally, the conclusion reached is that flow separation will take place if the particle Mach number is greater than 0.2 or 0.3 and, more significantly, if the included angle of the tapered pipe is greater than a critical value, typically reported widely in the literature as lying between 5 and 7°. The work to date (June 1994) at QUB would indicate that the angle is of very little significance but that gas particle Mach number alone is the important factor to monitor for flow separation. The current conclusion would be, phrased mathematically:

If $M_{s1} \geq 0.65$ employ the constant pressure equation, Eq. 2.15.4
If $M_{s1} < 0.65$ employ the momentum equation, Eq. 2.10.9 (2.15.5)

Future work on correlation of theory with experiment will shed more light on this subject, as can be seen in Sec. 2.19.7. Suffice it to say that there is sufficient evidence already to confirm that any computational method that universally employs the momentum equation for the solution of diffusing flow, in steeply tapered pipes where the Mach number is high, will inevitably produce a very inaccurate assessment of the unsteady gas flow behavior.

2.16 Reflection of pressure waves in pipes for outflow from a cylinder

This situation is fundamental to all unsteady gas flow generated in the intake or exhaust ducts of a reciprocating IC engine. Fig. 2.16 shows an exhaust port (or valve) and pipe, or the throttled end of an exhaust pipe leading into a plenum such as the atmosphere or a silencer box. Anywhere in an unsteady flow regime where a pressure wave in a pipe is incident on a pressure-filled space, box, plenum or cylinder, the following method is applicable to determine the magnitude of the mass outflow, of its thermodynamic state and of the reflected pressure wave. The theory to be generated is generally applicable to an intake port (or valve) and pipe for inflow into a cylinder, plenum, crankcase, or at the throttled end of an intake pipe from the atmosphere or a silencer box, but the subtle differences for this analysis are given in Sec. 2.17.

You may well be tempted to ask what then is the difference between this theoretical treatment and that given for the restricted pipe scenario in Sec. 2.12, for the drawings in Figs. 2.16 and 2.12 look remarkably similar. The answer is direct. In the theory presented here, the space from whence the particles emanate is considered to be sufficiently large and the flow so three-dimensional as to give rise to the fundamental assumption that the particle velocity within the cylinder is considered to be zero, i.e., c_1 is zero.

Design and Simulation of Two-Stroke Engines

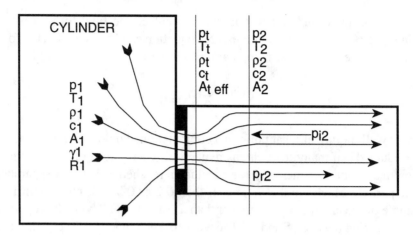

Fig. 2.16 Outflow from a cylinder or plenum to a pipe.

The solution of the gas dynamics of the flow must include separate treatments for subsonic outflow and sonic outflow. The first presentation of the solution for this type of flow was by Wallace and Nassif [2.5] and their basic theory was used in a computer-oriented presentation by Blair and Cahoon [2.6]. Probably the earliest and most detailed exposition of the derivation of the equations involved is that by McConnell [2.7]. However, while all of these presentations declared that the flow was analyzed non-isentropically, a subtle error was introduced within the analysis that negated that assumption. Moreover, all of the earlier solutions, including that by Bingham [2.19], used fixed values of the cylinder properties throughout and solved the equations with either the properties of air ($\gamma = 1.4$ and $R = 287$ J/kgK) or exhaust gas ($\gamma = 1.35$ and $R = 300$ J/kgK). The arithmetic solution was stored in tabular form and indexed during the course of a computation. Today, that solution approach is inadequate, for the precise equations in fully non-isentropic form must be solved at each instant of a computation for the properties of the gas which exists at that location at that juncture.

Since a more complex solution, i.e., that for restricted pipes in Sect. 2.12, has already been presented, the complete solution for outflow from a cylinder or plenum in an unsteady gas-dynamic regime will not pose any new theoretical difficulties.

The case of subsonic particle flow will be presented first and that for sonic flow is given in Sec. 2.16.1.

In Fig. 2.16 the expanding flow from the throat to the downstream superposition point 2 is seen to leave turbulent vortices in the corners of that section. That the streamlines of the flow give rise to particle flow separation implies a gain of entropy from the throat to area section 2. On the other hand, the flow from the cylinder to the throat is contracting and can be considered to be isentropic in the same fashion as the contractions debated in Secs. 2.11 and 2.12. This is summarized on the temperature-entropy diagram in Fig. 2.17 where the gain of entropy for the flow rising from pressure p_t to pressure p_{s2} is clearly visible. The isentropic nature of the flow from p_1 to p_t, a vertical line on Fig. 2.17, can also be observed.

The properties and composition of the gas particles are those of the gas at the <u>exit</u> of the cylinder to the pipe. The word "exit" is used most precisely. For most cylinders and plenums

Chapter 2 - Gas Flow through Two-Stroke Engines

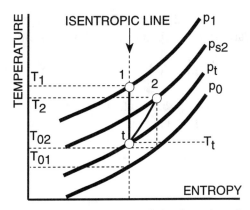

(a) temperature-entropy characteristics for subsonic outflow.

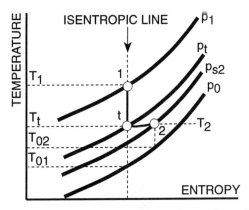

(b) temperature-entropy characteristics for sonic outflow.

Fig. 2.17 Temperature-entropy characteristics for cylinder or plenum outflow.

the process of flow within the cylinder is one of mixing. In which case the properties of the gas at the exit for an outflow process are that of the mean of all of the contents. Not all internal cylinder flow is like that. Some cylinders have a stratified in-cylinder flow process. A two-stroke engine cylinder would be a classic example of that situation. There the properties of the gas exiting the cylinder would vary from combustion products only at the commencement of the exhaust outflow to a gas which contains increasingly larger proportions of the air lost during the scavenge process; it would be mere coincidence if the exiting gas at any instant had the same properties as the average of all of the cylinder contents.

This is illustrated in Fig. 2.25 where there are stratified zones labeled as CX surrounding the intake and exhaust apertures. The properties of the gas in those zones will differ from the mean values for all of the cylinder, labeled in Fig. 2.25 as P_C, T_C, etc., and also the gas properties R_C and γ_C. In that case a means of tracking the extent of the stratification must be employed and these variables determined as P_{CX}, T_{CX}, R_{CX}, γ_{CX}, etc., and employed for those properties subscripted with a 1 in the text below. Further debate on this issue is found in Sec. 2.18.10.

Design and Simulation of Two-Stroke Engines

While this singularity of stratified scavenging should always be borne in mind, and dealt with should it arise, the various gas properties for cylinder outflow are defined as:

$$\gamma = \gamma_1 \quad R = R_1 \quad G_5 = G_{5_1} \quad G_7 = G_{7_1}, \quad \text{etc.}$$

As usual, the analysis of flow in this context uses, where appropriate, the equations of continuity, the First Law of Thermodynamics and the momentum equation.

The reference state conditions are:

density $\qquad \rho_{01} = \rho_{0t} = \dfrac{p_0}{RT_{01}} \qquad \rho_{02} = \dfrac{p_0}{RT_{02}}$ \hfill (2.16.1)

acoustic velocity $\quad a_{01} = a_{0t} = \sqrt{\gamma RT_{01}} \quad a_{02} = \sqrt{\gamma RT_{02}}$ \hfill (2.16.2)

The continuity equation for mass flow in previous sections is still generally applicable and repeated here, although the entropy gain is reflected in the reference acoustic velocity and density at position 2:

$$\dot{m}_t - \dot{m}_2 = 0 \qquad (2.16.3)$$

This equation becomes, where the particle flow direction is not conventionally significant:

$$\rho_t [C_c A_t][C_s c_t] - \rho_{02} X_{s2}^{G5} A_2 G_5 a_{02}(X_{i2} - X_{r2}) = 0 \qquad (2.16.4)$$

The above equation, for the mass flow continuity for flow from the throat to the downstream station 2, contains the coefficient of contraction on the flow area, C_c, and the coefficient of velocity, C_s. These are conventionally connected in fluid mechanics theory to a coefficient of discharge, C_d, to give an effective throat area, A_{teff}, as follows:

$$C_d = C_c C_s \quad \text{and} \quad A_{teff} = C_d A_t$$

This latter equation of mass flow continuity becomes:

$$C_d \rho_t A_t c_t - \rho_{02} X_{s2}^{G5} A_2 G_5 a_{02}(X_{i2} - X_{r2}) = 0$$

The First Law of Thermodynamics was introduced for such flow situations in Sec. 2.8. The analysis required here follows similar logical lines. The First Law of Thermodynamics for flow from the cylinder to superposition station 2 can be expressed as:

$$h_1 + \frac{c_1^2}{2} = h_{s2} + \frac{c_{s2}^2}{2}$$

130

or,
$$G_5 a_1^2 - \left(G_5 a_{s2}^2 + c_{s2}^2\right) = 0 \tag{2.16.5}$$

The First Law of Thermodynamics for flow from the cylinder to the throat can be expressed as:

$$h_1 + \frac{c_1^2}{2} = h_t + \frac{c_t^2}{2}$$

or,
$$C_p(T_1 - T_t) - \frac{c_t^2}{2} = 0 \tag{2.16.6}$$

The momentum equation for flow from the throat to superposition station 2 is expressed as:

$$A_2(p_t - p_{s2}) + \dot{m}_{s2}(c_t - c_{s2}) = 0 \tag{2.16.7}$$

The unknown values will be the reflected pressure wave at the boundary, p_{r2}, the reference temperature at position 2, namely T_{02}, and the pressure, p_t, and the velocity, c_t, at the throat. There are four unknowns, necessitating four equations, namely the mass flow equation in Eq. 2.16.4, the two First Law equations, Eq.2.16.5 and Eq.2.16.6, and the momentum equation, Eq.2.16.7. All other "unknown" quantities can be computed from these values and from the "known" values. The known values are the downstream pipe area, A_2, the throat area, A_t, the gas properties leaving the cylinder, and the incident pressure wave, p_{i2}.

Recalling that,

$$X_1 = \left(\frac{p_1}{p_0}\right)^{G17} \quad \text{and} \quad X_{i2} = \left(\frac{p_{i2}}{p_0}\right)^{G17} \quad \text{and setting} \quad X_t = \left(\frac{p_t}{p_0}\right)^{G17}$$

then due to isentropic flow from the cylinder to the throat, the temperature reference conditions are given by:

$$a_1 = a_{01} X_1 \quad \text{or} \quad T_{01} = \frac{T_1}{X_1^2}$$

As T_1 and X_1 are input parameters to any given problem, then T_{01} is readily determined. In which case, from Eqs. 2.16.1 and 2.16.2, so are the reference densities and acoustic velocities for the cylinder and throat conditions. As shown below, so too can the density and temperature at the throat be related to the reference conditions.

$$\rho_t = \rho_{01} X_t^{G5} \quad \text{and} \quad T_t = \frac{(a_{01} X_t)^2}{\gamma R}$$

The continuity equation set in Eq. 2.16.4 reduces to:

$$\rho_{01} X_t^{G5} C_d A_t c_t - \rho_{02}(X_{i2} + X_{r2} - 1)^{G5} A_2 G_5 a_{02}(X_{i2} - X_{r2}) = 0 \qquad (2.16.8)$$

The First Law of Thermodynamics in Eq.2.16.5 reduces to:

$$G_5(a_{01} X_1)^2 - \left[(G_5 a_{02}(X_{i2} - X_{r2}))^2 + G_5 a_{02}^2 (X_{i2} + X_{r2} - 1)^2 \right] \qquad (2.16.9)$$

The First Law of Thermodynamics in Eq. 2.16.6 reduces to:

$$G_5 \left[(a_{01} X_1)^2 - (a_{01} X_t)^2 \right] - c_t^2 = 0 \qquad (2.16.10)$$

The momentum equation, Eq. 2.16.7, reduces to:

$$\begin{aligned} & p_0 \left[X_t^{G7} - (X_{i2} + X_{r2} - 1)^{G7} \right] \\ & + \left[\rho_{02}(X_{i2} + X_{r2} - 1)^{G5} \times G_5 a_{02}(X_{i2} - X_{r2}) \right] \times \\ & \left[c_t - G_5 a_{02}(X_{i2} - X_{r2}) \right] = 0 \end{aligned} \qquad (2.16.11)$$

The five equations, Eqs.2.16.8 to 2.16.11, cannot be reduced any further as they are polynomial functions of the four unknown variables, X_{r2}, X_t, a_{02}, and c_t. These functions can be solved by a standard iterative method for such problems. I have determined that the Newton-Raphson method for the solution of multiple polynomial equations is stable, accurate and rapid in execution. The arithmetic solution on a computer is conducted by a Gaussian Elimination method.

2.16.1 Outflow from a cylinder where sonic particle velocity is encountered

In the above analysis of unsteady gas outflow from a cylinder the particle velocity at the throat will quite commonly be found to reach, or even attempt to exceed, the local acoustic velocity. This is not possible in thermodynamic or gas-dynamic terms. The highest particle velocity permissible is the local acoustic velocity at the throat, i.e., the flow is permitted to become choked. Therefore, during the mathematical solution of Eqs. 2.16.8 to 2.16.11, the local Mach number at the throat is monitored and retained at unity if it is found to exceed it.

As, $\qquad M_t = \dfrac{c_t}{a_{01} X_t} = 1 \quad \text{then} \quad c_t = a_{01} X_t \qquad (2.16.12)$

Chapter 2 - Gas Flow through Two-Stroke Engines

Also, contained within the solution of the First Law of Thermodynamics for outflow from the cylinder to the throat, in Eq. 2.16.10, is a direct solution for the pressure ratio from the cylinder to the throat. The combination of Eqs. 2.16.10 and 2.16.12 provides:

$$G_5\{(a_{01}X_1)^2 - (a_{01}X_t)^2\} - (a_{01}X_t)^2 = 0$$

Consequently, $\quad \dfrac{X_t}{X_1} = \sqrt{\dfrac{G_5}{G_5+1}} \quad$ or $\quad \dfrac{p_t}{p_1} = \left(\dfrac{2}{\gamma+1}\right)^{G35}$ (2.16.13)

The pressure ratio from the cylinder to the throat where the flow at the throat is choked, i.e., where the Mach number at the throat is unity, is known as "the critical pressure ratio." Its deduction is also to be found in many standard texts on thermodynamics or gas dynamics. It is applicable only if the upstream particle velocity is considered to be zero. Consequently it is not a universal "law" and its application must be used only where the thermodynamic assumptions used in its creation are relevant. For example, it is not employed in either Secs. 2.12.1 or 2.17.1.

This simplifies the entire procedure because it gives a direct solution for two of the unknowns and replaces two of the four equations employed above for the subsonic solution. It is probably easier and more accurate from an arithmetic standpoint to eliminate the momentum equation, use the continuity and the First Law of Eqs. 2.16.8 and 2.16.9, but it is more accurate thermodynamically to retain it!

The acquisition of all related data for pressure, density, particle velocity and mass flow rate at both superposition stations and at the throat follows directly from the solution of the two polynomials for X_{r2} and a_{02}.

2.16.2 Numerical examples of outflow from a cylinder

The application of the above theory is illustrated by the calculation of outflow from a cylinder using the data given in Table 2.16.1. The nomenclature for the data is consistent with the theory and the associated sketch in Fig. 2.17. The units of the data, if inconsistent with strict SI units, is indicated in the several tables. The calculation output is shown in Tables 2.16.2 and 2.16.3.

Table 2.16.1 Input data to calculations of outflow from a cylinder

No.	P_1	T_1 °C	Π_1	d_t mm	d_2 mm	C_d	P_{i2}	Π_2
1	5.0	1000	0.0	3	30	0.9	1.0	0.0
2	5.0	1000	1.0	3	30	0.9	1.0	1.0
3	1.8	500	0.0	25	30	0.75	1.0	0.0
4	1.8	500	0.0	25	30	0.75	1.1	0.0
5	1.8	500	0.0	25	30	0.75	0.9	0.0

Table 2.16.2 Output from calculations of outflow from a cylinder

No.	P_{r2}	P_{s2}	T_{s2} °C	P_t	T_t °C	$\dot{m}_{s2}$ g/s
1	1.0351	1.0351	999.9	2.676	805.8	3.54
2	1.036	1.036	999.9	2.641	787.8	3.66
3	1.554	1.554	486.4	1.319	440.0	85.7
4	1.528	1.672	492.5	1.546	469.5	68.1
5	1.538	1.392	479.9	1.025	392.9	94.3

Table 2.16.3 Further output from calculations of outflow from a cylinder

No.	c_t	M_t	c_{s2}	M_{s2}	a_{01} & a_{0t}	a_{02}
1	663.4	1.0	18.25	0.025	582.4	717.4
2	652.9	1.0	18.01	0.025	568.3	711.5
3	372.0	0.69	175.4	0.315	519.6	525.1
4	262.9	0.48	130.5	0.234	519.6	522.1
5	492.7	0.945	213.5	0.385	519.6	530.5

The input data for test numbers 1 and 2 are with reference to a "blowdown" situation from gas at high temperature and pressure with a small-diameter port simulating a cylinder port or valve that has just commenced its opening. The cylinder has a pressure ratio of 5.0 and a temperature of 1000°C. The exhaust pipe diameter is the same for all of the tests, at 30 mm. In tests 1 and 2 the port diameter is equivalent to a 3-mm-diameter hole and has a coefficient of discharge of 0.90. The gas in the cylinder and in the exhaust pipe in test 1 has a purity of zero, i.e., it is all exhaust gas.

The purity defines the gas properties as a mixture of air and exhaust gas where the air is assumed to have the properties of specific heats ratio, γ, of 1.4 and a gas constant, R, of 287 J/kgK. The exhaust gas is assumed to have the properties of specific heats ratio, γ, of 1.36 and a gas constant, R, of 300 J/kgK. For further explanation see Eqs. 2.18.47 to 2.18.50.

To continue, in test 1 where the cylinder gas is assumed to be exhaust gas, the results of the computation in Tables 2.16.2 and 2.16.3 show that the flow at the throat is choked, i.e., M_t is 1.0, and that a small pulse with a pressure ratio of just 1.035 is sent into the exhaust pipe. The very considerable entropy gain is evident by the disparity between the reference acoustic velocities at the throat and at the pipe, a_{0t} and a_{02}, at 582.4 and 717.4 m/s, respectively. It is clear that any attempt to solve this flow regime as an isentropic process would be very inaccurate.

The presentation here of a non-isentropic analysis with variable gas properties is unique and its importance can be observed by a comparison of the results of tests 1 and 2. Test data set 2 is identical to set number 1 with the exception that the purity in the cylinder and in the pipe is assumed to be unity, i.e., it is air. The mass flow rate from data set 1 is 3.54 g/s and it is 3.66 g/s when using data set 2; that is an error of 3.4%. Mass flow errors in simulation translate ultimately into errors in the prediction of air mass trapped in a cylinder, a value directly related to power output. This error of 3.4% is even more significant than it appears as the effect is compounded throughout the entire simulation of an engine when using a computer.

The test data sets 3 to 5 illustrate the ability of pressure wave reflections to dramatically influence the "breathing" of an engine. The situation is one of exhaust from a cylinder from gas at high temperature and pressure with a large-diameter port simulating a cylinder port or valve which is at a well-open position. The cylinder has a pressure ratio of 1.8 and a temperature of 500°C. The exhaust pipe diameter is the same for all of the tests, at 30 mm. The port diameter is equivalent to a 25-mm-diameter hole and has a typical coefficient of discharge of 0.75. The gas in the cylinder and in the exhaust pipe has a purity of zero, i.e., it is all exhaust gas. The only difference between these data sets 3 to 5 is the amplitude of the pressure wave in the pipe incident on the exhaust port at a pressure ratio of 1.0, i.e., undisturbed conditions, or at 1.1, i.e., providing a modest opposition to the flow, or at 0.9, i.e., a modest suction effect on the cylinder, respectively. The results show considerable variations in the ensuing mass flow rate exiting the cylinder, ranging from 85.7 g/s when the conditions are undisturbed in test 3, to 68.1 g/s when the incident pressure wave is of compression, to 94.3 g/s when the incident pressure wave is one of expansion. These swings of mass flow rate represent variations of −20.5% to +10%. It will be observed that test 4 with the lowest mass flow rate has the highest superposition pressure ratio, P_{s2}, at the pipe point, and test 5 with the highest mass flow rate has the lowest superposition pressure in the pipe. As this is the pressure that would be monitored by a fast response pressure transducer, one would be tempted to conclude that test 3 is the one with the stronger wave action. Such is the folly of casually examining measured pressure traces in the exhaust ducts of engines; this opinion has been put forward before in Sec. 2.2.1.

This illustrates perfectly both the advantages of utilizing pressure wave effects in the exhaust system of an engine to enhance the mass flow through it, and the disadvantages of poorly designing the exhaust system. These simple numerical examples reinforce the opinions expressed earlier in Sec. 2.8.1 regarding the effective use of reflections of pressure waves in exhaust pipes.

2.17 Reflection of pressure waves in pipes for inflow to a cylinder

This situation is fundamental to all unsteady gas flow generated in the intake or exhaust ducts of a reciprocating IC engine. Fig. 2.18 shows an inlet port (or valve) and pipe, or the throttled end of an intake pipe leading into a plenum such as the atmosphere or a silencer box. Anywhere in an unsteady flow regime where a pressure wave in a pipe is incident on a pressure-filled space, box, plenum or cylinder, the following method is applicable to determine the magnitude of the mass inflow, of its thermodynamic state and of the reflected pressure wave.

In the theory presented here, the space into which the particles disperse is considered to be sufficiently large, and also three-dimensional, to give the fundamental assumption that the particle velocity within the cylinder is considered to be zero.

Design and Simulation of Two-Stroke Engines

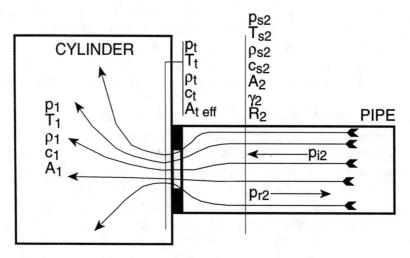

Fig. 2.18 Inflow from a pipe to a cylinder or plenum.

$$c_1 = 0 \qquad (2.17.1)$$

The case of subsonic particle flow will be presented first and that for sonic flow is given in Sec. 2.17.1.

In Fig. 2.18 the expanding flow from the throat to the cylinder gives pronounced turbulence within the cylinder. The traditional assumption is that this dissipation of turbulence energy gives no pressure recovery from the throat of the port or valve to the cylinder. This assumption applies only where subsonic flow is maintained at the throat.

$$p_t = p_1 \qquad (2.17.2)$$

On the other hand, the flow from the pipe to the throat is contracting and can be considered to be isentropic in the same fashion as other contractions debated in Secs. 2.11 and 2.12. This is summarized on the temperature-entropy diagram in Fig. 2.19 where the gain of entropy for the flow rising from pressure p_t to cylinder pressure p_1 is clearly visible. The isentropic nature of the flow from p_{s2} to p_t, a vertical line on Fig. 2.19, can also be observed.

The properties and composition of the gas particles are those of the gas at the superposition point in the pipe. The various gas properties for cylinder inflow are defined as:

$$\gamma = \gamma_2 \quad R = R_2 \quad G_5 = G_{5_2} \quad G_7 = G_{7_2}, \text{ etc.}$$

As usual, the analysis of flow in this context uses, where appropriate, the equations of continuity, the First Law of Thermodynamics and the momentum equation. However, the momentum equation is not employed in this particular analysis for subsonic inflow, as the constant pressure assumption used in Eq. 2.17.2 reflects an even higher gain of entropy, i.e., energy dissipation due to turbulence, than would be the case if the momentum equation were

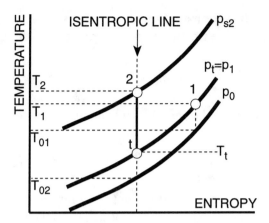

(a) temperature-entropy characteristics for subsonic

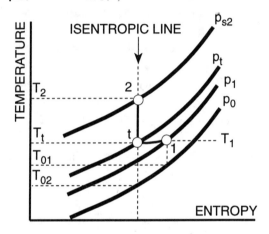

(b) temperature-entropy characteristics for sonic

Fig. 2.19 Temperature-entropy characteristics for cylinder or plenum inflow.

to be involved. The constant pressure assumption has been used in this regard before within this text, notably in the section dealing with diffusing flow in tapered pipes (Sec. 2.15.1).

The reference state conditions are:

density $\qquad \rho_{02} = \rho_{0t} = \dfrac{p_0}{RT_{02}} \qquad \rho_{01} = \dfrac{p_0}{RT_{01}}$ (2.17.3)

acoustic velocity $\quad a_{02} = a_{0t} = \sqrt{\gamma RT_{02}} \qquad a_{01} = \sqrt{\gamma RT_{01}}$ (2.17.4)

Design and Simulation of Two-Stroke Engines

The continuity equation for mass flow in previous sections is still generally applicable and repeated here, although the isentropic process from pipe to throat is reflected in the ensuing theory:

$$\dot{m}_t - \dot{m}_2 = 0 \tag{2.17.5}$$

This equation becomes, where the particle flow direction is not conventionally significant:

$$\rho_t [C_c A_t][C_s c_t] - \rho_{02} X_{s2}^{G5} A_2 G_5 a_{02} (X_{i2} - X_{r2}) = 0 \tag{2.17.6}$$

The above equation, for the mass flow continuity for flow from the throat to the downstream station 2, contains the coefficient of contraction on the flow area, C_c, and the coefficient of velocity, C_s. These are conventionally connected in fluid mechanics theory to a coefficient of discharge, C_d, to give an effective throat area, $A_{t\,eff}$, as follows:

$$C_d = C_c C_s \quad \text{and} \quad A_{t\,eff} = C_d A_t$$

This latter equation of mass flow continuity becomes:

$$C_d \rho_t A_t c_t - \rho_{02} X_{s2}^{G5} A_2 G_5 a_{02} (X_{i2} - X_{r2}) = 0$$

The First Law of Thermodynamics for flow from the pipe to the throat can be expressed as:

$$h_2 + \frac{c_2^2}{2} = h_t + \frac{c_t^2}{2}$$

or,

$$C_p(T_2 - T_t) + \left(\frac{c_2^2}{2} - \frac{c_t^2}{2}\right) = 0 \tag{2.17.7}$$

The unknown values will be the reflected pressure wave at the boundary, p_{r2}, and the velocity, c_t, at the throat. There are two unknowns, necessitating two equations, namely the mass flow equation in Eq. 2.17.6, and the First Law equation, Eq. 2.17.7. All other "unknown" quantities can be computed from these values and from the "known" values. The known values are the downstream pipe area, A_2, the throat area, A_t, the gas properties at the superposition position 2 in the pipe, and the incident pressure wave, p_{i2}.

It will be recalled that,

$$X_1 = \left(\frac{p_1}{p_0}\right)^{G17} \quad \text{and} \quad X_{i2} = \left(\frac{p_{i2}}{p_0}\right)^{G17} \quad \text{and setting} \quad X_t = \left(\frac{p_t}{p_0}\right)^{G17} = \left(\frac{p_1}{p_0}\right)^{G17}$$

The reference temperature for the cylinder is given by:

$$a_1 = a_{01}X_1 \quad \text{or} \quad T_{01} = \frac{T_1}{X_1^2}$$

As T_1 and X_1 are input parameters to any given problem, then T_{01} is readily determined. In which case, from Eqs. 2.17.1 and 2.17.2 so are the reference densities and acoustic velocities for the cylinder and throat conditions. As shown below, the density and temperature at the throat are related to their isentropic reference conditions and to the assumption regarding the throat pressure equality with cylinder pressure.

$$\rho_t = \rho_{02}X_t^{G5} = \rho_{02}X_1^{G5} \quad \text{and} \quad T_t = \frac{(a_{02}X_1)^2}{\gamma R}$$

The continuity equation set in Eq. 2.17.6 reduces to:

$$\rho_{02}X_1^{G5}C_dA_tc_t - \rho_{02}(X_{i2} + X_{r2} - 1)^{G5}A_2G_5a_{02}(X_{i2} - X_{r2}) = 0 \qquad (2.17.8)$$

The First Law of Thermodynamics in Eq. 2.17.7 reduces to:

$$\left[G_5(a_{02}(X_{i2} + X_{r2} - 1))^2 - G_5(a_{02}X_1)^2\right]$$
$$+\left[(G_5a_{02}(X_{i2} - X_{r2}))^2 - c_t^2\right] = 0 \qquad (2.17.9)$$

The two equations, Eqs. 2.17.8 and 2.17.9, cannot be reduced any further as they are polynomial functions of the two variables, X_{r1} and c_t. These functions can be solved by a standard iterative method for such problems. I have determined that the Newton-Raphson method for the solution of multiple polynomial equations is stable, accurate and rapid in execution. The arithmetic solution on a computer is conducted by a Gaussian Elimination method.

2.17.1 Inflow to a cylinder where sonic particle velocity is encountered

In the above analysis of unsteady gas outflow from a cylinder the particle velocity at the throat will quite commonly be found to reach, or even attempt to exceed, the local acoustic velocity. This is not possible in thermodynamic or gas-dynamic terms. The highest particle velocity permissible is the local acoustic velocity at the throat, i.e., the flow is permitted to become choked. Therefore, during the mathematical solution of Eqs. 2.17.8 and 2.17.9, the local Mach number at the throat is monitored and retained at unity if it is found to exceed it.

As,
$$M_t = \frac{c_t}{a_{02}X_t} = 1 \quad \text{then} \quad c_t = a_{02}X_t \tag{2.17.10}$$

The solution for the critical pressure ratio in Sec. 2.16.1 cannot be used here as the pipe, i.e., upstream, particle velocity is not zero.

This simplifies the entire procedure for this gives a direct relationship between two of the unknowns. However, this does not reduce the solution complexity as it also eliminates a previous assumption for the subsonic flow regime that the throat pressure is equal to the cylinder pressure, i.e., as shown in Eq. 2.17.2. Therefore, the previous equations, Eqs. 2.17.8 and 2.17.9, must be revisited and that assumption removed:

The continuity equation set in Eq. 2.17.8 becomes:

$$a_{02}\rho_{02}X_t^{G6}C_dA_t - \rho_{02}(X_{i2} + X_{r2} - 1)^{G5}A_2G_5a_{02}(X_{i2} - X_{r2}) = 0 \tag{2.17.11}$$

The First Law of Thermodynamics in Eq. 2.17.9 becomes:

$$\left[G_5(a_{02}(X_{i2} + X_{r2} - 1))^2 - G_5(a_{02}X_t)^2\right] + \left[(G_5a_{02}(X_{i2} - X_{r2}))^2 - (a_{02}X_t)^2\right] = 0 \tag{2.17.12}$$

The two equations, Eqs. 2.17.11 and 2.17.12, cannot be reduced any further as they are polynomial functions of the two variables, X_{r1} and X_t. These functions can be solved by a standard iterative method for such problems. I have determined that the Newton-Raphson method for the solution of multiple polynomial equations is stable, accurate and rapid in execution. The arithmetic solution on a computer is conducted by a Gaussian Elimination method.

The acquisition of all related data for pressure, density, particle velocity and mass flow rate at the pipe superposition station and at the throat follows directly from the solution of the two polynomials for X_{r2} and X_t.

2.17.2 Numerical examples of inflow into a cylinder

The application of the above theory is illustrated by the calculation of inflow into a cylinder using the data given in Table 2.17.1. The nomenclature for the data is consistent with the theory and the associated sketch in Fig. 2.18. The units of the data, if inconsistent with strict SI units, are indicated in the several tables. The calculation output is shown in Tables 2.17.2 and 2.17.3.

The input data for all of the tests are with reference to a normal intake situation from air at an above-atmospheric temperature and atmospheric pressure through a large port, equivalent to 20 mm diameter with a discharge coefficient of 0.75, simulating a cylinder port or valve near its maximum opening. The cylinder has a pressure ratio of 0.65 and a temperature of 70°C. The reference temperature at the pipe point is 70°C, an apparently high temperature

Chapter 2 - Gas Flow through Two-Stroke Engines

Table 2.17.1 Input data to calculations of inflow into a cylinder

No.	P_1	T_{02} °C	P_1	d_t mm	d_2 mm	C_d	P_{i2}	P_2
1	0.65	70	1.0	20	30	0.75	1.0	1.0
3	0.65	70	1.0	20	30	0.75	0.9	1.0
2	0.65	70	1.0	20	30	0.75	1.2	1.0
4	0.65	40	1.0	20	30	0.75	1.2	1.0

Table 2.17.2 Output from calculations of inflow into a cylinder

No.	P_{r2}	P_{it}	T_{i2} °C	P_t	T_t °C	$\dot{m}_{s2}$ g/s
1	0.801	0.801	48.9	0.65	30.3	36.0
3	0.779	0.698	36.5	0.65	30.3	21.1
2	0.910	1.095	78.9	0.65	30.3	57.3
4	0.910	1.095	48.2	0.65	3.75	60.0

Table 2.17.3 Further output from calculations of inflow into a cylinder

No.	c_t	M_t	c_{s2}	M_{s2}	a_{01} & a_{0t}
1	201.9	0.578	58.0	0.161	371.2
3	118.2	0.338	37.4	0.106	371.2
2	321.4	0.921	73.8	0.196	371.2
4	307.0	0.921	70.5	0.196	354.6

that would be quite normal for an intake system, the walls of which have been heated by conduction from the rest of the power unit. The intake pipe diameter is the same for all of the tests—30 mm. The gas in the cylinder and in the induction pipe has a purity of 1.0, i.e., it is air.

The test data sets 1, 2 and 3 illustrate the ability of pressure wave reflections to influence the "breathing" of an engine. The difference between the data sets 1 to 3 is the amplitude of the pressure wave in the pipe incident on the intake port at a pressure ratio of 1.0, i.e., undisturbed conditions, or at 0.9, i.e., providing a modest opposition to the flow, or at 1.2, i.e., a good ramming effect on the cylinder, respectively. The results show very considerable variations in the ensuing mass flow rate entering the cylinder, ranging from 36.0 g/s when the conditions are undisturbed in test 1, to 21.1 g/s when the incident pressure wave is one of expansion, to 57.3 g/s when the incident pressure wave is one of compression. These variations in mass flow rate represent changes of −41.4% to +59.2%. "Intake ramming," first discussed in Sec. 2.8.2, has a profound effect on the mass of air which can be induced by an engine. Equally, if the intake system is poorly designed, and provides expansion wave reflections at the intake valve or port during that process, then the potential for the deterioration of induction of air is equally self-evident.

The test data set 4 is almost identical to that of test data set 2, except that the intake system has been cooled, as in "intercooled" in a modern turbo-diesel automobile engine, to give a reference air temperature at the pipe point of 40°C in comparison to the hotter air at 70°C in test data set 2. The density of the intake air has been increased by 8.8%. This does not translate directly into the same order of increase of air mass flow rate. The computation shows that the air mass flow rate rises from 57.1 to 60.0 g/s, which represents a gain of 5.1%. However, as intake air flow and torque are almost directly related, the charge cooling in test data set 4 would indicate a useful gain of torque and power of the latter percentage.

2.18 The simulation of engines by the computation of unsteady gas flow

Many computational methods have been suggested for the solution of this theoretical situation, such as Riemann variables [2.10], Lax-Wendroff [2.42] and other finite difference procedures [2.46], and yet others [2.12, 2.49]. The basic approach adopted here is to re-examine the fundamental theory of pressure wave motion and adapt it to a mesh method interpolation procedure. At the same time the boundary conditions for inflow and outflow, such as the filling and emptying of engine cylinders, are resolved for the generality of gas properties and in terms of the unsteady gas flow that controls those processes. The same generality of gas property and composition is traced throughout the pipe system. This change of gas property is very significant in two-stroke engines where the exhaust blowdown is followed by short-circuited scavenge air. It is also very significant at varying load levels in diesel engines. Vitally important in this context is the solution for the continual transmission and reflection of pressure waves as they encounter both differing temperature gradients and gas properties, and both gradual and sudden changes of area throughout the engine ducting. Of equal importance is the ability of the calculation to predict the effect of internal heat generation within the duct or of external heat transfer with respect to it, and to be able to trace the effect of the ensuing gas temperature change on both the pressure wave system and the net gas flow.

The computation of unsteady gas flow through the cylinders of reciprocating internal-combustion engines is a technology which is now nearly forty years old. The original paper by Benson *et al.* [2.10] formalized the use of Rieman variables as a technique for tracing the motion of pressure waves in the ducts of engines and the effect on the filling and emptying of the cylinders attached to them. Subsequently, other models have been introduced using finite difference techniques [2.29] and Lax-Wendroff methods [2.42, 2.46, 2.49]. Many technical papers and books have been published on this subject, including some of mine, which are listed in the references.

For a more complete discussion of this subject, and in particular for the description of the previous computational procedures produced by authors from the University of Manchester (UMIST) and QUB, the textbooks by Benson [2.4] and Blair [2.25], and the papers by Chen [2.47] and Kirkpatrick *et al.* [2.41] may be studied.

This introduction is somewhat brief, for a full discussion of the computational procedures developed by others would fill, indeed they have filled, many a textbook, let alone an introduction to this chapter. Consequently, many scores of worthy contributors to the literature have not been cited in the references and I hope that those who are not mentioned here will not feel slighted in any way.

I started working in the era when the graphical method of characteristics [2.8] was the only means of computing unsteady gas flow in pipes. While cumbersome in the arithmetic extreme before the advent of the digital computer, and hence virtually ineffective as a design tool, it had the considerable merit of giving an insight into unsteady gas flow in a manner which the Riemann variable, the finite difference or Lax-Wendroff methods have never provided. Doubtless the programmers of such computational techniques have gained these insights, but this rarely, if ever, applied to the users of the software, i.e., the engineers and engine designers who had to employ it. As one who has taught such computational techniques to several generations of students, it can be confirmed that the insights gained by students always occurred most readily from lectures based on the papers produced by Wallace *et al.* [2.5] or the notes produced by Bannister [2.2]. These latter papers were all based on the fundamentals of pressure wave motion. Actually one author, Jones [2.9], succeeded in solving the graphical method of characteristics by a computational procedure.

For some time, I have been re-examining many of these computational procedures to improve the quality of the design tool emanating from them and have found that, for one reason or another, all of the available methods fall short of what is required for engine design needs. In particular, as two-stroke cycle engines appear to have considerable potential as future automobile powerplants, or require to be redesigned to incorporate stratified charging for the reduction of emissions for the simpler engines used in lawnmowers, mopeds, and chainsaws, a design model firmly based on the motion of pressure waves is absolutely essential. At the same time the computational model must be able to trace with great accuracy the rapidly changing gas properties in the cylinder and in the exhaust pipe, where the exhaust gas from blowdown is rapidly followed by large amounts of fresh air, or fresh air and fuel, into the exhaust pipe and onward through an exhaust catalyst or silencer. Equally, for outboard engines, where the exhaust is often internally water-cooled, unless the model can incorporate the thermodynamic effects of that water "injection," it will be less than useful as a design tool. Diesel engines, be they two- or four-stroke cycle units, also have gas properties in the exhaust system that vary considerably as a function of the load level.

Unlike the four-stroke cycle engine, the two-stroke engine is heavily dependent on pressure wave effects for the scavenging and charging processes and the model therefore must be based on their transmission, propagation and reflections. It must not be thought that the theoretical model described below is uniquely applicable to a two-stroke cycle engine, for such is not the case. Indeed, the many recent papers on inlet manifold "tuning," employed as a means of modifying the volumetric efficiency-speed curve as a precursor to altering the torque curve, indicates the growing interest in the prediction of unsteady gas motion by the designers of four-stroke engines.

It should not be assumed that the four-stroke spark-ignition engine is immune from these effects. The common occurrence of blowback into the inlet tract during the valve overlap period at the end of the exhaust stroke necessitates the tracking of variable gas properties within that duct. If this does not happen then a very inaccurate assessment of the ensuing air flow into the cylinder and of its trapped charge purity will result.

Design and Simulation of Two-Stroke Engines

A further requirement is also essential for a simulation model that describes the complexities of pressure wave motion in an engine; its principles must be capable of being understood by the users, otherwise the process of optimizing a given design will not only be long-winded and tedious, but prone to decision making on geometrical aspects of the engine which are illogically correct at one engine speed and load, yet disastrously wrong at another. Such dichotomies must be thoroughly checked out by a designer who understands the fundamental principles on which the model is based.

The basis of the theoretical model employed here, dubbed the GPB model, fulfills all of the criteria specified above.

2.18.1 The basis of the GPB computation model

The unsteady gas flow process being modeled is illustrated in Fig. 2.20. It is also the same process that has been debated throughout this chapter. The computational procedure is somewhat similar to that found in other characteristics solutions. The pipe or pipes in the ducts of the engine being modeled are divided into meshes of a given length, L. The pressure waves propagating leftward and rightward are shown in Fig. 2.20 and at the instant that snapshot is taken for a particular mesh labeled as J, the pressure values at the left end are p_R and p_L and at the right end are p_{R1} and p_{L1}. The gas in the mesh space has properties of gas constant, R, and specific heats ratio, γ, which are assumed to be known at any instant of time. The reference density, ρ_0, and temperature, T_0, are also assumed to be known at any instant. As the diameter, d, and the volume, V, of the mesh space are a matter of geometrical fact, the mass in the mesh space can be determined at that instant. The average pressure throughout the mesh

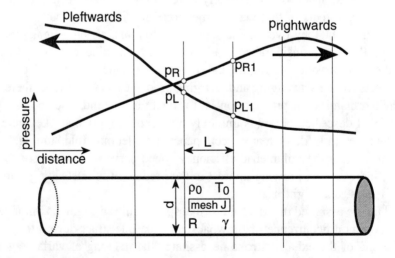

Fig. 2.20 Meshing of the duct for pressure wave propagation.

space can be considered to be the mean of the superposition pressures at each end of the mesh. The average superposition pressure amplitude ratio is given by X_J:

$$X_J = \frac{(X_R + X_L - 1) + (X_{R1} + X_{L1} - 1)}{2} \qquad (2.18.1)$$

Consequently the average pressure, p_J, density, ρ_J, and temperature, T_J, are found from:

$$p_J = p_0 X_J^{G7} \qquad (2.18.2)$$

$$\rho_J = \rho_0 X_J^{G5} \qquad (2.18.3)$$

$$T_J = T_0 X_J^2 \qquad (2.18.4)$$

The acoustic velocity, a_J, and mass, m_J, in the mesh are obtained by:

$$a_J = a_0 X_J \qquad (2.18.5)$$

$$m_J = \rho_J V_J \quad \text{where} \quad V_J = \frac{\pi}{4} d^2 L \qquad (2.18.6)$$

What is required of the GPB computation model is to determine:
(a) The effect of the motion of the pressure waves at either end of the mesh during some suitable time interval, dt, on the thermodynamics of the gas in the mesh space, J.
(b) The effect of the motion of the pressure waves at the right-hand end of the mesh, J–1, or the left-hand end of mesh, J+1, during some suitable time interval, dt, on the thermodynamics of the gas in the mesh space, J.
(c) The effect of the motion of the pressure waves propagating during time, dt, through mesh space, J, on their alteration in amplitude due to friction or area change.
(d) The effect on the pressure waves after arriving at the right-hand end of the mesh J, or at the left-hand end of mesh J, and encountering differing gas properties in adjacent mesh spaces, J+1 and J–1, respectively.
(e) The effect on the pressure waves after arriving at the right-hand end of the mesh J, or at the left-hand end of mesh J, and encountering a geometrical discontinuity such as a throttle, the valve or port of an engine cylinder, a branch in the pipe system, or an increase or decrease in duct area.

The first and most basic problem in any such computation model is to select the time interval at each step of the computation.

2.18.2 Selecting the time increment for each step of the calculation

It is not essential that the mesh lengths in the inlet or exhaust pipes be equal, or be equal in any given section of an inlet or exhaust system. The reasoning behind these statements is found in the arithmetic nature of the ensuing iterative procedures, where interpolation of values is permissible, but extrapolation is arithmetically unstable [2.14]. This is best seen in Fig. 2.21. The calculation is to be advanced in discrete time steps. The value of the time step, dt, is obtained by sweeping each mesh space throughout the simulated pipe geometry and determining the "fastest" propagation velocity in the system. In the example illustrated in Fig. 2.21 this happens to be at some other mesh than mesh J. At either end of the mesh the pressure waves, p_R and p_L and p_{R1} and p_{L1}, induce left and right running superposition velocities, α_R and α_L. These superposition velocities can be determined from Eqs. 2.2.8 and 2.2.9.

It is assumed that there are linear variations within the mesh length, L, of pressure and propagation velocity.

Consequently, the time increment, dt, at mesh J is the least value of time taken to traverse the mesh J, where L is the mesh length peculiar to mesh J, by the fastest of any one of the four propagation modes at either end of mesh J:

$$dt = \frac{L}{\alpha_{sL}} \quad \text{or} \quad dt = \frac{L}{\alpha_{sL1}} \quad \text{or} \quad dt = \frac{L}{\alpha_{sR}} \quad \text{or} \quad dt = \frac{L}{\alpha_{sR1}} \qquad (2.18.7)$$

This ensures that all subsequent iterative procedures are by interpolation, thus satisfying the Courant, Friedrich and Lewy "stability criterion" [2.14].

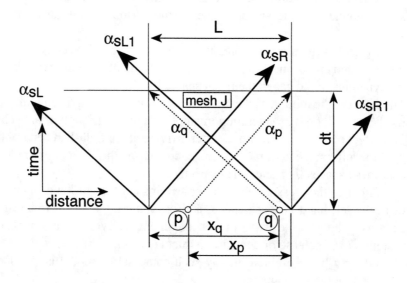

Fig. 2.21 The propagation of the pressure waves within mesh J.

Chapter 2 - Gas Flow through Two-Stroke Engines

2.18.3 The wave transmission during the time increment, dt

This is illustrated in Fig. 2.21. Consider the mesh J of length L. At the left end of the mesh the rightward-moving pressure wave, p_R, is not propagating fast enough at α_{SR} to reach the right end of the mesh in time dt. Consequently, a value of superposition propagation velocity, α_p, from a wave point of pressure amplitude ratio, X_p, linearly related to its physical position p and the p_R and p_{R1} values, will just index the right end of the time-distance point at time dt, while it is being mutually superposed upon a leftward pressure wave, α_q, emanating from physical position q. This leftward-propagating pressure wave point of amplitude X_q will just indent the left-hand intersection of the time-distance mesh during the time increment dt. The values of X_q and α_q are also linearly related to their physical position and in terms of the leftward wave pressures, X_L and X_{L1}, at either end of mesh J.

In short, the calculation presumption is that between any two meshes there is a linear variation of wave pressure, wave superposition pressure and superposition propagation velocity, both leftward and rightward, and that the values of X_p and X_q will, should no other effect befall them, become the new values of rightward and leftward pressure wave at either end of mesh J at the conclusion of time increment dt.

2.18.4 The interpolation procedure for wave transmission through a mesh

Having determined the time increment for a calculation step, and knowing the gas properties within any mesh volume for that transmission, the simulation must now determine the values of X_p and X_q, within the terms outlined above. The situation is as sketched in Fig. 2.21.

The propagation of rightward wave X_p through leftward wave X_q is conducted at superposition velocities, α_p and α_q, respectively. Retaining the sign convention that rightward motion is positive, then from Eqs. 2.2.9 and 2.2.10, these values of propagation velocity are determined as,

$$\alpha_p = a_0 \left(G_6 X_p - G_4 X_q - 1 \right) \tag{2.18.8}$$

$$\alpha_q = -a_0 \left(G_6 X_q - G_4 X_p - 1 \right) \tag{2.18.9}$$

The time taken from their respective dimensional starting points, p and q, is the same, dt, where dt is equal to the minimum time step inferred from the application of the stability criterion in Sec. 2.18.2.

Therefore, and determining the arithmetic values of the lengths x_p and x_q,

$$x_p = \alpha_p dt \tag{2.18.10}$$

$$x_q = |\alpha_q| dt \tag{2.18.11}$$

The dimensional values x_p and x_q also relate to the numeric values of X_p and X_q as linear variations of the change of wave pressure between the two ends of the mesh J boundaries.

Thus,

$$X_p = X_R + (X_{R1} - X_R)\frac{L - x_p}{L} \qquad (2.18.12)$$

$$X_q = X_{L1} + (X_L - X_{L1})\frac{L - x_q}{L} \qquad (2.18.13)$$

Eliminating x_p and x_q from Eqs. 2.8.10-13, produces two further equations,

$$\frac{X_{R1} - X_p}{X_{R1} - X_R} = \frac{a_0 dt}{L}(G_6 X_p - G_4 X_q - 1) \qquad (2.18.14)$$

$$\frac{X_L - X_q}{X_L - X_{L1}} = \frac{a_0 dt}{L}(G_6 X_q - G_4 X_p - 1) \qquad (2.18.15)$$

By defining the following groupings of variables as terms A to E, Eqs. 2.18.14 and 2.18.15 become Eqs. 2.18.16 and 2.18.17. The variables in the terms A-E are ones which would be "known" quantities at the commencement of any mesh calculation.

$$A = E(X_{R1} - X_R)$$

$$B = E(X_L - X_{L1})$$

$$C = \frac{X_{R1}}{A}$$

$$D = \frac{X_L}{B}$$

$$E = \frac{a_0 dt}{L}$$

$$X_p\left(G_6 + \frac{1}{A}\right) - G_4 X_q - C - 1 = 0 \qquad (2.18.16)$$

$$X_q\left(G_6 + \frac{1}{A}\right) - G_4 X_p - D - 1 = 0 \qquad (2.18.17)$$

These simultaneous equations can be solved for the unknown quantities, X_p and X_q, after further collections of known terms within F_R and F_L are made for simplification:

$$F_R = \frac{G_6 + \dfrac{1}{A}}{G_4}$$

$$F_L = \frac{G_6 + \dfrac{1}{B}}{G_4}$$

The final outcome is that

$$X_p = \frac{1 + D + F_L + F_L C}{G_4(F_R F_L - 1)} \qquad (2.18.18)$$

$$X_q = \frac{1 + C + F_R + F_R D}{G_4(F_R F_L - 1)} \qquad (2.18.19)$$

Thus, assuming that the value of wave pressure is going to be modified by friction or area change during its travel during the time step, the new values of leftward and rightward wave pressures at the left- and right-hand ends of mesh J, X_p and X_q, at the conclusion of the time step, dt, are going to be given by:

$$X_{R1new} = X_p + \{\pm \text{ friction effects } (\pm) \text{ area change effects}\} \qquad (2.18.20)$$

$$X_{Lnew} = X_q + \{\pm \text{ friction effects } (\pm) \text{ area change effects}\} \qquad (2.18.21)$$

Each particle within the mesh space is assumed to experience this superposition process involving the pressure waves, p_p and p_q. Consequently the new values of the unsteady gas-dynamic parameters attributed to the particles undergoing this superposition effect are:

pressure amplitude ratio $\qquad X_s = X_p + X_q - 1 \qquad (2.18.22)$

pressure $\qquad p_s = p_0 X_s^{G7} \qquad (2.18.23)$

density $\qquad \rho_s = \rho_0 X_s^{G5} \qquad (2.18.24)$

temperature $\qquad T_J = T_0 X_J^2 \qquad$ (2.18.25)

particle velocity $\qquad c_s = G_5 a_0 (X_p - X_q) \qquad$ (2.18.26)

2.18.5 Singularities during the interpolation procedure

It is obvious that arithmetic problems could arise with the procedure given in Sec. 2.18.4 if the values of X_R and X_{R1}, or the values of X_L and X_{L1}, are equal. This would certainly be true in a model start-up situation, where the pipes would be "dead" and all of the array elements of pressure amplitude ratio X_R, X_{R1}, X_L and X_{L1} would be unity. In that situation the values of A and B would be zero, C and D would be infinity and the calculation would collapse. Therefore, separate solutions are required for the wave pressures, X_p and X_q, in these unique situations, either at start-up or as cover during the course of a complete calculation. The solution is fairly straightforward using Eqs. 2.18.16 and 2.18.17, for the three possibilities (a)-(c) involved.

In case (i) the following is the solution if it is true that X_R is equal to X_{R1}:

$$X_p = X_{R1} \qquad (2.18.27)$$

Hence,

$$X_q = \frac{1 + D + G_4 X_p}{G_6 + \dfrac{1}{B}} \qquad (2.18.28)$$

In case (ii) the following is the solution if it is true that X_L is equal to X_{L1}:

$$X_q = X_{L1} \qquad (2.18.29)$$

Hence,

$$X_p = \frac{1 + C + G_4 X_q}{G_6 + \dfrac{1}{A}} \qquad (2.18.30)$$

In case (iii) the following is the solution if it is true that X_R is equal to X_{R1} and also that X_L is equal to X_{L1}:

$$X_q = X_{L1} \qquad (2.18.31)$$

and,

$$X_p = X_{R1} \qquad (2.18.32)$$

Chapter 2 - Gas Flow through Two-Stroke Engines

2.18.6 Changes due to friction and heat transfer during a computation step

The theoretical treatment for this is dealt with completely in Secs. 2.3 and 2.4. The value of the new wave pressure amplitude ratios at the mesh boundary, X_p and X_q, computed for mesh J are used as being representative of the superposition process experienced by all of the particles in the mesh space during the time step, dt. The pressure drop is computed as in Sec. 2.3 and the total heat transfer due to friction, dQ_f, can be computed for all of the particles in mesh space J.

The surface area of the mesh J is given by A_J, where,

$$A_J = \pi dL \tag{2.18.33}$$

At this point the warning given in the latter part of Sec. 2.3 should be heeded for the computation of the heat generated in the mesh space due to friction and Eqs. 2.3.15 and 2.3.16 modified accordingly. The correct solution for the friction force, F, for all of the particles within the mesh space is:

$$F = A_J \left[C_f \frac{\rho_s c_s^2}{2} \right] \tag{2.18.34}$$

However, as each particle is assumed to move at superposition particle velocity, c_s, during time, dt, the work, δW_f, resulting in the heat generated, δQ_f, can be calculated by:

$$\delta W_f = F dx = F c_s dt = \left| \frac{A_J C_f \rho_s c_s^3 dt}{2} \right| = \delta Q_f \tag{2.18.35}$$

The sign of this equation must always be positive as the heat due to friction is additive to the system comprising the gas particles within the mesh.

A similar process is adopted for the calculation of the heat transfer effect to the wall. It is discussed fundamentally in Sec. 2.4. It is assumed that the wall temperature, T_w, is a known quantity at the location of mesh J. The convection heat transfer coefficient, C_h, can be computed using the theory in Sec. 2.4. Consequently the heat transfer to or from the pipe walls at the mesh J, with respect to the system comprising all of the particles within the boundaries of mesh J, is:

$$\delta Q_h = A_J C_h (T_w - T_s) dt \tag{2.18.36}$$

2.18.7 Wave reflections at the inter-mesh boundaries after a time step

The above sections, which discuss the simulation technique using the GPB method, have given the theory of deduction of the amplitudes of the leftward and righward pressure waves, p_p and p_q, arriving at the left and right end boundaries of a mesh J. The calculation method is employed similarly for all other meshes in all of the ducts of the engine. All meshes are assumed to have a constant diameter appropriate to that mesh. That applies even to tapered

Design and Simulation of Two-Stroke Engines

sections within any of the ducts as a glance at Sec. 2.15 will confirm. However, the ability to ascertain the values of p_p and p_q is but half of the story. Consider the sketch of the model at this juncture as given in Fig. 2.22. It deals with parallel ducting and tapered ducting in sections (a) and (b), respectively. These are discussed in the sub-sections below.

(a) parallel ducting

Fig. 2.22(a) is an illustration of two adjacent meshes, labeled 1 and 2. After the application of the preceding theory in this section the GPB model now has the information that pressure waves with pressure amplitude ratio X_{p1} and X_{q1}, and X_{p2} and X_{q2}, have arrived at the left- and right-hand boundaries of meshes 1 and 2, respectively. The pressures are the

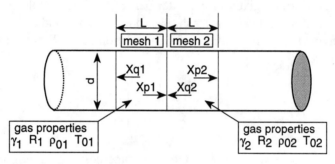

(a) two adjacent meshes in a parallel pipe

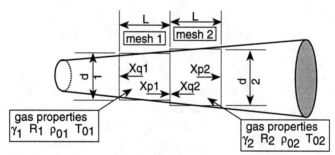

(b) two adjacent meshes in a tapered pipe

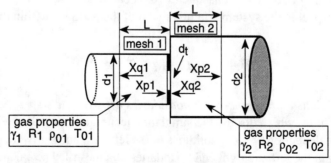

(c) two adjacent meshes in a restricted pipe

Fig. 2.22 Adjacent meshes in pipes of differing discontinuities.

"new" values, i.e., the values of X_p and X_q for each mesh having endured the loss of wave energy due to friction as described in Sec. 2.18.6. Consider the inter-mesh boundary between mesh 1 and 2 as being representative of all other inter-mesh boundaries within any one of the pipes in the entire ducting of the engine. At the commencement of the computation for the time step, the information for mesh 1 had available the left- and rightward pressure waves for that mesh at the left- and right-hand boundaries; as seen in Sec. 2.18.2 they were labeled as p_R and p_L and p_{R1} and p_{L1}, and to denote that they are attached to mesh they will be relabeled here as $_1p_R$ and $_1p_L$ and $_1p_{R1}$ and $_1p_{L1}$. The equivalent values for mesh 2 at the commencement of the computation time step would be $_2p_R$ and $_2p_L$ and $_2p_{R1}$ and $_2p_{L1}$. To start a second time step this same set of information must be updated and this discussion elucidates the connection between the new values of X_{p1} and X_{q2}, i.e., two variables, for the inter-mesh boundary between mesh 1 and 2 and the updated values required for $_1p_{R1}$ and $_1p_{L1}$ and $_2p_R$ and $_2p_L$, i.e., four unknown values, at the same physical position.

If the gas properties and reference gas state are identical in meshes 1 and 2 then the solution is trivial, as follows:

$$_1X_{L1} = {_2X_{q2}} \quad \text{and} \quad {_2X_R} = {_1X_{p1}} \qquad (2.18.37)$$

If, as is almost inevitable, the gas properties and reference gas state are not identical in meshes 1 and 2, then a temperature discontinuity exists at the inter-mesh boundary and the solution is conducted exactly as set out in Sec. 2.5. The similarity of the required solution and of the nomenclature is evident by comparison of Fig. 2.22(a) and Fig. 2.5.

(b) tapered pipes

The same approach to this computation situation within the GPB model exists for two of the four unknowns, i.e., $_1p_{L1}$ and $_2p_R$, at the inter-mesh boundary in the tapered section of any pipe. In this situation the known quantities are the values of $_1p_{R1}$ and $_2p_L$, as:

$$_1p_{R1} = p_0 X_{p1}^{G7} \qquad (2.18.38)$$

and

$$_2p_L = p_0 X_{q2}^{G7} \qquad (2.18.39)$$

The similarity of the sketch in Fig. 2.22(b) and Fig. 2.15 indicates that the fundamental theory for this solution is given in Sec. 2.15 with the base theory as given in Secs. 2.10 and 2.11. As related there, the basis of the method employed is to recognize whether the particle flow is diffusing or contracting. If it is contracting then the flow is considered to be isentropic and the thermodynamics of the solution is given in Sec. 2.11 for subsonic particle flow and in Sec. 2.11.1 for sonic flow. The Benson "constant pressure" criterion gives an excellent initial guess for the values of the unknown quantities and considerably reduces the number of iterations required for the application of the more complete gas-dynamic theory in Sec. 2.11. If the particle flow is recognized to be diffusing then the flow is considered to be non-isentropic and the solution is given in Sec. 2.10 for subsonic particle flow and in Sec. 2.10.1 for sonic flow.

As with contracting flow, the Benson "constant pressure" criterion gives an excellent initial guess for the values of the three remaining unknown quantities in diffusing flow. Remember that as the flow is non-isentropic, the third unknown quantity is the reference temperature, incorporating the entropy gain, on the downstream side of the expansion. For diffusing flow, special attention should be paid to the information on flow separation given in Sec. 2.15.1 where the pipe may be steeply tapered or the particle Mach number is high.

2.18.8 Wave reflections at the ends of a pipe after a time step

During computer modeling, the duct of an engine is meshed in distance terms and referred to within the computer program as a pipe which may have a combination of constant and gradually varying area sections between discontinuities. Thus, a tapered pipe is not a discontinuity in those terms, but a sudden change of section is treated as such. So too is the end of a pipe at an engine port or valve, or at a branch. This meshing nomenclature is shown in Fig. 2.22(c), with a restricted pipe employed as the example.

At first glance, the "joint" between mesh 1 and mesh 2 in Fig. 2.22(c), with the change of diameter from d_1 to d_2 from mesh 1 to mesh 2, appears to be no different from the previous case of tapered pipes discussed in Sec. 2.18.7. However, the fact that there is an orifice of diameter d_t between the two mesh sections, and that the basic theory for a restricted pipe is presented in a separate section, Sec. 2.12, prevents this situation from being discussed as just another inter-mesh boundary problem. It is in effect a pipe discontinuity.

All engine configurations being modeled contain pipes, plenums, cylinders and are fed air from, and exhaust gas to, the atmosphere. The atmosphere is nothing but another form of plenum. A cylinder during the open cycle is nothing but another form of plenum in which one of the walls moves and holes, of varying size and shape, open and close as a function of time. It is the pipes that connect these discontinuities, i.e., cylinders, plenums and the atmosphere, together. However, pipes are connected to other pipes, either at branches or at the type of junction such as shown in Fig. 2.22(c). This latter case of the restricted pipe cannot be treated as an inter-mesh problem as it is simply easier, from the viewpoint of the overall organization of the logic and structure of the computer software, to consider it as a junction at the ends of two separate pipes.

The point has already been made in Sec. 2.18.7 that each mesh space is an "island" of information and that the results of the pipe analyses in Secs. 2.18.1 to 2.18.6 lead only to the determination of the new values of left and right moving pressure waves at the left- and right-hand boundaries of the mesh space, i.e., p_p and p_q. In Sec. 2.18.7, the analysis focused on an inter-mesh boundary and the acquisition of the remaining two unknown values for the pressure waves at each boundary of every mesh, except for that at the left-hand end of the left-hand mesh of any pipe, and also for the right-hand end of the right-hand mesh of any pipe.

For example, imagine in Fig. 2.22(a) that mesh 1 is at the extreme left-hand end of the pipe. The value of X_{q1} automatically becomes the required value of X_L for mesh 1. To proceed with the next time step of the calculation, the value of X_R is required. What then is the value of X_R? The answer is that it will depend on what the left-hand end of the pipe is connected to, i.e., a cylinder, a plenum, etc. Equally, imagine in Fig. 2.22(a) that mesh 2 is at the extreme right-hand end of the pipe. The value of X_{p1} automatically becomes the required

Chapter 2 - Gas Flow through Two-Stroke Engines

value of X_{R1} for mesh 1. To proceed with the next time step of the calculation, the value of X_{L1} is required. What then is the value of X_{L1}? The answer, as before, is that it will depend on what the right-hand end of the pipe is connected to, i.e., a branch, a restricted pipe, the atmosphere, etc.

The GPB modeling method stores information regarding the "hand" of the two end meshes in every pipe, and must be able to index the name of that "hand" at its connection to its own discontinuity, so that the numerical result of the computation of boundary conditions at the ends of every pipe is placed in the appropriate storage location within the computer.

Consider each boundary condition in turn, a restricted pipe, a cylinder, plenum or atmosphere, and a branch.

(a) a restricted pipe, as sketched in Fig. 2.22(c)
Imagine, just as it is sketched, that mesh 1 is at the right-hand end of pipe 1 and that mesh 2 is at the left-hand end of pipe 2. Thence:

$$_1X_{R1} = X_{p1} \quad \text{and} \quad _2X_L = X_{q2} \qquad (2.18.40)$$

The unknown quantities are the values of the reflected pressure waves, $_1p_{L1}$ and $_2p_R$, represented here by their pressure amplitude ratios, namely $_1X_{L1}$ and $_2X_R$. The analytical solution for these, and also for the entropy gain on the downstream side of whichever direction the particle flow takes, is to be found in Sec. 2.12. In terms of the notation in Fig. 2.8, which accompanies the text of Sec. 2.12, the value referred to as X_{p1} is clearly the incident wave X_{i1}, and the value of X_{q2} is obviously the incident pressure wave X_{i2}.

(b) a cylinder, plenum, or the atmosphere as sketched in Figs. 2.16 and 2.18
A cylinder, plenum, or the atmosphere is considered to be a large "box," sufficiently large to consider the particle velocity within it to be effectively zero. For the rest of this section, a cylinder, plenum, or the atmosphere will be referred to as a "box." In the theory given in Secs. 2.16 and 2.17, the entire analysis is based on knowing the physical geometry at any instant and, depending on whether the flow is inflow or outflow, either the thermodynamic state conditions of the pipe or the "box" are known values.

As previously in this section, the "hand" of the incident wave at the mesh in the pipe section adjacent to the cylinder is an important element of the modeling process. Imagine that mesh 1, as shown in Fig. 2.22(a), is that mesh at the left-hand end of the pipe and attached to the cylinder exactly as it is sketched in either Fig. 2.16 or 2.18. In which case at the end of the time step in computation, to implement the theory given in Secs. 2.16 and 2.17, the following is the nomenclature interconnection for that to occur:

$$_1X_L = X_{q1} \qquad (2.18.41)$$

and

$$p_{i2} = p_{01} X_{q1}^{G7} \qquad (2.18.42)$$

The properties of the gas subscripted as 2 in Fig. 2.16 or 2.18 are the properties of the gas in the mesh 1 used here as the illustration.

At the conclusion of the computation for boundary conditions at the "box," using the theory of Secs. 2.16 and 2.17, the amplitude of the reflected pressure wave, p_{r2}, is provided. So too for outflow is the entropy gain in the form of the reference temperature, T_{02}. In nomenclature terms, this reflected pressure wave is the "missing" pressure wave value at the left-hand of mesh 1 to permit the computation to proceed to the next time step.

$$_1X_R = \left(\frac{p_{r2}}{p_0}\right)^{G17} \tag{2.18.43}$$

One important final point must be made in this section regarding the rather remote possibility of encountering supersonic particle velocity in any of the pipe systems during the superposition process at any mesh point. By the end of this section you have been brought to the point where the amplitudes of all of the left and right moving pressure waves have been established at both ends of all meshes within the ducting of the engine. At this juncture it is necessary to search each of these mesh positions for the potential occurrence of supersonic particle velocity in the manner described completely in Sec. 2.2.4 and, if found, apply the corrective action of a weak shock to give the necessary subsonic particle velocity (Secs. 2.15-2.17). Unsteady gas flow does not permit supersonic particle velocity. It is self-evident that the particles cannot move faster than the disturbance which is giving them the signal to move.

2.18.9 Mass and energy transport along the duct during a time step

In the real flow situation the energy and mass transport is conducted at the molecular level. Computational facilities are not large enough to accomplish this, so it is carried out in a mesh grid spacing of some 10 to 25 mm, a size commonly used in automotive engines. The size of mesh length is deduced by making the simple assumption that the calculation time step, dt, should translate to an advance of about 1° or 2° crank angle for any engine design. Within each mesh space, as shown in Fig. 2.20, the properties of the gas contained are assumed to be known at the start of a time step. During the subsequent time step, dt, as a result of the wave transmission, particles and energy are going to be transported from mesh space to mesh space, and heat transfer is going to occur by internal means, such as friction or a catalyst, or by external means through the walls of the duct.

To determine the effect of all of these mechanisms, the First Law of Thermodynamics is employed, and the situation is sketched in Figs. 2.23 and 2.24. Here the energy transport across the boundaries for the mesh space J will have unique values depending on the particle directions of this transport. This is illustrated in Fig. 2.23, where four different cases are shown to be possible.

The four cases will be described below. All pressures, particle velocities and densities are, by definition, for the superposition situation at the physical location at the appropriate end of the mesh J. The left-hand end of any mesh is denoted as the "in" side and the right-hand end as the "out" side for flow of mass and energy and as a sign convention when apply-

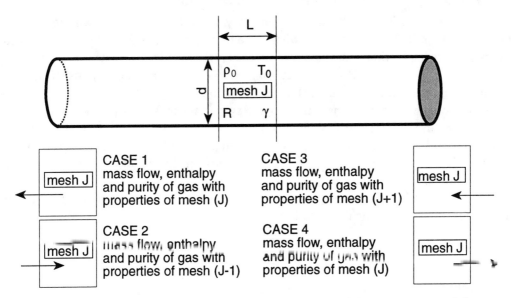

Fig. 2.23 Mass, energy and gas species transport at mesh J during time interval dt.

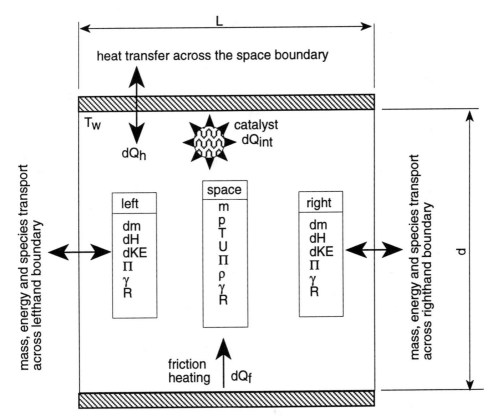

Fig. 2.24 The mesh J which encounters mass and energy transfer during time dt.

ing the First Law of Thermodynamics. Manifestly, as with case 2, there will be situations where the gas is flowing outward at what is nominally the "in" side of the mesh but the arithmetic sign convention of the pressure wave analysis takes care of that problem automatically. Remember that the GPB computer simulation must interrogate each mesh boundary and apply with precision the result of that interrogation as cases 1 to 4.

At the commencement of the time step, the system for mesh J has a known mass, pressure, temperature, and volume. The notation describes the "before" and "after" situation during the time step, dt, for the mass, pressure, and temperature by a "b" and "a" prefix for the system properties. This gives a symbolism for mesh J for the "before" conditions of mass, $_bm_J$, pressure, $_bP_J$, and temperature, $_bT_J$.

As each mesh has a constant flow area, A_J, volume, V_J, and a stagnation temperature, $_bT_J$, the four cases can be analyzed as follows:

(i) case 1

pressure $\quad\quad\quad\quad\quad\quad _JX_{in} = {}_JX_R + {}_JX_L - 1$

particle velocity $\quad\quad\quad _Jc_{in} = {}_JG_5 \, {}_Ja_0({}_JX_R - {}_JX_L)$

density $\quad\quad\quad\quad\quad\quad _J\rho_{in} = {}_J\rho_0 \, {}_JX_S^{{}_JG_5}$

specific enthalpy $\quad\quad\quad _Jdh_{in} = {}_JC_P \, {}_bT_J + \dfrac{{}_Jc_{in}^2}{2}$

mass flow increment $\quad\quad _Jdm_{in} = {}_J\rho_{in} \, A_J \, {}_Jc_{in} \, dt$

enthalpy increment $\quad\quad _JdH_{in} = {}_Jdh_{in} \, {}_Jdm_{in}$

air flow increment $\quad\quad\quad _Jd\Pi_{in} = {}_J\Pi \, {}_Jdm_{in}$

(ii) case 2

pressure $\quad\quad\quad\quad\quad\quad _JX_{in} = {}_JX_R + {}_JX_L - 1$

particle velocity $\quad\quad\quad _Jc_{in} = {}_{J-1}G_5 \, {}_{J-1}a_0({}_JX_R - {}_JX_L)$

density $\quad\quad\quad\quad\quad\quad _J\rho_{in} = {}_{J-1}\rho_0 \, {}_JX_S^{{}_{J-1}G_5}$

specific enthalpy $\quad\quad\quad _Jdh_{in} = {}_{J-1}C_P \, {}_bT_{J-1} + \dfrac{{}_Jc_{in}^2}{2}$

mass flow increment		$_Jdm_{in} = {_J}\rho_{in}\, A_J\, {_J}c_{in}\, dt$
enthalpy increment		$_JdH_{in} = {_J}dh_{in}\, {_J}dm_{in}$
air flow increment		$_Jd\Pi_{in} = {_{J-1}}\Pi\, {_J}dm_{in}$

(iii) case 3

pressure		$_JX_{out} = {_J}X_{R1} + {_J}X_{L1} - 1$
particle velocity		$_Jc_{out} = {_{J+1}}G_5\, {_{J+1}}a_0({_J}X_{R1} - {_J}X_{L1})$
density		$_J\rho_{out} = {_{J+1}}\rho_0\, {_J}X_{out}^{{_{J+1}}G_5}$
specific enthalpy		$_Jdh_{out} = {_{J+1}}C_P\, {_b}T_{J+1} + \dfrac{{_J}c_{out}^2}{2}$
mass flow increment		$_Jdm_{out} = {_J}\rho_{out}\, A_J\, {_J}c_{out}\, dt$
enthalpy increment		$_JdH_{out} = {_J}dh_{out}\, {_J}dm_{out}$
air flow increment		$_Jd\Pi_{out} = {_{J+1}}\Pi\, {_J}dm_{out}$

(iv) case 4

pressure		$_JX_{out} = {_J}X_{R1} + {_J}X_{L1} - 1$
particle velocity		$_Jc_{out} = {_J}G_5\, {_J}a_0({_J}X_{R1} - {_J}X_{L1})$
density		$_J\rho_{out} = {_J}\rho_0\, {_J}X_{out}^{{_J}G_5}$
specific enthalpy		$_Jdh_{out} = {_J}C_P\, {_b}T_J + \dfrac{{_J}c_{out}^2}{2}$
mass flow increment		$_Jdm_{out} = {_J}\rho_{out}\, A_J\, {_J}c_{out}\, dt$
enthalpy increment		$_JdH_{out} = {_J}dh_{out}\, {_J}dm_{out}$
air flow increment		$_Jd\Pi_{out} = {_J}\Pi\, {_J}dm_{out}$

Design and Simulation of Two-Stroke Engines

For the end meshes the required information for cases 1 to 4 is deduced from the boundary conditions of the flow which have been applied appropriately at the left- or right-hand edge of a mesh space. This means that the "hand" of the flow has to be taken into account when transferring the information from the solution emanating from the particular boundary condition for inflow or outflow with respect to the mesh space as distinct from "inflow" or "outflow" from a cylinder, plenum, restricted pipe or branch. However, the logic of the sign convention is quite straightforward in practice.

The acquisition of the numerical information regarding heating effects due to friction and heat transfer in any time step, δQ_f and δQ_h, have already been dealt with in Sec. 2.18.6. Should a catalyst or some similar internal heating device be present within the mesh space and in operation, during the time increment dt, then it is assumed that the quantity of heat which emanates from it, δQ_{int}, can be determined and used as numerical input to the analysis below. It should be pointed out that catalysts are a very common component within exhaust systems at this point in history as a means of reducing exhaust emissions. However, a cooling device could be employed instead to give this internal heat transfer effect, and water injection would be one such example. In both examples, the chemical composition of the gas will also change and this requires a further extension to the analysis given below.

During the time step, and from the continuity equation, we derive the new system mass, $_am_J$:

$$_am_J = {_bm_J} + {_Jdm_{in}} - {_Jdm_{out}} \tag{2.18.44}$$

The First Law of Thermodynamics for the system which is the mesh space is:

heat transfer + energy in = change of system state + energy out + work done

$$\left(\delta Q_{int} + \delta Q_f + \delta Q_h\right)_J + {_JdH_{in}} = dU_J + {_JdH_{out}} + P_J dV_J \tag{2.18.45}$$

The work term is clearly zero. All of the terms except that for the change of system state are already known through the theory given above in this section. Expansion of this unknown term reveals:

$$dU_J = \left[{_am_J}\left({_au_J} + \frac{{_ac_J^2}}{2}\right)\right]_J - \left[{_bm_J}\left({_bu_J} + \frac{{_bc_J^2}}{2}\right)\right]_J \tag{2.18.46}$$

As the velocities at either end of a mesh are almost identical, the difference between the kinetic energy terms is insignificant, so they can be neglected. This reduces Eq. 2.18.46 to:

$$dU_J = {_JC_V}\left({_am_J}\,{_aT_J} - {_bm_J}\,{_bT_J}\right) \tag{2.18.47}$$

This can be solved directly for the system temperature, $_aT_J$, after the time step.

The gas properties in the mesh space will have changed due to the mass transport across its boundaries. As with the case of mass and energy transport, direction is a vital consideration and the four cases presented in Fig. 2.23 are applicable to the discussion. For almost all engine calculations the gases within it are either exhaust gas or air. This situation will be debated here, as it is normality, but the more general case of a multiplicity of gases being present throughout the system can be handled with equal simplicity. After all, air and exhaust gas are composed of a multiplicity of gases.

Consider a mixture of air and exhaust gas with a purity, Π, defined as:

$$\Pi = \frac{\text{mass of air}}{\text{total mass}} \qquad (2.18.48)$$

Hence the reasoning for the inclusion of the air flow increment in the four cases of mass and energy transport presented above and in Fig. 2.23.

The new purity, $_a\Pi_J$, in mesh space J is found simply as follows:

$$_a\Pi_J = \frac{_bm_J \; _b\Pi_J + _Jd\Pi_{in} - _Jd\Pi_{out}}{_am_J} \qquad (2.18.49)$$

If the gas properties of air and exhaust gas are denoted by their respective gas constants and specific heat ratios as R_{air} and R_{exh} and γ_{air} and γ_{exh}, then the new properties of the gas in the mesh space after the time step are:

gas constant
$$_aR_J = {_a\Pi_J} \, R_{air} + (1 - {_a\Pi_J}) \, R_{exh} \qquad (2.18.50)$$

specific heats ratio
$$_a\gamma_J = \frac{_aC_{P_J}}{_aC_{V_J}} = \frac{_a\Pi_J C_{P_{air}} + (1 - {_a\Pi_J}) C_{P_{exh}}}{_aC_{P_J} - {_aR_J}} \qquad (2.18.51)$$

It should be noted that all of the gas properties employed in the theory must be indexed for their numerical values based on the gas composition and temperature at every step in time and at every location, using the theoretical approach given in Sec. 2.1.6.

From this point it is possible, using the theory given in Sec. 2.18.3, to establish the remaining properties in space J, in particular the reference acoustic velocity, density and temperature. The average superposition pressure amplitude ratio, $_aX_J$, in the duct is derived using Eq. 2.18.1 with the updated values of the pressure amplitude ratios at either end of the mesh space. The connection for the new reference temperature, $_aT_0$, is given by Eq. 2.18.3 as:

$$_aT_0 = \frac{_aT_J}{_aX_J^2} \qquad (2.18.52)$$

Consequently the other reference conditions are:

reference acoustic velocity $\quad _a a_0 = \sqrt{_a\gamma_J \, _a R_J \, _a T_0}$ (2.18.53)

reference density $\quad _a\rho_0 = \dfrac{p_0}{_a R_J \, _a T_0}$ (2.18.54)

All of the properties of the gas at the conclusion of a time step, dt, have now been established at all of the mesh boundaries and in all of the mesh volumes.

2.18.10 The thermodynamics of cylinders and plenums during a time step

During a time step in calculation the pipes of the engine being simulated are connected to cylinders and plenums. For simplicity, as in Sec. 2.18.8, an engine cylinder or a plenum will be referred to occasionally and collectively as a "box." A plenum is in reality a cylinder of constant volume in calculation terms. During the time step, due to mass flow entering or leaving the box, the state conditions of the box will change. For example, during exhaust outflow from an engine cylinder, the pressure and the temperature fall and its mass is reduced. In the GPB simulation method proceeding in small time steps of 1° or 2° of crankshaft angle at some rotational speed, the situation is treated as quasi-steady flow for that period of time. To proceed to the next time step of the simulation the new state conditions in all cylinders and plenums must be determined. It should be recalled that the whole point of the simulation process is to predict the mass and state conditions of the gas in the cylinder at the conclusion of the open cycle, and as influenced by the pressure wave action in the ducting, so that a closed cycle computation may provide the requisite data of power, torque, fuel consumption or emissions.

The application of the boundary conditions given in Secs. 2.16 and 2.17 provides all of the information on mass, energy and air flow at the mesh boundaries adjacent to the cylinder or plenum. Fig. 2.25 shows the cylinder with state conditions of pressure, P_C, temperature, T_C, volume, V_C, mass, m_C, etc., at the commencement of a time step. The gas properties are defined by the purity, Π_C, the data value of which leads directly to the gas constant, R_C, and specific heats ratio, γ_C, in the manner shown by Eqs. 2.18.50 and 2.18.51 and from the application of the theory in Sec. 2.1.6 regarding gas properties.

It will be noted that the box has "inflow" and "outflow" apertures, which implies that there are only two apertures. The values of mass flow increment during the time step, shown as either dm_I and dm_E, are in fact the combined total of all of those ports or valves designated as being "inflow" or "outflow." The same reasoning applies to the energy and air flow terms, dH_I and dP_I, or dH_E and dP_E; they are the combined totals of the energy and air flow terms of the perhaps several intake or exhaust ducts at a cylinder. All of these terms are the direct equivalents of, indeed those at the boundary edges of a pipe mesh adjacent to a box are identical to, those quoted as cases 1 to 4 in Sec. 2.18.9.

Note that the direction arrow at either "inflow" or "outflow" in Fig. 2.25 is bidirectional. In short, simply because a port is designated as inflow does not mean that the flow is always

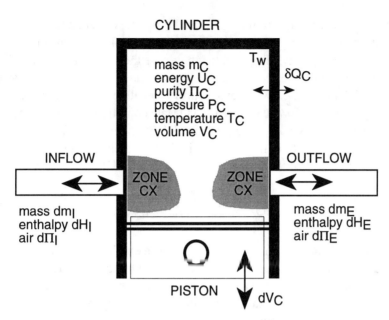

Fig. 2.25 The thermodynamics of open cycle flow through a cylinder.

toward the cylinder. All ports and valves experience backflow at some point during the open cycle. The words "inflow" and "outflow" are, in the case of an engine cylinder, a convenient method of denoting ports and valves whose nominal job is to supply air into the cylinder of an engine. When dealing with an intake plenum, or an exhaust silencer box, "inflow" and "outflow" become a directionality denoted at the whim of the modeler; but having made a decision on the matter, the ensuing sign convention must be adhered to rigidly. The sign convention, common in engineering thermodynamics, is that inflow is "positive" and that outflow is also "positive"; this sign convention is employed not only in the theory below, but throughout this text. Backflow, by definition, is opposite to that which is decreed as positive and is then a negative quantity. The mesh computation must then reorient in sign terms the numerical values for the right- and left-hand ends of pipes during inflow and outflow boundary calculations as appropriate to those junctions defined as "inflow" or "outflow" at the cylinders or plenums. While the computer software logic for this is trivial, care must be taken not to confuse the thermodynamic needs of the mesh spaces defined in Sec. 2.18.9 with those of the cylinder or plenum being examined here.

Heat transfer is defined as positive for heat added to a system and work out is also a positive action. Employing this convention, the First Law of Thermodynamics reads as:

heat transfer + energy in = change of system state + energy out + work done

The entire computation at this stage makes the assumption that the previous application of the boundary conditions, using the theory of Secs. 2.16 and 2.17, has produced the correct values of the terms dm_I, dH_I, dP_I, dm_E, dH_E and dP_E. To reinforce an important point which has been made before, during the application of the boundary conditions in the case of a cylinder where the outflow is stratified, the local properties of zone CX, which are pressure, P_{CX}, temperature, T_{CX}, purity, Π_{CX}, etc., are those that replace the mean values of pressure, P_{CX}, temperature, T_{CX}, purity, Π_{CX}, etc. In which case the solution of the continuity equation and the First Law of Thermodynamics is as accurate and as straightforward as it was in Sec. 2.18.9 in Eqs. 2.18.44 to 2.18.49. I provide further debate on this topic in Ref. [2.32].

The subscript notation for the properties and state conditions within the box after the time step, dt, is C1, i.e., the new values are pressure, P_{C1}, temperature, T_{C1}, purity, Π_{C1}, etc.

The heat transfer, δQ_C, to or from the box in the time step is given by the local convection heat transfer coefficient, C_h, the total surface area, A_C, and the average wall temperature of the box, T_w.

$$\delta Q_C = C_h A_C (T_w - T_C) dt \tag{2.18.55}$$

The value of the heat transfer coefficient, C_h, to be employed during the open cycle is the subject of much research, of which the work by Annand [2.58] is noteworthy. The approach by Annand is recommended for the acquisition of heat transfer coefficients for both the open and the closed cycle within the engine cylinder. For all engines, it should be noted that at some period during the closed cycle an allowance must be made for the cooling of the cylinder charge due to the vaporization of the fuel. For spark-ignition engines it is normal to permit this to happen linearly from the trapping point to the onset of ignition. For compression-ignition units it is conventional to consider that this occurs simultaneously with each packet of fuel being burned during a computational time step. The work of Woschni [2.60] has also provided significant contributions to this thermodynamic field.

The continuity equation for the process during the time step, dt, is given by:

$$m_{C1} = m_C + dm_I - dm_E \tag{2.18.56}$$

The First Law of Thermodynamics for the cylinder or plenum system is:

$$\delta Q_C + dH_I = dU_C + dH_E + \frac{P_C + P_{C1}}{2} dV_C \tag{2.18.57}$$

The work term is clearly zero for a plenum of constant volume. All of the terms except that for the change of system state, dU_C, and cylinder pressure, P_{C1}, are already known through the theory given above in this section. Expansion of one unknown term reveals:

$$dU_C = m_{C1} u_{C1} - m_C u_C \tag{2.18.58}$$

This reduces to:

$$dU_C = m_{C1} C_{V_{C1}} T_{C1} - m_C C_{V_C} T_C \tag{2.18.59}$$

where the value of the specific heat at constant volume is that appropriate to the properties of the gas within the cylinder at the beginning and end of the time step:

$$C_{V_C} = \frac{R_C}{\gamma_C - 1} \quad \text{and} \quad C_{V_{C1}} = \frac{R_{C1}}{\gamma_{C1} - 1}$$

Normally, as with the debate on the gas constant, R, below, the values of C_V and γ should be those at the beginning and end of the time step. However, little inaccuracy ensues, as does considerable algebraic simplification, by taking the known values, C_{VC}, γ_C and R_C, at the commencement of the time step and assuming that they persist for the duration of that time step. The exception to this is during a combustion process where the temperature changes during a given time step are so extreme that the gas properties must be indexed correctly using the theory of Sec. 2.1.6 and, if necessary, an iteration undertaken for several steps to acquire sufficient accuracy.

At the conclusion of the time step the cylinder volume is V_{C1} caused by the piston movement and this is:

$$V_{C1} = V_C + dV_C \tag{2.18.60}$$

As the mass of the cylinder, m_{C1}, is given by Eq. 2.18.56 and the new cylinder pressure and temperature are related by the state equation:

$$P_{C1}V_{C1} = m_{C1}R_C T_{C1} \tag{2.18.61}$$

Eqs. 2.18.56 to 2.18.61 may be combined to produce a direct solution for T_{C1} for a cylinder or plenum as:

$$T_{C1} = \frac{\delta Q_C + dH_I - dH_E + m_C C_V T_C - \dfrac{P_C dV_C}{2}}{(m_C + dm_I - dm_E)\left(C_V + \dfrac{R_C dV_C}{2V_C}\right)} \tag{2.18.62}$$

This can be solved directly for the system temperature, T_{C1}, after the time step, and with dV_C as zero in the event that any plenum or cylinder has no volume change. The cylinder pressure is found from Eq. 2.18.61.

The gas properties in the box will have changed due to the mass transport across its boundaries. For almost all engine calculations the gases within the box are either exhaust gas or air. This situation will be debated here, as it is normality, but the more general case of a multiplicity of gases being present throughout the system can be handled with equal simplicity. After all, air and exhaust gas are composed of a multiplicity of gases. This argument, with the same words, is precisely that mounted in the previous section Sec. 2.18.9 for flow through the mesh spaces.

Design and Simulation of Two-Stroke Engines

The new purity, Π_{C1}, in the box is found simply as follows:

$$\Pi_{C1} = \frac{m_C \Pi_C + d\Pi_I - d\Pi_E}{m_{C1}} \qquad (2.18.63)$$

The new properties of the gas in the box after the time step are:

gas constant $\qquad R_{C1} = \Pi_{C1} R_{air} + (1 - \Pi_{C1}) R_{exh} \qquad (2.18.64)$

specific heats ratio $\quad \gamma_{C1} = \dfrac{C_{P_{C1}}}{C_{V_{C1}}} = \dfrac{\Pi_{C1} C_{P_{air}} + (1 - \Pi_{C1}) C_{P_{exh}}}{C_{P_{C1}} - R_{C1}} \qquad (2.18.65)$

From this point it is possible, using the theory given in Secs. 2.18.3 and 2.1.6, to establish the remaining properties in the box, in particular the reference acoustic velocity, density and temperature. The connection for the new reference temperature, T_0, is given by the isentropic relationship between pressure and temperature as:

$$\frac{T_{C1}}{T_0} = \left(\frac{p_{C1}}{p_0}\right)^{\frac{\gamma_{C1}}{\gamma_{C1}-1}} \qquad (2.18.66)$$

Consequently the other reference conditions to be employed at the commencement of the next time step are:

reference acoustic velocity $\qquad a_0 = \sqrt{\gamma_{C1} R_{C1} T_0} \qquad (2.18.67)$

reference density $\qquad \rho_0 = \dfrac{p_0}{R_{C1} T_0} \qquad (2.18.68)$

All of the properties of the gas at the conclusion of a time step, dt, have now been established at all of the mesh boundaries, in all of the mesh volumes and in all cylinders and plenums of the engine being modeled. It becomes possible to proceed to the next time step and continue with the GPB simulation method, replacing all of the "old" values with the "new" ones acquired during the progress described in entirety in this Sec. 2.18. A new time step may now commence with all data stores refreshed with the updated numerical information.

Information on the effect of the results of the modeling process of the open cycle and in all of the ducts may need to be collected. This is discussed in the next section.

2.18.11 Air flow, work, and heat transfer during the modeling process

The modeling of an engine, or the modeling of any device, that inhales and exhales in an unsteady fashion, is oriented toward the determination of the effect of that unsteady process

Chapter 2 - Gas Flow through Two-Stroke Engines

on many facets of the operation. For example, the engine designer will wish to know the totality of the air flow into the engine as well as the quantification of it with respect to time or crankshaft angle, i.e., to determine the engine *delivery ratio* as well as the extent, or lack, of backflow at certain periods of crankshaft rotation which may deteriorate the overall value. Thus, during a calculation, summations of quantities will be made by the modeler to aid the design process. To illustrate this, the example of delivery ratio will be used in the first instance.

Air flow into an engine

The air flow into an engine is the summation of all of the increments of air flow at each time step at any point in the intake tract. Any mesh point can be selected for this purpose, as the net mass of air flow should be identical at every mesh point over a long time period such as a complete engine cycle. Consider the general case first. A parameter B is required to be assessed for its mean value $\overline{B}$ over a period of time t at a particular location J. The time period starts at t_1 and ends at t_2. Equally well, for an engine running at engine speed N this can be carried out over, and is more informative during, specific periods of crankshaft angle θ ranging from θ_1 to θ_2. This is effected by:

$$\overline{B} = \frac{\sum_{t=t_1}^{t=t_2} B\,dt}{\sum_{t=t_1}^{t=t_2} dt} = \frac{\sum_{\theta=\theta_1}^{\theta=\theta_2} B\,d\theta}{\sum_{\theta=\theta_1}^{\theta=\theta_2} d\theta} \qquad (2.18.69)$$

The relationship between crankshaft angle, in degree units, and time for an engine is:

$$\theta = \frac{60}{360N} = \frac{1}{6N} \qquad (2.18.70)$$

For the specific case of air flow into the engine, the modeler will select a mesh J for the assessment point, and the shrewd modeler will select mesh J as being that value beside the intake valves or ports of the engine. The total mass of air flow, m_a, passing this point will then be:

$$m_a = \frac{\sum_{\theta=0}^{\theta=720} \dot{m}_J \Pi_J\,d\theta}{\sum_{\theta=0}^{\theta=720} d\theta} \qquad (2.18.71)$$

The crankshaft period selected will be noted as 720°. This would be correct for a four-stroke cycle engine for that is its total cyclic period. On the other hand, for a two-stroke cycle engine the cyclic period is 360° and the modeler would perform the summation appropriately. If the engine were a multi-cylinder unit of n cylinders then the accumulation process would be repeated at all of the intake ports of the unit. To transfer that data to a delivery ratio, DR, value the conventional theory is used where ρ_{ref} is the reference density for the particular industry standard employed and V_{SV} is the swept volume of all of the cylinders of the engine:

$$\text{Delivery Ratio} \qquad DR = \frac{\sum_{cylinder=1}^{cylinder=n} m_a}{\rho_{ref} V_{SV}} \qquad (2.18.72)$$

The term for cyclic air flow rate for a two-stroke engine is called *scavenge ratio*, SR, where the reference volume term is the entire cylinder volume. This is given by:

$$\text{Scavenge Ratio} \qquad SR = \frac{\sum_{cylinder=1}^{cylinder=n} m_a}{\rho_{ref}(V_{SV} + V_{CV})} \qquad (2.18.73)$$

The above example is for one of the many such terms required for design assessment during modeling. The procedure is identical for the others. Some more will be given here as further examples.

Work done during an engine cycle

The work output for an engine is that caused by the cylinder pressure, p_C, acting on the piston(s) of the power unit. The in-cylinder work is known as the indicated work and is often reduced to a pseudo-dimensionless value called *mean effective pressure*; in this case it is the *indicated mean effective pressure*. Let us assume that there are n cylinders each with an identical bore area, A, but a total swept volume, V_{SV}. The piston movement in any one cylinder at each time step is a variable, dx, as is the time step, dt, and the volume change, dV.

The work done, δW, at each time step in each cylinder is,

$$\delta W = p_C A dx = p_C dV \qquad (2.18.74)$$

The total work done, W_C, for that cylinder over a cycle is given by the summation of all such terms for that period. Remember that the thermodynamic cycle period is θ_C, and for a four-stroke engine is 720° and for a two-stroke engine is 360°. Thus this accumulated work done is,

$$W_C = \frac{\sum_{\theta=0}^{\theta=\theta_C} \delta W d\theta}{\sum_{\theta=0}^{\theta=\theta_C} d\theta} \qquad (2.18.75)$$

The total work done by the engine, W_E, over a cycle is given by,

$$W_E = \sum_{cylinder=1}^{cylinder=n} W_C \qquad (2.18.76)$$

The indicated power output of the engine, which is calculated by the simulation, is the rate of execution of this work at N rpm and is:

four-stroke engine $\qquad \dot{W}_I = W_E \dfrac{N}{120} \qquad (2.18.77)$

two-stroke engine $\qquad \dot{W}_I = W_E \dfrac{N}{60} \qquad (2.18.78)$

The indicated mean effective pressure, imep, for the engine is that pressure which would act on the pistons throughout the thermodynamic cycle and produce the same work output, thus:

$$imep = \frac{W_E}{V_{SV}} \qquad (2.18.79)$$

The brake values, i.e., those which would be measured on a "brake" or dynamometer, are any or all of the above values of indicated performance multiplied by the mechanical efficiency, η_m.

Other work and heat-transfer-related terms, such as pumping mean effective pressure during the intake stroke and also during the exhaust stroke of a four-stroke engine, or pumping mean effective pressure for the induction into the crankcase of a two-stroke engine, or heat loss during the open cycle of any engine, can be determined throughout the GPB modeling process, either cycle by cycle or cumulatively over many cycles, in exactly the same fashion as for imep. Similarly, the designer may wish to know, and relate to measured terms, the mean pressures and temperatures at significant locations throughout the engine; such mean values are found using the methodology given above.

2.18.12 The modeling of engines using the GPB finite system method

In the literature there has been reported correlation of measurement and this theory in several international conferences and meetings. The engines modeled and compared with experiments include two-stroke and four-stroke power units. One reference in particular [2.32] describes the latest developments in the inclusion within the GPB modeling process of the scavenging of a two-stroke engine. The present method employed for the closed cycle period of the modeling process is as I describe [2.25, Chapter 4] using a rate of heat release approach for the combustion period. The references list these publications [2.31, 2.32, 2.33, 2.34, 2.35, 2.40 and 2.41].

2.19 The correlation of the GPB finite system simulation with experiments

A theoretical simulation process in design engineering which has not been checked for accuracy against relevant experiments is, depending on the purposes for which it is required, at best potentially misleading and at worst potentially dangerous. In the technology allied to the simulation of unsteady gas flow, many experiments have been carried out by the researchers involved. Virtually every reference in the literature cited here carries evidence, relevant or irrelevant, of experimentation designed to test the validity of the theories presented by their author(s).

I am closely associated with a new series of experiments, reported by Kirkpatrick *et al.* [2.41, 2.65, 2.66], designed specifically to test the validity of the theories of unsteady gas flow, and in particular those presented here, and to compare and contrast the GPB finite system simulation with both the experiments and with other simulation methods such as Riemann characteristics, Lax-Wendroff, Harten-Lax-Leer, etc. The experimental apparatus is quite unique and is detailed fully by Kirkpatrick [2.41]. It will be described here sufficiently well so that the presentation of the experimental test results may be understood fully. Although the main purpose is to determine the extent of the accuracy of the GPB simulation method, the test method and the experimental results illustrate also many of the contentions in the theory presented above.

2.19.1 The QUB SP (single pulse) unsteady gas flow experimental apparatus

Most experimenters and modelers in unsteady gas dynamics have correlated the measured pressure-time diagrams in the ducts of engines, firing or motored, against their theoretical contentions. As all unsteady gas flow within the ducts of engines is in a state of superposition, this makes the process of correlation very difficult indeed. It is almost impossible to tell which wave is traveling in which direction. While fast-response pressure transducers are still the best experimental tool that the theoretician in this subject possesses, the simple truth is that they are totally directionally insensitive. That much is manifestly clear in just about every numerical example quoted up to this point in the text. Worse, the correlation of mass flow is in an even more parlous state when working with engines, either motored or firing. While the cylinder pressure may be recorded accurately, the density record in the same place is non-existent since a temperature, or purity, or density transducer with a sufficiently fast response has yet to be invented. Thus, while the experimenter may infer the mass of trapped charge, or as a matter of even greater necessity the mass of trapped air charge, in the cylinder from the overall engine air consumption and the cylinder pressure transducer record, the blunt truth is that he is "whistling in the wind." For those who are my contemporaries in this technology, let them be assured that I readily admit to being as guilty of perfidy in my technical publications as they are. It was this "guilt complex" which led to the design of the QUB SP apparatus. The nomenclature "SP" stands for "single pulse."

The criteria for the design of the apparatus are straightforward:

(i) Base it on the assumption that a fast-response pressure transducer is the only accurate experimental tool readily available.

(ii) The pipe(s) attached to the device must be sufficiently long as to permit visibility of a pressure wave traveling left or right without undergoing superposition in the plane of the transducer while recording some particular phenomenon of interest.

(iii) The cylinder of the device must be capable of having the mass and purity of its contents recorded with absolute accuracy.

(iv) The cylinder of the device must be capable of containing any gas desired, and at a wide variety of state conditions, prior to the commencement of the experiment which could be the simulation of either an exhaust or an induction process.

(v) The pipes attached to the cylinder must be capable of containing any gas desired, over a wide variety of state conditions, prior to the commencement of the experiment which could be the simulation of either an exhaust or an induction process.

(vi) The pipes attached to the device must be capable of holding any of the discontinuities known in engine technology, i.e., diffusers, cones, bends, branches, sudden expansions and contractions and restrictions, throttles, carburetors, catalysts, silencers, air filters, poppet valves, etc.

These design criteria translate into the single pulse device shown in Fig. 2.26. The cylinder is a rigid cast iron container of 912 cm^3 volume and can contain gas up to a pressure of 10 bar and a temperature of 500°C. The gas can be heated electrically (H). The cylinder has valves, V, which permit the charging of gas into the cylinder at sub-atmospheric or supra-atmospheric pressure conditions. The port, P, at the cylinder has a 25-mm-diameter hole that mates with a 25-mm-diameter aluminum exhaust pipe. The valve mechanism, S, is a flat polished nickel-steel plate with a 25-mm-diameter hole that mates perfectly with the port and pipe at maximum opening. It is actuated by a pneumatic impact cylinder and its movement is recorded by an attached 2-mm pitch comb sensed by an infrared source and integral photodetector. Upon impact the valve slider opens the port from a "perfect" sealing of the cylinder, gas flows from (or into in an induction process) the cylinder and seals it again upon the conclusion of its passing. A damper, D, decelerates the valve to rest after the port has already been closed. An exhaust pulse generated in this manner is typical in time and amplitude of, say, a two-stroke engine at 3000 rpm. A typical port opening lasts for 0.008 second.

The cylinder gas properties of purity, pressure and temperature are known at commencement and upon conclusion of an event; in the case of purity, by chemical analysis through a valve, V, if necessary. The pressure is recorded by a fast-response pressure transducer. The temperature is known at commencement and at conclusion, without concern about the time response of that transducer, so that the absolute determination of cylinder mass can be conducted accurately.

The coefficients of discharge, C_d, of the cylinder port, under wide variations of cylinder-to-pipe pressure ratio giving rise to inflow or outflow and valve opening as port-pipe area

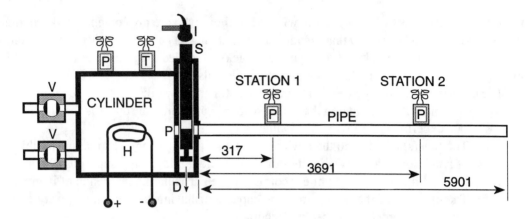

Fig. 2.26 QUB SP apparatus with a straight pipe attached.

ratio, are determined under steady flow conditions with air. They are presented in Fig. 2.27 and are a classic picture of the variation of C_d with respect to such parameters. Their measurement is necessary for the accurate correlation of experiment with theory, as would be the case for any engine simulation to be accurate, as inspection of Eq. 2.16.4 will indicate. In this context, the discussion in Appendix A2.3 requires careful study, for the traditional methods employed for the reduction of the measured data for the coefficients of discharge, C_d, have been determined to be inadequate for use in conjunction with a theoretical engine simulation.

It is acknowledged that the initial design and development work on the QUB SP rig was carried out by R.K. McMullan and finalized by S.J. Kirkpatrick. All of the pressure diagrams and discharge coefficient data presented in Figs. 2.27-2.53 were recorded by S.J. Kirkpatrick, and those in Figs. 2.55-2.57 were measured by D.O. Mackey. These individuals carried out this work as part of their research work for a Ph.D. at The Queen's University of Belfast.

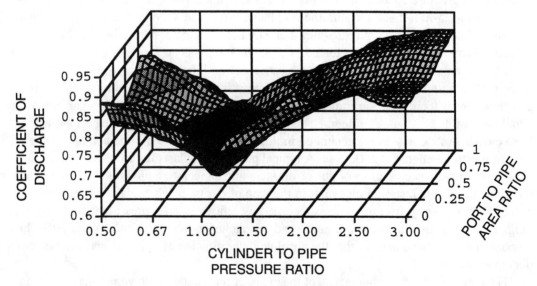

Fig. 2.27 Coefficient of discharge for the QUB SP (single pulse) apparatus.

2.19.2 A straight parallel pipe attached to the QUB SP apparatus

The experiments simulate both exhaust and intake processes with a straight pipe attached to the cylinder. Fig. 2.26 shows a straight aluminum pipe of 5.9 m and 25-mm internal diameter attached to the port and ending at the atmosphere. The pipe has a relatively smooth bore typical of the quality of ducting common in an engine design. There are pressure transducers attached to the pipe at stations 1 and 2 at the length locations indicated.

(i) the outflow process producing compression pressure waves in the exhaust pipe

The cylinder was filled with air and the initial cylinder pressure and temperature were 1.5 bar and 293 K. In Figs. 2.28 to 2.30 are the pressure-time records in the cylinder and at stations 1 and 2. The result of the computations using the GPB modeling method are shown in the same figures. The results are so close that it is sometimes difficult to distinguish between theory and experiment, but where differences can be discerned they are indicated on the figure.

The correlation of measured and calculated mass is virtually exact. The criterion used is the ratio of the final to the initial cylinder mass, and in this case both the calculated and the measured values were 0.866. As can be seen from the figures, no wave action impinged on the

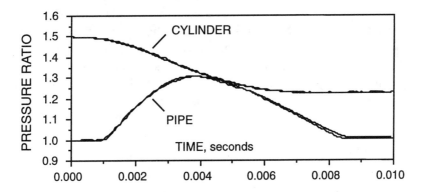

Fig. 2.28 Measured and calculated cylinder and pipe pressures.

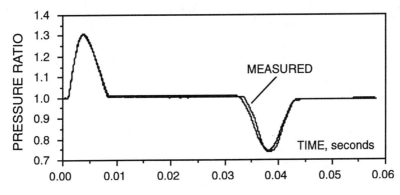

Fig. 2.29 Measured and calculated pressures at station 1.

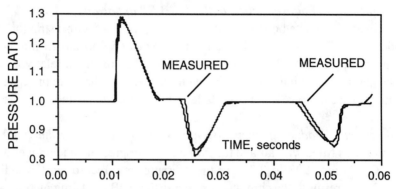

Fig. 2.30 Measured and calculated pressures at station 2.

port to influence the cylinder emptying process, so this is more of a tribute to the accuracy of the cylinder to pipe boundary conditions of Sec. 2.16, rather than any comment on the GPB finite system modeling of the wave action in the pipe system!

The main event is the creation of the exhaust pulse as a wave of compression, the peak of which is passing station 1 at 0.00384 second. It reflects at the open end as an expansion wave and the peak returns to station 1 at 0.0382 second. The basic theory of the motion of finite amplitude waves is discussed in Sec. 2.1.4 and of reflections of compression waves at a plain open end in Sec. 2.8.1. As the reference acoustic velocity in air at 20°C is 343 m/s, which is estimated to be the average propagation velocity on the simplistic grounds that the expansion wave travels as much below sonic velocity as the compression wave is supersonic above it, the expected return point in time by a very simple computation is:

$$\text{time } t = 0.00384 + \frac{2(5901 - 317)}{1000 \times 343} = 0.0364 \text{ second}$$

This approximate calculation can be seen to give an answer which is close to reality.

In Figs. 2.28 to 2.30 the individual waves can be seen clearly. The creation of the exhaust pulse is accurate as can be seen in Figs. 2.28 and 2.29. Later it passes station 2 after 0.01 second. Steep-fronting of the compression wave has occurred as discussed in Sec. 2.1.5. In Fig. 2.30 the steep-fronted compression wave reflects at the plain open end as an expansion wave. The profile has changed little by the time it passes station 2 going leftward, but upon returning to station 2 it has steepened at the rear of the expansion wave, again as discussed in Sec. 2.1.5. It will be seen that a shock did not develop on the wave at any time even though the long pipe runs provided the waves with the space and time to so comply.

Although the size and scale of the diagrams make it difficult to observe, nevertheless it can be seen that the pressure at 0.01 second in Fig. 2.28 and at 0.02 second in Fig. 2.29 is slightly above atmospheric. It can be seen that the pressure at 0.04 second in Fig. 2.30 is slightly below atmospheric. These arise due to the main compression wave, and the expansion wave reflection of it, sending their continual reflections due to friction in the opposite direction to their propagation. As this is an important topic, a separate graph is presented in

Fig. 2.58(a) which expands the time scale for the time period between 0.007 and 0.01 second. The close correlation between the measured and calculated pressures, and the residual waves attributable to the friction reflections after 0.0085 second, is now observable by zooming in on this time period. As for friction causing deterioration of the peak of the pressure wave, as it passed station 1 it had a measured peak pressure ratio of 1.308 (calculated at 1.307) and when it passed station 2 it had reduced to 1.290 (calculated at 1.273). The theoretical discussion on this subject is contained in Sec. 2.3.

It can be seen that the GPB modeling method provides a very accurate pressure-time history of the recorded events for the motion of compression waves and of the exhaust process from a cylinder.

(ii) the inflow process producing expansion pressure waves in the intake pipe

The cylinder was filled with air and the initial cylinder pressure and temperature were 0.8 bar and 293 K. In Figs. 2.31 to 2.33 are the pressure-time records in the cylinder and at stations 1 and 2. The result of the computations using the GPB modeling method are shown in the same figures. The results are sufficiently close to warrant the description of being good, but where differences can be discerned they are indicated in the figure.

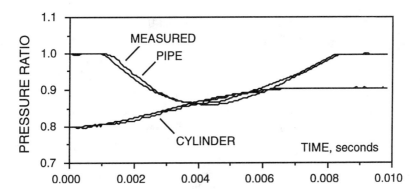

Fig. 2.31 *Measured and calculated cylinder and pipe pressures.*

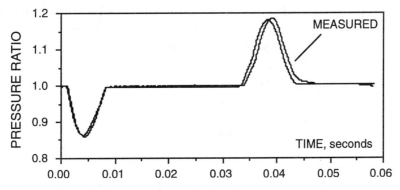

Fig. 2.32 *Measured and calculated pressures at station 1.*

Design and Simulation of Two-Stroke Engines

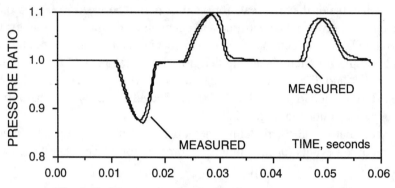

Fig. 2.33 Measured and calculated pressures at station 2.

The correlation of measured and calculated mass is also good. The criterion used is the ratio of the final to the initial cylinder mass. The calculated and the measured values are 1.308 and 1.311, respectively, which is an error of 0.2%. As with the exhaust process in (i) above, no wave action other than friction reflections impinged on the port to influence the cylinder emptying process.

The main event is the creation of the intake pulse as a wave of expansion, the peak of which is passing station 1 at 0.004 second. It reflects at the open end as a compression wave and the peak returns to station 1 at 0.0383 second. The basic theory of the motion of finite amplitude waves is discussed in Sec. 2.1.4 and of reflections of expansion waves at a plain open end in Sec. 2.8.3. As the reference acoustic velocity in air at 20°C is 343 m/s, which is estimated to be the average propagation velocity on the simplistic grounds that the expansion wave travels as much below sonic velocity as the compression wave is supersonic above it, the expected return point in time by a very simple computation is:

$$\text{time } t = 0.004 + \frac{2(5901 - 317)}{1000 \times 343} = 0.0366 \text{ second}$$

This approximate calculation can be seen to give an answer which is similar to the experimental value.

In Figs. 2.31 to 2.33 the individual waves can be seen clearly. The creation of the induction pulse is accurate as can be seen in Figs. 2.31 and 2.32. Later it passes station 2 after 0.01 second. Steepening of the tail of the expansion wave has occurred as discussed in Sec. 2.1.5. It will be seen that a shock did not develop on the wave at any time.

The effect of friction is presented in Fig. 2.58(b) which expands the time scale for the time period between 0.007 and 0.01 second. The close correlation between the measured and calculated pressures, and the residual waves attributable to the friction reflections after 0.0085 second causing the sub-atmospheric pressure on the traces, is now observable by zooming in on this time period. As for friction causing deterioration of the peak of the pressure wave, as it passed station 1 it had a measured peak pressure ratio of 0.864 (calculated at 0.859) and when it passed station 2 it had increased to 0.871 (calculated at 0.876). The theoretical discussion on this subject is contained in Sec. 2.3.

Chapter 2 - Gas Flow through Two-Stroke Engines

It can be seen that the GPB modeling method provides a very accurate pressure-time history of the recorded events for the motion of expansion waves and of the induction process into a cylinder. However, it was the "less than perfect" theoretical correlation with this experiment which led to the study on discharge coefficients [5.25].

2.19.3 A sudden expansion attached to the QUB SP apparatus

The experiment simulates an exhaust process with a straight pipe incorporating a sudden expansion attached to the cylinder. Fig. 2.34 shows a straight aluminum pipe of 3.394 m and 25-mm internal diameter attached to the port, followed by a 2.655 m length of 80-mm-diameter pipe ending at the atmosphere. There are pressure transducers attached to the pipe at stations 1, 2 and 3 at the length locations indicated. The basic theory of pressure wave reflections at sudden expansions and contractions is given in Secs. 2.10 and 2.12.

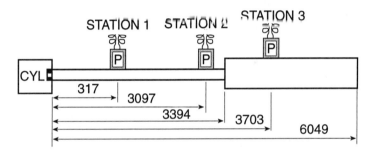

Fig. 2.34 QUB SP rig with a sudden expansion attached.

The cylinder was filled with air and the initial cylinder pressure and temperature were 1.5 bar and 293 K. In Figs. 2.35 to 2.37 are the measured pressure-time records in the cylinder and at stations 1, 2 and 3. The results of the computations using the GPB modeling method are shown in the same figures, and the correlation is very good. Differences that can be discerned between measurement and computation are indicated on the figure.

In Fig. 2.35 the basic action of reflection at a sudden expansion is observed. The exhaust pulse passes the pressure transducer at station 1 at 0.004 second and the sudden expansion in

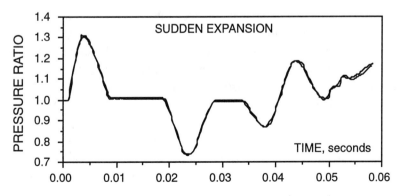

Fig. 2.35 Measured and calculated pressures at station 1.

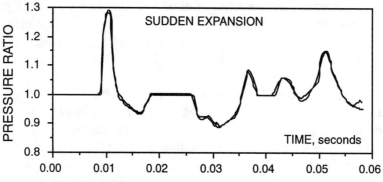

Fig. 2.36 Measured and calculated pressures at station 2.

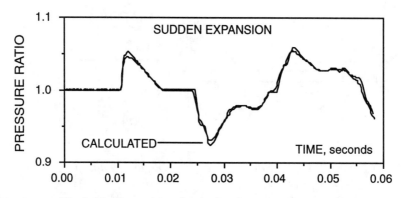

Fig. 2.37 Measured and calculated pressures at station 3.

pipe area sends an expansion wave reflection back to arrive at 0.024 second. It is echoed off the closed end, is reflected again at the sudden expansion, but this time as a compression wave which returns to station 1 at 0.044 second preceded by an expansion wave at 0.038 second. From whence comes this expansion wave?

In Fig. 2.37, the pressure record at station 3 at 0.012 second shows the onward transmission of the residue of the original exhaust pulse having traversed the sudden expansion in pipe area. In short the original exhaust pulse commenced its journey with a pressure ratio of 1.3 which is reduced to 1.05 in the 80-mm pipe when passing station 2. This onward transmitted wave reflects off the open end as an expansion wave, returns to the sudden area change which it now sees as a sudden contraction. The onward transmission leftward of that reflection process is seen to proceed past the transducer at station 2 in Fig. 2.36 at 0.03 second and arrive at station 1 at 0.038 second, thereby answering the query posed in the previous paragraph.

In Fig. 2.36 the pressure transducer is sufficiently close to the sudden change of pipe area that all pressure records are of the superposition process. This is evident between 0.01 and 0.02 second, where the rightward propagating exhaust pulse is partially superposed on the leftward travel of its own reflection from the sudden expansion in pipe area. These measured superposition processes are followed closely in amplitude and phase by the theory.

Chapter 2 - Gas Flow through Two-Stroke Engines

It can be observed that the GPB finite system modeling by computer provides a very accurate pressure-time history of the recorded events for the reflection of compression pressure waves at sudden increases in pipe area. There is also evidence, as the second-order reflections are a mixture of expansion and compression waves which see the sudden area change as both an expansion and a contraction in pipe area, that the theory is generally applicable to all such boundary conditions.

2.19.4 A sudden contraction attached to the QUB SP apparatus

The experiment simulates an exhaust process with a straight pipe incorporating a sudden contraction attached to the cylinder. Fig. 2.38 shows a straight aluminum pipe of 108 mm and 25-mm internal diameter attached to the port, followed by a 2.346 m length of 80-mm-diameter pipe; the final length to the open end to the atmosphere is a 2.21 m length of pipe of 25-mm internal diameter. There are pressure transducers attached to the pipe at stations 1, 2 and 3 at the length locations indicated. The basic theory of pressure wave reflections at sudden expansions and contractions is given in Secs. 2.10 and 2.12.

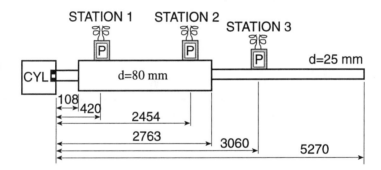

Fig. 2.38 QUB SP rig with a sudden contraction attached.

The cylinder was filled with air and the initial cylinder pressure and temperature were 1.5 bar and 293 K. In Figs. 2.39 to 2.41 are the measured pressure-time records in the cylinder and at stations 1, 2 and 3. The results of the computations using the GPB modeling method are shown on the same figures, and the correlation is very good. Where differences can be observed between measurement and computation are indicated on the figure.

The short 108 mm length means that the pressure wave encounters a sudden expansion before proceeding to the point of conduction of this particular test, i.e., the sudden contraction posed to the exhaust pressure wave. In the previous section this was found to be accurate and so it is no surprise to see that the measured and computed pressure waves at station 1 are in close correlation at 0.05 second in Fig. 2.39. The amplitude of the exhaust pressure wave passing station 1 has a pressure ratio of 1.065.

The sudden contraction sends a compression wave reflection back from it and can be seen in a superposition condition at station 1 at 0.018 second and at station 2 earlier at 0.011 second. The clean and non-superposed onward transmission of the exhaust pulse can be found passing station 3 at 0.012 second with a pressure ratio of 1.12. This is a higher amplitude than

Design and Simulation of Two-Stroke Engines

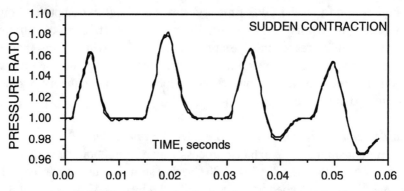

Fig. 2.39 *Measured and calculated pressures at station 1.*

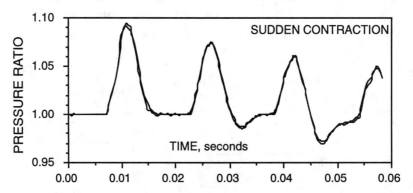

Fig. 2.40 *Measured and calculated pressures at station 2.*

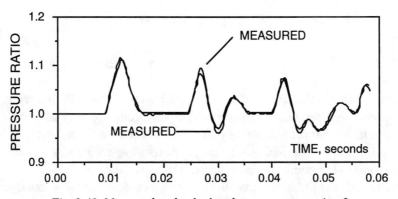

Fig. 2.41 *Measured and calculated pressures at station 3.*

the original exhaust pulse of 1.06 atm, but the wave proceeding to the outlet has been contracted from a 80-mm-diameter pipe into a 25-mm tail-pipe.

It can be observed that the GPB finite system modeling gives a very accurate representation of the measured events for the reflection of compression pressure waves at sudden decreases in pipe area.

2.19.5 A divergent tapered pipe attached to the QUB SP apparatus

The experiment simulates an exhaust process with a straight pipe incorporating a divergent taper attached to the cylinder. Fig. 2.42 shows a straight aluminum pipe of 3.406 m and 25-mm internal diameter attached to the port, followed by a 195 mm length of steeply tapered pipe at 12.8° included angle to 68-mm diameter; the final length to the open end to the atmosphere is a 2.667 m length of pipe of 68-mm internal diameter. This form of steep taper is very commonly found within the ducting of IC engines, including the diffuser sections of highly tuned two-stroke engines. There are pressure transducers attached to the pipe at stations 1, 2 and 3 at the length locations indicated. The basic theory of pressure wave reflections in tapered pipes is given in Sec. 2.15.

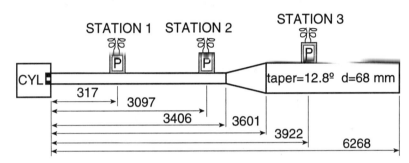

Fig. 2.42 QUB SP rig with a divergent taper attached.

The cylinder was filled with air and the initial cylinder pressure and temperature were 1.5 bar and 293 K. In Figs. 2.43 to 2.45 are the measured pressure-time records in the cylinder and at stations 1, 2 and 3. The results of the computations using the GPB modeling method are shown on the same figures, and the correlation is very good. Differences that can be observed between measurement and computation are indicated on the figure.

The tapered pipe acts as an expansion to the area of the pipe system and sends an expansion wave reflection back from it. It is observed arriving back at station 1 at 0.024 second in Fig. 2.43, albeit in a superposition condition from the nearby closed end.

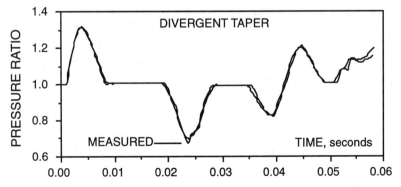

Fig. 2.43 Measured and calculated pressures at station 1.

Design and Simulation of Two-Stroke Engines

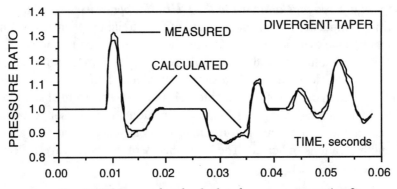

Fig. 2.44 Measured and calculated pressures at station 2.

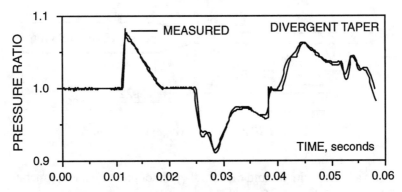

Fig. 2.45 Measured and calculated pressures at station 3.

Actually, as it is a short, steeply tapered pipe there is a close resemblance between the behavior of this pipe and the sudden contraction discussed in Sec. 2.19.3 and illustrated in Figs. 2.34 to 2.37. The cylinder release conditions were identical in both cases so the exhaust pulses which arrived at the discontinuity in areas was the same. Consequently there are great similarities between Figs. 2.35 and 2.43 for the pressure transducer at station 1, and between Figs. 2.36 and 2.44 for the pressure transducer at station 2. The records for the tapered pipe look slightly more "peaky" in certain places. However, that effect is seen more strongly in Fig. 2.45 where the onward transmission of the residual of the exhaust pulse is observed to have developed a shock front as it passes the pressure transducer at station 3. In other words, the tapered pipe has contributed to, and exaggerated, the normal distortion process of a compression wave profile. It is also interesting to compare the amplitudes of the transmitted and reflected waves, as the tapered pipe should be more efficient at this than a sudden expansion. The peak of the suction wave from the tapered pipe undergoing superposition at 0.024 second at station 1 is 0.68 atm; for the sudden expansion it is 0.74 atm or less strength of expansion wave reflection. The peak of the transmitted wave from the tapered pipe passing station 3 undisturbed at 0.011 second is 1.08 atm; for the sudden expansion it is 1.05 atm or an exhaust pulse of reduced strength is delivered farther down the pipe system.

Chapter 2 - Gas Flow through Two-Stroke Engines

It can be observed that the GPB finite system modeling gives a very accurate representation of the measured events for the reflection of compression pressure waves at a steeply tapered pipe segment within ducting.

2.19.6 A convergent tapered pipe attached to the QUB SP apparatus

The experiment simulates an exhaust process with a straight pipe incorporating a convergent taper attached to the cylinder. Fig. 2.46 shows a straight aluminum pipe of 108 mm and 25-mm internal diameter attached to the port, followed by a 2.667 m length of 68-mm parallel section pipe, then a steeply tapered pipe at 12.8° included angle convergent to 25-mm diameter over a 195 mm length; the final length to the open end to the atmosphere is a 2.511 m length of pipe of 25-mm internal diameter. This form of steep taper is very commonly found within the ducting of IC engines, including the rear cone sections of expansion chambers of highly tuned two-stroke engines. There are pressure transducers attached to the pipe at stations 1, 2 and 3 at the length locations indicated. The basic theory of pressure wave reflections in tapered pipes is given in Sec. 2.15.

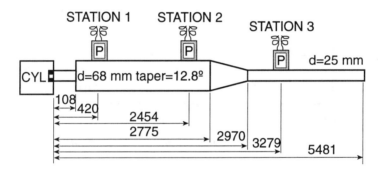

Fig. 2.46 QUB SP rig with a convergent cone attached.

The cylinder was filled with air and the initial cylinder pressure and temperature were 1.5 bar and 293 K. In Figs. 2.47 to 2.49 are the measured pressure-time records in the cylinder and at stations 1, 2 and 3. The result of the computations using the GPB modeling method are shown on the same figures, and the correlation is very good. Where differences can be observed between measurement and computation are indicated on the figure.

The converging tapered pipe acts as a contraction to the area of the pipe system and sends a compression wave reflection back from the compression wave of the exhaust pulse. It is observed arriving back at station 1 at 0.02 second in Fig. 2.47, albeit in a superposition condition from the nearby contraction.

Actually, as it is a short, steeply tapered pipe there is a close resemblance between the behavior of this pipe and the sudden contraction discussed in Sec. 2.19.4 and illustrated in Figs. 2.38 to 2.41. The cylinder release conditions were identical in both cases so the exhaust pulses which arrived at the discontinuity in areas was the same. Consequently there are great similarities between Figs. 2.39 and 2.47 for the pressure transducer at station 1, and between Figs. 2.40 and 2.48 for the pressure transducer at station 2. This close comparison also extends to the pressure transducer at station 3, in the tapered cone version in Fig. 2.49 and in

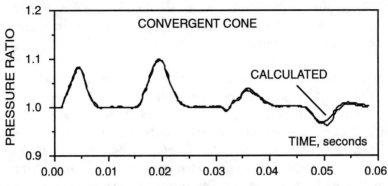

Fig. 2.47 Measured and calculated pressures at station 1.

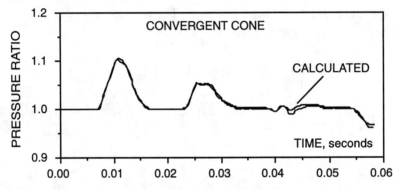

Fig. 2.48 Measured and calculated pressures at station 2.

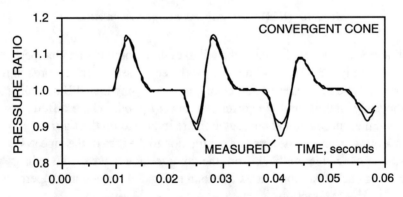

Fig. 2.49 Measured and calculated pressures at station 3.

Fig. 2.41 for the sudden expansion. This was not the case in the previous section, when it was observed that the tapered diffuser pipe had distorted the profile of the exhaust pulse into a shock front as it reached the station 3 pressure transducer. It is germane to this discussion, and it is fully covered in the theoretical analysis in Secs. 2.10, 2.12 and 2.15, that a process in a diffuser is non-isentropic whereas that in a nozzle is isentropic. Shock formation in Fig. 2.45

is clearly a manifestation of non-isentropic behavior while the undistorted profile emanating from the nozzle in Fig. 2.49 is obviously more efficient. Even the profile in Fig. 2.37 of the transmitted exhaust pulse onward from the sudden expansion exhibits elements of steep-front formation.

It is also interesting to compare the amplitudes of the transmitted and reflected waves, as the nozzle should be more efficient at this than a sudden contraction. The peak of the compression wave from the tapered pipe undergoing superposition at 0.02 second at station 1 is 1.105 atm; for the sudden contraction it is weaker at 1.08 atm. The peak of the transmitted wave from the converging tapered pipe passing station 3 undisturbed at 0.012 second is 1.155 atm; for the sudden contraction it is 1.12 atm or an exhaust pulse of reduced strength is delivered onward down the pipe system.

One final word on the above sections, which reiterates many of the theoretical considerations from Secs. 2.10, 2.12 and 2.15, is that area changes, which in discussion are labeled loosely as "cones," "nozzles," "diffusers," etc., are only accurate as nomenclature when taken in the context of the direction of the particle velocity at that instant of the discussion. In short, what is a nozzle for a particle traveling rightward at one instant in time is a diffuser to another traveling leftward in another time frame.

It can be observed that the GPB finite system modeling gives a very accurate representation of the measured events for the reflection of compression pressure waves at a steeply tapered pipe segment within ducting. Taking heed of the warning in the previous paragraph, note that the word nozzle is not mentioned in this last statement.

2.19.7 A longer divergent tapered pipe attached to the QUB SP apparatus

The experiment simulates an exhaust process with a straight pipe attached to the cylinder and incorporating a long divergent taper attached at the end of that straight pipe. Fig. 2.50 shows a straight aluminum pipe of 3.417 m and 25-mm internal diameter attached to the port, followed by a 600 mm length of tapered pipe at 8° included angle to 109-mm diameter at the open end to the atmosphere. This form of steep taper is very commonly found within the ducting of IC engines, including the diffuser outlet sections of highly tuned four-stroke engines. They are often referred to as "megaphones" for obvious reasons. There is a pressure transducer attached to the pipe at the length location indicated. The basic theory of pressure wave reflections in tapered pipes is presented in Sec. 2.15.

The cylinder was filled with air and the initial cylinder pressure and temperature were 2.0 bar and 293 K. In Figs. 2.51 to 2.53 are the (same) measured pressure-time record at the pressure transducer location. The megaphone acts as an expansion to the area of the pipe system and sends a strong expansion wave reflection back from it. This is chased back along the pipe system by the reflection at the atmosphere of the residue of the exhaust pulse which makes it to the open end. It returns as an expansion wave, but now strengthening as it nozzles its way back to the smaller area of the parallel pipe leading back to the cylinder. It is observed arriving back at the pressure transducer at 0.021 second in Figs. 2.51 to 2.53 in a virtually undisturbed condition apart from residual friction waves. The original exhaust pulse has a pressure ratio of 1.55 and the suction reflection has a peak of 0.63 atm. The suction reflection has a shock wave at its tail, a feature already observed regarding the action of tapered pipes in Sec. 2.19.5. It will be seen from the discussion below that the superposition of oppositely

Design and Simulation of Two-Stroke Engines

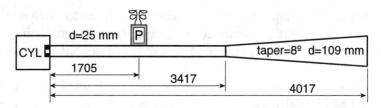

Fig. 2.50 QUB SP rig with a megaphone exhaust attached.

moving expansion and compression waves gives very high particle velocities within the pipe and diffuser system, in excess of a Mach number of 0.7. Consider the strong suction wave evident in the figures at 0.02 second; if it arrives at the exhaust valves or ports of an engine, particularly a four-stroke engine during the long valve overlap period of a high-performance unit, it can induce a considerable flow of air through the combustion chamber, removing the residue of the combustion products, and initiate a high volumetric efficiency. The earlier discussion in Sec. 2.8.1 alludes to this potential for the enhanced cyclic charging of an engine with air.

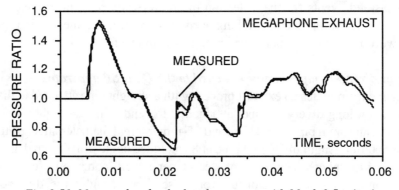

Fig. 2.51 Measured and calculated pressures with Mach 0.5 criterion.

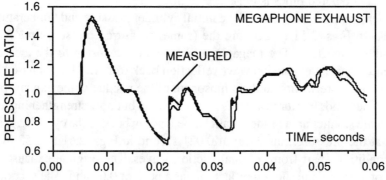

Fig. 2.52 Measured and calculated pressures with Mach 0.6 criterion.

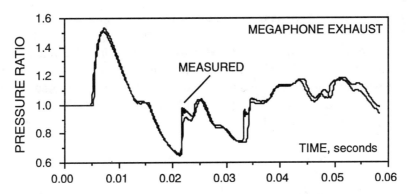

Fig. 2.53 Measured and calculated pressures with Mach 0.7 criterion.

The result of the computations using the GPB modeling method are shown on the same figures with the theoretical criterion for flow separation from the walls of a diffuser, as debated in Sec. 2.15.1, differing in each of the three figures. Recall from the discussion in that section, and by examining the criterion declared in Eq. 2.15.5, that gas particle flow separation from the walls of a diffuser will induce deterioration of the amplitude of the reflection of a compression wave as it traverses a diffuser. The taper of 8° included angle employed here would be considered in steady gas flow to be sufficiently steep as to give particle flow separation from the walls. The theory used to produce the computations in Figs. 2.51 to 2.53 was programmed to record the gas particle velocity at every mesh within the diffuser section. The Eq. 2.15.3 statement was implemented at every mesh at every time step, except that the computational switch was set at a Mach number of 0.5 when computing the theory shown in Fig. 2.51, at a Mach number of 0.6 for the theory plotted in Fig. 2.52, and at a Mach number of 0.7 for the theory presented in Fig. 2.53. It will be seen that the criterion presented in Eq. 2.15.5 provides the accuracy required. It is also clear that any computational method which cannot accommodate such a fluid mechanic modification of its thermodynamics will inevitably provide considerable inaccuracy. Total reliance on the momentum equation alone for this calculation gives a reflected wave amplitude at the pressure transducer of 0.5 atm. It is also clear that flow separation from the walls occurs only at very high Mach numbers. As the criterion of Eq. 2.15.5 is employed for the creation of the theory in Figs. 2.43 to 2.45, where the taper angle is at 12.8° included, it is a reasonable assumption that wall taper angle in unsteady gas flow is not the most critical factor.

The differences between measurement and computation are indicated on each figure. It can be observed that the GPB finite system modeling gives a very accurate representation of the measured events for the reflection of compression pressure waves at a tapered pipe ending at the atmosphere.

2.19.8 A pipe with a gas discontinuity attached to the QUB SP apparatus

The experiment simulates an exhaust process with a straight pipe attached to the cylinder. Fig. 2.54 shows a straight aluminum pipe of 5.913 m and 25-mm internal diameter attached to the port and ending at a closed end with no exit to the atmosphere. There are pressure transducers attached to the pipe at stations 1, 2 and 3 at the length locations indicated. However, at

3.401 m from the cylinder port there is a sliding valve S with a 25-mm circular aperture which, when retracted, seals off both segments of the exhaust pipe. Prior to the commencement of the experiment, the cylinder was filled with air and the initial cylinder pressure and temperature were 1.5 bar and 293 K. The segment of the pipe betwen the cylinder and the valve S is filled with air at a pressure and temperature of 1.5 bar and 293 K, and the segment between the valve S and the closed end is filled with carbon dioxide at the same state conditions. The experiment is conducted by retracting the valve S to the fully open position and impacting the valve at the cylinder port open at the same instant, i.e., the pneumatic impact cylinder I in Fig. 2.26 opens the cylinder port P and then closes it in the manner described in Sec. 2.19.1. An exhaust pulse is propagated into the air in the first segment of the pipe and encounters the carbon dioxide contained in the second segment before echoing off the closed end. In the quiescent conditions for the instant of time between retracting the valve S and sending a pressure wave to arrive at that position some 0.015 second later, it is not anticipated that either the carbon dioxide or the air will have migrated far from their initial positions, if at all. This is a classic experiment examining the boundary conditions for pressure wave reflections at discontinuities in gas properties in engine ducting as described in Sec. 2.5. The gas properties of carbon dioxide are significantly different from air as the ratio of specific heats, γ, and the gas constant, R, are 1.28 and 189 J/kgK, respectively. The reference densities in the two segments of the pipe in this experiment, and employed in the GPB modeling process, are then given by:

$$\text{reference density of air} \qquad _{\text{air}}\rho_0 = \frac{p_0}{R_{\text{air}}T_0} = \frac{101325}{287 \times 293} = 1.205 \text{ kg/m}^3$$

$$\text{reference density of } CO_2 \qquad _{CO_2}\rho_0 = \frac{p_0}{R_{CO_2}T_0} = \frac{101325}{189 \times 293} = 1.830 \text{ kg/m}^3$$

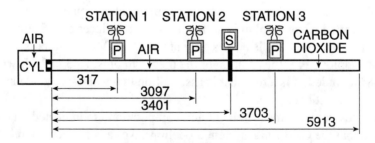

Fig. 2.54 QUB SP rig with pipe containing a gas discontinuity.

Observe that the reference density of carbon dioxide is significantly higher than air by some 52%. The reference acoustic velocities, a_0, which profoundly affect the propagation and the particle velocities, are also very different as:

$$a_0 \text{ in air} \qquad _{\text{air}}a_0 = \sqrt{\gamma_{\text{air}}R_{\text{air}}T_0} = \sqrt{1.4 \times 287 \times 293} = 343.1 \text{ m/s}$$

a_0 in CO_2 $\quad _{CO_2}a_0 = \sqrt{\gamma_{CO_2}R_{CO_2}T_0} = \sqrt{1.28 \times 189 \times 293} = 266.2$ m/s

The reference acoustic velocity for air is some 28.9% higher than that in carbon dioxide.

At equal values of pressure wave amplitude this provides significant alterations to the motion of the pressure wave in each segment of the gas in the pipe. Consider the theory of Sec. 2.1.4 for a compression wave with a pressure ratio of 1.3 at the above state conditions in a 25-mm-diameter pipe, i.e., a pipe area, A, of 0.000491 m².

For air the results for pressure amplitude ratio, X, particle velocity, c, propagation velocity, α, density, ρ, and mass flow rate, $\dot{m}$, would be:

X $\qquad X = P^{G17} = 1.3^{\frac{1}{7}} = 1.0382$

c $\qquad c = G_5 a_0 (X - 1) = 5 \times 343.1 \times 0.0382 = 65.5$ m/s

α $\qquad \alpha = a_0(G_6 X - G_5) = 343.1 \times (6 \times 1.0382 - 5) = 421.7$ m/s

ρ $\qquad \rho = \rho_0 X^{G5} = 1.205 \times 1.0382^5 = 1.453$ kg/m³

$\dot{m}$ $\qquad \dot{m} = \rho A c = 1.453 \times 0.000491 \times 65.5 = 0.0467$ kg/s

For carbon dioxide the results for pressure amplitude ratio, X, particle velocity, c, propagation velocity, α, density, ρ, and mass flow rate, $\dot{m}$, would be:

X $\qquad X = P^{G17} = 1.3^{0.1094} = 1.0291$

c $\qquad c = G_5 a_0 (X - 1) = 7.143 \times 266.2 \times 0.0291 = 55.3$ m/s

α $\qquad \alpha = a_0(G_6 X - G_5) = 266.2 \times (8.143 \times 1.0291 - 7.143) = 329.3$ m/s

ρ $\qquad \rho = \rho_0 X^{G5} = 1.830 \times 1.0291^{7.143} = 2.246$ kg/m³

$\dot{m}$ $\qquad \dot{m} = \rho A c = 2.246 \times 0.000491 \times 55.3 = 0.061$ kg/s

The GPB finite system simulation method has no difficulty in modeling a pipe system to include these considerable disparities in gas properties and the ensuing behavior in terms of wave propagation. The computation is designed to include the mixing and smearing of the gases at the interface between mesh systems which have different properties at every mesh within the computation, and the reflections of the pressure waves at the inter-mesh boundaries as a function of those differing gas properties. The theory for this is presented in Sec. 2.18.

In Figs. 2.55 to 2.57 are the measured pressure-time records in the cylinder and at stations 1, 2 and 3. The result of the computations using the GPB modeling method are shown on the same figures. The pressure traces are so close together that it is difficult to distinguish between theory and experiment, but where differences can be discerned they are indicated on the figure.

The first point of interest to observe is in Fig. 2.55, where there is a "bump" of pressure at 0.023 second. This is the echo of the initial exhaust pulse, having propagated through the air in the first pipe segment off the more dense carbon dioxide, which commences at position S in the second pipe segment, returning to station 1. At that point it is observed as a superposition process bouncing off the closed end at the port.

The second point of interest is the overall accuracy of the computation. The phasing of the pressure waves is very accurate. Considering the disparity of the propagation velocities of waves of equal pressure ratio of 1.3, illustrated above, i.e., some 28% faster in air, the phasing error if the computation had been carried out with air only would have been very considerable. In Fig. 2.55 the return time, peak to peak of the exhaust pulse, passing station 1 to return

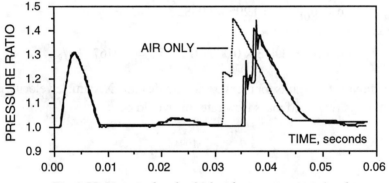

Fig. 2.55 *Measured and calculated pressures at station 1.*

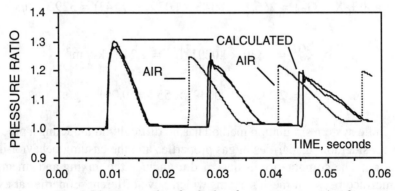

Fig. 2.56 *Measured and calculated pressures at station 2.*

Chapter 2 - Gas Flow through Two-Stroke Engines

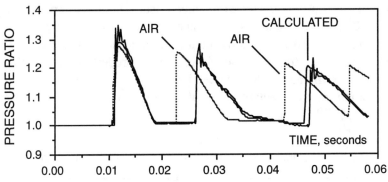

Fig. 2.57 Measured and calculated pressures at station 3.

to station 1, in 0.03 second. Very approximately and ignoring interference due to superposition as in Eqs. 2.2.9 and 2.2.10, in a pipe filled with air only, that return time would have been

$$\text{return time, } t \qquad t = \frac{2(5.913 - 0.317)}{421.7} = 0.0265 \text{ second}$$

As the peak of the exhaust pulse passed station 1 at 0.004 second originally, it would have returned there at 0.0265 + 0.004, or 0.0305 second, and not at the observed value of 0.035 second. The phase error emanating from a simulation process detailing a completely air-filled pipe would have been highly visible in Fig. 2.55.

In fact, in Figs. 2.55 to 2.57 is a third pressure trace where the computation has been conducted with the pipe filled only with air. The traces are marked as "air" or "air only." The phase error deduced very simply above is seen to be remarkably accurate, for in Fig. 2.55 the "air only" returning reflection does arrive at station 1 at 0.031 second. On the other graphs the "air only" computation reveals the serious error that can occur in UGD simulation when the correct properties of the actual gases involved are not included. Needless to add, in Fig. 2.55 there is no sign of the pressure wave "bump" at 0.025 second for there is no CO_2 in that computation!

From the measured and computed pressure traces it can be seen that the waves have steep-fronted and that the computation follows that procedure very accurately. Some phase error is seen by 0.047 second at station 3. This is after some 15.5 meters of pressure wave propagation.

It can be observed that the GPB finite system modeling gives a very accurate representation of the measured events for the reflection of compression pressure waves at gas property variations within ducting.

2.20 Computation time

One of the important issues for any computer code is the speed of its operation. Kirkpatrick et al. [2.41] conclude that the GPB finite system simulation method and the Lax-Wendroff (+Flux Corrected Transport) have equality of computational speed and both are several times faster than the non-homentropic method of characteristics. Of great importance is the inherent ability of the GPB code to automatically take into account the presence of variable gas properties and variable gas species throughout a duct. This is not the case with the Lax-Wendroff (+FCT) code, and the inclusion of variable gas properties and variable gas species within a duct for that computer code has been reported to slow it down by anything from a factor of 1.6 for a minimum acceptable level of accuracy to a factor of 5 to attain complete accuracy [2.62, 2.63].

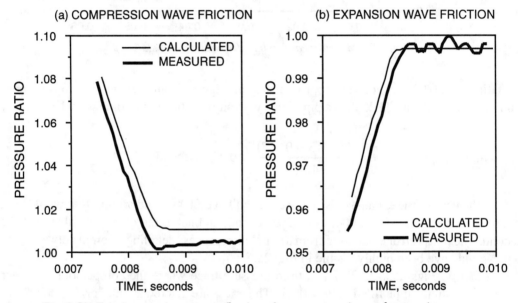

Fig. 2.58 Friction causing wave reflections from compression and expansion waves.

2.21 Concluding remarks

The GPB finite system simulation technique can be used with some confidence regarding its accuracy for the thermodynamics and unsteady gas dynamics of flow from and into engine cylinders, and throughout the ducting attached to the internal-combustion engine.

It is an interesting commentary on the activities of those who fund research in engineering, in this case the Science and Engineering Research Council of the UK, that when they were approached some years ago for the funding of the theoretical and experimental work described in this text, a "committee of peers" rejected the application for such funding on the grounds that it had "all been done before." The last line of the *Second Mulled Toast* should be item 1 on any agenda for meetings of organizations that fund research in engineering.

References for Chapter 2

2.1 S. Earnshaw, "On the Mathematical Theory of Sound," *Phil.Trans.Roy.Soc.*, Vol 150, p133, 1860.

2.2 F.K. Bannister, "Pressure Waves in Gases in Pipes," *Ackroyd Stuart Memorial Lectures*, University of Nottingham, 1958.

2.3 G. Rudinger, <u>Wave Diagrams for Non-Steady Flow in Ducts</u>, Van Nostrand, New York, 1955.

2.4 R.S. Benson, <u>The Thermodynamics and Gas Dynamics of Internal Combustion Engines</u>, Volumes 1 and 2, Clarendon Press, Oxford, 1982.

2.5 F.J. Wallace, M.H. Nassif, "Air Flow in a Naturally Aspirated Two-Stroke Engine," *Proc.I. Mech.E.*, Vol 1B, p343, 1953.

2.6 G.P. Blair, W.L. Cahoon,"A More Complete Analysis of Unsteady Gas Flow Through a High Specific Output Two-Cycle Engine," SAE Paper No.720156, Society of Automotive Engineers, Warrendale, PA, 1972.

2.7 J.H. McConnell, "Unsteady Gas Flow Through Naturally Aspirated Four-Stroke Cycle Internal Combustion Engines," Doctoral Thesis, The Queen's University of Belfast, March, 1974.

2.8 P. De Haller, "The Application of a Graphic Method to some Dynamic Problems in Gases," *Sulzer Tech. Review*, Vol 1, p6, 1945.

2.9 A. Jones, Doctoral Thesis, University of Adelaide, Australia, 1979.

2.10 R.S. Benson, R.D. Garg, D. Woollatt, "A Numerical Solution of Unsteady Flow Problems," *Int.J.Mech.Sci.*, Vol 6, pp117-144, 1964.

2.11 D.R. Hartree, "Some Practical Methods of Using Characteristics in the Calculation of Non-Steady Compressible Flow," US Atomic Energy Commission Report, AECU-2713, 1953.

2.12 M. Chapman, J.M. Novak, R.A. Stein, "A Non-Linear Acoustic Model of Inlet and Exhaust Flow in Multi-Cylinder Internal Combustion Engines," Winter Annual Meeting, ASME, Boston, Mass., 83-WA/DSC-14, November, 1983.

2.13 P.J. Roache, <u>Computational Fluid Dynamics</u>, Hermosa Publishers, USA, 1982.

2.14 R. Courant, K. Friedrichs, H. Lewy, Translation Report NYO-7689, *Math. Ann*, Vol 100, p32, 1928.

2.15 G.P. Blair, J.R. Goulburn, "The Pressure-Time History in the Exhaust System of a High-Speed Reciprocating Internal Combustion Engine," SAE Paper No. 670477, Society of Automotive Engineers, Warrendale, PA, 1967.

2.16 G.P. Blair, M.B. Johnston, "Unsteady Flow Effects in the Exhaust Systems of Naturally Aspirated, Crankcase Compression, Two-Stroke Cycle Engines," SAE Paper No. 680594, Society of Automotive Engineers, Warrendale, PA, 1968.

2.17 G.P. Blair, J.R. Goulburn, "An Unsteady Flow Analysis of Exhaust Systems for Multi-Cylinder Automobile Engines," SAE Paper No. 690469, Society of Automotive Engineers, Warrendale, PA, 1969.

2.18 G.P. Blair, M.B. Johnston, "Simplified Design Criteria for Expansion Chambers for Two-Cycle Gasoline Engines," SAE Paper No. 700123, Society of Automotive Engineers, Warrendale, PA, 1970.

2.19   J.F. Bingham, G.P. Blair, "An Improved Branched Pipe Model for Multi-Cylinder Automotive Engine Calculations," *Proc.I.Mech.E.*, Vol. 199, No. D1, 1985.

2.20   A.J. Blair, G.P. Blair, "Gas Flow Modelling of Valves and Manifolds in Car Engines," *Proc.I.Mech.E.*, International Conference on Computers in Engine Technology, University of Cambridge, April, 1987, C11/87.

2.21   J.S. Richardson, "Investigation of the Pulsejet Engine Cycle," Doctoral Thesis, The Queen's University of Belfast, March, 1981.

2.22   M. Kadenacy, British Patents 431856, 431857, ...,484465, 511366.

2.23   E. Giffen, "Rapid Discharge of Gas from a Vessel into the Atmosphere," *Engineering*, August 16, 1940, p134.

2.24   G.P. Blair, "Unsteady Flow Characteristics of Inward Radial Flow Turbines," Doctoral Thesis, The Queen's University of Belfast, May 1962.

2.25   G.P. Blair, "The Basic Design of Two-Stroke Engines," SAE R-104, Society of Automotive Engineers, Warrendale, PA, 1990, ISBN 1-56091-008-9.

2.26   R. Fleck, R.A.R. Houston, G.P. Blair, "Predicting the Performance Characteristics Of Twin Cylinder Two-Stroke Engines for Outboard Motor Applications," SAE Paper No. 881266, Society of Automotive Engineers, Warrendale, PA, 1988.

2.27   F.A. McGinnity, R. Douglas, G.P. Blair, "Application of an Entropy Analysis to Four-Cycle Engine Simulation," SAE Paper No. 900681, Society of Automotive Engineers, Warrendale, PA, 1990.

2.28   R. Douglas, F.A. McGinnity, G.P. Blair, "A Study of Gas Temperature Effects on the Prediction of Unsteady Flow," Conference on Internal Combustion Engine Research in Universities and Polytechnics, *Proc.I.MechE.*, C433/036, 30-31 January 1991, London, pp47-55.

2.29   M. Chapman, J.M. Novak, R.A. Stein, "Numerical Modelling of Inlet and Exhaust Flows in Multi-Cylinder Internal Combustion Engines," Proc.ASME Winter Meeting, Phoenix, November 1982.

2.30   D.E. Winterbone, "The Application of Gas Dynamics for the Design of Engine Manifolds," IMechE Paper No. CMT8701, 1987.

2.31   G.P. Blair, "An Alternative Method for the Prediction of Unsteady Gas Flow through the Reciprocating Internal Combustion Engine," SAE Paper No. 911850, Society of Automotive Engineers, Warrendale, PA, 1991.

2.32   G.P. Blair, "Correlation of an Alternative Method for the Prediction of Engine Performance Characteristics with Measured Data," SAE Paper No. 930501, Society of Automotive Engineers, Warrendale, PA, 1993.

2.33   G.P. Blair, "Correlation of Measured and Calculated Performance Characteristics of Motorcycle Engines," Funfe Zweiradtagung, Technische Universität, Graz, Austria, 22-23 April 1993, pp5-16.

2.34   G.P. Blair, R.J. Kee, R.G. Kenny, C.E. Carson, "Design and Development Techniques Applied to a Two-Stroke Cycle Automobile Engine," International Conference on Comparisons of Automobile Engines, Verein Deutscher Ingenieuer (VDI-GFT), Dresden, 3-4 June 1993, pp77-103.

2.35 G.P. Blair, S.J. Magee, "Non-Isentropic Analysis of Varying Area Flow in Engine Ducting," SAE Paper No. 932399, Society of Automotive Engineers, Warrendale, PA, 1993.
2.36 R.S. Benson, D. Woollatt, W.A. Woods, "Unsteady Flow in Simple Branch Systems," *Proc.I.Mech.E.*, Vol 178, Pt 3I(iii), No 10, 1963-64.
2.37 B.E.L. Deckker, D.H. Male, "Unsteady Flow in a Branched Duct," *Proc.I.Mech.E.*, Vol 182, Pt 3H, No 10, 1967-68.
2.38 D.S. Miller, Internal Flow Systems, BHRA, Fluid Engineering Series, No.5, ISBN 0 900983 78 7, 1978.
2.39 F.A. McGinnity, "The Effect of Temperature on Engine Gas Dynamics," Ph.D. thesis, The Queen's University of Belfast, December 1989.
2.40 G.P. Blair, "Non-Isentropic Analysis of Branched Flow in Engine Ducting," SAE Paper No. 940395, Society of Automotive Engineers, Warrendale, PA, 1994.
2.41 S.J. Kirkpatrick, G.P. Blair, R. Fleck, R.K. McMullan, "Experimental Evaluation of 1-D Computer Codes for the Simulation of Unsteady Gas Flow Through Engines - A First Phase," SAE Paper No. 941685, Society of Automotive Engineers, Warrendale, PA, 1994.
2.42 P.D. Lax, B. Wendroff, "Systems of Conservation Laws," *Communs. Pure Appl. Math.*, Vol 15, 1960, pp217-237.
2.43 R.D. Richtmyer, K.W. Morton, Difference Methods for Initial Value Problems, 2nd Ed., John Wiley & Sons, Inc., 1967.
2.44 H. Niessner, T. Bulaty, "A Family of Flux-Correction Methods to Avoid Overshoot Occurring With Solutions of Unsteady Flow Problems," Proceedings of GAMM 4th Conf., 1981, pp241-250.
2.45 D.L. Book, J.P. Boris, K. Hain, "Flux-Corrected Transport II: Generalization of the Method," *J. Comp. Phys.*, Vol 18, 1975, pp248-283.
2.46 A. Harten, P.D. Lax, B. Van Leer, "On Upstream Differencing and Gudunov type Schemes for Hyperbolic Conservation Laws," *SIAM Rev.*, Vol 25, 1983, pp35-61.
2.47 C. Chen, A. Veshagh, F.J. Wallace, "A Comparison Between Alternative Methods for Gas Flow and Performance Prediction of Internal Combustion Engines," SAE Paper No. 921734, Society of Automotive Engineers, Warrendale, PA, 1992.
2.48 T. Ikeda, T. Nakagawa, "On the SHASTA FCT Algorithm for the Equation," *Maths. Comput.*, Vol 33, No 148, October 1979, pp1157-1169.
2.49 S.K. Gudunov, "A Difference Scheme for Numerical Computation of Discontinuous Solutions of Equations of Fluid Dynamics," *Mat. Sb.*, Vol 47, 1959, pp271-290.
2.50 S.C. Low, P.C. Baruah, "A Generalised Computer Aided Design Package for an I.C. Engine Manifold System," SAE Paper No. 810498, Society of Automotive Engineers, Warrendale, PA, 1981.
2.51 G.P. Blair, R. Fleck, "The Unsteady Gas Flow Behaviour in a Charge Cooled Rotary Piston Engine," SAE Paper No. 770763, Society of Automotive Engineers, Warrendale, PA, 1977.
2.52 F.K. Bannister, G.F. Mucklow, "Wave Action Following the Release of Compressed Gas from a Cylinder," *Proc. I. Mech. E.*, Vol 159, 1948, p269.

2.53 R.S. Benson, "Influence of Exhaust Belt Design on the Discharge Process in Two-Stroke Engines," Proc. I. Mech. E., Vol 174, No 24, 1960, p713.

2.54 W.A. Woods, "On the Formulation of a Blowdown Pulse in the Exhaust System of a Two-Stroke Cycle Engine," *Int. J. Mech. Sci.*, Vol 4, 1962, p259.

2.55 W.L. Cahoon, "Unsteady Gas Flow Through a Naturally Aspirated Two-Stroke Internal Combustion Engine," Ph.D. Thesis, Dept. Mech. Eng., The Queen's University of Belfast, Nov. 1971.

2.56 F.J. Wallace, R.W. Stuart-Mitchell, "Wave Action following the Sudden Release of Air Through an Engine Port System," Proc. I. Mech. E., Vol 1B, 1953, p343.

2.57 F.J. Wallace, M.H. Nassif, "Air Flow in a Naturally Aspirated Two-Stroke Engine," *Proc.I.Mech.E.*, Vol 168, 1954, p515.

2.58 W.J.D. Annand, "Heat Transfer in the Cylinders of Reciprocating Internal Combustion Engines," *Proc.I.Mech.E.*, Vol 177, 1963, p973.

2.59 G.P. Blair, W.L. Cahoon, C.T. Yohpe, "Design of Exhaust Systems for V-twin Motorcycle Engines," SAE Paper No. 942514, Society of Automotive Engineers, Warrendale, PA, 1994.

2.60 G. Woschni, "A Universally Applicable Equation for Instantaneous Heat Transfer in the Internal Combustion Engine," SAE Paper No. 670931, Society of Automotive Engineers, Warrendale, PA, 1967.

2.61 D.D. Agnew, "What is Limiting Engine Air Flow. Using Normalised Steady Air Flow Bench Data," SAE Paper No. 942477, Society of Automotive Engineers, Warrendale, PA, 1994.

2.62 D.E. Winterbone, R.J. Pearson, "A Solution of the Wave Equation using Real Gases," *Int.J.Mech.Sci.*, Vol 34, No 12, 1992, pp917-932.

2.63 R.J. Pearson, D.E. Winterbone, "Calculating the Effects of Variations in Composition on Wave Propagation in Gases," *Int.J.Mech.Sci.*, Vol 35, No 6, 1993.

2.64 J.F. Bingham, "Unsteady Gas Flow in the Manifolds of Multicylinder Automotive Engines," Ph.D. Thesis, Dept. Mech. Eng., The Queen's University of Belfast, Oct. 1983.

2.65 G.P. Blair, S.J. Kirkpatrick, R. Fleck, "Experimental Evaluation of a 1D Modelling Code for a Pipe Containing Gas of Varying Properties," SAE Paper No. 950275, Society of Automotive Engineers, Warrendale, PA, 1995.

2.66 G.P. Blair, S.J. Kirkpatrick, D.O. Mackey, R. Fleck, "Experimental Evaluation of a 1D Modelling Code for a Pipe System Containing Area Discontinuities," SAE Paper No. 950276, Society of Automotive Engineers, Warrendale, PA, 1995.

Appendix A2.1 The derivation of the particle velocity for unsteady gas flow

This section owes much to the text of Bannister [2.2], which I found to be a vital component of my education during the period 1959-1962. His lecture notes have remained a model of clarity and a model of the manner in which matters theoretical should be written by those who wish to elucidate others.

The exact differential equations employed by Earnshaw [2.1] in his solution of the propagation of a wave of finite amplitude are those established using the notation of Lagrange.

Fig. A2.1(a) shows a frictionless pipe of unit cross-section, containing gas at reference conditions of ρ_0 and p_0. The element AB is of length, dx, and at distance, x, from an origin of time and distance. Fig. A2.1(b) shows the changes that have occurred in the same element AB by time, t, due to the influence of a pressure wave of finite amplitude. The element face, A, has now been displaced to a position, L, farther on from the initial position. Thus, at time t, the distances of A and B from the origin are no longer separated by dx but by a dimension which is a function of that very displacement; this is shown in Fig. A2.1(b). The length of the element is now $\left(1 + \dfrac{\partial L}{\partial x}\right)dx$. The density, ρ, in this element at this instant is related by the fact that the mass in the element is unchanged from its initial existence at the reference conditions and that the pipe area, A, is unity;

$$\rho_0 A dx = \rho A\left(1 + \frac{\partial L}{\partial x}\right)dx \tag{A2.1.0}$$

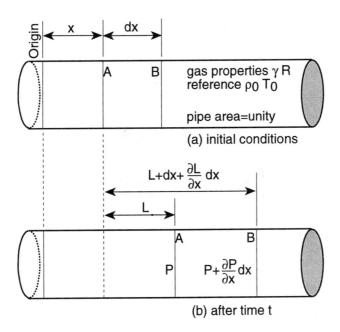

Fig. A2.1 Lagrangian notation for a pressure wave.

Hence,
$$\frac{\rho_0}{\rho} = 1 + \frac{\partial L}{\partial x} \qquad (A2.1.1)$$

Since the process is regarded as isentropic,

$$\frac{p}{p_0} = \left(\frac{\rho}{\rho_0}\right)^\gamma$$

Then,
$$\frac{p}{p_0} = \left(1 + \frac{\partial L}{\partial x}\right)^{-\gamma} \qquad (A2.1.2)$$

Partial differentiaton of this latter expression gives:

$$\frac{\partial p}{\partial x} = -\gamma p_0 \left(1 + \frac{\partial L}{\partial x}\right)^{-\gamma-1} \frac{\partial^2 L}{\partial x^2} \qquad (A2.1.3)$$

The accelerating force applied to the gas element in the duct of unity area, A, at time, t, is:

$$A\left(p - \left(p + \frac{\partial p}{\partial x} dx\right)\right) \quad \text{or} \quad -\frac{\partial p}{\partial x} A dx$$

The mass in the element is $\rho_0 A dx$ from Eq. A2.1.1 and from Newton's Laws where force equals mass times acceleration:

$$-\frac{\partial p}{\partial x} A dx = \rho_0 A dx \frac{\partial^2 L}{\partial t^2} \qquad (A2.1.4)$$

Eliminating the area A and substituting from Eq. A2.1.4 gives:

$$\frac{\gamma p_0}{\rho_0}\left(1 + \frac{\partial L}{\partial x}\right)^{-\gamma-1} \frac{\partial^2 L}{\partial x^2} = \frac{\partial^2 L}{\partial t^2} \qquad (A2.1.5)$$

As the reference acoustic velocity, a_0, can be stated as:

$$a_0 = \sqrt{\gamma R T_0} = \sqrt{\frac{\gamma p_0}{\rho_0}}$$

Chapter 2 - Gas Flow through Two-Stroke Engines

and replacing the distance from the origin as y, i.e., replacing (x+L), then Eq. A2.1.5 becomes:

$$a_0^2 \left(\frac{\partial y}{\partial x}\right)^{-\gamma-1} \frac{\partial^2 y}{\partial x^2} = \frac{\partial^2 y}{\partial t^2} \qquad (A2.1.6)$$

This is the fundamental thermodynamic equation and the remainder of the solution is merely mathematical "juggling" to effect a solution. This is carried out quite normally by making logical substitutions until a solution emerges. Let $\frac{\partial y}{\partial t} = f\left(\frac{\partial y}{\partial x}\right)$, in which case the following are the results of this substitution:

$$\frac{\partial^2 y}{\partial t^2} = f'\left(\frac{\partial y}{\partial x}\right)\frac{\partial}{\partial t}\left(\frac{\partial y}{\partial x}\right) \quad \text{where} \quad f'\left(\frac{\partial y}{\partial x}\right) = \frac{d\left(f\left(\frac{\partial y}{\partial x}\right)\right)}{d\left(\frac{\partial y}{\partial x}\right)}$$

Thus, by transposition:

$$\frac{\partial^2 y}{\partial t^2} = f'\left(\frac{\partial y}{\partial x}\right)\frac{\partial}{\partial x}\left(\frac{\partial y}{\partial t}\right) = f'\left(\frac{\partial y}{\partial x}\right)\frac{\partial}{\partial x}\left(f\left(\frac{\partial y}{\partial x}\right)\right)$$

Hence
$$\frac{\partial^2 y}{\partial t^2} = f'\left(\frac{\partial y}{\partial x}\right) f'\left(\frac{\partial y}{\partial x}\right)\frac{\partial^2 y}{\partial x^2} = \left(f'\left(\frac{\partial y}{\partial x}\right)\right)^2 \frac{\partial^2 y}{\partial x^2}$$

This relationship is substituted into Eq. A2.1.6 which produces:

$$a_0^2 \left(\frac{\partial y}{\partial x}\right)^{-\gamma-1} \frac{\partial^2 y}{\partial x^2} = \left(f'\left(\frac{\partial y}{\partial x}\right)\right)^2 \frac{\partial^2 y}{\partial x^2}$$

or
$$\left(\frac{\partial y}{\partial x}\right)^{\frac{-\gamma-1}{2}} = \pm\frac{1}{a_0} f'\left(\frac{\partial y}{\partial x}\right)$$

Integrating this expression introduces an integration constant, k:

$$\left(\frac{\partial y}{\partial x}\right)^{\frac{1-\gamma}{2}} = \pm\frac{\gamma-1}{2a_0} f\left(\frac{\partial y}{\partial x}\right) + k$$

As y = x + L, then

$$\frac{\partial y}{\partial x} = 1 + \frac{\partial L}{\partial x} = \frac{\rho_0}{\rho} = \left(\frac{p}{p_0}\right)^{\frac{-1}{\gamma}}$$

and from the original substitution, the fact that x is a constant, and that the gas particle velocity is c and is also the rate of change of dimension L with time:

$$f\left(\frac{\partial y}{\partial x}\right) = \frac{\partial y}{\partial t} = \frac{\partial(x+L)}{\partial t} = \frac{\partial L}{\partial t} = c$$

Therefore

$$\left(\frac{p}{p_0}\right)^{\frac{\gamma-1}{2\gamma}} = \pm\frac{(\gamma-1)c}{2a_0} + k$$

If at the wave head the following facts are correct, the integration constant, k, is:

$$p = p_0 \quad c = 0 \quad \text{then} \quad k = 1$$

whence the equation for the particle velocity, c, in unsteady gas flow is deduced:

$$c = \pm\frac{2a_0}{\gamma-1}\left[\left(\frac{p}{p_0}\right)^{\frac{\gamma-1}{2\gamma}} - 1\right]$$

The positive sign is the correct one to adopt because if $p > p_0$ then c must be a positive value for a compression wave.

Appendix A2.2 Moving shock waves in unsteady gas flow

The text in this section owes much to the often-quoted lecture notes by Bannister [2.2]. The steepening of finite amplitude waves is discussed in Sec. 2.1.5 resulting in a moving shock wave. Consider the case of the moving shock wave, AB, illustrated in Fig. A2.2. The propagation velocity is α and it is moving into stationary gas at reference conditions, ρ_0 and p_0. The pressure and density behind the shock front are p and ρ, while the associated gas particle velocity is c. Imagine imposing a mean gas particle velocity, a, on the entire system illustrated in Fig. A2.2(a) so that the regime in Fig. A2.2(b) becomes "reality." This would give a stationary shock, AB, i.e., the moving front would be brought to rest and the problem is now reduced to one of steady flow. Consider that the duct area is A and is unity.

The continuity equation shows across the now stationary shock front:

$$(\alpha - c)\rho A = \alpha \rho_0 A \tag{A2.2.1}$$

The momentum equation gives, where force is equal to the rate of change of momentum,

$$(\alpha - (\alpha - c))\alpha \rho_0 A = (p - p_0)A$$

or
$$c\alpha\rho_0 = p - p_0 \tag{A2.2.2}$$

This can be rearranged as:

$$\frac{p}{\rho} = \frac{p_0}{\rho} + \frac{c\alpha\rho_0}{\rho} \tag{A2.2.3}$$

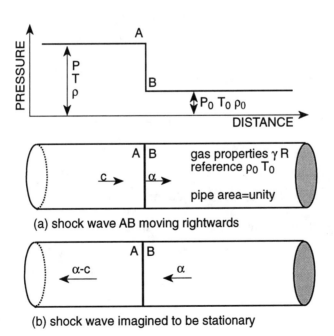

Fig. A2.2 *The moving shock wave.*

The First Law of Thermodynamics across the now stationary shock front shows:

$$C_p T_A + \frac{(\alpha - c)^2}{2} = C_p T_B + \frac{c^2}{2}$$

or

$$\frac{\gamma R}{\gamma - 1} T_A + \frac{(\alpha - c)^2}{2} = \frac{\gamma R}{\gamma - 1} T_B + \frac{c^2}{2}$$

or

$$\frac{1}{\gamma - 1} \frac{p}{\rho} + \frac{p}{\rho} + \frac{(\alpha - c)^2}{2} = \frac{1}{\gamma - 1} \frac{p_0}{\rho_0} + \frac{p_0}{\rho_0} + \frac{c^2}{2}$$

or

$$-\alpha c + \frac{c^2}{2} = \frac{1}{\gamma - 1} \frac{p}{\rho} + \frac{p}{\rho} - \frac{1}{\gamma - 1} \frac{p_0}{\rho_0} - \frac{p_0}{\rho_0}$$

Substituting from Eq. A2.2.3 for $\frac{p_0}{\rho}$ and writing $a_0^2 = \frac{\gamma p_0}{\rho_0}$ produces:

$$\alpha^2 - \frac{\gamma + 1}{2} \alpha c - a_0^2 = 0 \qquad (A2.2.4)$$

Therefore
$$c = \frac{2 a_0}{\gamma + 1} \left(\frac{\alpha}{a_0} - \frac{a_0}{\alpha} \right) \qquad (A2.2.5)$$

Combining Eqs. A2.2.2 and A2.2.4:

$$\alpha^2 - a_0^2 = \frac{\gamma + 1}{2} \left(\frac{p - p_0}{\rho_0} \right)$$

Dividing throughout by a_0^2 and substituting $\frac{\gamma p_0}{\rho_0}$ for it provides the relationship for the propagation velocity of a moving compression shock wave as:

$$\alpha = a_0 \sqrt{ \frac{\gamma + 1}{2\gamma} \frac{p}{p_0} + \frac{\gamma - 1}{2\gamma} } \qquad (A2.2.6)$$

Substituting this latter expression for the shock propagation velocity, α, into Eq. A2.2.5 gives a direct relationship for the gas particle velocity, c:

$$c = \frac{\dfrac{a_0}{\gamma}\left(\dfrac{p}{p_0} - 1\right)}{\sqrt{\dfrac{\gamma+1}{2\gamma}\dfrac{p}{p_0} + \dfrac{\gamma-1}{2\gamma}}} \quad \text{(A2.2.7)}$$

The temperature and density relationships behind the shock are determined as follows, first from the equation of state:

$$\frac{p}{p_0} = \frac{\rho R T}{\rho_0 R T_0} = \frac{\rho T}{\rho_0 T_0} \quad \text{(A2.2.8)}$$

which when combined with Eq. A2.2.1 gives:

$$\frac{T}{T_0} = \frac{p}{p_0}\left(\frac{\alpha - c}{\alpha}\right)$$

which when combined with Eq. A2.2.5 gives:

$$\frac{T}{T_0} = \frac{p}{p_0}\left(\frac{\gamma-1}{\gamma+1} + \frac{2}{\gamma+1}\frac{a_0^2}{\alpha^2}\right)$$

and in further combination with Eq. A2.2.6 reveals the temperature relationship:

$$\frac{T}{T_0} = \frac{p}{p_0}\left(\frac{\dfrac{\gamma-1}{\gamma+1}\dfrac{p}{p_0} + 1}{\dfrac{p}{p_0} + \dfrac{\gamma-1}{\gamma+1}}\right) \quad \text{(A2.2.9)}$$

and in further combination with Eq. A2.2.8 reveals the density relationship:

$$\frac{\rho}{\rho_0} = \frac{\dfrac{p}{p_0} + \dfrac{\gamma-1}{\gamma+1}}{\dfrac{\gamma-1}{\gamma+1}\dfrac{p}{p_0} + 1} \quad \text{(A2.2.10)}$$

Design and Simulation of Two-Stroke Engines

The functions in Eqs. 2.2.9 and 2.2.10 reveal the non-isentropic nature of the flow. For example, an isentropic compression would give the following relation between pressure and density:

$$\frac{p}{p_0} = \left(\frac{\rho}{\rho_0}\right)^\gamma$$

This is clearly quite different from that deduced for the moving shock wave in Eq. A2.2.10. The non-isentropic functions relating pressure, temperature and density for a moving shock wave are often named in the literature as the Rankine-Hugoniot equations. They arise again in the discussion in Sec. 2.2.4.

Chapter 2 - Gas Flow through Two-Stroke Engines

Appendix A2.3 Coefficients of discharge in unsteady gas flow

In various sections of this chapter, such as Secs. 2.12, 2.16, 2.17, and 2.19, a coefficient of discharge, C_d, is employed to describe the effective area of flow through a restriction encountered at the throat of a valve or port at a cylinder, or of flow past a throttle or venturi section in an inlet duct. Indeed a map of measured coefficients of discharge, C_d, is displayed in Fig. 2.27. It will be observed from this map that it is recorded over a wide range of pressure ratios and over the full range of port-to-pipe-area ratio, k, from zero to unity. The area ratio, k, is defined as:

$$\text{Area ratio, } k = \frac{\text{throat area}}{\text{pipe area}} = \frac{A_t}{A_p} \qquad (A2.3.1)$$

Where the geometry under scrutiny is a port in the cylinder wall of a two-stroke engine and is being opened or closed by the piston moving within that cylinder, it is relatively easy to determine the effective throat and pipe areas in question. All such areas are those regarded as normal to the direction of the particle flow. In some circumstances these areas are more difficult to determine, such as a poppet valve in an engine cylinder, and this matter will be dealt with more completely below. A poppet valve is not a valving device found exclusively in a four-stroke engine, as many two-stroke engines have used them for the control of both inlet and exhaust flow.

Measurement of coefficient of discharge

The experimental set-up for these measurements varies widely but the principle is illustrated in Fig. A2.3. The cylinder of the engine is mounted on a steady flow rig and the experimentally determined air mass flow rate, $\dot{m}_{ex}$, is measured at various pressure drops from the cylinder to the pipe for outflow, or vice-versa for inflow, through the aperture of area, A_t, leading from the cylinder to the pipe of area, A_p. The illustration shows a two-stroke engine cylinder with the piston held stationary giving a geometric port area, A_t, feeding an exhaust pipe with an inner diameter, d_{is}, and where the flow is coming from, or going to, a pipe where the diameter is d_p. It involves the measurement of the pressures and temperatures, p_0, T_0, within the cylinder and at the pipe point, p_2, T_2, respectively. In the illustration, the flow is suction from the atmosphere, giving "exhaust" flow to the exhaust pipe, and so the cylinder pressure and temperature are the atmospheric conditions, p_0 and T_0. The values of pressure and temperature within the cylinder are regarded as stagnation values, i.e., the particle velocity is so low as to be considered zero, and at the pipe point they are the static values. The mass rate of flow of the gas, $\dot{m}_{ex}$, for the experiment is measured by a meter, such as a laminar flow meter or an orifice designed to ISO, BS, DIN or ASME standards. The experiment is normally conducted using air as the flow medium.

The theoretical mass flow rate, $\dot{m}_{is}$, is traditionally determined [2.7] using isentropic nozzle theory between the cylinder and the throat for outflow, or pipe and throat for inflow, using the measured state conditions in the cylinder, at the throat, or in the pipe, which are p_1 and T_1, p_t and T_t, or p_2 and T_2, respectively, and using the nomenclature associated with

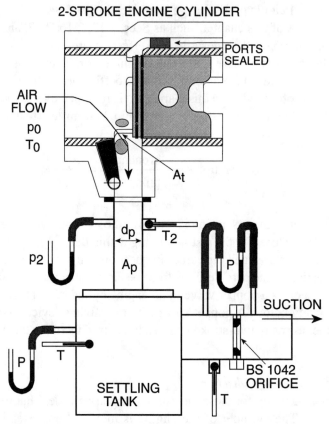

Fig. A2.3 *Experimental apparatus for C_d measurement.*

either Fig. 2.16 or Fig. 2.18. The theory is normally modified to incorporate the possibility of sonic flow at the throat, and the careful researcher [2.7] ensures that the critical pressure ratio is not applied to the inflow condition for flow from pipe to cylinder. Some [2.61] employ overall pressure ratios sufficiently low so as to not encounter this theoretical problem.

The coefficient of discharge, C_{dis}, is then determined as:

$$C_{dis} = \frac{\dot{m}_{ex}}{\dot{m}_{is}} \tag{A2.3.2}$$

The extra appellation of "is" to the subscript for C_d refers to the fact that the measured mass flow rate is compared with that calculated isentropically.

If the objective of measuring the coefficient of discharge, C_d, is simply as a comparator process for the experimental improvement of the flow in two-stroke engine porting, or of poppet valving in the cylinder heads of two- or four-stroke engines, then this classic method is completely adequate for the purpose. The only caveats offered here for the enhancement of that process is that the correct valve curtain area for poppet valves should be employed (see

Appendix A5.1), and that the range of pressure ratios used experimentally should be much greater than the traditional levels [2.61].

The determination of C_d for accurate application within an engine simulation

However, if the object of the exercise is the very necessary preparation of C_d maps of the type displayed by Kirkpatrick [2.41], for employment within engine simulation and modeling, then the procedure described above for the deduction of the coefficient of discharge, C_d, is totally inaccurate.

The objective is to measure and deduce C_d in such a manner that when it is "replayed" in a computer simulation of the engine, at identical values of pressure, temperature and area ratio, the calculation would predict exactly the mass flow rate that was measured. At other thermodynamic state conditions, the discharge coefficient is characterized by the pressure ratio and the area ratio and fluid mechanic similarity is assumed. There may be those who will feel that they could postulate and apply more sophisticated similarity laws, and that is a legitimate aspiration.

For the above scenario to occur, it means that the theoretical assessment of the ideal mass flow rate during the experimental deduction of C_d must not be conducted by a simple isentropic analysis, but with exactly the same set of thermodynamic software as is employed within the actual computer simulation. In short, the bottom line of Eq. A2.3.2 must be acquired using the non-isentropic theory, described earlier, applied to the geometry and thermodynamics of the steady flow experiment which mimics the outflow or inflow at a cylinder, or some similar plenum to duct boundary. This gives a significantly different answer for C_d by comparison with that which would be acquired by an isentropic analysis of the ideal mass flow rate. To make this point completely, this gives an "ideal" discharge coefficient, C_{di}, and is found by:

$$C_{di} = \frac{\dot{m}_{ex}}{\dot{m}_{nis}} \qquad (A2.3.3)$$

where $\dot{m}_{nis}$ is the theoretical non-isentropic mass flow rate described above.

However, even with this modification to the classic analytical method, the value of "ideal" discharge coefficient, C_{di}, is still not the correct value for accurate use within a computer engine simulation. Taking cylinder outflow as the example, if Eqs. 2.16.8 and 2.17.8 are examined it can be seen that the actual coefficient of discharge arises from the need for the prediction of an effective area of the throat of the restriction, formed by the reality of gas flow through the aperture of the cylinder port. Hence, the value of port area, A_{eff}, is that value which, when presented into the relevant thermodynamic software for the analysis of a particular flow regime at the <u>measured</u> values of upstream and downstream pressure and temperature, will calculate the <u>measured</u> value of mass flow rate, $\dot{m}_{ex}$. This involves an iterative process within the theory for flow to or from the measured value of pipe area, A_2, until the experimentally measured values of p_1, T_1, p_2, T_2, and $\dot{m}_{ex}$ coincide for a unique numerical value of effective throat to pipe area ratio, k_{eff}. The relevant value of C_d to be employed with

Design and Simulation of Two-Stroke Engines

an engine simulation value, i.e., the "actual" coefficient of discharge, C_{da}, is then determined as:

$$k_{eff} = \frac{A_{eff}}{A_2} = C_{da}\frac{A_t}{A_2} = C_{da}k \tag{A2.3.4}$$

The algebraic solution to this iterative procedure (for it requires a solution for the several unknowns from an equal number of simultaneous polynomial equations) is not trivial. The number of unknowns depends on whether the flow regime is inflow or outflow, and it can be subsonic or sonic flow for either flow direction; the number of unknowns can vary from two to five, depending on the particular flow regime encountered. The iterative procedure is completed until a satisfactory error band has been achieved, usually 0.01% for any one unknown variable.

Then, and only then, with the incorporation of the actual discharge coefficient, C_{da}, at the same cylinder-to-pipe pressure ratio, P, and geometric area ratios, k, into the simulation will the correct value of mass flow rate and pressure wave reflection and formation be found in the replay mode during an unsteady gas-dynamic and thermodynamic engine computer simulation.

Apart from some discussion in a thesis by Bingham [2.64], I am unaware of this approach to the determination of the actual coefficient of discharge, C_{da}, being presented in the literature until recent times [5.25]. I have published a considerable volume of C_d data relating to both two- and four-stroke engines, but all of it is in the format of C_{di} and all of it is in the traditional format whereby the ideal mass flow rate was determined by an isentropic analysis. Where the original measured data exist in a numeric format, that data can be re-examined and the required C_{da} determined. Where it does not, and the majority of it no longer exists as written records, then that which I have presented is well nigh useless for simulation purposes. Furthermore, it is the complete digitized map, such as in Fig. 2.27, that is needed for each and every pipe discontinuity to provide accurate simulation of unsteady gas flow through engines.

Some measurements of C_d at the exhaust port of an engine

In Figs. A2.4 to A2.7 are the measured discharge coefficients for both inflow and outflow at the exhaust port of a 125 cm^3 Grand Prix racing motorcycle engine, as shown in Fig. A2.3 [5.25]. These figures plot the discharge coefficients with respect to pressure ratio from the cylinder to the pipe, P, and for geometrical area ratios, k, of 0.127, 0.437, 0.716 and 0.824, respectively. On each figure is shown both the actual discharge coefficient, C_{da}, and the ideal coefficient, C_{di}, as given by Eq. A2.3.3. It should be noted that any difference between these two numbers, if replayed back into an engine simulation at identical k and P values, will give precisely that ratio of mass flow rate difference.

At a low area ratio, k, there is almost no difference between C_{di} and C_{da}. The traditional analysis would be equally effective here. As the area ratio increases, and where the mass flow rate is, almost by definition, increasing, then the mass flow rate error through the application of a C_{di} value also rises. The worst case is probably inflow at high area ratios, where it is seen to be 20%.

Chapter 2 - Gas Flow through Two-Stroke Engines

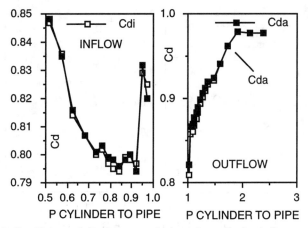

Fig. A2.4 Coefficient of discharge variations for cylinder inflow and outflow.

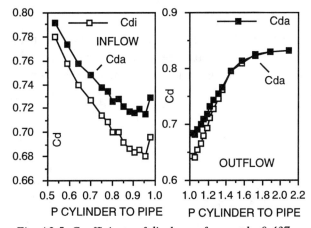

Fig. A2.5 Coefficients of discharge for port $k=0.437$.

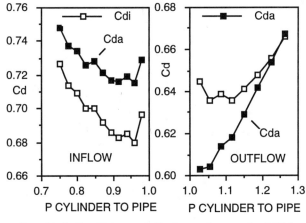

Fig. A2.6 Coefficients of discharge for port $k=0.716$.

Design and Simulation of Two-Stroke Engines

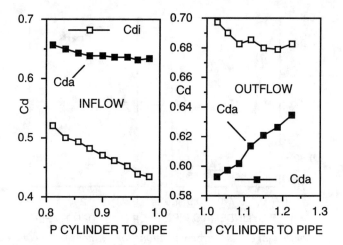

Fig. A2.7 Coefficients of discharge for port k=0.824.

Further discussion is superfluous. The need to accurately measure and, even more important, to correctly reduce the data to obtain the actual discharge coefficient, C_{da}, is obvious. A fuller discussion of this topic, together with more extensive measurements, is to be found in Ref. [5.25].

Chapter 3

Scavenging the Two-Stroke Engine

3.0 Introduction

In Chapter 1, particularly Sec. 1, there is a preliminary description of scavenging for the various types of two-stroke engines. This chapter continues that discussion in greater depth and provides practical advice and references to computer design programs to aid the design process for particular types of engine.

3.1 Fundamental theory

The original paper on scavenging flow was written by Hopkinson [3.1] in 1914. It is a classic paper, written in quite magnificent English, and no serious student of the two-stroke engine can claim to be such if that paper has not been read and absorbed. It was Hopkinson who conceived the notion of "perfect displacement" and "perfect mixing" scavenge processes. Benson [1.4] also gives a good account of the theory and expands it to include the work of himself and Brandham. The theory used is quite fundamental to our conception of scavenging, so it is repeated here in abbreviated form. Later it will be shown that there is a problem in correlation of these simple theories with measurements.

The simple theories of scavenging all postulate the ideal case of scavenging a cylinder which has a constant volume, V_{cy}, as shown in Fig. 3.1, with a fresh air charge in an isothermal, isobaric process. It is obvious that the real situation is very different from this idealized postulation, for the reality contains gas flows occurring at neither constant cylinder volume, constant cylinder temperature, nor constant cylinder pressure. Nevertheless, it is always important, theoretically speaking, to determine the ideal behavior of any system or process as a marker of its relationship with reality.

In Fig. 3.1 the basic elements of flow are represented. The incoming scavenge air can enter either a space called the "displacement zone" where it will be quite undiluted with exhaust gas, or a "mixing zone" where it mixes with the exhaust gas, or it can be directly short-circuited to the exhaust pipe providing the worst of all scavenging situations.

In this isothermal and isobaric process, the incoming air density, ρ_a, the cylinder gas density, ρ_c, and the exhaust gas density, ρ_{ex}, are identical. Therefore, from the theory previously postulated in Sec. 1.5, the values of purity, scavenge ratio, scavenging efficiency, trapping efficiency, and charging efficiency in this idealized simulation become functions of the volume of the several components, rather than the mass values as seen in Eqs. 1.5.1-1.5.9. The following equations illustrate this point.

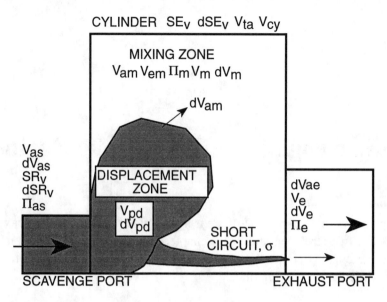

Fig. 3.1 Physical representation of isothermal scavenge model.

In terms of Fig. 3.1, the volume of scavenge flow which makes up the total quantity of air supplied is V_{as} with a purity value of Π_{as}. The purity of the incoming scavenge flow is unity as it is presumed to be air only. Purity in this idealized process is defined volumetrically as:

$$\Pi = \frac{\text{volume of air}}{\text{total volume}}$$

The first is for scavenge ratio, SR, subscripted as SR_v to make the point precisely that it is now a volumetrically related parameter, derived from Eq. 1.5.7:

$$SR_v = \frac{V_{as}}{V_{cy}} \qquad (3.1.1)$$

The cylinder reference volume, V_{cy}, does not have to be the swept volume, V_{sv}, but it is clearly the first logical option for an idealized flow regime.

The second is for scavenging efficiency, SE, derived from Eq. 1.5.9, where the volume of air trapped is V_{ta} and the volume of exhaust gas trapped is V_{ex}. It is also denoted as SE_v to illustrate that the ideal scavenge process is conducted volumetrically.

$$SE_v = \frac{\rho_a V_{ta}}{\rho_a V_{ta} + \rho_a V_{ex}} = \frac{V_{ta}}{V_{ta} + V_{ex}} \qquad (3.1.2)$$

Hence as
$$V_{cy} = V_{ta} + V_{ex}$$

Chapter 3 - Scavenging the Two-Stroke Engine

this simplifies to:
$$SE_v = \frac{V_{ta}}{V_{cy}} \tag{3.1.3}$$

The relationship for trapping efficiency, TE, denoted as TE_v in this ideal concept, follows from Eq. 1.5.13 as:

$$TE_v = \frac{V_{ta}}{V_{as}} \tag{3.1.4}$$

The expression for the ideal charging efficiency by volume, CE_v, follows from Eq. 1.5.15 as:

$$CE_v = \frac{V_{ta}}{V_{cy}} \tag{3.1.5}$$

In this ideal scavenge process, it is clear from manipulation of the above equations that the charging and scavenging efficiencies are identical:

$$CE_v = SE_v \tag{3.1.6}$$

and
$$TE_v = \frac{SE_v}{SR_v} \tag{3.1.7}$$

3.1.1 Perfect displacement scavenging

In the perfect displacement process from Hopkinson [3.1], all fresh charge entering the cylinder is retained and "perfectly displaces" the exhaust gas. This air enters the perfect scavenge volume, V_{pd}, shown in Fig. 3.1. Corresponding to the present theoretical presumption of perfect displacement scavenging the value of the short-circuit proportion, σ, is zero, and the mixing volume illustrated contains only exhaust gas. In other words, the volume of air in the mixing zone is V_{am} and is zero; the quantity of air entering the mixing zone is dV_{am} and is zero; the quantity of air entering the displacement zone at any instant is dV_{pd} and is equal to dV_{as}; the quantity of air entering the exhaust at any instant is dV_{ae} and is equal to zero. In short, in perfect displacement scavenging the air can fill the cylinder until it is filled completely, in which case it then spills into the exhaust pipe.

Therefore, if all entering volumes of fresh charge, V_{as}, are less than the cylinder volume, V_{cy}, or:

if $\qquad\qquad V_{as} \leq V_{cy}$

then $\qquad\qquad V_{ta} = V_{as} \tag{3.1.8}$

consequently, $\qquad\qquad SE_v = SR_v \tag{3.1.9}$

and $\qquad\qquad TE_v = 1 \tag{3.1.10}$

If the scavenge ratio, SR_v, exceeds unity, and clearly SE_v cannot do so, then:

if
$$V_{as} > V_{cy}$$

then
$$V_{ta} = V_{cy} \qquad (3.1.11)$$

consequently,
$$SE_v = 1 \qquad (3.1.12)$$

and
$$TE_v = \frac{SE_v}{SR_v} = \frac{1}{SR_v} \qquad (3.1.13)$$

3.1.2 Perfect mixing scavenging

This was Hopkinson's second concept. In this scenario the entering air has no perfect displacement characteristic, but upon arrival in the cylinder proceeds to mix "perfectly" with the exhaust gas. The perfect displacement zone in Fig. 3.1 does not exist and V_{pd} is zero. The resulting increment of exhaust gas effluent, dV_e, is composed solely of the mixed cylinder charge at that particular instant. As V_{pd} is zero, then the volume of trapped air is composed of V_{am} less that which has been lost to the exhaust system. To analyze this process, consider the situation at some point in time where the instantaneous values of air supplied have been V_{as}, and the scavenge ratio and scavenging efficiency values to date are SR_v and SE_v, respectively. Upon the supply of a further increment of air, dV_{as}, this will induce an exhaust flow of equal volume, dV_e, containing an increment of air, dV_{ae}. The volume of air retained in the cylinder, dV_{ta}, due to this flow increment is given by:

$$dV_{ta} = dV_{as} - dV_{ae}$$

or
$$dV_{ta} = dV_{as} - dV_e \Pi_m$$

However, in this idealized concept, as the cylinder purity is numerically equal to the scavenging efficiency, then:

$$dV_{ta} = dV_{as} - dV_e SE_v \qquad (3.1.14)$$

Differentiation of Eq. 3.1.3 for scavenging efficiency gives:

$$dSE = \frac{dV_{ta}}{V_{cy}} \qquad (3.1.15)$$

Substituting dV_{ta} from Eq. 3.1.15 into Eq. 3.1.14, and taking into account that dV_{ex} is numerically equal to dV_{as}, produces:

$$V_{cy} dSE = dV_{as} - dV_{as} SE_v \qquad (3.1.16)$$

Rearranging this differential equation so that it may be integrated, and using a differential form of Eq. 3.1.1 for SR_v yields, employing the same assumption regarding a reference volume:

as
$$dSR_v = \frac{dV_{as}}{V_{cy}}$$

then
$$\frac{dSE_v}{1 - dSE_v} = \frac{dV_{as}}{V_{cy}} = dSR_v \qquad (3.1.17)$$

This equation may be integrated on the left-hand side from the integration limits of 0 to SE_v, and on the right-hand side from 0 to SR_v, to give:

$$\log_e(1 - SE_v) = -SR_v \qquad (3.1.18)$$

The further manipulation of this equation produces the so-called "perfect mixing" equation:

$$SE_v = 1 - e^{-SR_v} \qquad (3.1.19)$$

From Eq. 3.1.7, the trapping efficiency during this process is:

$$TE_v = \frac{1 - e^{-SR_v}}{SR_v} \qquad (3.1.20)$$

3.1.3 Combinations of perfect mixing and perfect displacement scavenging

Benson and Brandham [3.2] suggested a two-part model for the scavenging process. The first part is to be perfect displacement scavenging until the air flow has reached a volumetric scavenge ratio value of SR_{pd}. At that point, when the theoretical situation in the cylinder is considered to be much like that in Fig. 3.1, the perfect scavenge volume and the exhaust gas are "instantaneously" mixed together. From that point on, until the fresh air flow process is concluded, a perfect mixing process takes place. This is described in great detail by Benson in his book [1.4]. He also presents the theoretical analysis of such a flow, conducted in a very similar manner to that above. The result of that analysis reveals:

when
$$0 < SR_v < SR_{pd}$$

then
$$SE_v = SR_v \qquad (3.1.21)$$

and if in excess or
$$SR_v > SR_{pd}$$

then
$$SE_v = 1 - (1 - SR_{pd})e^{(SR_{pd} - SR_v)} \qquad (3.1.22)$$

Design and Simulation of Two-Stroke Engines

In other words, in terms of the symbolism shown in Fig. 3.1, in the first part of the Benson two-part model, the volume of air, dV_{am}, supplied into the mixing zone is zero. In the second part of the Benson two-part model, there is no further air supplied into the perfect displacement zone, i.e., the value of dV_{pd} is zero.

3.1.4 Inclusion of short-circuiting of scavenge air flow in theoretical models

In the book by Benson [1.4], the theory for the Benson-Brandham two-part model described in the preceding section is extended to include short-circuiting of the flow directly to the exhaust, as illustrated in Fig. 3.1. A proportion of the incoming scavenge flow, σ, is diverted into the exhaust duct without scavenging exhaust gas or mixing with it in the cylinder. This results in modifications to Eqs. 3.1.21 and 22 to account for the fact that any cylinder scavenging is being conducted by an air flow of reduced proportions, numerically $(1 - \sigma)SR_v$. After such modifications, Eqs. 3.1.21 and 22 become:

when $\qquad 0 < SR_v \leq (1 - \sigma)SR_{pd}$

then $\qquad SE_v = (1 - \sigma)SR_v \qquad (3.1.23)$

and when $\qquad (1 - \sigma)SR_v > SR_{pd}$

then $\qquad SE_v = 1 - (1 - SR_{pd})\,e^{(SR_{pd} - (1-\sigma)SR_v)} \qquad (3.1.24)$

This two-part <u>volumetric</u> scavenge model has been widely quoted and used in the literature. Indeed the analytical approach has been extended in many publications to great complexity [3.3].

3.1.5 The application of simple theoretical scavenging models

Eqs. 3.1.1-24 are combined within a simple package, included in the Appendix Listing of Computer Programs as Prog.3.1, BENSON-BRANDHAM MODEL. This and all of the programs listed in the Appendix are available separately from SAE. It is self-explanatory in use, with the program prompting the user for all of the data values required. As an example, the output plotted in Figs. 3.2 and 3.3 is derived using the program. Apart from producing the data to plot the perfect displacement scavenging and perfect mixing scavenging lines, two further examples are shown for a perfect scavenging period, SR_{pd}, of 0.5, but one is with zero short-circuiting and the second is with σ equal to 10%. These are plotted in Figs. 3.2 and 3.3 as SE_v-SR_v and TE_v-SR_v characteristics. To calculate a perfect mixing characteristic, all that needs to be specified is that SR_{pd} and σ are both zero, because that makes Eq. 3.1.24 identical to Eq. 3.1.19 within the program.

From an examination of the two figures, it is clear that a TE_v-SR_v characteristic provides a better visual picture for the comparison of scavenging behavior than does a SE_v-SR_v graph. This is particularly evident for the situation at light load, or low SR_v levels, where it is nearly impossible to tell from Fig. 3.2 that the 10% short-circuit line has fallen below the perfect mixing line. It is very clear, however, from Fig. 3.3 that such is the case. Further, it is easier to

Chapter 3 - Scavenging the Two-Stroke Engine

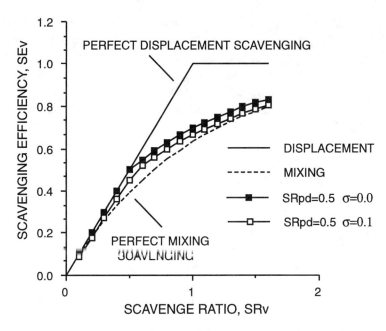

Fig. 3.2 Benson-Brandham model of scavenging characteristics.

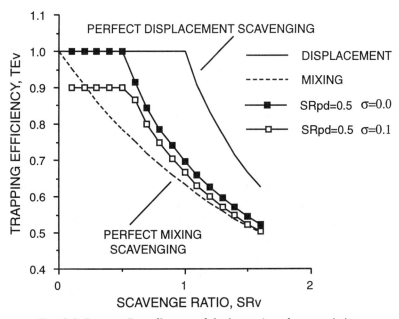

Fig. 3.3 Benson-Brandham model of trapping characteristics.

217

come to quantitative judgments from TE_v-SR_v characteristics. Should the two lines in Fig. 3.3 be the TE_v-SR_v behavior for two real, rather than theoretically ideal, engine cylinders, then it would be possible to say positively that the bmep or power potential of one cylinder would be at least 10% better than the other. This could be gauged visually as the order of trapped mass improvement, for that is ultimately what trapping efficiency implies. It would not be possible to come to this judgment so readily from the SE_v-SR_v graph.

On the other hand, the SE_v-SR_v characteristic also indicates the purity of the trapped charge, as that is the alternative definition of scavenging efficiency, and in this case one that is totally accurate. As flammability limits of any charge are connected with the level of trapped exhaust gas, the SE value in mass terms indicates the potential for the cylinder to fire that charge upon ignition by the spark plug. Should that not happen, then the engine will "four-stroke," or "eight-stroke" until there has been sufficient scavenging processes for the SE value to rise to the appropriate level to fire. In very broad terms, it is unlikely that a cylinder homogeneously filled with air-fuel mixture and exhaust gas to a SE value of less than 0.4 will be successfully fired by a spark plug. This is bound to be the case at light loads or at low SR levels, and is a common characteristic of carburetted two-stroke SI engines. A more extended discussion of this topic takes place in Sec. 4.1.3. In such a situation, where the engine commences to fire unevenly, there will be a considerable loss of air-fuel mixture through the exhaust port, to the considerable detriment of specific fuel consumption and the level of exhaust emissions of unburned hydrocarbons. To further emphasize this important point, in the case of the differing cylinder scavenging characteristics cited in Fig. 3.3, there would be a small loss of fresh charge with one of them, and at least 10% loss with the other, should the SR value for each scavenge process be less than 0.5 while the engine was in this "four-stroking" mode. This makes it very important to be able to design or develop cylinders with high trapping efficiencies at light loads and low SR values. Remember that the SR axis is, in effect, the same axis as throttle opening. Care must be taken to differentiate between mass-related SE-SR and volume-related SE_v-SR_v characteristics; this matter will be dealt with later in this chapter.

The literature is full of alternative theoretical models to that suggested by Benson and Brandham. Should you wish to study this matter further, the technical publications of Baudequin and Rochelle [3.3] and Changyou and Wallace [3.4] should be perused, as well as the references contained within those papers.

This subject will be covered again later in this chapter, as it is demonstrated that these simple two-part models of scavenging provide poor correlation with experimentally derived results, even for experiments conducted volumetrically.

You are probably wondering where, in all of the models postulated already in this chapter, does loop or cross or uniflow scavenging figure as an influence on the ensuing SE-SR or TE-SR characteristics. The answer is that they do in terms of SR_{pd} and σ values, but none of the literature cited thus far provides any numerical values of use to the engine designer or developer. The reason is that this requires correlation of the theory with relevant experiments conducted on real engine cylinders and until recent times this had not occurred. The word "relevant" in the previous sentence is used very precisely. It means an experiment conducted as the theory prescribes, by an isobaric, isothermal and isovolumic process. Irrelevant experiments, in that context, are quite common, as the experimental measurement of SE-SR behavior by me [3.7] and [3.13], Asanuma [3.8] and Booy [3.9] in actual firing engines will demonstrate. Such experiments are useful and informative, but they do not assist with the assignment of theoretical SR_{pd} and σ values to the many engine cylinders described in those publications. An assignment of such parameters is vital if the results of the experiments are to be used to guide engine simulations employing mass-based thermodynamics and gas dynamics.

3.2 Experimentation in scavenging flow

Since the turn of the 20th Century, the engineers involved with the improvement of the two-stroke engine took to devolving experimental tests aimed at improving the scavenging efficiency characteristics of the engines in their charge. This seemed to many to be the only logical methodology because the theoretical route, of which Sec. 3.1 could well be described as the knowledge base pre-1980, did not provide specific answers to specific questions. Many of these test methods were of the visualization kind, employing colored liquids as tracers in "wet" tests and smoke or other visible particles in "dry" tests. I experimented extensively with both methods, but always felt the results to be more subjective than conclusive.

Some of the work was impressive in its rigor, such as that of Dedeoglu [3.10] or Rizk [3.11] as an example of liquid simulation techniques, or by Ohigashi and Kashiwada [3.12] and Phatak [3.19] as an example of gas visualization technology.

The first really useful technique for the improvement of the scavenging process in a particular engine cylinder, be it a loop- or cross- or uniflow-scavenged design, was proposed by Jante [3.5]. Although the measurement of scavenging efficiency in the firing engine situation [3.7, 3.8, 3.9] (Plate 3.2) is also an effective development and research tool, it comes too late in terms of the time scale for the development of a particular engine. The cylinder with its porting has been designed and constructed. Money has been spent on casting patterns and on the machining and construction of a finished product or prototype. It is somewhat late in the day to find that the SE-SR characteristic is, possibly, less than desirable! Further, the testing process itself is slow, laborious and prone to be influenced by extraneous factors such as the effect of dissimilarly tuned exhaust pipes or minor shifts in engine air-fuel ratio. What Jante [3.5] proposed was a model test on the actual engine cylinder, or a model cylinder and piston capable of being motored, which did not have the added complexity of confusing the scavenging process with either combustion behavior or the unsteady gas dynamics associated with the exhaust tuning process.

3.2.1 The Jante experimental method of scavenge flow assessment

The experimental approach described by Jante [3.5] is sketched in Fig. 3.4. A photograph of an experimental apparatus for this test employed at QUB some years ago is shown in Plate 3.1. It shows an engine, with the cylinder head removed, which is being motored at some constant speed. The crankcase provides the normal pumping action and a scavenging flow exits the transfer ports and flows toward, and out of, the open cylinder bore. At the head face is a comb of pitot tubes, which is indexed across the cylinder bore to provide a measured value of vertical velocity at various points covering the entire bore area. Whether the pitot tube comb is indexed radially or across the bore to give a rectangular grid pattern for the recording of the pitot tube pressures is immaterial. The use of pitot tubes for the recording of gas velocities is well known, and the theory or practice can be found in any textbook on fluid mechanics [3.14]. The comb of pitot tubes shown in Fig. 3.4 can be connected to a variety of recording devices, ranging from the simplest, which would be an inclined water manometer bank, to some automatic data logging system through a switching valve from a pressure transducer. Irrespective of the recording method, the velocity of gas, c, in line with the pitot tube is given by:

$$c \approx 4\sqrt{\Delta p_{water}} \quad m/s \qquad (3.2.1)$$

where Δp_{water} is the pitot-static pressure difference in units of mm of water. The engine can be motored at a variety of speeds and the throttle set at various opening levels to vary the value of scavenge ratio, SR. The most important aspect of the resulting test is the shape of the velocity contour map recorded at the open cylinder bore. Jante [3.5] describes various significant types of patterns, and these are shown in Fig. 3.5. There are four velocity contour patterns, (a)-(d), shown as if they were recorded from various loop-scavenged engines with two side transfer ports and one exhaust port. The numerical values on the velocity contours are in m/s units.

Jante describes pattern (a) as the only one to produce good scavenging, where the flow is ordered symmetrically in the cylinder about what is often referred to as the "plane of symme-

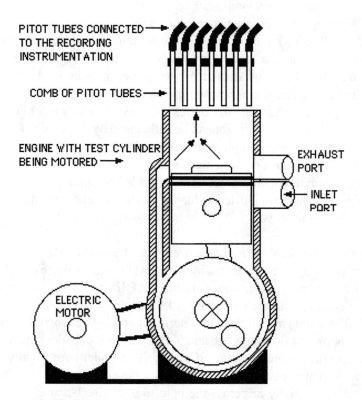

Fig. 3.4 Experimental configuration for Jante test.

Chapter 3 - Scavenging the Two-Stroke Engine

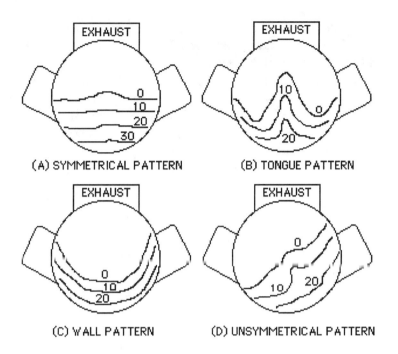

Fig. 3.5 Typcial velocity contours observed in the Jante test.

Plate 3.1 A comb of pitot tubes for the Jante test on a motorcycle engine.

try." The velocity contours increase in strength from zero at the center of the cylinder to a maximum at that side of the cylinder opposite to the exhaust port; in the "trade" this side is usually referred to as the "back of the cylinder." (As with most professional specializations, the use of jargon is universal!) The other patterns, (b)-(d), all produce bad scavenging characteristics. The so-called "tongue" pattern, (b), would give a rapid and high-speed flow over the cylinder head face and proceed directly to, and presumably out of, the exhaust port. The pattern in (c), dubbed the "wall" pattern, would ultimately enfold large quantities of exhaust gas and become a classic mixing process. So, too, would the asymmetrical flow shown in (d).

I, at one time, took a considerable interest in the use of the Jante test method as a practical tool for the improvement of the scavenging characteristics of engines. At QUB there is a considerable experimental and theoretical effort in the development of actual engines for industry. Consequently, up to 1980, this method was employed as an everyday development tool. At the same time, a research program was instituted at QUB to determine the level of its effectiveness, and the results of those investigations are published in technical papers [1.23, 3.13]. In an earlier paper [3.6] I had published the methodology adopted experimentally at QUB for the recording of these velocity contours, and had introduced several analytical parameters to quantify the order of improvement observed from test to test and from engine to engine. It was shown in Refs. [1.23] and [3.13] that the criterion of "mean velocity," as measured across the open cylinder bore, did determine the ranking order of scavenging efficiency in a series of cylinders for a 250cc Yamaha motorcycle engine. In that work the researchers at QUB also correlated the results of the Jante testing technique with scavenging efficiency measurements acquired under firing conditions with the apparatus shown in Plate 3.2. The work of Nuti and Martorano [3.33] confirmed that cylinders tested using the Jante simulation

Plate 3.2 QUB apparatus for the measurement of scavenging efficiency under firing conditions.

Chapter 3 - Scavenging the Two-Stroke Engine

method correlated well with scavenging efficiency measurements acquired under firing conditions by cylinder gas sampling. In particular, they agreed that the "mean velocity" criterion that I proposed [3.6] was an accurate indicator of the scavenging behavior under firing conditions.

The main advantages of the Jante test method are that (a) it is a test conducted under dynamic conditions, (b) it is a test which satisfies the more important, but not all, of the laws of fluid mechanics regarding similarity (see Sec. 3.2.2), (c) it is a test on the actual hardware relevant to the development program for that engine, (d) it does not require expensive instrumentation or hardware for the conduct of the test, and (e) it has been demonstrated to be an effective development tool.

The main disadvantages of the Jante test method are that (a) it is a test of scavenging flow behavior conducted without the presence of the cylinder head, which clearly influences the looping action of the gas flow in that area, (b) it is difficult to relate the results of tests on one engine to another engine even with an equal swept volume, or with an unequal swept volume, or with a differing bore-stroke ratio, (c) it is even more difficult to relate "good" and "bad" velocity contour patterns from loop- to cross-scavenged engines, (d) it has no relevance as a test method for uniflow-scavenged engines, where the flow is deliberately swirled by the scavenge ports at right angles to the measuring pitot tubes, (e) because it is not an absolute test method producing SE-SR or TE-SR characteristics, it provides no information for the direct comparison of the several types of scavenging flow conducted in the multifarious geometry of engines in existence, and (f) because it is not an absolute test method, no information is provided to assist the theoretical modeler of scavenging flow during that phase of a mathematical prediction of engine performance.

3.2.2 Principles for successful experimental simulation of scavenging flow

Some of the points raised in the last paragraph of Sec. 3.2.1 require amplification in more general terms.

Model simulation of fluid flow is a science [3.14] developed to deal with, for example, the realistic testing of model aircraft in wind tunnels. Most of the rules of similarity that compare the model to be tested to reality are expressed in terms of dimensionless quantities such as Reynolds, Froude, Mach, Euler, Nusselt, Strouhal, and Weber numbers, to name but a few. Clearly, the most important dimensionless quantity requiring similarity is the Reynolds number, for that determines whether the test is being conducted in either laminar or turbulent flow conditions, and, as the real scavenging flow is demonstrably turbulent, at the correct level of turbulence. To be pedantic, all of the dimensionless quantities should be equated exactly for precision of simulation. In reality, in any simulation procedure, some will have less relevance than others. Dedeoglu [3.10], in examining the applicability of these several dimensionless factors, maintained that the Strouhal number was an important similarity parameter.

Another vitally important parameter that requires similarity in any effective simulation of scavenging flow is the motion of the piston as it opens and closes the transfer ports. The reason for this is that the gas flow commences as laminar flow at the port opening, and becomes fully developed turbulent flow shortly thereafter. That process, already discussed in Chapters 1 and 2, is one of unsteady gas-dynamic flow where the particle velocity varies in time from zero to a maximum value and then returns to zero at transfer port closure. If one attempts to simulate the scavenging flow by holding the piston stationary, and subjects the cylinder to steady flow of either gas or liquid, then that flow will form attachments to either the piston crown or the cylinder walls in a manner which could not take place in the real velocity-time situation. This effect has been investigated by many researchers, Rizk [3.11], Percival [3.16], Sammons [3.15], Kenny [3.18] and Smyth [3.17], and they have concluded that, unless the test is being conducted in steady flow for some specific reason, this is not a realistic simulation of the scavenging flow. As more recent publications show Jante test results conducted in steady flow it would appear that not all are yet convinced of that conclusion [3.50].

3.2.3 Absolute test methods for the determination of scavenging efficiency

To overcome the disadvantages posed by the Jante test method, it is preferable to use a method of assessment of scavenging flow that will provide measurements of scavenging, trapping and charging efficiency as a function of scavenge ratio. It would also be preferable to conduct this test isolated from the confusing effects of combustion and exhaust pipe tuning characteristics. This implies some form of model test, but this immediately raises all of the issues regarding similarity discussed in Sec. 3.2.2. A test method which does not satisfy these criteria, or at least all of the vitally important dimensionless criteria, is unacceptable.

It was Sammons [3.15] who first postulated the use of a single-cycle apparatus for this purpose. Because of the experimental difficulty of accurately measuring oxygen concentrations by gas analysis in the late 1940s (a vital element of his test method), his proposal tended to be forgotten. It was revived in the 1970s by Sanborn [3.13], who, together with researchers at QUB, investigated the use of a single-cycle apparatus using liquids to simulate the flow of fresh charge and exhaust gas. Sanborn continued this work [3.21] and other researchers investigated scavenging flow with this experimental methodology [3.22].

At QUB there was a growing realization that the occasional confusing result from the liquid-filled apparatus was due to the low Reynolds numbers found during the experiment, and that a considerable period of the simulated flow occurred during laminar or laminar-turbulent transition conditions. Consequently, the QUB researchers reverted to the idea postulated by Sammons [3.15], for in the intervening years the invention of the paramagnetic analyzer, developed for the accurate determination of oxygen proportions in gas analysis, had realized the experimental potential inherent in Sammons' original technique. (As a historical note, for it illustrates the frailty of the human memory, I had forgotten about, or had relegated to the subconscious, the Sammons paper and so reinvented his experimental method.)

However, as will become more evident later, the QUB apparatus incorporates a highly significant difference from the Sammons experimental method, i.e., the use of a constant volume cylinder during the scavenge process.

The QUB apparatus sketched in Fig. 3.6, and appearing in Plate 3.3, is very thoroughly described and discussed by Sweeney [3.20]. The salient features of its operation are a constant volume crankcase and a constant volume cylinder during the single cycle of operation from tdc to tdc at a known turning rate. The equivalent rotational rate is 700 rev/min. The cylinder is filled with air to represent exhaust gas, and the crankcase is filled with carbon dioxide to represent the fresh air charge. Here, one similarity law is being well satisfied, in that the typical density ratio of crankcase air charge to cylinder exhaust gas is about 1.6, which corresponds almost exactly with the density ratio of 1.6 for the carbon dioxide to air used in the experiment. The Reynolds numbers are well into the turbulent zone at the mid-flow position, so yet another similarity law is satisfied.

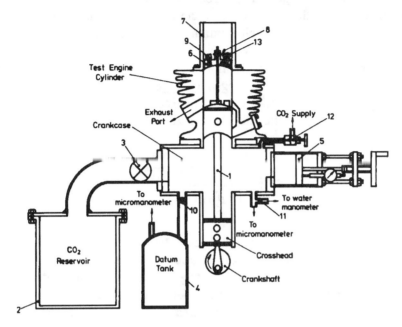

Fig. 3.6 QUB single-cycle gas scavenging apparatus.

Plate 3.3 The single-cycle gas scavenging experimental apparatus at QUB.

The gases in the cylinder and the crankcase are at atmospheric temperature at the commencement of the process. The crankcase pressure is set at particular levels to produce differing values of scavenge ratio, SR_v, during each experiment. At the conclusion of a single cycle, the crankshaft is abruptly stopped at tdc by a wrap-spring clutch brake, retaining under the movable cylinder head the trapped charge from the scavenging flow. The movable cylinder head is then released from the top piston rod and depressed so that more than 75% of the trapped gas contents are forced through a paramagnetic oxygen analyzer, giving an accurate measurement from a representative gas sample.

If υ_{O_2} is the measured oxygen concentration (expressed as % by volume) in the trapped charge, M_{on} is the molecular ratio of nitrogen to oxygen in air, and C_{pm} is a correction coefficient for the slight paramagnetism exhibited by carbon dioxide, then Sweeney [3.20] shows that the volumetric scavenging efficiency, SE_v, is given by:

$$SE_v = \frac{1-(1+M_{on})\frac{\upsilon_{O_2}}{100}}{1-C_{pm}(1+M_{on})} \tag{3.2.2}$$

The correction coefficient for carbon dioxide in the presence of oxygen, C_{pm}, is a negative number, -0.00265. The value of M_{on} is traditionally taken as 79/21 or 3.762.

The scavenge ratio, SR_v, is found by moving the piston, shown as item 5 in Fig. 3.6, inward at the conclusion of the single cycle experiment until the original crankcase state conditions of pressure and temperature are restored. A dial gauge accurately records the piston movement. As the piston area is known, the volume of charge which left the crankcase, V_{cc}, is readily determined. The volume of charge, V_c, which scavenged the cylinder, is then calculated from the state equation:

$$\frac{p_{cc}V_{cc}}{T_{cc}} = \frac{p_{cy}V_c}{T_{cy}} \tag{3.2.3}$$

The temperatures, T_{cy} and T_{cc}, are identical and equal to the atmospheric temperature, T_{at}. The cylinder pressure, p_{cy}, is equal to the atmospheric pressure, p_{at}.

The cylinder volume can be set to any level by adjusting the position of the cylinder head on its piston rod. The obvious value at which to set that volume is the cylinder swept volume, V_{sv}.

The scavenge ratio, SR_v, is then:

$$SR_v = \frac{V_c}{V_{sv}} \tag{3.2.4}$$

In the final paragraph of Sec. 3.1, the desirability was emphasized of conducting a relevant scavenging experiment in an isothermal, isovolumic and isobaric fashion. This experiment satisfies these criteria as closely as any experiment ever will. It also satisfies all of the relevant criteria for dynamic similarity. It is a dynamic experiment, for the piston moves and provides the realistic attachment behavior of scavenge flow as it would occur in the actual process in the firing engine. Sweeney [3.20, 3.23] demonstrates the repeatability of the test method and of its excellent correlation with experiments conducted under firing conditions. Some of those results are worth repeating here, for that point cannot be emphasized too strongly.

The experimental performance characteristics, conducted at full throttle for a series of modified Yamaha DT 250 engine cylinders, are shown in Fig. 3.7. Each cylinder has identical engine geometry so that the only modifications made were to the directioning of the transfer ports and the shape of the transfer duct. Neither port timings nor port areas were affected so that each cylinder had almost identical SR characteristics at any given rotational speed. Thus, the only factor influencing engine performance was the scavenge process. That this is significant is clearly evident from that figure, as the bmep and bsfc behavior is affected by as much as 15%. When these same cylinders were tested on the single-cycle gas scavenging rig, the SE_v-SR_v characteristics were found to be as shown in Fig. 3.8. The figure needs closer examination, so a magnified region centered on a SR_v value of 0.9 is shown in Fig. 3.9. Here, it can be observed that the ranking order of those same cylinders is exactly as in Fig. 3.7, and so too is their relative positions. In other words, Cylinders 14 and 15 are the best and almost as effective as each other, so too are cylinders 9 and 7 but at a lower level of scavenging efficiency and power. The worst cylinder of the group is cylinder 12 on all counts. The "double indemnity" nature of a loss of 8% SE_v at a SR_v level of 0.9, or a TE_v drop of 9%, is translated into the bmep and bsfc shifts already detailed above at 15%.

A sustained research and development effort has taken place at QUB in the experimental and theoretical aspects of scavenging flow. For the serious student of the subject, the papers published from QUB form a series, each using information and thought processes from the preceding publication. That reading list, in consecutive order, is [3.6], [1.23], [3.13], [3.20], [1.10], [3.23], [1.11], and [3.17].

3.2.4 Comparison of loop, cross and uniflow scavenging

The QUB single-cycle gas scavenge experiment permits the accurate and relevant comparison of SE_v-SR_v and TE_v-SR_v characteristics of different types of scavenging. From some of those previous papers, and from other experimental work at QUB hitherto unpublished, test results for uniflow-, loop- and cross-scavenged engine cylinders are presented to illustrate the several points being made. At this stage the most important issue is the use of the experimental apparatus to compare the various methods of scavenging, in order to derive some fundamental understanding of the effectiveness of the scavenging process conducted by these several methodologies. In Sec. 3.3 the information gained will be used to determine the theoretical relevance of this experimental data in the formulation of a model of scavenging to be incorporated within a complete theoretical model of the firing engine.

Figs. 3.10-3.13 give the scavenging and trapping characteristics for eight engine cylinders, as measured on the single-cycle gas scavenging rig. It will be observed that the test results fall between the perfect displacement line and perfect mixing line from the theories of Hopkinson [3.1]. By contrast, as will be discussed in Sec. 3.3, some of the data presented by others [3.3, 3.4] lie below the perfect mixing line.

Design and Simulation of Two-Stroke Engines

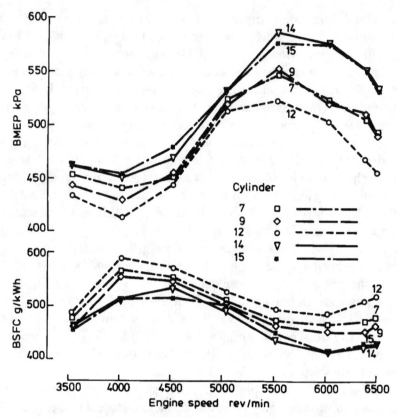

Fig. 3.7 *Measured performance characteristics.*

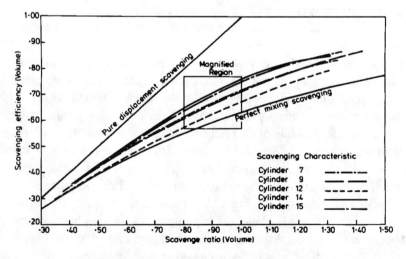

Fig. 3.8 *Measured isothermal scavenging characteristics.*

Chapter 3 - Scavenging the Two-Stroke Engine

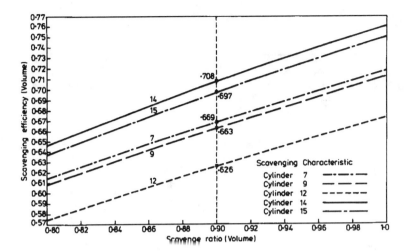

Fig. 3.9 Magnified region of SE-SR characteristics.

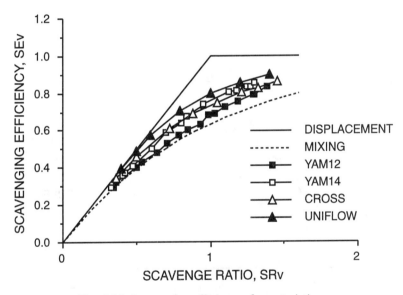

Fig. 3.10 Scavenging efficiency characteristics.

The cylinders used in this study, virtually in order of their listing in Figs. 3.10-3.13, are as follows, first for those shown in Figs. 3.10 and 3.11:
(a) A 250 cm^3 loop-scavenged cylinder, modified Yamaha DT 250 cylinder no.12, and called here YAM12, but previously discussed in [1.23]. The detailed porting geometry is drawn in that paper, showing five scavenge ports.

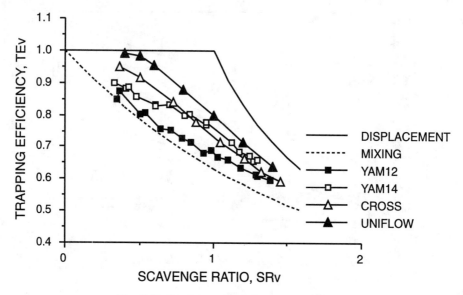

Fig. 3.11 Trapping efficiency characteristics.

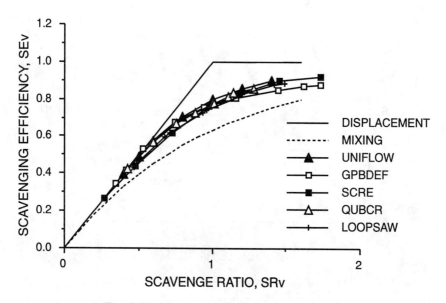

Fig. 3.12 Scavenging efficiency characteristics.

(b) A 250 cm³ loop-scavenged cylinder, modified Yamaha DT 250 cylinder no.14, and called here YAM14, but previously discussed in [1.23]. The detailed porting geometry is drawn in that paper, showing five scavenge ports.

(c) A 409 cm³ classic cross-scavenged cylinder, called CROSS. It is a cylinder from an outboard engine. The detailed porting geometry and the general layout is virtually as illustrated in Fig. 1.3. However, the engine details are proprietary and further technical

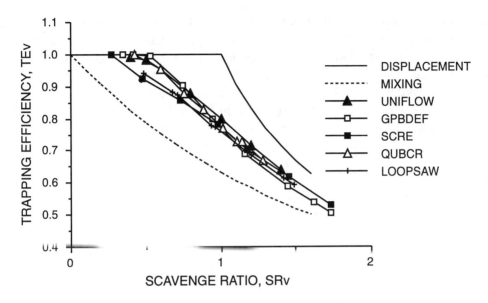

Fig. 3.13 Trapping efficiency characteristics.

description is not possible. The design approach is discussed in Sec. 3.5.2 and a sketch of the layout is to be found in Fig. 3.32(a).

(d) A ported uniflow-scavenged cylinder of 302 cm³ swept volume, called UNIFLOW. It has a bore stroke ratio of 0.6, and the porting configuration is not dissimilar to that found in the book by Benson [1.4, Vol 2, Fig. 7.7, p213]. The engine details are proprietary and further technical description is not possible.

Then there are those cylinders shown in Figs. 3.12 and 3.13, with the UNIFLOW cylinder repeated to provide a standard of comparison:

(e) A 409 cm³ cross-scavenged cylinder, called GPBDEF. It is a prototype cylinder designed to improve the scavenging of the same outboard engine described above as CROSS. The technique used in the design is given Sec. 3.5.3. The detailed porting geometry and the general layout is illustrated in Figs. 3.32(b) or 3.34(a).

(f) A loop-scavenged cylinder of 375 cm³ swept volume, called SCRE. This engine has three transfer ports, after the fashion of Fig. 3.38.

(g) A 250 cm³ QUB cross-scavenged cylinder, called QUBCR, and previously described in [1.10] in considerable detail. The detailed porting geometry is drawn in that paper. The general layout is almost exactly as illustrated in Figs. 1.4 or 3.34(b).

(h) A loop-scavenged cylinder of 65 cm³ swept volume, called LOOPSAW. This engine has two transfer ports, after the fashion of the upper left sketch in Fig. 1.2 or Fig. 3.35. The engine is designed for use in a chainsaw and is very typical of all such cylinders designed for small piston-ported engines employed for industrial or outdoor products.

The disparate nature of the scavenging characteristics of these test cylinders is clearly evident. Note that they all fall within the envelope bounded by the lines of "perfect displacement scavenging" and "perfect mixing scavenging." By the end of 1994 at QUB, a large library of information on this subject has been amassed with up to four QUB single-cycle gas rigs conducting tests on over 1300 differing cylinder geometries. During the course of this accumulation of data, there have been observed test points which have fallen below the "perfect mixing" line. Remember when examining these diagrams that the higher the scavenge ratio then so too is the potential bmep or torque that the engine may attain. An engine may be deliberately designed to produce a modest bmep, such as a chainsaw or a small outboard, and hence the quality of its scavenging characteristics above, say, a SR_V value of 1.0 is of no consequence. If an engine is to be designed to produce high specific power and torque, such as a racing outboard or a motocross engine, then the quality of its scavenging characteristics below, say, a SR_V value of 1.0 is equally of little consequence.

Figs. 3.10 and 3.11 present what could be loosely described as the best and worst of scavenging behavior. As forecast from the historical literature, the UNIFLOW cylinder has the best scavenging characteristic from the lowest to the highest scavenge ratios. It is not the best, however, by the margin suggested by Changyou [3.4], and can be approached by optimized loop and cross scavenging as shown in Figs. 3.12 and 3.13. The two Yamaha DT 250 cylinders, discussed in Sec. 3.2.2, are now put in context, for the firing performance parameters given in Fig. 3.7 are obtained at scavenge ratios by volume in excess of unity. This will be discussed in greater detail later in Chapter 5.

It will be observed from the trapping efficiency diagram, Fig. 3.11, that the superior scavenging of the CROSS engine is more readily observed by comparison with the equivalent SE-SR graph in Fig. 3.10. This is particularly so for scavenge ratios less than unity. While the trapping of the UNIFLOW engine at low SR_V values is almost perfect, that for the CROSS engine is also very good, making the engine a good performer at idle and light load; that indeed is the experience of both the user and the researcher. The CROSS engine does not behave so well at high scavenge ratios, making it potentially less suitable as a high-performance unit.

That it is possible to design loop scavenging poorly is evident from the characteristic shown for YAM12, and the penalty in torque and fuel consumption as already illustrated in Fig. 3.7.

Figs. 3.12 and 3.13 show that it is possible to design loop- and cross-scavenged engines to approach the "best in class" scavenging of the UNIFLOW engine. Indeed, it is arguable that the GPBDEF and QUBCR designs are superior to the UNIFLOW at scavenge ratios by volume below 0.5. Put very crudely, this might translate to superior performance parameters in a firing engine at bmep levels of 2.5 or 3.0 bar, assuming that the combustion efficiency and related characteristics are equivalent.

Observe in Figs. 3.12 and 3.13 that both the LOOPSAW and SCRE designs, both loop scavenged, are superior to either YAM12 or YAM14, and approach the UNIFLOW engine at high SR values above unity. For the SCRE, a large-capacity cylinder designed to run at high bmep and piston speed, that is an important issue and indicates the success of the optimization to perfect that particular design. On the other hand, the LOOPSAW unit is designed to run at low bmep levels, i.e., 4 bar and below, and so the fact that it has good scavenging at high scavenge ratios is not significant in terms of that design requirement. To improve its firing performance characteristics, it can be observed that it should be optimized, if at all possible, to approach the GPBDEF scavenging characteristic at SR_V values below unity.

The inferior nature of the scavenging of the cross-scavenged engine, CROSS, at full throttle or a high SR$_V$ value, is easily seen from the diagrams. It is for this reason that such power units have generally fallen from favor as outboard motors. However, in Sec. 3.5.3 it is shown that classical cross scavenging can be optimized in a superior manner to that already reported in the literature [1.10] and to a level somewhat higher than a mediocre loop-scavenged design. It is easier to develop a satisfactory level of scavenging from a cross-scavenged design, by the application of the relatively simple empirical design recommendations of Sec. 3.5, than it is for a loop-scavenged engine.

The modified Yamaha cylinder, YAM12, has the worst scavenging overall. Yet, if one merely examined by eye the porting arrangements of the loop-scavenged cylinders, YAM14, YAM12, LOOPSAW and SCRE, prior to the conduct of a scavenge test on a single-cycle apparatus, it is doubtful if the opinion of any panel of "experts" would be any more unanimous on the subject of their scavenging ability than they would be on the quality of the wine being consumed with their dinner! This serves to underline the importance of an absolute test for scavenging and trapping efficiencies of two-stroke engine cylinders.

It is at this juncture that the first whiff of suspicion appears that the Benson-Brandham model, or any similar theoretical model, is not going to provide sensible data for the prediction of scavenging behavior. In Fig. 3.2 the scavenging exhibited by that theory for a cylinder with no short-circuiting and a rather generous allowance of a 50% period of perfect displacement prior to any mixing process, appears to have virtually identical scavenging characteristics at high scavenge ratios to the very worst cylinder experimentally, YAM12, in Fig. 3.10. As all of the other cylinders in Figs. 3.10 and 3.11 have much better scavenging characteristics than the cylinder YAM12, clearly the Benson-Brandham model would have great difficulty in simulating them, if at all. In the next section this theoretical problem is investigated.

3.3 Comparison of experiment and theory of scavenging flow
3.3.1 Analysis of experiments on the QUB single-cycle gas scavenging rig

As has already been pointed out, the QUB single-cycle gas scavenging rig is a classic experiment conducted in an isovolumic, isothermal and isobaric fashion. Therefore, one is entitled to compare the measurements from that apparatus with the theoretical models of Hopkinson [3.1], Benson and Brandham [3.2], and others [3.3], to determine how accurate they may be for the modeling of two-stroke engine scavenging. Eq. 3.1.19 for perfect mixing scavenging is repeated here:

$$SE_V = 1 - e^{-SR_V} \qquad (3.1.19)$$

Manipulation of this equation shows:

$$\log_e(1 - SE_V) = -SR_V \qquad (3.3.1)$$

Consideration of this equation for the analysis of any experimental SE$_V$ and SR$_V$ data should reveal a straight line of traditional equation format, with a slope m and an intercept on the y axis of value c when x is zero.

straight line equation: $\quad y = mx + c$

The Benson-Brandham model contains a perfect scavenging period, SR_{pd}, before total mixing occurs, and a short-circuited proportion, σ. This resulted in Eqs. 3.1.23 and 24, repeated here:

when $\quad 0 < SR_v \leq (1 - \sigma)SR_{pd}$

then $\quad SE_v = (1 - \sigma)SR_v \quad$ (3.1.23)

and when $\quad (1 - \sigma)SR_v > SR_{pd}$

then $\quad SE_v = 1 - (1 - SR_{pd}) e^{(SR_{pd} - (1-\sigma)SR_v)} \quad$ (3.1.24)

Manipulation of this latter equation reveals:

$$\log_e(1 - SE_v) = (\sigma - 1)SR_v + \log_e(1 - SR_{pd}) + SR_{pd} \quad (3.3.2)$$

Consideration of this linear equation as a straight line shows that test data of this type should give a slope m of $(\sigma - 1)$ and an intercept of $\log_e(1 - SR_{pd}) + SR_{pd}$. Any value of short-circuiting σ other than 0 and the maximum possible value of 1 would result in a line of slope m.

where $\quad 0 > m > -1$

The slope of such a line could not be less than -1, as that would produce a negative value of the short-circuiting component, σ, which is clearly theoretically impossible.

Therefore it is vital to examine the experimentally determined data presented in Sec. 3.2.4 to acquire the correlation of the theory with the experiment. The analysis is based on plotting $\log_e(1 - SE_v)$ as a function of SR_v from the experimental data for two of the cylinders, YAM14 and YAM12, as examples of "good" and "bad" loop scavenging, and this is shown in Fig. 3.14. From a correlation standpoint, it is very gratifying that the experimental points fall on a straight line. In Refs. [1.11, 3.34] many of the cylinders shown in Figs. 3.10 to 3.13 are analyzed in this manner and are shown to have a similar quality of fit to a straight line. What is less gratifying, in terms of an attempt at correlation with a Benson and Brandham type of theoretical model, is the value of the slope of the two lines. The slopes lie numerically in the region between -1 and -2. The better scavenging of YAM14 has a value which is closer to -2. Not one cylinder ever tested at QUB has exhibited a slope greater than -1 and would have fallen into a category capable of being assessed for the short-circuit component σ in the manner of the Benson-Brandham model. Therefore, there is no correlation possible with any of the "traditional" models of scavenging flow, as all of those models would seriously underestimate the quality of the scavenging in the experimental case, as alluded to in the last paragraph of Sec. 3.2.4.

Chapter 3 - Scavenging the Two-Stroke Engine

It is possible to take the experimental results from the single-cycle rig, plot them in the logarithmic manner illustrated and derive mathematical expressions for SE_v-SR_v and TE_v-SR_v characteristics representing the scavenging flow of all significant design types, most of which are noted in Fig. 3.16. This is particularly important for the theoretical modeler of scavenging flow who has previously been relying on models of the Benson-Brandham type.

The straight line equation, as seen in Fig. 3.14, and in Refs. [1.11, 3.34], is not adequate to represent the scavenging behavior sufficiently accurately for modeling purposes and has to be extended as indicated below:

$$\log_e(1 - SE_v) = \kappa_0 + \kappa_1 SR_v + \kappa_2 SR_v^2 \qquad (3.3.3)$$

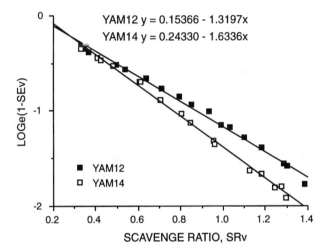

Fig. 3.14 Logarithmic SE_v-SR_v characteristics.

Manipulating this equation shows:

$$SE_v = 1 - e^{\left(\kappa_0 + \kappa_1 SR_v + \kappa_2 SR_v^2\right)} \qquad (3.3.4)$$

Fig. 3.15 shows the numerical fit of Eq. 3.3.3 to the experimental data for two cylinders, SCRE and GPBDEF. The SCRE cylinder has a characteristic on the log SE-SR plot which approximates a straight line but it is demonstrably fitted more closely by the application of Eq. 3.3.3. The GPBDEF cylinder could not sensibly be fitted by a straight line equation; this is a characteristic exhibited commonly by engines with high trapping efficiency at low scavenge ratios.

Therefore, the modeler can take from Fig. 3.16 the appropriate κ_0, κ_1 and κ_2 values for the type of cylinder being simulated and derive realistic values of scavenging efficiency, SE_v, and trapping efficiency, TE_v, at any scavenge ratio level, SR_v. What this analysis fails to do is to provide the modeler with any information regarding the influence of "perfect displacement" scavenging, or "perfect mixing" scavenging, or "short-circuiting" on the experimental

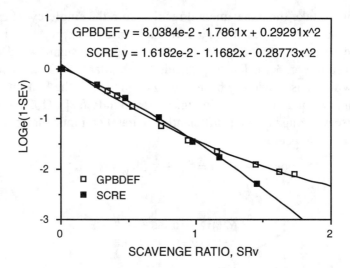

Fig. 3.15 Logarithmic SE_v-SR_v characteristics

or theoretical scavenging behavior of any of the cylinders tested. The fact is that these concepts are valuable only as an aid to understanding, for the real scavenge process does not proceed with an abrupt transition from displacement to mixing scavenging, and any attempt to analyze it as such has been demonstrated above to result in failure.

NAME	TYPE	κ_0	κ_1	κ_2
UNIFLOW	PORTED UNIFLOW	3.3488E-2	−1.3651	−0.21735
LOOPSAW	2 PORT LOOP	2.6355E-2	−1.2916	−0.12919
SCRE	3 PORT LOOP	1.6182E-2	−1.1682	−0.28773
CROSS	4 PORT + DEFLECTOR	1.2480E-2	−1.2289	−0.090576
QUBCR	QUB CROSS	2.0505E-2	−1.3377	−0.16627
GPBDEF	GPB CROSS	8.0384E-2	−1.7861	+0.29291
YAM1	5 PORT LOOP	1.8621E-2	−0.91737	−0.46621
YAM6	5 PORT LOOP	3.1655E-2	−1.0587	−0.16814
YAM12	5 PORT LOOP	−1.4568E-2	−0.84285	−0.28438
YAM14	5 PORT LOOP	2.9204E-2	−1.0508	−0.34226

Fig. 3.16 Experimental values of scavenging coefficients.

Modelers of scavenging flow would do well to remember that the scavenging characteristics being encapsulated via Fig. 3.16 from the experimental results are scavenge ratio, scavenging efficiency and trapping efficiency __by volume__ and not by mass. This is a common error to be found in the literature where scavenging characteristics are provided without specifying precisely the reference for those values. Further discussion and amplification of this point will be found in Chapter 5 on engine modeling.

3.3.2 A simple theoretical scavenging model which correlates with experiments

The problem with all of the simple theoretical models presented in previous sections of this chapter is that the theoretician involved was under some pressure to produce a single mathematical expression or a series of such expressions. Much of this work took place in the pre-computer age, and those who emanate from those slide rule days will appreciate that pressure. Consequently, even though Benson or Hopkinson knew perfectly well that there could never be an abrupt transition from perfect displacement scavenging to perfect mixing scavenging, this type of theoretical "fudge" was essential if Eqs. 3.1.8 to 3.1.24 were to ever be realized and be "readily soluble," arithmetically speaking, on a slide rule. Today's computer- and electronic calculator-oriented engineering students will fail to understand the sarcasm inherent in the phrase, "readily soluble." It should also be said that there was no experimental evidence against which to judge the validity of the early theoretical models, for the experimental evidence contained in paper [3.20], and here as Figs. 3.10 to 3.13, has only been available since 1985.

The need for a model of scavenging is vital when conducting a computer simulation of an engine. This will become much more evident in Chapter 5, but already a hint of the complexity has been given in Chapter 2. In that chapter, in Secs. 2.16 and 2.18.10, the theory under consideration is outflow from a cylinder of an engine, exactly as would be the situation during a scavenge process. Naturally, inflow of fresh air would be occurring at the same juncture through the scavenge or transfer ports from the crankcase or from a supercharger. However, in the time element under consideration for a scavenge process, the precise cylinder properties are indexed for the computation of the outflow from the cylinder. These cylinder properties are pressure, temperature, and the cylinder gas properties so that the outflow from the cylinder is calculated correctly and gas of the appropriate purity is delivered into the exhaust pipe. What then is the appropriate purity to use in the computational step? By definition a scavenging process is one that is stratified. If the scavenging could be carried out by a perfect displacement process then the gas leaving the cylinder would be exhaust gas while the cylinder purity and scavenging efficiency approached unity. If the short-circuiting was disastrously complete it would be air leaving the cylinder and the cylinder contents would remain as exhaust gas! The modeling computation requires information regarding the properties of the gas in the plane of the exhaust port for its purity and temperature while assuming that the pressure throughout the cylinder is uniform. The important theoretical step is to be able to characterize the behavior of scavenging of any cylinder as observed volumetrically on the QUB single-cycle gas scavenging rig and connect it to the mass- and energy-based thermodynamics in the computer simulation of a firing engine [3.35].

Consider the situation in Fig. 3.17, where scavenging is in progress in a constant volume cylinder under isobaric and isothermal conditions with both the scavenge and the exhaust ports open. In other words, exactly as carried out experimentally in a QUB single-cycle gas scavenge rig, or as closely as any experiment can ever mimic an idealized concept. The volume of air retained within the cylinder during an incremental step in scavenging is given by, from the volumes of air entering by the scavenge ports and leaving the cylinder through the exhaust port,

$$dV_{ta} = dV_{as} - dV_{ae} \qquad (3.3.5)$$

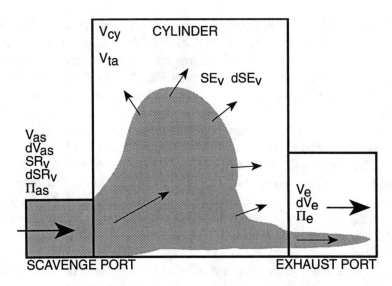

Fig. 3.17 Model of scavenging for simulation.

Dividing across by the instantaneous cylinder volume, V_{cy}, and assuming correctly that the incremental volume of charge supplied has the same volume which exits the exhaust ports, this gives,

$$dSE_{cy} = \Pi_{as}dSR_v - \Pi_e dSR_v \tag{3.3.6}$$

as
$$dSE_v = dSE_{cy} = dSR_v(\Pi_{as} - \Pi_e) \tag{3.3.7}$$

for
$$dV_{as} = \Pi_{as}V_{cy}dSR_v \tag{3.3.8}$$

and
$$dV_{ae} = \Pi_e dV_e \tag{3.3.9}$$

Thence, the essential data for the purity leaving the cylinder is given by rearranging Eq. 3.3.6:

$$\Pi_e = \Pi_{as} - \frac{dSE_v}{dSR_v} \tag{3.3.10}$$

In most circumstances, as the purity, Π_{as}, is unity for it is fresh charge, Eq. 3.3.10 can be reduced to:

$$\Pi_e = 1 - \frac{dSE_v}{dSR_v} \tag{3.3.11}$$

From the measured SE_v-SR_v curves for any cylinder, such as those in Figs. 3.10 and 3.12, it is possible to determine the charge purity leaving the exhaust port at any instant during the ideal volumetric scavenge process. At any given value of scavenge ratio, from Eq. 3.3.11 it is deduced from this simple function involving the tangent to the SE_v-SR_v curve at that point.

Chapter 3 - Scavenging the Two-Stroke Engine

In Fig. 3.15, for two engine cylinders GPBDEF and SCRE, the SE_v-SR_v characteristics have been curve fitted with a line of the form of Eq. 3.3.3:

$$\log_e(1-SE_v) = \kappa_0 + \kappa_1 SR_v + \kappa_2 SR_v^2$$

The purity of the charge leaving the exhaust port can be deduced from the appropriate differentiation of the relationship in Eq. 3.3.3, or its equivalent in the format of Eq. 3.3.4, and inserting into Eq. 3.3.11 resulting in:

$$\Pi_e = \Pi_{as} + (\kappa_1 + 2\kappa_2 SR_v)e^{\kappa_0 + \kappa_1 SR_v + \kappa_2 SR_v^2} \qquad (3.3.12)$$

The application of Eq. 3.3.12 to the experimental scavenging data presented in Fig. 3.16 for the curve fitted coefficients, κ_0, κ_1 and κ_2, for many of the cylinders in that table is given in Figs. 3.18 and 3.19. Here the values of purity, Π_e, at the entrance to the exhaust port are plotted against scavenge ratio, SR_v. The presentation is for eight of the cylinders and they are split into Figs. 3.18 and 3.19 for reasons of clarity.

The characteristics that emerge are in line with predictions of a more detailed nature, such as the simple computational gas-dynamic model suggested by Sher [3.24, 3.25] and the more complex CFD computations by Blair *et al.* [3.37]. In one of these papers, Sher shows that the shape of the charge purity characteristic at the exhaust port during the scavenge process should have a characteristic profile. Sher suggests that the more linear the profile the worse is the scavenging, the more "S"-like the profile the better is the scavenging. The worst scavenging, as will be recalled, is from one of the loop-scavenged cylinders, YAM12. The best overall scavenging is from the loop-scavenged SCRE and UNIFLOW cylinders and, at low scavenge ratios, the cross-scavenged cylinder GPBDEF.

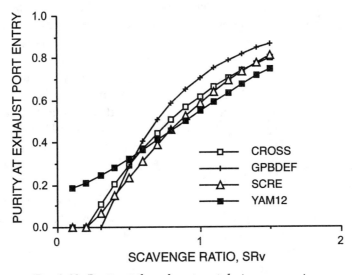

Fig. 3.18 Purity at the exhaust port during scavenging.

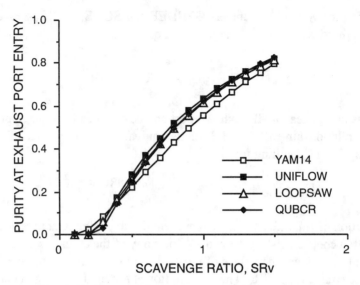

Fig. 3.19 Purity at the exhaust port during scavenging.

In Fig. 3.18 the poor scavenging of cylinder YAM12 is clearly evident. At low scavenge ratios the purity at the exhaust port entry is actually higher than the SR_v value. This indicates very significant short-circuit losses. At the highest SR_v values an apparent contradiction occurs in that it appears to be the best cylinder of that group. The fact is that so much fresh charge has been lost by YAM12 by that point that the cylinder purity (SE_v) is the lowest of that group of cylinders (see Fig. 3.10), and therefore the exhaust port purity is also the lowest as a consequence!

This simple model of scavenging has good correlation with experimentation and more complex theory. It provides mathematical relationships for the interconnection between scavenge ratio, scavenging efficiency and exhaust port purity for the measurements taken over the wide spectrum of scavenging variations seen in practice in the design of two-stroke engines. The employment of Eqs. 3.3.4 and 3.3.12, in terms of the experimentally determined coefficients, κ_0, κ_1 and κ_2, permits the ready assessment of the potential scavenging performance of any measured cylinder or, as an estimation, of any particular type of cylinder, or the characterization of the effect of "good" or "bad" scavenging on the performance potential of an engine. It allows the modeler to interrelate the measured scavenging characteristics of real engine cylinders into an engine simulation, but this will be more fully described in Chapter 5.

The final cautionary words of reminder for you are that the scavenging characteristics, derived by the above analysis of measured data from a QUB single-cycle gas scavenge rig, are by volume and not by mass. Always remember that any simulation of an actual process is just that, for only the real process brings into play the full gamut of changes of thermodynamic state and gas properties which influence the final outcome so heavily. The great benefit of the single-cycle gas rig is that it is a relatively simple experiment in scavenging flow that has been proven to give excellent results in comparison with firing engine tests, but at the same time, due to the thermodynamic and fluid mechanic idealism of that experiment, provides vital information without which simulation of scavenging within a computer model could not be conducted.

3.3.3 Connecting a volumetric scavenging model with engine simulation

The engine simulation by computer, as presented in Chapters 2 and 5, traces the thermodynamics and gas dynamics of flow in, out and through the engine cylinder. The previous section, and in particular Eq. 3.3.12, provided the vital piece of information regarding the value of the purity of the charge leaving the cylinder. It is not an unreasonable assumption that the gas leaving the cylinder is sufficiently well mixed that the exiting gas has a common temperature, T_e. It is also a very reasonable assumption that the pressure throughout the cylinder is uniform and that the exiting charge has a pressure which is the same as the mean value for all of the cylinder contents. However, the temperature of the exiting charge of a stratified process must be composed of exhaust gas and fresh charge elements which are a function of their physical position within the cylinder.

There are two ways of dealing with this problem of the determination of the exiting charge temperature, one simple and one more complex.

The most complex method is actually the easiest to debate, for it consists of conducting a CFD computation using a proprietary computer code, such as Phoenics or StarCD, and obtaining the information very accurately and very completely for the entire scavenge operation. The effectiveness of such a code is described in Sec. 3.4. At the same time, it must be stated that CFD computations are slow, even on the fastest of computers, and the incorporation of a CFD computation of scavenging into the 1D engine simulation as described in Chapters 2 and 5 would change its computation time from several minutes into many days.

Thus, until the great day comes when CFD computations can be accomplished in minutes and not days within an engine simulation model a simpler criterion is required to estimate the temperatures of the air and exhaust components of the charge lost from the cylinder during scavenging. I have presented such a simple criterion [3.35] consisting of the realistic assumption that the temperature differential ΔT_{ax} at any instant between the air and the exhaust is linked to the mean temperature in the cylinder. In other words, as the scavenge flow rate rises, this reduces both the mean temperature within the cylinder, T_{cy}, and the temperature of the exhaust gas, T_{ex}; but also there is an increase of the temperature of the trapped air, T_{ta}. The arithmetic value of the temperature differential, ΔT_{ax} which links these literal statements together is defined by a factor, C_{temp}, thus:

$$T_{ex} = T_{ta} + \Delta T_{ax} = T_{ta} + C_{temp}T_{cy} \qquad (3.3.13)$$

The temperature differential factor, C_{temp}, is clearly a complex function of time, i.e., the engine speed, the displacement/mixing ratio of the fluid mechanics of the particular scavenge process, the heat transfer characteristics of the geometry under investigation, etc. Thus no simple empirical criterion is ever going to satisfy the needs of the simulation for accuracy. My best estimation for the arithmetic value of C_{temp} is given by the following function based on its relationship to (a) the attained value of cylinder scavenging efficiency, SE_{cy}, computed as being trapped at the conclusion of a scavenging process, and (b) the mixing caused by time and piston motion, the effects of which are expressed by the mean piston speed, c_p. The estimated value of C_{temp} would then be employed throughout the following events during the open cycle period for the scavenging of that particular cylinder. The functions with respect to attained scavenging efficiency and mean piston speed, κ_{se} and κ_{cp}, are relative, i.e., they are maximized at unity, so that the relationship for C_{temp} is observed in Eq. 3.3.14 to have an upper bound of 0.65.

$$C_{temp} = 0.65\kappa_{cp}\kappa_{se} \qquad (3.3.14)$$

I have observed from practice that the best estimation of the relationships for κ_{se} and κ_{cp} are:

$$\kappa_{se} = -16.12 + 100.15 SE_{cy} - 229.72 SE_{cy}^2 + 229.56 SE_{cy}^3 - 82.898 SE_{cy}^4$$

$$\kappa_{cp} = -0.34564 + 0.17968 c_p - 8.0818 \times 10^{-3} c_p^2 + 1.2161 \times 10^{-4} c_p^3$$

The value of mean piston speed is obtained from engine geometry using Eq. 1.7.2.

The actual values of trapped air temperature and trapped exhaust gas temperature are then found from a simple energy balance for the cylinder at the instant under investigation, thus:

$$m_{ta} C_{v\,ta} T_{ta} + m_{ex} C_{v\,ex} T_{ex} = m_{cy} C_{v\,cy} T_{cy}$$

or, dividing through by the cylinder mass, m_{cy},

$$SE_{cy} C_{v\,ta} T_{ta} + (1 - SE_{cy}) C_{v\,ex} T_{ex} = C_{v\,cy} T_{cy} \qquad (3.3.15)$$

The arithmetic values for the temperatures of the air and the exhaust gas is obtained by solution of the linear simultaneous equations, Eqs. 3.3.13 and 3.3.15.

3.3.4 Determining the exit properties by mass

It was shown in Sec. 3.3.2 that the exit purity by volume can be found by knowing the volumetric scavenging characteristics of the particular cylinder employed in a particular engine simulation model. The assumption is that the exiting charge purity, Π_e, is determined as a volumetric value via Eq. 3.3.12, and further assumed to be so mixed by that juncture as to have a common exit temperature, T_e. The important issue is to determine the exiting charge purity by mass at this common temperature. It will be observed that Eq. 3.3.12 refers to the assessment of the volumetric related value of the scavenge ratio in the cylinder, SR_v (which should be noted is by volume). The only piece of information available at any juncture during the computation, of assistance in the necessary estimation of SR_v, is either the scavenging efficiency of the cylinder or the scavenge ratio applied into the cylinder, both of which are mass related within the computer simulation. The latter is of little help as the expansion, or contraction, of the air flow volumetrically into the cylinder space is controlled by the cylinder thermodynamics. The instantaneous cylinder scavenging efficiency can be converted from a mass-related value to a volumetric-related value once the average temperatures of the trapped air and the exhaust gas, T_{ta} and T_{ex}, have been found from the above empirical theory. This is found by:

$$SE_v = \frac{V_{ta}}{V_{cy}} = \frac{v_{ta} m_{ta}}{v_{cy} m_{cy}} = \frac{R_a T_{ta}}{R_{cy} T_{cy}} SE_{cy} \qquad (3.3.16)$$

The relevant value of the instantaneous scavenge ratio by volume, SR_v, is found by the insertion of SE_v into Eq. 3.3.3, taking into account the κ coefficients for the particular engine cylinder. Having determined SR_v, the ensuing solution for the exiting charge purity by volume, Π_{ev}, follows directly from Eq. 3.3.12. To convert this value to one which is mass related, Π_{em}, at equality of temperature, T_e, for air and exhaust gas at the exit point, the following theoretical consideration is required of the thermodynamic connection between the mass, m_{ae}, and the volume, V_{ae}, of air entering the exhaust system. From the fundamental definitions of purity by mass and volume:

$$\frac{\Pi_{em}}{\Pi_{ev}} = \frac{m_{ae}/m_e}{V_{ae}/V_e} = \frac{\rho_{ae}}{\rho_e} = \frac{R_a}{R_e}$$

In the unlikely event that the gas constants for air and the exiting charge are identical, the solution is trivial. The gas constants of the mixture are related as follows, by simple proportion:

$$R_e = \Pi_{em} R_a + (1 - \Pi_{em}) R_{ex}$$

Combining these two equations gives a quadratic equation for Π_{em}. Whence the solution for the exiting purity by mass, Π_{em}, is reduced as:

$$\Pi_{em} = \frac{-\dfrac{R_{ex}}{R_a} + \sqrt{\left(\dfrac{R_{ex}}{R_a}\right)^2 + 4\Pi_{ev}\left(1 - \dfrac{R_{ex}}{R_a}\right)}}{2\left(1 - \dfrac{R_{ex}}{R_a}\right)} \qquad (3.3.17)$$

This exit purity, Π_{em}, is, of course, that value required for the engine simulation model to feed the cylinder boundary equations as the apparent cylinder conditions for the next stage of cylinder outflow, rather than the mean values within the cylinder. Further study of Sec. 2.16 or 2.17 will elaborate on this point.

The other properties of the exiting cylinder gas are found as a mixture of Π_{em} proportions of air and exhaust gas, where the pressure is the mean box pressure, p_{cy}, but the temperature

Design and Simulation of Two-Stroke Engines

is T_e which is found from the following approximate consideration of the enthalpy of the exiting mixture of trapped air, T_{ta}, and the exhaust gas, T_{ex}.

$$T_e = \frac{\Pi_{em} C_{p\,ta} T_{ta} + (1 - \Pi_{em}) C_{p\,ex} T_{ex}}{\Pi_{em} C_{p\,ta} + (1 - \Pi_{em}) C_{p\,ex}} \qquad (3.3.18)$$

Further properties of the gas mixture are required and are determined using the theory for such mixtures, which is found in Sec. 2.1.6.

Incorporation of the theory into an engine simulation
This entire procedure for the simulation of scavenging is carried out at each step in the computation by a 1D computer model of the type discussed in Chapter 2.

In Chapter 5, Sec. 5.5.1, is a detailed examination of every aspect of the theoretical scavenging model on the in-cylinder behavior of a chainsaw engine. In particular, Figs. 5.16 to 5.21 give numerical relevance to the theory in the above discussion.

Further discussion on the scavenge model, within a simulation of the same chainsaw engine, is given in Chapter 7. Examples are provided of the effect of widely differing scavenging characteristics on the ensuing performance characteristics of power, fuel economy and exhaust emissions; see Figs. 7.19 and 7.20 in Sec. 7.3.1.

3.4 Computational fluid dynamics

In recent years a considerable volume of work has been published on the use of Computational Fluid Dynamics, or CFD, for the prediction of in-cylinder and duct flows in (four-stroke cycle) IC engines. Typical of such publications are those by Brandstatter [3.26], Gosman [3.27] and Diwakar [3.28].

At The Queen's University of Belfast, much experience has been gained from the use of general-purpose CFD codes called PHOENICS and StarCD. The structure and guiding philosophy behind this theoretical package has been described by their originators, Spalding [3.29] or Gosman [3.27]. These CFD codes are developed for the simulation of a wide variety of fluid flow processes. They can analyze steady or unsteady flow, laminar or turbulent flow, flow in one, two, or three dimensions, and single- or two-phase flow. The program divides the control volume of the calculated region into a large number of cells, and then applies the conservation laws of mass, momentum and energy, i.e., the Navier-Stokes equations, over each of these regions. Additional conservation equations are solved to model various flow features such as turbulent fluctuations. One approach to the solution of the turbulent flow is often referred to as a k-epsilon model to estimate the effective viscosity of the fluid. The mathematical intricacies of the calculation have no place in this book, so the interested reader is referred to the publications of Spalding [3.29] or Gosman [3.27], and the references those papers contain.

A typical computational grid structure employed for an analysis of scavenging flow in a two-stroke engine is shown in Fig. 3.20. Sweeney [3.23] describes the operation of the program in some detail, so only those matters relevant to the current discussion will be dealt with here.

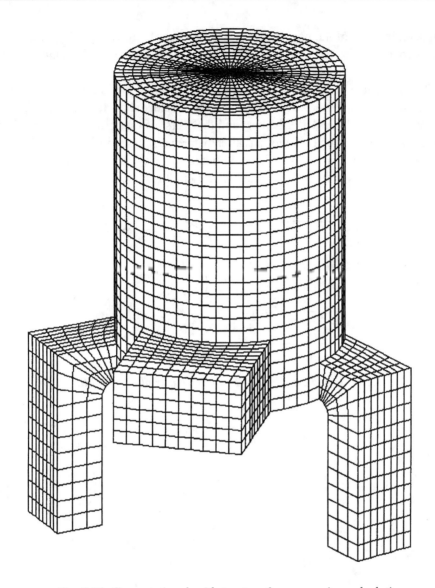

Fig. 3.20 Computational grid structure for scavenging calculation.

In the preceding sections, there has been a considerable volume of information presented regarding experimentally determined scavenging characteristics of engine cylinders on the single-cycle gas scavenging rig. Consequently, at QUB it was considered important to use the PHOENICS CFD code to simulate those experiments and thereby determine the level of accuracy of such CFD calculations.

Further gas-dynamics software was written at QUB to inform the PHOENICS code as to the velocity and state conditions of the entering and exiting scavenged charge at all cylinder port boundaries, or duct boundaries if the grid in Fig. 3.20 were to be employed. The flow entering the cylinder was assumed by Sweeney [3.23] to be "plug flow," i.e., the direction of flow of the scavenge air at any port through any calculation cell at the cylinder boundary was

Design and Simulation of Two-Stroke Engines

in the designed direction of the port. A later research paper by Smyth [3.17] shows this to be inaccurate, particularly for the main transfer ports in a loop-scavenged design, and there is further discussion of that in Sec. 3.5.4.

Examples of the use of the calculation are given by Sweeney [3.23] and a precis of the findings is shown here in Figs. 3.21-30. The examples selected as illustrations are modified Yamaha cylinders, Nos. 14 and 12, the best and the worst of that group whose performance characteristics and scavenging behavior has already been discussed in Sec. 3.2.4. The computation simulates the flow conditions of these cylinders on the QUB single-cycle test apparatus so that direct comparison can be made with experimental results.

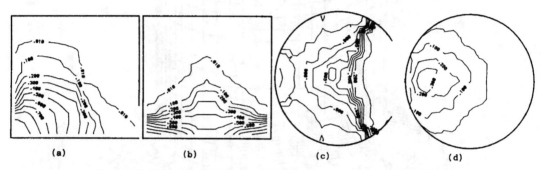

Fig. 3.21 Yamaha cylinder No. 14 charge purity plots at 39° BBDC.

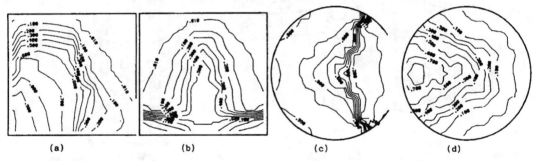

Fig. 3.22 Yamaha cylinder No. 14 charge purity plots at 29° BBDC.

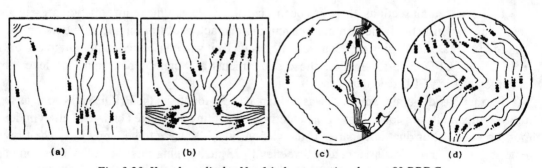

Fig. 3.23 Yamaha cylinder No. 14 charge purity plots at 9° BBDC.

Chapter 3 - Scavenging the Two-Stroke Engine

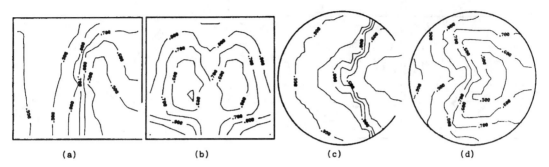

Fig. 3.24 Yamaha cylinder No. 14 charge purity plost at 29° ABDC.

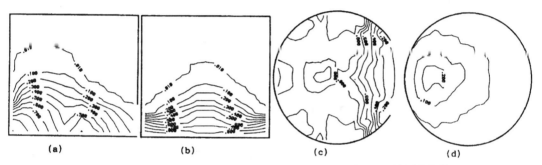

Fig. 3.25 Yamaha cylinder No. 12 charge purity plots at 39° BBDC.

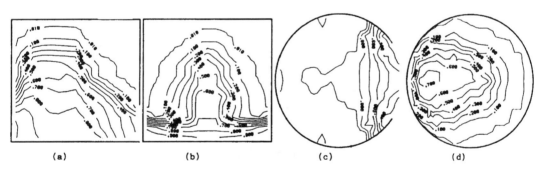

Fig. 3.26 Yamaha cylinder No. 12 charge purity plots at 29° BBDC.

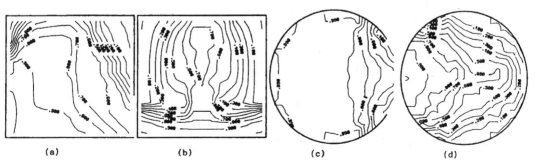

Fig. 3.27 Yamaha cylinder No. 12 charge purity plots at 9° BBDC.

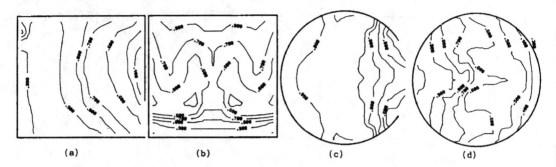

Fig. 3.28 Yamaha cylinder No. 12 charge purity plots at 29° ABDC.

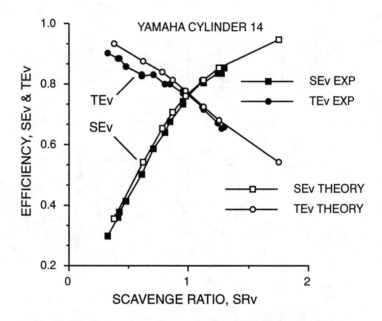

Fig. 3.29 Comparison of experiment and CFD computation.

Comparisons of the measured and calculated SE_v-SR_v and TE_v-SR_v profiles are given in Figs. 3.29 and 3.30. It will be observed that the order of accuracy of the calculation, over the entire scavenging flow regime, is very high indeed. For those who may be familiar with the findings of Sweeney *et al.* [3.23], observe that the order of accuracy of correlation of the theoretical predictions with the experimental data is considerably better in Figs. 3.29 and 30 than it was in the original paper. The reason for this is that the measured data for the variance of the flow from the port design direction acquired by Smyth [3.17] and Kenny [3.31] was applied to the CFD calculations for those same modified Yamaha cylinders to replace the plug flow assumption originally used. It is highly significant that the accuracy is improved, for this moves the CFD calculation even closer to becoming a proven design technique.

Chapter 3 - Scavenging the Two-Stroke Engine

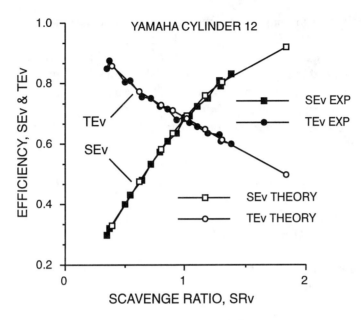

Fig. 3.30 Comparison of experiment and CFD computation.

In terms of insight, Figs. 3.21-3.24 and 3.25-3.28 show the in-cylinder charge purity contours for cylinders 14 and 12, respectively. In each figure are four separate cylinder sections culled from the cells in the calculation. The cylinder section (a) is one along the plane of symmetry with the exhaust port to the right. The cylinder section (b) is at right angles to section (a). The cylinder section (c) is on the surface of the piston with the exhaust port to the right. The cylinder section (d) is above section (c) and halfway between the piston crown and the cylinder head. In Fig. 3.21(a) and Fig. 3.25(a) it can be seen from the flow at 39° bbdc that the short-circuiting flow is more fully developed in cylinder 12 than in cylinder 14. The SE_v level at the exhaust port for cylinder 14 is 0.01 whereas it is between 0.1 and 0.2 for cylinder 12. This situation gets worse by 29° bbdc, when the SE_V value for cylinder 14 is no worse than 0.1, while for cylinder 12 it is between 0.1 and 0.4. This flow characteristic persists at 9° bbdc, where the SE_V value for cylinder 14 is between 0.1 and 0.2 and the equivalent value for cylinder 12 is as high as 0.6. By 29° abdc, the situation has stabilized with both cylinders having SE_V values at the exhaust port of about 0.6. For cylinder 12, the damage has already been done to its scavenging efficiency before the bdc piston position, with the higher rate of fresh charge flow to exhaust by short-circuiting. This ties in precisely with the views of Sher [3.24] and also with my simple scavenge model in Sec. 3.3.2, regarding the Π_e-SR_v profile at the exhaust port, as shown in Fig. 3.18.

Typical of a good scavenging cylinder such as cylinder 14 by comparison with bad scavenging as in cylinder 12 is the sharp boundary between the "up" and the "down" parts of the looping flow. This is easily seen in Figs. 3.23 and 3.27. It is a recurring feature of all loop-scavenged cylinders with good scavenging behavior, and this has been observed many times by the research team at QUB.

At the same time, examining those same figures but in cylinder section (d) closer to the cylinder head level, there are echos of Jante's advice on good and bad scavenging. Cylinder 12 would appear to have a "tongue," whereas cylinder 14 would conform more to the "good" Jante pattern as sketched in Fig. 3.5(a). You may wish to note that I give the measured Jante velocity profiles for cylinders 12 and 14 in Ref. [1.23], and cylinder 12 does have a pronounced "tongue" pattern at 3000 rev/min.

CFD calculations use considerable computer time, and the larger the number of cells the longer is the calculation time. To be more precise, with the simpler cell structure used by Sweeney [3.23] than that shown in Fig. 3.20, and with 60 time steps from transfer port opening to transfer port closing, the computer run time was 120 minutes on a VAX 11/785 mainframe computer. As one needs seven individual calculations at particular values of scavenge ratio, SR, from about 0.3 to 1.5, to build up a total knowledge of the scavenging characteristic of a particular cylinder, that implies about 14 hours of computer run time.

A further technical presentation in this field, showing the use of CFD calculations to predict the scavenging and compression phases of in-cylinder flow in a firing engine, has been presented by Ahmadi-Befrui *et al.* [3.30]. In that paper, the authors also used the assumption of plug flow for the exiting scavenge flow into the cylinder and pre-calculated the state-time conditions at all cylinder-port boundaries using an unsteady gas-dynamic calculation of the type described in Chapter 2 and shown in Chapter 5. Nevertheless, their calculation showed the effect on the in-cylinder flow behavior of a varying cylinder volume during the entire compression process up to the point of ignition.

However, the insight which can be gained from this calculation technique is so extensive, and the accuracy level is also so very impressive, there is every likelihood that it will eventually become the standard design method for the optimization of scavenging flow in two-stroke engines in the years ahead; this opinion takes into account the caveats expressed in Sec. 3.6.

3.5 Scavenge port design

By this time any of you involved in actual engine design will be formulating the view that, however interesting the foregoing parts of this chapter may be, there has been little practical advice on design dispensed thus far. I would contend that the best advice has been presented, which would be to construct a single-cycle gas test rig, at best, or a Jante test apparatus if R&D funds will not extend to the construction of the optimum test method. Actually, by comparison with many test procedures, neither is an expensive form of experimental apparatus. However, the remainder of this chapter is devoted to practical advice to the designer or developer of scavenge ports for two-stroke engines.

3.5.1 Uniflow scavenging

The only type which will be discussed is that for scavenging from the scavenge ports which are deliberately designed to induce charge swirl. This form of engine design is most commonly seen in diesel power units such as the Rootes TS3 or the Detroit Diesel truck engines. Many uniflow-scavenged engines have been of the opposed piston configuration. The basic layout would be that of Fig. 1.5(a), where the scavenge air enters around the bdc position. One of the simplest port plan layouts is shown as a section through the scavenge

Chapter 3 - Scavenging the Two-Stroke Engine

ports in Fig. 3.31. The twelve ports are configured at a swirl angle, ϕ_p, of some 10° or 15°. The port guiding edge is a tangent to a swirl circle of radius, r_p. A typical value for r_p is related to the bore radius, r_{cy}, and would be in the range:

$$\frac{r_{cy}}{2} < r_p < \frac{r_{cy}}{1.25}$$

The port width, x_p, is governed by the permissible value of the inter-port bar, x_b. This latter value is determined by the mechanical strength of the liner, the stiffness of the piston rings in being able to withstand flexure into the ports, and the strong possibility of being able to dispense with pegging the piston rings. Piston rings are normally pegged in a two-stroke engine to ensure that the ends of the ring do not catch in any port during rotation. Because of the multiplicity of small ports in the uniflow case, the potential for the elimination of pegging greatly reduces the tendency for the piston rings to stick, as will tend to be the case if they are held to a fixed position or if they are exposed to hot exhaust gas during passage over the exhaust ports. The piston ring pegs can be eliminated if the angle subtended by the ports is less than a critical value, θ_{crit}. Although this critical port angle q_{crit} is affected by the radial stiffness of the piston rings, nevertheless it is possible to state that it must be less than 23° for

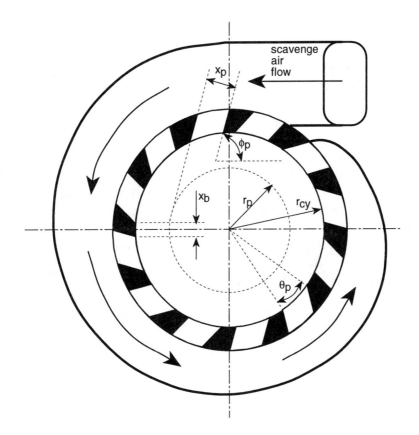

Fig. 3.31 Scavenge port plan layout for uniflow scavenging.

Design and Simulation of Two-Stroke Engines

absolute security, but can be as high as 30° if the piston rings are sufficiently stiff radially and, perhaps more importantly, the ports have adequate top and bottom corner radii to assist with the smooth passage of the rings past the top and bottom edges of the port.

It is clear that uniflow scavenging utilizes the maximum possible area of the cylinder for scavenge porting by comparison with any variant of loop or cross scavenging, as shown in Figs. 1.2-1.4.

The actual design shown in Fig. 3.31, where the ports are at 15° and the radius, r_p, value is sketched at 66% of the cylinder radius dimension, r_{cy}, is a very useful starting point for a new and unknown design.

The use of splitters in the "scavenge belt," to guide the flow into a swirling mode, may not always be practical, but it is a useful optimum to aim at, for the splitter helps to retain the gas motion closer to the design direction at the lower values of scavenge ratio, when the gas velocities are inherently reduced. The use of a scroll, as seen in Fig. 3.31, is probably the best method of assisting the angled ports to induce swirling flow. It is difficult to incorporate into multi-cylinder designs with close inter-cylinder spacing.

The optimum value of the swirl orientation angle, ϕ_p, here shown as 15°, is very much affected by many factors, principal among which is the bore-stroke ratio. It is difficult to be didactic about limits for this criterion, but conventional wisdom, based upon experiment, decrees:

$$5° < \phi_p < 20°$$

If the number of ports are N, then a useful guide to the designer is given by the ratio of the total effective port width to the bore dimension, C_{pb}, where:

$$C_{pb} = \frac{Nx_p}{2r_{cy}} \qquad (3.5.1)$$

The ratio, C_{pb}, is defined as the *port width ratio*. As will be evident in the following sections, it is difficult to raise this ratio much in excess of 1.25 for loop- and cross-scavenged engines. The higher this ratio, the lower can be the port timings to flow a desired scavenge ratio, SR, at the design speed. It is clear from Fig. 3.31 that the value of C_{pb} is considerably greater than unity, as the uniflow engine does not share this cylinder plane with the exhaust port. If the engine is an opposed piston uniflow design, the port timings for both scavenge and exhaust ports can be modest indeed, even when high scavenge ratios are required as is the case for diesel engines. In actual fact, the C_{pb} value for Fig. 3.31 is exactly 2.0.

It is possible to deduce from simple geometry some potential values for the value of the port-width-to-bore ratio.

The effective width of the port, x_p, can be shown to be:

$$x_p = 2r_{cy} \sin\frac{\theta_p}{2} \cos\phi_p$$

Consequently, the port-width-to-bore ratio is deduced as:

$$C_{pb} = N \sin \frac{\theta_p}{2} \cos \phi_p$$

Simply, if one has a design with 12 ports made up of each port segment being 30°, composed of a port angle, θ_p, at 25° and a port bar which subtends 5°, and if the port swirl angle, ϕ_p, is 15°, then the port-width-to-bore ratio, C_{pb}, from the above equation is calculated at 2.51. It is obvious that the absolute maximum which this value can have is π, or 3.14.

It is generally considered that uniflow scavenging provides the optimum in scavenging efficiency at any given scavenge ratio. Examination of the experimental results in Figs. 3.12 and 3.13 would show that this statement is true, but not by the margin stated to be conventional in a literature not overly endowed with experimental back-up to the debate. The fact is that uniflow scavenging is best suited to very long stroke engines, such as marine engines with a bore-stroke ratio of 0.5 or less, where it is undoubtedly the best scavenging method. For automotive engines, gasoline or diesel, with bore-stroke ratios normally close to unity, loop and cross scavenging can be equally effective. Here another factor enters and that is that optimization of uniflow scavenging is much more difficult experimentally than loop or cross scavenging. A swirling uniflow process at low air flow rates has a bad habit of forming a weak central spiral vortex leaving the outer radii unscavenged and vice-versa at high flow rates, where the vortex is concentrated more at the bore circumference leaving the center of the cylinder poorly scavenged. The optimization of uniflow scavenging is much more difficult than the design of a simple "end-to-end" process.

3.5.2 Conventional cross scavenging

The cross-scavenged engine usually has a fairly simple port layout and the scavenging of it is quite easily optimized if the simple design rules set out below are followed. This is in marked contrast to loop and uniflow scavenging, where empiricism as a design guide for the optimization of scavenging is not so effective nor so clear cut. The fundamental reason for this is that the deflector provides the directional guidance for the scavenge flow, whereas in loop and uniflow scavenging the guidance to the flow is based on the interaction of jets of air emanating from the ports. In cross scavenging, the scavenge ports and the exhaust ports are often drilled from the exhaust port side in a single operation, thus simplifying the manufacturing process. Also, if this is carried out with some precision, it increases the accuracy of the port timing events over ports which are merely cast into position.

One of the great advantages of cross-scavenged engines, which is rarely exploited in practice, is that it is relatively easy to design the engine with many narrow ports and eliminate the need for piston ring pegging. This greatly enhances piston ring life and durability, and the lack of excessive wear on the same piece of piston ring passing over the exhaust port improves engine compression retention over longer time periods. When a piston ring can rotate freely in the absence of a peg, there is a self-cleaning action in the ring grooves and more even wear on the rings, and this too enhances the resistance of the engine to piston ring sticking and seizure. The criterion for port width, in order to eliminate ring pegging, is the same as that debated for uniflow scavenging above, i.e., the port angle shown in Fig. 3.32 as the subtended value, θ_p, should be less than 25° in most circumstances.

The basic layout is shown in Fig. 3.32(a) with further amplification in Fig. 1.3. Fig. 3.32(a) contains a plan section with a view on the piston crown and also an elevation section through the scavenge and exhaust ports at the bdc position. Further amplification is also given by Plates 1.6 and 4.2. Fig. 3.32 shows no splitters in the scavenge duct separating the scavenge ports and which would give directional guidance to the scavenge flow. While this is the preferred geometry so that the gas flow follows the design directions, it is rarely seen in practice and the open transfer duct from crankcase to the port belt, which is illustrated here, is the common practice in the industry. The "ears" of the deflector are usually centered about a diameter or are placed slightly toward the scavenge side. The edge of the ears are normally chamfered at 45° and given some (0.5 to 1.0 mm) clearance, x_c, from the bore so that they may enter the combustion chamber above the tdc point without interference. Contrary to popular belief, I [1.10] showed that while this apparently gives a direct short-circuit path to the exhaust port, the actual SE-SR characteristic was hardly affected. Consequently, any loss of engine performance caused by chamfering 50% off the ears of the deflector, and giving them the necessary side clearance, must be due to a further deterioration of the combustion efficiency by introducing even longer and more tortuous flame paths. The "ears" of the solid deflector are chamfered to prevent the corners from glowing and inducing pre-ignition (see Chapter 4).

In Chapters 6 and 7 there will be discussions on the determination of port timings and port heights for all types of engines. In terms of Fig. 3.32(a), it will be assumed that the scavenge port height at bdc is h_s, and there are a number of scavenge ports, N_s. The most common value for N_s is three, but multiple port designs with five or more scavenge ports have been seen in practice. If the ports are rectangular in profile, the maximum scavenge flow area, A_{sp}, and including corner radii, r_c, is given by:

$$A_{sp} = N_s\left(h_s x_s - r_c^2(4 - \pi)\right) \tag{3.5.2}$$

Where the maximum area of the ports at bdc are segments of drilled holes, the designer is forced to calculate that area for use in Eq. 3.5.2. There are two important design criteria, one being the match of the maximum scavenge flow area, A_{sp}, to the deflector flow area, A_{def}, the other being the deflector height. The area in Fig. 3.32(a) enclosed by the deflector radius and the cylinder bore, showing the scavenge flow area past the deflector, is that which is under scrutiny and is defined as A_{def}.

The value of maximum scavenge flow area, A_{sp}, determines the first of the two important design criteria, i.e., the value for r_d which is the radius of the deflector. So that the flow exits the scavenge ports, makes as smooth a transition as possible through 90° and goes toward the cylinder head, the value of the deflector flow area, A_{def}, should approximate the area of the scavenge ports. The best deflection process would be where a restriction took place, but that would reduce the scavenge ratio, SR. The rule of thumb for guidance in this matter can be

Chapter 3 - Scavenging the Two-Stroke Engine

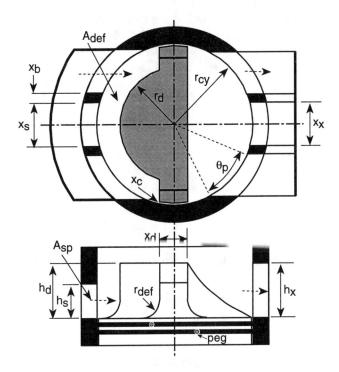

Fig. 3.32(a) Port and deflector layout for conventional cross scavenging.

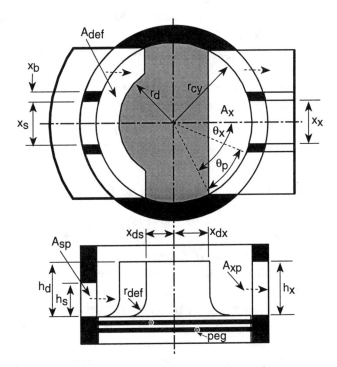

Fig. 3.32(b) Port and deflector layout for unconventional cross scavenging.

expressed as a *deflection ratio*, C_A, which is the ratio of the scavenge flow area, A_{sp}, to the deflector flow area, A_{def}, where the deflection ratio is given by:

$$C_A = \frac{A_{def}}{A_{sp}} \tag{3.5.3}$$

The values of C_A for the best possible answer for conventional cross scavenging is where:

$$1.1 < C_A < 1.2$$

Remember from Sec. 3.2.4 that the reported behavior of traditional cross scavenging is not particularly good, especially at high scavenge ratios. By optimizing such factors as deflection ratio, it is possible to improve upon that level of scavenging effectiveness. Using a C_A factor within the range indicated, it becomes possible to compute a value for the deflector radius, r_d, assuming that the deflector ears are approximately on a diameter of the cylinder bore:

$$r_d = \sqrt{r_{cy}^2 - \frac{2 C_A A_s}{\pi}} \tag{3.5.4}$$

The vital importance of the deflection ratio in the design of the piston crown and the porting of a cross-scavenged engine is illustrated in Fig. 3.33. This graph shows the results of two scavenge tests for a cross-scavenged engine with a deflector of the traditional type, but where the deflection ratios are markedly different at 1.65 and 1.15. The experimental results in Fig. 3.33 are presented as TE_v-SR_v characteristics, and it is observed that the behavior of the classic type of cross scavenging is greatly improved by the optimization of the deflection ratio.

The second and equally important design criterion is for the height of the deflector, h_d. This is defined in the form of a *deflector height ratio*, which is the ratio of port height to deflector height, C_H, where:

$$C_H = \frac{h_d}{h_s} \tag{3.5.5}$$

The values of C_H for the best possible scavenging lie in the band given by:

$$1.4 < C_H < 1.6$$

There are profound disadvantages in designing too tall a deflector. The taller the deflector, the more exposed it will be to the combustion process and the hotter it will become. Worse, the taller the deflector the more complex the combustion space becomes which further deteriorates combustion efficiency. As there is no scavenging advantage in exceeding a C_H value of 1.6, it is often wiser to keep the deflector as low as is practical.

The transition radius at the base of the deflector, r_{def}, is not a particularly critical dimension as far as scavenging is concerned [1.10], and it should be designed as approximately one-third of the port height, h_s.

Chapter 3 - Scavenging the Two-Stroke Engine

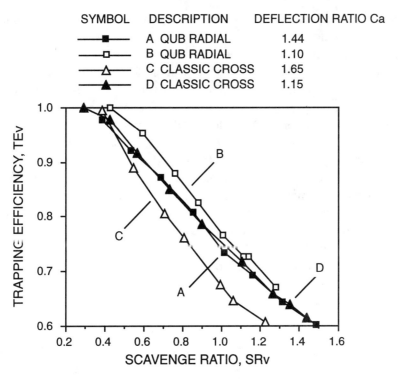

Fig. 3.33 Effect of deflection ratio on scavenging of cross-scavenged engines.

The scavenge port width ratio, C_{pb}, for this type of engine is usually somewhat less than unity, particularly in the straight-drilled port design configuration shown in Fig. 3.32(a). In Fig. 3.32(a) the numerical value is approximately 0.80, which is typical of outboard motor practice.

3.5.3 Unconventional cross scavenging
Fig. 3.32(b) illustrates an alternative design which provides some significant advantages over the conventional design approach. The unconventional design is characterized by:
(a) A deflector which fills the cylinder bore in exactly the same fashion as the QUB type discussed in the Sec. 3.5.4. This permits the deflector to run cooler because heat transfer takes place to the cylinder walls by conduction and from a deflector which is sufficiently wide so that a significant amount of cooling by convection to the crankcase gases can occur on the underside of the piston. With no side clearance on the deflector, the deflection of air charge is more positive and short-circuiting to the exhaust ports from that source is eliminated. While experiments on the single-cycle gas scavenge rig show that a deflector side clearance does not adversely affect the scavenge characteristics at high scavenge ratios, it is at low scavenge ratios that the effect is significant in that the trapping efficiency can be reduced from a "perfect" 100% to a "near perfect" 95 to 96%. This may sound like pedantry, but if the aim of an engine design at low load is zero emission of hydrocarbons at low scavenge ratios

Design and Simulation of Two-Stroke Engines

then this could well amount to an engine design which satisfies, or does not satisfy, emissions legislation.

(b) A deflector which has a discernable downward path for the exhaust side flow so that the particle motion for the in-cylinder gas is more forcibly directed toward the cylinder head as a more pronounced loop into and over the combustion chamber.

(c) A deflector which has a significantly more regular shape than is conventional, and a larger flat area on the top of the piston, so that a more regular combustion chamber shape may be designed to improve the efficiency of that process.

The design procedure follows that of conventional cross scavenging, as given in Sec. 3.5.2, except that the deflector edge on the exhaust side, displaced at x_{dx} from the piston centerline, must also be incorporated into the design loop. The design proposition is that the deflector flow area on the exhaust side, A_x, should equal the total flow area of the exhaust ports at the bdc position, A_{xp}; in other words, the deflector on the exhaust side should guide but not restrict. Simplistically, but including port corner radii, for a design which has N_x rectangular exhaust ports with corner radii, r_c,

$$A_{xp} = N_x \left(h_x x_x - r_c^2 (4 - \pi) \right)$$

As the deflector area, A_x, on the exhaust side is the segment of a circle of radius, r_{cy}, subtended by the angle $2\theta_x$ (in radians), less the area of the triangle back to the centerline, this reduces to:

$$A_x = \frac{2\theta_x}{2\pi} \left(\pi r_{cy}^2 \right) - \frac{1}{2} \left(2 r_{cy} \sin \theta_x \right) \left(r_{cy} \cos \theta_x \right)$$

$$= \frac{r_{cy}^2}{2} \left(2\theta_x - \sin 2\theta_x \right)$$

In this geometrical context it should be noted that:

$$x_{dx} = r_{cy} \cos \theta_x \quad \text{or} \quad \theta_x = \cos^{-1} \frac{x_{dx}}{r_{cy}}$$

This introduces the concept of a deflection ratio for the exhaust side of the piston, C_X, which is defined as:

$$C_X = \frac{A_x}{A_{xp}}$$

and where acceptable design values range from:

$$1.1 < C_X < 1.2$$

This type of cross scavenging, colloquially referred to at QUB as being given by a GPB deflector, has near perfect displacement scavenging up to modest scavenge ratios of about 0.5

or 0.6. Above fresh charge SR_v values of 1.0, it would not be as good as well-optimized loop scavenging. In Fig. 3.12 or 3.13, the measured scavenge lines coded as GPBDEF are from just such a design and the characteristics described above are clearly evident. For an engine being designed to have a low specific power output, say between 3 and 4 bar as a maximum bmep, then a design with optimized porting including this form of scavenging but held at a low scavenge ratio, i.e., no higher than 0.4 to 0.5, could well have the best trapping efficiency, the lowest hydrocarbon emissions, and the least specific fuel consumption for a simple carburetted design.

From a mechanical design standpoint it should be noted that while the engine "height" is no greater than a conventional cross-scavenged engine, the cylinder liner is longer by at least the dimension of the deflector height, h_d, and possibly also the deflector height plus the squish clearance, x_{sq}, depending on the cylinder head layout being employed.

3.5.3.1 The use of Prog.3.3(a) GPB CROSS PORTS

Fig. 3.34(a) shows the screen output of Prog.3.3(a) GPB CROSS PORTS, available from SAE. It is for a small engine with unconventional cross scavenging as debated in Sec. 3.5.3. However, the same program can also assist with the vast majority of the design process for conventional cross scavenging as the basic piston crown layout is virtually the same, apart from the piston crown profile on the exhaust side.

In terms of the relationship of the data input values for the program to the text in Secs. 3.5.2 and 3.5.3, and to Figs. 3.32(a) and (b), the following explanation is offered to the designer. Let each line of input data be examined in turn.

The terms for bore, stroke, con-rod length and engine speed, i.e., BO, ST, CRL and RPM are self-explanatory and, as with all of the units in the programs, they are in mm dimensions unless otherwise stated.

The trapped compression ratio is TCR; the radius in the corner of the combustion chamber is RCV; the number of cylinders of the engine is NCY; the bmep expected at the RPM is BMEP in bar units.

The exhaust ports and transfer ports open at EPO and TPO °atdc, respectively; the number of exhaust and transfer ports are NEXP and NTRP, respectively.

The dimensions, x_{dx} and x_{ds}, are the deflector widths on the exhaust and transfer sides of the piston centerline and are coded as CEX and CTR, respectively; the values of the bars of the exhaust and transfer ports, x_b, are EBAR and TBAR, respectively.

The dimensions, x_x and x_s, are the widths of the exhaust and transfer ports and are coded as WEP and WTP, respectively; the deflection ratios, C_A, and deflector height ratio, C_H, are shown as CDA and CDH, respectively.

The squish clearance, the piston lengths to crown and skirt, and the gudgeon pin offset (+ in direction of rotation) are shown as SQCL, GPC, GPS and GPOFF, respectively.

The port corner radii, r_c, are presented as RTE, RBE, RTT and RBT, the top and bottom corner radii of the exhaust and transfer ports, respectively.

The output data are also fairly self-explanatory; however, each line shows: the engine total swept volume (cm^3) and power output (kW); the subtended port angles, θ_p, of the various exhaust ports; the subtended port angles, θ_p, of the various transfer ports; the exhaust side

Design and Simulation of Two-Stroke Engines

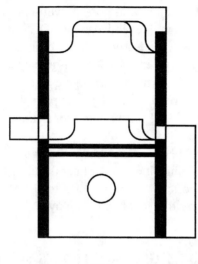

INPUT DATA and data coding for change
BO,ST,CRL,RPM   48.0 36.0 65.0 6500.0
TCR,RCV,NCY,BMEP 7.0 4.0 1.0 4.0
EPO,TPO,NEXP,NTRP 112.0 125.0 4.0 4.0
CEX,CTR,EBAR,TBAR 12.0 12.0 3.0 3.0
WEP,WTP,CDA,CDH 9.0 9.0 1.1 1.4
SQCL,GPC,GPS,GPOFF 1.2 21.0 21.0 0.0
RTE,RBE,RTT,RBT 3.0 3.0 1.0 1.0

OUTPUT DATA
Swept Volume & Power (kW) 65.1   2.8
Exhaust Port angles, º 22.4 25.8
Transfer Port Angles, º 22.4 25.8
Area ratio, piston to exhaust ports 1.20
Blow, Ex &Tr TASV xE4 2.4 108.1 62.9
Defl. Rcy-Rd, & Rd 5.7 18.3
Clearance Volume & Chamber Height 8.1 5.6
Heights, Hd, Hex & Htr ports 8.7 9.1 6.0
Areas, Ex & Tr ports 295.9 211.9

Fig. 3.34(a) GPB deflector cross scavenging through program Prog.3.3(a), GPB CROSS PORTS.

CURRENT INPUT DATA FOR DEFLECTOR PISTON CROWN AND PORTING
(B) BORE,mm= 70   (S) STROKE,mm= 70   (C) CON-ROD,mm= 140
(T) TRANSFER OPENS,deg atdc= 118   (SM) SIDE/MIDDLE PORT WIDTH RATIO = 1
(CR) PORT CORNER RADII,mm= 3   (BC) PORT BAR CHORD WIDTH,mm= 5
(WD) DEFLECTOR WIDTH,mm= 65   (DR) DEFLECTOR RADIUS,mm= 120

OUTPUT DATA
SQUISH AREA RATIO=0.680
PORT WIDTH/BORE RATIO=1.03
DEFLECTOR/PORTS AREA RATIO=1.16
TRANSFER HEIGHT & AREA=   15.1 1062.8
SIDE, MIDDLE PORT WIDTHS,mm= 23.9 23.9
SIDE PORT ANGLE,deg=40.0
MIDDLE PORT ANGLE,deg=40.0
DEFLECTOR ANGLE,deg=136.4
DEFLECTION RATIO,DY/BORE=0.378
DEFLECTOR POSITION,mm,DX & DY=43.5 26.5
PORT-DEFLECTOR ANGLE,deg= 6.1
SQUISH AREA,mm2,=2617
DEFLECTOR FLOW AREA,mm2,=1231

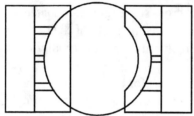

TYPE CODE FOR NEW DATA AND A RECALCULATION
OR TYPE CODE FOR PRINTING(P), FILING(F) OR QUITTING(Q) ??

Fig. 3.34(b) QUB-type cross scavenging through program Prog.3.3(b), QUB CROSS PORTS.

deflection ratio, C_X; the blowdown, exhaust and transfer port time areas of those ports (vide Sec. 6.1); the deflector side clearance, r_{cy}-r_d, and the deflector radius, r_d; the clearance volume (cm^3) and the chamber height above the piston at tdc; the heights, h_d, h_x and h_s, of the deflector, the exhaust ports and transfer ports, respectively; the areas, A_{xp} and A_{sp}, of the exhaust ports and transfer ports, respectively.

A change of data is invited by the programmer for any or all of the input data, as is their recomputation. The code for the data change is the name of that input data parameter. A printout of the entire screen display is available, which is drawn to scale from the input and output data. The design process is aided by the computer program immediately drawing the sketch of the piston and the porting to scale on the computer screen as a function of the input data.

As a supplement to the general discussion on unconventional cross scavenging in Sec. 3.5.3, in Fig. 3.34(a) the breadth of the deflector at 24 mm for a bore of 48 mm means that the piston can be cast with some 6 or 7 mm thick walls all over the crown without encouraging the thick, solid deflector which accompanies the design of the piston in conventional cross scavenging. The availability of a more compact combustion chamber is also evident, albeit with some complexity in the design of the cylinder head spigoted into the cylinder bore. The lack of the scavenge leakage path to the exhaust port, of dimension x_c, is also to be observed. Just as important is the fact that this design will scavenge well at high bore-stroke ratios as the scavenge characteristic called GPBDEF in Figs. 3.12 and 3.13 is for an engine with a bore-stroke ratio greater than 1.3.

3.5.4 QUB-type cross scavenging

A preliminary discussion of this type of scavenging was made in Sec. 1.2.2 and shown in Fig. 1.4. It is clear that the radial disposition of the transfer ports will raise the port width ratio, C_{pb}, and the value for such designs is usually about 1.0. While one can use the straight-in port arrangement, it has been demonstrated that the scavenging of such a design is actually worse than conventional cross scavenging [1.10]. There are further, unpublished, engine tests from QUB which would substantiate that statement. The radial arrangement of scavenge ports has been shown [1.10] to have excellent scavenging characteristics by comparison with good quality loop scavenging and uniflow scavenging, as demonstrated in Figs. 3.12 and 3.13.

The criterion used by the designer to optimize the scavenging process, prior to conducting a confirmatory test on a single-cycle test apparatus, is very similar to that followed for conventional cross scavenging. The deflection ratio, C_A, calculated from Eq. 3.5.3 is still pertinent, and the determination of it, for this more complex geometry, is aided by computer program, Prog.3.3(b).

3.5.4.1 The use of Prog.3.3(b), QUB CROSS PORTS

The design process is best illustrated by the example shown in Fig. 3.34(b) for a 70 mm bore and stroke "square" engine. The figure shows the screen display from the program, slightly accentuated to show the deflector flow area by shading that portion of the piston crown. By requiring input values for cylinder bore, stroke, con-rod length, transfer port timing and transfer port corner radii, the program accurately calculates the total scavenge port height and area, h_s and A_{xp}, as 15.1 mm and 1062.8 mm^2, respectively. The deflector width and radius input data values of 65 and 120, respectively, give the deflector flow area, A_{def}, as 1231 mm². Consequently, the C_A ratio is calculated as 1.16, which would be regarded as a satisfactory value. The value of deflector radius, selected here as 120 mm, and known to provide good scavenging for this particular bore dimension of 70 mm [1.10], can vary over a wide range from 30% to 80% greater than the cylinder bore dimension without unduly affecting the quality of scavenging.

It is difficult to over-emphasize the importance of the deflection ratio in the design of cross-scavenged engines, and further information with respect to the QUB type of cross scavenging is presented in Fig. 3.33. This shows the experimental results for trapping efficiency for two QUB-type engines, one with a deflection ratio of 1.15 and the other with 1.44. The one with the C_A value of 1.15 is the QUBCR cylinder discussed in Sec. 3.2.4 and Figs. 3.12 and 3.13. The higher C_A ratio lowers the trapping efficiency of the QUB type of cross-scavenged cylinder down to a level which places it equal to an optimized example of a classic cross-scavenged design. As further evidence, the QUB type with the higher C_A ratio of 1.44 was built as a firing engine, not just as a model cylinder for experimentation on the single-cycle scavenge rig, and the resulting performance characteristics fell at least 10% short of the behavior reported for the QUB 400 loop-scavenged research engine in Chapters 5 and 7; in other words, they were somewhat disappointing. To emphasize the point, a QUB-type deflector piston engine of the square dimensions of 70 mm bore and stroke and with a deflection ratio of 1.16, with a transfer port layout very similar to that shown in Fig. 3.34(b), was constructed and experimentally produced very similar levels of brake specific fuel consumption, bmep, trapping efficiency and exhaust emission characteristics to the QUB 400 loop-scavenged engine [3.39]. The application of the design into other areas ranging from direct-injection to mopeds and biogas engines can be found in other publications [3.39-3.44].

As described previously for the classic cross-scavenged cylinder, the deflection ratio criterion for the QUB type is the same:

$$1.1 < C_A < 1.2$$

Because the design of this deflector piston influences both scavenge and combustion behavior, the value of *squish area ratio*, C_{sq}, is an important design parameter. This is defined in the same manner as for any engine, namely:

$$C_{sq} = \frac{\text{piston area inducing squish flow}}{\text{cylinder bore area}} \qquad (3.5.6)$$

The squish area is indicated in Fig. 3.34(b), and the value is calculated as 0.680 in the output data table. For this type of piston, acceptable C_{sq} values fall in the range:

$$0.65 < C_{sq} < 0.75$$

The criterion above is not universally applicable, as will be seen when further discussion of this matter takes place when combustion chamber design is debated in Chapter 4.

The detailed selection of the transfer port geometry occurs with the input data values of side-to-middle port width ratio, C_{sm}, and the inter-port bar width, x_b. If at all possible, it is recommended that C_{sm} be equal to unity [1.10]. The data value for the port bar width, x_b, is determined from the mechanical considerations of the piston ring passing over the bar and the ensuing requirement for the rings not to trap in the port nor to excessively wear either the ring or the port bar during the required engine life span. To put an approximate number on the value of x_b, a 3 mm bar for a 40 mm bore and a 6 mm bar for a 90 mm bore would be regarded as typical custom and practice. From all such considerations, the C_A value for deflection ratio is produced with the individual port widths and angles subtended at the bore center. The port width ratio, C_{pb}, for this particular design is 1.03, and is usually some 20% higher than a conventional cross-scavenged engine. The position of the deflector center on the bore centerline, listed on Fig. 3.34(b) as DX and DY, at 43.5 and 26.5 mm, respectively, completes all of the required geometry of the piston crown, in port plan dimensions at least. Like the criterion for the conventional deflector piston given in Eq. 3.5.5, the C_H range for deflector height lies between 1.4 and 1.6, and if this were followed up for a combustion chamber design in Chapter 4, one would expect to see a deflector height, h_d, between 21 and 24 mm.

The design process is aided by the computer program immediately drawing the sketch of the piston and the porting to scale on the computer screen. Input data change on the computer is carried out by typing in the code for the relevant data required; it is the letter in brackets shown in front of the data value in the input data section on Fig. 3.34(b).

Most designers (and I am no exception to this rule) make their final decisions and acceptances by eye, having satisfied themselves that the numerical values of the design criteria fall within acceptable boundaries. Such boundaries are dictated by custom, practice, past experience, new information and, most important of all, the instinct for innovation. In the case of the QUB cross-scavenging engine, such a design type is relatively new, and further design information and experience has been gained since 1990. Therefore, the design criteria quoted numerically have been confirmed with the passage of time and further experience, now that several such engines have been in production and further units have been designed and developed [3.39-3.44].

3.5.5 Loop scavenging

The design process for a loop-scavenged two-stroke engine to ensure good scavenging characteristics is much more difficult than for either the uniflow- or the cross-scavenged engine. In the first place, the number of potential porting layouts is much greater, as inspection of Figs. 1.2 and 3.37 will prove, and these are far from an exhaustive list of the geometrical combinations which have been tried in practice. In the second place, the difference between a successful transfer port geometry and an unsuccessful one is barely discernible by eye. If you care to peruse the actual port plan layouts of Yamaha DT 250 enduro motorcycle engine cylinders 1, 2, 3, 4, 5, 6, 7, 8, 9, 10, 12, 14, 15, presented in papers from QUB [1.23, 3.13 and 3.23], it is doubtful if the best scavenging cylinders could have been predicted in advance of conducting either the firing tests or the single-cycle experimental assessment of their SE_v-SR_v behavior.

In a loop-scavenge design, it is the main port that controls the scavenge process and it is necessary to orient that port correctly. The factors that influence that control are illustrated in Fig. 3.35. These geometrical values are the angles, AM1 and AM2, and their "target" points

Design and Simulation of Two-Stroke Engines

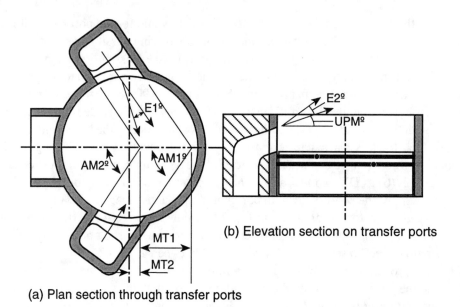

(a) Plan section through transfer ports

(b) Elevation section on transfer ports

Fig. 3.35 Nomenclature for design for Prog.3.4, LOOP SCAVENGE DESIGN.

on the plane of symmetry defined by the distances, MT1 and MT2, respectively. A further factor is the upsweep angle of the main port, UPM. Until the publication of the papers from QUB by Smyth *et al.* [3.17] and Kenny *et al.* [3.31], it was suspected that the exiting gas flow from the main ports did not follow the design direction of the port, but deviated from it in some manner. In particular, it was speculated that the plan angle of the exiting fresh charge would be deviated by an angle, E1, from the design direction, due to the entrainment or short-circuiting of the flow with the exiting exhaust gas flow. That the flow might have a deviation of angle, E2, in the vertical direction was regarded as more problematic. Smyth [3.17] showed that the value of E1 varied from 24.5° to 10.4° from the port opening point to full port opening at bdc, and that the value of E2 varied from 34° to 14° over the same range. As these measurements were taken in a particular engine geometry, the answer from Smyth's experiments cannot be regarded as universal. Nevertheless, as his engine had rather good scavenging for a simple two-port design, it is of the type sketched in Fig. 3.35, with AM1 and AM2 values of 50°, and as the values of gas flow deviation quoted are considerable in magnitude, it must be assumed that most engines have somewhat similar behavior. The actual experimental data presented by Kenny *et al.* [3.31], which was measured by laser doppler velocimetry in an optical access cylinder, are shown in Fig. 3.36. A more extensive discussion of research into this same topic, and for this same engine cylinder, can be found in the paper by McKinley *et al.* [3.45].

As no designer can predict such gas flow deviation behavior, the recommendations for angles and targets for the main port, which have been demonstrated to provide successful scavenging characteristics, should be followed. In the not too distant future, accurate CFD calculations will replace much of the empirical advice in the succeeding sections. Neverthe-

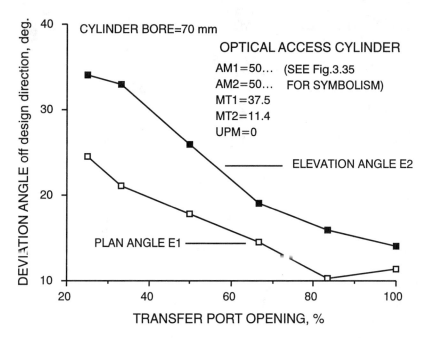

Fig. 3.36 Deviation angles from the design direction for loop scavenging.

less, as CFD calculations are expensive in terms of computer time, designers will always have a need for some better starting point based more on experimental evidence than intuition.

Finally, for those who wish to examine the full extent of porting design seen in two-stroke engines, albeit mostly loop-scavenged two-stroke motorcycle engines, the comprehensive design compendia [3.32] emanating from the Technical University of Graz deserve study as a source of reference on what has been produced.

3.5.5.1 The main transfer port

The orientation of the main port, which is the scavenge port immediately beside the exhaust port, is already stated as the designer's first priority and this has some potential for empirical guidance. If one examines the port plan layouts of all of the cylinders known to provide successful scavenging, several factors stand out quite noticeably:

(a) The upsweep angle of the main port, UPM, is rarely larger than 10°.
(b) The value of AM2 is usually between 50° and 55°.
(c) The target point for MT2 is usually between 10% and 15% of the cylinder bore dimension, BO.
(d) The target point for MT1 is approximately on the edge of the cylinder bore.
(e) The port is tapered to provide an accelerating flow though the port, i.e., AM1 is greater than AM2, and AM1 is rarely larger than 70°.
(f) The larger the angle, AM1, the more the target point, MT1, is inside the cylinder bore. The narrower the angle, AM1, the farther outside the cylinder bore is the target

Design and Simulation of Two-Stroke Engines

point, MT1. The target length, MT1, is a function of the angle, AM1, and a useful empirical relationship is:

$$\text{MT1} = \text{BO}\left(\frac{50 - \text{AM1}}{275} + 0.55\right) \qquad (3.5.7)$$

To formalize some of the criteria stated above, the range of values for AM1 is usually in the band:

$$50 < \text{AM1} < 70$$

Eq. 3.5.7 shows a bore edge intersection for MT1 at a value for AM1 of 63.75°.

(g) In multiple transfer port layouts, i.e., greater than three, it is not uncommon to find the main port with parallel sides.

3.5.5.2 Rear ports and radial side ports

Such ports are sketched in Fig. 3.37, and the main design feature is that they should have upsweep angles between 50° and 60° to ensure attachment of the flow to the cylinder wall opposite to the exhaust port. Jante [3.5] gives a good discussion of such an attachment phenomenon, often known as the *Coanda effect*.

3.5.5.3 Side ports

Such ports are sketched in the bottom half of Figs. 1.2 and 3.37. They can have straight sides or, more usually, the side nearest to the main port has a similar slope to the main port, AM1. The objective is to have the opposing flow paths meet at the back wall, attach to it, and flow up to the cylinder head in a smooth manner. Consequently, an upsweep angle of between 15° and 25° helps that process to occur. Such ports, when employed in conjunction with a rear port, assist the rear port scavenge flow to attach to the back wall by providing a greater pressure differential across the face of the rear port forcing it upward rather than toward the exhaust port.

3.5.5.4 Inner wall of the transfer ports

The transfer ports in many engines, particularly the main transfer ports, have an inner wall of the type seen in Fig. 3.35. The inner and outer walls form a profiled duct from the crankcase exit to the transfer port. This has been discussed in many publications from QUB and the general conclusion is that loop scavenging is always improved by the provision of a contoured duct featuring an inner wall, particularly for the main transfer port. The effective duct profile in elevation is often referred to as resembling a (tea) "cup handle." It is not sufficient to leave this to be formed by the liner, assuming that the design features such, and that the bore surface is not simply plated by hard chromium or a silicon-based material, as this leaves but a vertical duct normally of constant area. The photograph in Plate 3.4 shows the profile of machined scavenge duct cores very clearly and the presence of the inner wall, and its profile, is quite visible.

Chapter 3 - Scavenging the Two-Stroke Engine

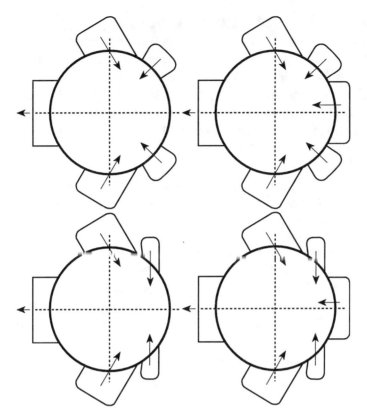

Figure 3.37 Together with Fig. 1.2 shows the port layouts for design by Prog.3.4, LOOP SCAVENGE DESIGN.

Put another way, small and inexpensive engines for industrial applications are often designed with "finger" ports which have no inner walls at all and which are open to the moving piston surface. The normal experience is that they exhibit a very indifferent quality of scavenging when measured on a single-cycle gas scavenge rig.

3.5.5.5 Effect of bore-to-stroke ratio on loop scavenging
The tradition has it that loop scavenging is most effective in any cylinder size when the engine is of "square" dimensions. Experiments at QUB have shown that the most effective band for loop scavenging is from a bore-stroke ratio between unity and 1.2.

3.5.5.6 Effect of cylinder size on loop scavenging
The tradition has it that the higher the engine swept volume, the better the scavenging quality for loop scavenging. Experimental experience at QUB over a thousand plus experiments shows that virtually all sizes of engine can be made to scavenge equally well if the R&D effort is put into so doing. However, it is probably true to say that it is much more difficult to make a small cylinder, say of 50 cm^3, scavenge well by design or by development than one of, say, 500 cm^3.

3.5.5.7 The use of Prog.3.4, LOOP SCAVENGE DESIGN

A useful design program for all of the porting layouts illustrated in Figs. 1.2 and 3.37 is available from SAE. The program is self-explanatory and the typical outputs shown in Figs. 3.38 and 3.39 are exactly what you see on the computer screen. The opening screen display asks you to decide on the number of transfer ports in the design, i.e., two, three, four or five, and from then on the design procedure will take the form of changing a particular data value until you are satisfied that the aims and objectives have been achieved. Such objectives would typically be the satisfaction of the criteria in Secs. 3.5.5.1 to 3.5.5.3 and to achieve the desired value of port width ratio, C_{pb}.

Upon completion of the design process, the printer will provide hard copy of the screen and, as can be seen from Fig. 3.39, give some useful manufacturing data for the cylinder liner. This becomes even more pertinent for four and five port layouts, where the structural strength of some of the port bars can become a major factor.

It is interesting to note that in the actual design shown for a 70-mm-bore cylinder in this three transfer port layout, the C_{pb} value of 1.06 is about the same as for the QUB deflector piston engine; it was 1.03. It is clear that it would be possible to design a four or five port layout to increase the C_{pb} value to about 1.3. Remember that the uniflow engine had an equivalent value of 2.5, although if that uniflow engine is of the long stroke type, or is an opposed piston engine, it will require a C_{pb} value at that level to achieve equality of flow area with the other more conventional bore-stroke ratio power units.

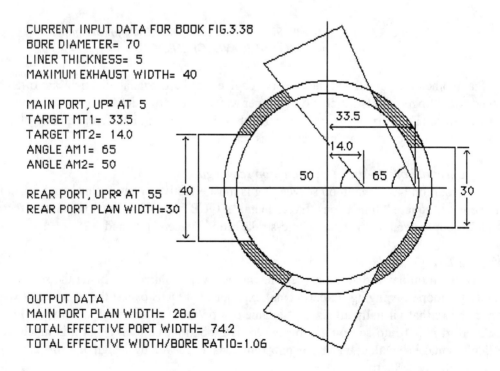

Fig. 3.38 First page of output from Prog.3.4, giving the scavenge geometry of the ports.

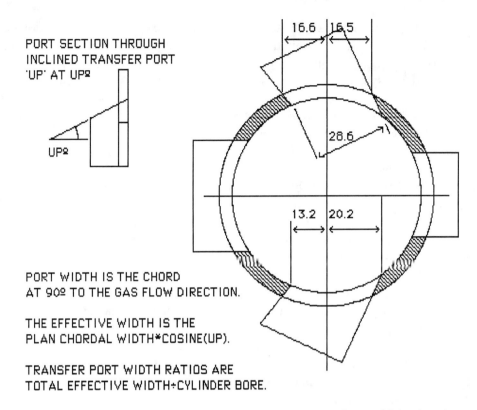

Fig. 3.39 *Second page of output from Prog.3.4, giving port edge machining locations.*

3.5.6 Loop scavenging design for external scavenging

The design of engines for the direct injection of fuel, be they compression-ignition or spark-ignition units, is often carried out by pressure charging the cylinder [3.47-3.49]. This has already been discussed in a preliminary manner in Sec. 1.2.4. In other words, the supply of fresh air for the scavenge process is by a supercharger or by a turbocharger directly to the scavenge (transfer) ports without the use of the crankcase as an air pumping system. Naturally, if a turbocharger is employed, which is driven by exhaust gas energy entering the turbine part of the device then the engine has no self-starting capability. Irrespective of how the engine is started, the air is supplied directly to the scavenge ports by an external air pump. If it is a supercharger, often referred to as a "blower," it can be of the Roots or Eaton type, a screw type, or a centrifugal design. If it is a turbocharger, then for the automotive engine this is conventionally an inward radial flow design for the turbine and an outward radial flow type for the compressor. For the marine engine, axial flow turbines and compressors are the most common units.

Externally scavenged engines are usually of a multi-cylinder design, in which case the trapping of the cylinder charge is of paramount importance as an excess of air supply leads to high pumping losses in the form of the power required to drive the supercharger. Obviously this is less vital for the turbocharged case where the air is supplied by exhaust gas energy to the turbine. Nevertheless, the engine performance is characterized by the cylinder trapping characteristics which are dominated by the optimization of the scavenging characteristics and also the unsteady gas dynamics within the exhaust manifold; a comprehensive debate on the latter topic is in Chapter 5. While the decision on the number of cylinders which will best optimize the exhaust system gas dynamics is left to the debate in Chapter 5, nevertheless the trapping characteristics to be optimized by scavenging are found by the enhancement of the in-cylinder flow process at the minimum port heights allied to maximized port flow areas.

The optimization of the in-cylinder flow process can be conducted experimentally using the single-cycle testing method [3.48] or theoretically using CFD [3.45]. Before embarking on either of these routes, some basic principles for the design should be established.

Bear in mind that the external "blower," of whatever configuration, is going to force air into the cylinder virtually throughout the entire scavenge flow period from the opening of the scavenge ports to their closing. This is in marked contrast to the action of the crankcase pump in the crankcase compression engine where the majority of the scavenge flow is concentrated around the bdc period, and very little of it occurs in the latter third of the scavenge period as the crankcase pressure has decayed considerably by that point. Therefore, for the "blown" engine, at equality of air flow rate and port area by comparison with a crankcase compression engine, the scavenging process is conducted with an air flow which has a longer duration but at velocities which are more constant and of a lower amplitude.

To achieve optimization of external scavenging, in terms of maximized port areas and minimized port heights, utilization of the majority of the port circumference assumes a high priority. A sketch of just such an approach is shown in Fig. 3.40, and is easily recognized from the history of loop-scavenged two-stroke diesel engines [3.49]. As the need for a transfer port from the crankcase has been eliminated, it becomes possible to direct the scavenge flow from the exhaust port side of the cylinder, and as close as possible to the same plane as the exhaust ports. This can reduce, and potentially eliminate, the deleterious plan and elevation angles of deviation for the flow of the first scavenge port, shown as angles E1 and E2 in Figs. 3.35 and 3.36. Many designs [3.49] have employed varying angles of upsweep on the ports, UPM° in Fig. 3.35, but my view is that this complicates the liner design and manufacture, and deteriorates the magnitude of the effective port areas. However, this advantage in gas-dynamic terms must not result in a sacrifice of good scavenging characteristics. In short, it is recommended that the optimization on the QUB single-cycle test apparatus is conducted logically using various test liners machined where the focus of the port edges is positioned at values dictated by their proportion of the cylinder radius, r_{cy}, along the axis of the cylinder on the scavenge side.

If the design is for a multi-cylinder unit, then the inter-cylinder disposition of the scavenge duct as a siamesed layout serving two adjacent cylinders is a straightforward procedure.

The design process for such a cylinder is geometrically complex and tedious. This tedium is greatly eased by the employment of the simple computer program, Prog.3.5 BLOWN PORTS.

3.5.6.1 The use of Prog.3.5, BLOWN PORTS

Fig. 3.41 shows a screen print of the display. In terms of the relationship of the data input values for the program to the text in Sec. 3.5.6, and to Fig. 3.41, the following explanation is offered. Let each line of input data be examined in turn.

Chapter 3 - Scavenging the Two-Stroke Engine

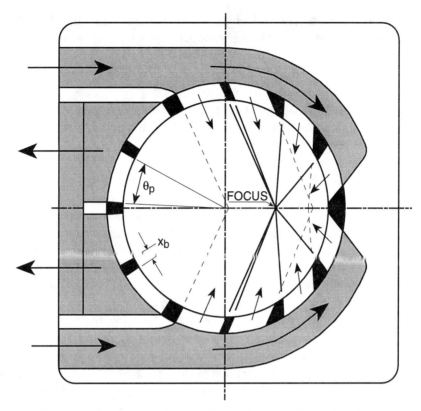

Fig. 3.40 Loop scavenging port layout for supercharged engines.

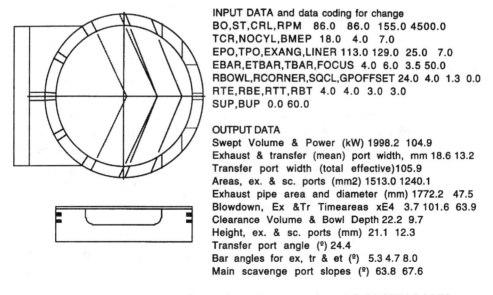

INPUT DATA and data coding for change
BO,ST,CRL,RPM   86.0 86.0 155.0 4500.0
TCR,NOCYL,BMEP   18.0 4.0 7.0
EPO,TPO,EXANG,LINER 113.0 129.0 25.0 7.0
EBAR,ETBAR,TBAR,FOCUS 4.0 6.0 3.5 50.0
RBOWL,RCORNER,SQCL,GPOFFSET 24.0 4.0 1.3 0.0
RTE,RBE,RTT,RBT 4.0 4.0 3.0 3.0
SUP,BUP 0.0 60.0

OUTPUT DATA
Swept Volume & Power (kW) 1998.2 104.9
Exhaust & transfer (mean) port width, mm 18.6 13.2
Transfer port width (total effective)105.9
Areas, ex. & sc. ports (mm2) 1513.0 1240.1
Exhaust pipe area and diameter (mm) 1772.2 47.5
Blowdown, Ex &Tr Timeareas xE4 3.7 101.6 63.9
Clearance Volume & Bowl Depth 22.2 9.7
Height, ex. & sc. ports (mm) 21.1 12.3
Transfer port angle (º) 24.4
Bar angles for ex, tr & et (º) 5.3 4.7 8.0
Main scavenge port slopes (º) 63.8 67.6

Fig. 3.41 Loop-scavenging design through program Prog.3.5, BLOWN PORTS.

The terms for bore, stroke, con-rod length and engine speed, i.e., BO, ST, CRL and RPM are self-explanatory and, as with all of the units in the programs, they are in mm dimensions unless otherwise stated.

The trapped compression ratio is TCR; the number of cylinders of the engine is NCY; the bmep expected at the RPM is BMEP in bar units.

The exhaust ports and transfer ports open at EPO and TPO °atdc, respectively; the subtended port angles, θ_p, of the exhaust ports is EXANG; the radial thickness of the cylinder liner in the scavenge port belt is LINER.

The chordal values of the bars of the exhaust and scavenge (transfer) ports, x_b, are EBAR and TBAR, while that for the bar between the last exhaust port and the first scavenge port is coded as ETBAR, respectively. The value of FOCUS is the proportion of the cylinder radius, from the cylinder center to the edge of the bore on the scavenge side, where the guiding edges of the main scavenge ports index the centerline. The guiding edges in question are the trailing edges of the first port, and the leading edge of the second backswept port. This is shown even more clearly in Fig. 3.40, where they are reproduced in the same manner as the screen display of the computer program. The side port is at right angles to the line of symmetry and the back ports are aligned with it, facing the exhaust port. Naturally, it is essential in this design configuration to elevate the back ports, usually by 50-60°. To allow the designer this flexibility, both the side and the back ports can be elevated by setting angles SUP and BUP between 0 and (a maximum of) 60°.

The plan radius of a recessed cylinder bowl in the fashion of a DI diesel engine is RBOWL; the corner radius of that bowl is RCV; the squish clearance, the piston lengths to crown and skirt, and the gudgeon pin offset (+ in direction of rotation) are shown coded as SQCL, GPC, GPS and GPOFF, respectively.

The port corner radii, r_c, are presented as RTE, RBE, RTT and RBT, the top and bottom corner radii of the exhaust and scavenge (transfer) ports, respectively.

The output data are also fairly self-explanatory. However, each line shows: the engine total swept volume (cm^3) and power (kW); the mean widths of all of the exhaust and scavenge ports; the port width ratio C_A; the areas A_{xp} and A_{sp} of the exhaust ports and scavenge ports, respectively; the exhaust pipe area and its diameter; the blowdown, exhaust and transfer port time areas of those ports (vide Sec. 6.1); the blowdown, exhaust and transfer port time areas of those ports (vide Sec. 6.1); the combustion chamber clearance volume (cm^3) and the depth of that chamber; the heights, h_x and h_s, of the exhaust ports and scavenge ports, respectively; the subtended port angles, θ_p, of each of the scavenge ports for they are specifically designed to be identical; the subtended angles, θ_b, of each of the bars of the exhaust ports, between the exhaust and the scavenge ports, and between each of the scavenge ports, respectively; the slopes (deg) of each of the focusing edges of the main scavenge ports with respect to the "horizontal" centerline of scavenge symmetry.

A change of data is invited by the program for any or all of the input data, as is their recomputation. The code for that data change is the name of that particular input data parameter, to be typed in by the user. A hard copy, i.e., a printout, is available of the entire screen display, and this is drawn to scale from the input and output data. The "by eye" design process is aided by the computer program immediately drawing the sketch of the piston and the porting to scale on the computer screen as a function of the input data.

The input and output data in Fig. 3.41 detail the geometry of a four-cylinder DI diesel engine. Those dedicated to the design of the spark-ignition engine should note that the expected power output of this two-liter diesel automobile engine is 104.9 kW at 4500 rpm, which is well up to the performance levels of any gasoline engine of the same swept volume. The design has been conducted so that the maximum port angle, θ_p, of any one port does not exceed 25°, permitting the use of unpegged rings on the piston. Even though the back ports are elevated at 60°, while the side ports are at 0°, the effective port width ratio, C_{pb}, for the scavenge ports is high at 1.23 and reflects the design criterion of maximized port area for the given bore circumference. The equivalent C_{pb} number for the exhaust ports is (4 times 18.6 divided by 86) 0.865, also a high number by loop-scavenging design standards. The exhaust and scavenge port timings are low at opening values of 113° and 129° atdc, which mild timing still gives acceptably high values of time-areas (see Sec. 6.1) in order to pass the requisite exhaust and scavenge flows in the allotted time span at 4500 rpm. The low port timings are required to give good cylinder trapping conditions for a four-cylinder engine from both a scavenging standpoint and from that of the exhaust system gas dynamics.

A fuller debate on the latter topic is in Chapter 5, Sec. 5.5.3, where this program is used to provide the porting data for the simulation of a four-cylinder, externally scavenged, spark-ignition, direct-injection engine for automobile applications.

3.6 Scavenging design and development

The development of good scavenging is essential for the two-stroke engine to provide the ever more sophisticated performance characteristics required of engines to meet international legislation for exhaust emissions and fuel consumption levels. In Fig. 3.42 is a tiny sample of the range of scavenging characteristics found experimentally from some 1300 differing geometries tested in recent times. They are for uniflow engines (triangle symbol), cross-scav-

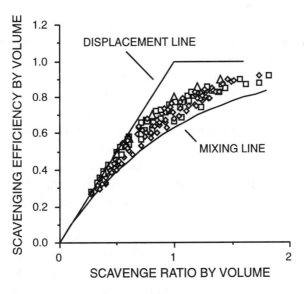

Fig. 3.42 Scatter of experimental SE_v-SR_v characteristics.

enged units (square symbol) and loop-scavenged designs (round symbol) and it is obvious that there is no clearcut winner in the race for perfect scavenging and that many, indeed the majority, of the units tested could withstand improvement by either experimental or by theoretical methods.

You can observe from reading this chapter that the knowledge base on scavenging has been greatly enlarged in the past two decades. As with many other branches of science and engineering this can be attributed to the ever wider use of computation which has moved from the mainframe to the desktop and laboratory bench level. Nowhere is this more true than for R&D related to scavenging in two-stroke engines.

The increasing use of computational fluid dynamics holds out the hope for a complete design process of the in-cylinder flow regime in terms of an accurate assessment of the SE-SR characteristics, either prior to the manufacture of the cylinder or as part of the onward development of existing engine designs.

The single-cycle gas testing apparatus, an experimental tool which became possible by the development of the accurate paramagnetic oxygen analyzer, is actually devoid of any complex computer control technology and could well have been employed for R&D much earlier in history. However, the preparation of model cylinders for experimental work on this rig has been greatly assisted by modern CAD/CAM techniques. Since 1990 it has been possible to machine scavenge duct profiles directly in a low-melting-point alloy using surface (or solid) modeling CAD software. The machined transfer duct core is immersed into a cold-setting plastic compound in a mold. When the plastic has set, the machined duct core can be "melted" out by immersion in hot water and the material itself recycled. A photograph of the machined cores and the finished test cylinder are shown in Plate 3.4. The important issue here is that when the test sequence on the single-cycle rig has been completed, the information on the scavenge duct profiles which produced the best SE_v-SR_v characteristics is not a drawing which can lead to misinterpretation, but a file of CAM data to permit the accurate transmission of the finalized scavenge duct geometry into the manufacturing process. At QUB it is possible to produce test cylinders and their scavenge ducts in approximately one week of CAD and CAM activity, and another week to conduct the experimental optimization process on the single-cycle test apparatus.

More advanced techniques are now available which speed up this process. The advent of "rapid prototyping," otherwise known as stereo lithography, means that the entire cylinder including its scavenge ducting can be modeled in 3D CAD software and manufactured by a laser etching process in a resin which hardens to the point where it can be used directly on the QUB single-cycle test apparatus. Plate 3.5 shows a prototype cylinder produced by stereo lithography at QUB for use on the single-cycle test apparatus. It is possible to conduct this entire prototype cylinder design and manufacturing process in two or three days and conclude the entire optimization process for the scavenging in ten days total. The data file for the "rapid prototyping" stage, and indeed the solid cylinder model it produces, can be used in the ensuing manufacturing process to create cores or molds for the production castings.

Which then is the optimum method to employ for the development of optimized scavenging for the two-stroke engine cylinder: single-cycle gas experiment or CFD theory, or should both techniques be applied during R&D in series or parallel? What does the word optimum

Chapter 3 - Scavenging the Two-Stroke Engine

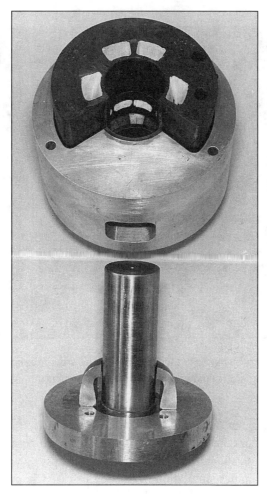

Plate 3.4 CAD/CAM techniques at QUB enhance the accuracy of transfer of cylinder design to experimental scavenging test.

mean? Is it simply the best answer? Is it the best answer in the shortest time or is it simply the least money expended in the search for the answer. In most realms of R&D activity, money expended tends to equate with time spent.

While computer technology has provided the CFD theory with which to perform the theoretical calculations, it has also provided a problem in terms of data preparation prior to that computation. Fig. 3.20 shows the computation grid required to perform the CFD calculation for flow from cell to cell within a two-stroke cycle engine cylinder. The computational meshing procedure is very time consuming; it can occupy several weeks of concentrated activity by a very experienced programmer. The computation process after that can take days on a fast computer, perhaps even requiring the use of a supercomputer. Thus, while the ensuing answer is highly effective and full of information for the designer and the developer, it takes time and is expensive. What is required to significantly advance the art and science of CFD is the ability to accurately and rapidly mesh the structure and space being investigated.

Plate 3.5 Stereo lithography techniques at QUB provide accurate prototype cylinders for scavenging experiments.

Thus, advanced computer technology has provided sophisticated tools for the optimization of scavenging both theoretically and experimentally. In this context remember that either approach must permit the acquisition of SE-SR characteristics, by volume or by mass, which are required for a 1D computer simulation of the engine (see Chapter 5 or Sec. 3.3). At this moment in time, the fastest and most dependable optimization method is actually the experimental one using the single-cycle gas rig. If the theoretical technology using CFD matures, with its accuracy certified by much more extensive correlation with experiments than is current and, more importantly in time and cost terms, automatic meshing becomes a practical reality, then that situation could well change.

References for Chapter 3

3.1 B. Hopkinson, "The Charging of Two-Cycle Internal Combustion Engines," *Trans. NE Coast Instn. Engrs. Shipbuilders*, Vol 30, 1914, p433.

3.2 R.S. Benson, P.J. Brandham, "A Method for Obtaining a Quantitative Assessment of the Influence of Charge Efficiency on Two-Stroke Engine Performance," *Int.J.Mech.Sci.*, Vol 11, 1969, p303.

3.3 F. Baudequin, P. Rochelle, "Some Scavenging Models for Two-Stroke Engines," *Proc.I.Mech.E.*, Vol 194, 1980, pp203-210.

3.4 C. Changyou, F.J. Wallace, "A Generalized Isobaric and Isochoric Thermodynamic Scavenging Model," SAE Paper No. 871657, Society of Automotive Engineers, Warrendale, Pa. 1987.

3.5 A. Jante, "Scavenging and Other Problems of Two-Stroke Spark-Ignition Engines," SAE Paper No. 680468, Society of Automotive Engineers, Warrendale, Pa. 1968.

3.6 G.P. Blair, "Studying Scavenge Flow in a Two-Stroke Cycle Engine," SAE Paper No. 750752, Society of Automotive Engineers, Warrendale, Pa. 1975.

3.7 G.P. Blair, M.C. Ashe, "The Unsteady Gas Exchange Characteristics of a Two-Cycle Engine," SAE Paper No. 760644, Society of Automotive Engineers, Warrendale, Pa. 1976.

3.8 T. Asanuma, S. Yanigahara, "Gas Sampling Valve for Measuring Scavenging Efficiency in High Speed Two-Stroke Engines," *SAE Transactions*, Vol 70, p.420, Paper T47, 1962.

3.9 R.R. Booy, "Evaluating Scavenging Efficiency of Two-Cycle Gasoline Engines," SAE Paper No. 670029, Society of Automotive Engineers, Warrendale, Pa., 1967.

3.10 N. Dedeoglu, "Model Investigations on Scavenging and Mixture Formation in the Dual-Fuel or Gas Engine," *Sulzer Tech Review*, Vol 51, No 3, 1969, p133.

3.11 W. Rizk, "Experimental Studies of the Mixing Processes and Flow Configurations in Two-Cycle Engine Scavenging," *Proc.I.Mech.E.*, Vol 172, 1958, p417.

3.12 S. Ohigashi, Y. Kashiwada, "A Study on the Scavenging Air Flow through the Scavenging Ports," *Bull.J.S.M.E.*, Vol 9, No 36, 1966.

3.13 D.S. Sanborn, G.P. Blair, R.G. Kenny, A.H. Kingsbury, "Experimental Assessment of Scavenging Efficiency of Two-Stroke Cycle Engines," SAE Paper No. 800975, Society of Automotive Engineers, Warrendale, Pa., 1980.

3.14 V.L. Streeter, Fluid Mechanics, McGraw-Hill, 3rd Edition, Tokyo, 1962.

3.15 H. Sammons, "A Single-Cycle Test Apparatus for Studying ÔLoop-Scavenging' in a Two-Stroke Engine," *Proc.I.Mech.E.*, Vol 160, 1949, p233.

3.16 W.H. Percival, "Method of Scavenging Analysis for Two-Stroke Cycle Diesel Cylinders," *SAE Transactions*, Vol 62, 1954, p737.

3.17 J.G. Smyth, R.G. Kenny, G.P. Blair, "Steady Flow Analysis of the Scavenging Process in a Loop Scavenged Two-Stroke Cycle Enginea Theoretical and Experimental Study," SAE Paper No. 881267, Society of Automotive Engineers, Warrendale, Pa., 1988.

3.18 R.G. Kenny, "Scavenging Flow in Small Two-Stroke Cycle Engines," Doctoral Thesis, The Queen's University of Belfast, May 1980.

3.19 R.G. Phatak, "A New Method of Analyzing Two-Stroke Cycle Engine Gas Flow Patterns," SAE Paper No. 790487, Society of Automotive Engineers, Warrendale, Pa., 1979.

3.20 M.E.G. Sweeney, R.G. Kenny, G.B. Swann, G.P. Blair, "Single Cycle Gas Testing Method for Two-Stroke Engine Scavenging," SAE Paper No. 850178, Society of Automotive Engineers, Warrendale, Pa., 1985.

3.21 D.S. Sanborn, W.M. Roeder, "Single Cycle Simulation Simplifies Scavenging Study," SAE Paper No. 850175, Society of Automotive Engineers, Warrendale, Pa., 1985.

3.22 C. Mirko, R. Pavletic, "A Model Method for the Evaluation of the Scavenging System in a Two-Stroke Engine," SAE Paper No. 850176, Society of Automotive Engineers, Warrendale, Pa., 1985.

3.23 M.E.G. Sweeney, R.G. Kenny, G.B. Swann, G.P. Blair, "Computational Fluid Dynamics Applied to Two-Stroke Engine Scavenging," SAE Paper No. 851519, Society of Automotive Engineers, Warrendale, Pa., 1985.

3.24 E. Sher, "An Improved Gas Dynamic Model Simulating the Scavenging Process in a Two-Stroke Cycle Engine," SAE Paper No. 800037, Society of Automotive Engineers, Warrendale, Pa., 1980.

3.25 E. Sher, "Prediction of the Gas Exchange Performance in a Two-Stroke Cycle Engine," SAE Paper No. 850086, Society of Automotive Engineers, Warrendale, Pa., 1985.

3.26 W. Brandstatter, R.J.R. Johns, G. Wrigley, "The Effect of Inlet Port Geometry on In-Cylinder Flow Structure," SAE Paper No. 850499, Society of Automotive Engineers, Warrendale, Pa., 1985.

3.27 A.D. Gosman, Y.Y. Tsui, C. Vafidis, "Flow in a Model Engine with a Shrouded Valve. A Combined Experimental and Computational Study," SAE Paper No. 850498, Society of Automotive Engineers, Warrendale, Pa., 1985.

3.28 R. Diwakar, "Three-Dimensional Modelling of the In-Cylinder Gas Exchange Processes in a Uniflow-Scavenged Two-Stroke Engine," SAE Paper No. 850596, Society of Automotive Engineers, Warrendale, Pa., 1985.

3.29 D.B. Spalding, "A General Purpose Computer Program for Multi-Dimensional One and Two-Phase Flow," *Mathematics and Computers in Simulation*, Vol 23, 1981, pp267-276.

3.30 B. Ahmadi-Befrui, W. Brandstatter, H. Kratochwill, "Multidimensional Calculation of the Flow Processes in a Loop-Scavenged Two-Stroke Cycle Engine," SAE Paper No. 890841, Society of Automotive Engineers, Warrendale, Pa., 1989.

3.31 R.G. Kenny. J.G. Smyth, R. Fleck, G.P. Blair, "The Scavenging Process in the Two-Stroke Engine Cylinder," Dritte Grazer Zweiradtagung, Technische Universitat, Graz, 13-14 April, 1989.

3.32 F. Laimbock, "Kenndaten und Konstruktion Neuzeitlicher Zweiradmotoren," Heft 31a, Heft 31b, Heft 31c, Institut for Thermodynamics and Internal Combustion Engines, Technische Universitat Graz, Austria, 1989.

3.33 M. Nuti, L. Martorano, "Short-Circuit Ratio Evaluation in the Scavenging of Two-Stroke S.I. Engine," SAE Paper No. 850177, Society of Automotive Engineers, Warrendale, Pa., 1985.

3.34 G.P. Blair, <u>The Basic Design of Two-Stroke Engines</u>, SAE R-104, Society of Automotive Engineers, Warrendale, Pa., ISBN 1-56091-008-9, February 1990.

3.35 G.P. Blair, "Correlation of an Alternative Method for the Prediction of Engine Performance Characteristics with Measured Data," SAE Paper No. 930501, Society of Automotive Engineers, Warrendale, Pa., 1993.

3.36 J.G. Smyth, R.G. Kenny, G.P. Blair, "Motored and Steady Flow Boundary Conditions Applied to the Prediction of Scavenging Flow in a Loop-Scavenged Two-Stroke Cycle Engine," SAE Paper No. 900800, Society of Automotive Engineers, Warrendale, Pa., 1990.

3.37 G.P. Blair, R.J. Kee, R.G. Kenny, C.E. Carson, "Design and Development Techniques Applied to a Two-Stroke Cycle Automobile Engine," International Conference on

Comparisons of Automobile Engines, Verein Deutscher Ingenieuer (VDI-GFT), Dresden, 3-4 June 1993.

3.38 J.R. Goulburn, G.P. Blair, P.M.D. Donohoe, "A Natural Gas Fired Two-Stroke Engine Designed for High Thermal Efficiency and Low Environmental Impact," Seminar on Gas Engines and Co-Generation, Institution of Mechanical Engineers, The National Motorcycle Museum, Solihull, Birmingham, 10-11 May 1990.

3.39 R.J. Kee, G.P. Blair, R. Douglas, "Comparison of Performance Characteristics of Loop and Cross Scavenged Two-Stroke Engines," SAE Paper No. 901666, Society of Automotive Engineers, Warrendale, Pa., 1990, pp113-129.

3.40 D. Campbell, G.P. Blair, "Anordnung von kleinvolumigen Zweitaktmotoren um gute Kraft und arme Emissionen zu erhalten," 4th International Motorcycle Conference, Verein Deutscher Ingenieure, Munich, 5-7 March 1991.

3.41 G.P. Blair, R.J. Kee, R.G. Kenny, C.E. Carson, "Design and Development Techniques Applied to a Two-Stroke Cycle Automobile Engine," International Conference on Comparisons of Automobile Engines, Verein Deutscher Ingenieuer (VDI-GFT), Dresden, 3-4 June 1993.

3.42 S.J. Magee, R. Douglas, G.P.Blair, J-P Cressard, "Reduction of Fuel Consumption and Emissions for a Small Capacity Two-Stroke Engine," SAE Paper No. 932393, Society of Automotive Engineers, Warrendale, Pa., 1993.

3.43 R.G. Kenny, R.J. Kee, C.E. Carson, G.P. Blair, "Application of Direct Air-Assisted Fuel Injection to SI Cross-Scavenged Two-Stroke Engine," SAE Paper No.932396, Society of Automotive Engineers, Warrendale, Pa., 1993, pp37-50.

3.44 K. Doherty, R. Douglas, G.P.Blair, R.J. Kee, "The Initial Development of a Two-Stroke Cycle Biogas Engine," SAE Paper No. 932398, Society of Automotive Engineers, Warrendale, Pa., 1993, pp69-76.

3.45 N.R. McKinley, R.G. Kenny, R. Fleck, "CFD Prediction of a Two-Stroke, In-Cylinder Steady Flow Field and Experimental Validation," SAE Paper No. 940399, Society of Automotive Engineers, Warrendale, Pa., 1994, pp125-137.

3.46 N.R. McKinley, R. Fleck, R.G. Kenny, "LDV Measurement of Transfer Efflux Velocities in a Motored Two-Stroke Cycle Engine," SAE Paper No. 921694, Society of Automotive Engineers, Warrendale, Pa., 1992.

3.47 D.Thornhill, R. Fleck, "Design of a Blower Scavenged, Piston Ported, V6 Two-Stroke Automotive Engine," SAE Paper No. 930980, Society of Automotive Engineers, Warrendale, Pa., 1993.

3.48 R. Fleck, D.Thornhill, "Single Cycle Scavenge Testing a Multi-Cylinder Externally Scavenged, Two-Stroke Engine with a Log Intake Manifold," SAE Paper No. 941684, Society of Automotive Engineers, Warrendale, Pa., 1994.

3.49 P.H. Schweitzer, <u>Scavenging of Two-Stroke Cycle Diesel Engines</u>, MacMillan, New York, 1949, p268.

3.50 D.W. Montville, B.A. Jahad, "An Extended Jante Test Method for Two-Stroke Piston-Ported Engine Development," SAE Paper No. 941679, Society of Automotive Engineers, Warrendale, Pa., 1994.

Chapter 4

Combustion in Two-Stroke Engines

4.0 Introduction

This chapter will mainly deal with combustion in the spark-ignition engine, but a significant segment of the discussion and the theory is about compression-ignition engines. The subject will be dealt with on a practical design basis, rather than as a fundamental treatise on the subject of combustion. It is not stretching the truth to say that combustion, and the heat transfer behavior which accompanies it, are the phenomena least understood by the average engine designer. The study of combustion has always been a specialized topic which has often been treated at a mathematical and theoretical level beyond that of all but the dedicated researcher. Very often the knowledge garnered by research has not been disseminated in a manner suitable for the designer to use. This situation is changing, for the advent of CFD design packages will ultimately allow the engine designer to predict combustion behavior without being required to become a mathematics specialist at the same time. For those who wish to study the subject at a fundamental level, and to be made aware of the current state of the science, Refs. [1.3-1.4, 4.1-4.5 and 4.28-4.31] will help provide a starting point for that learning process. There seems little point in repeating the fundamental theory of combustion in this book, because such theory is well covered in the literature. However, this chapter will cover certain aspects of combustion theory that are particularly applicable to the design and development of two-stroke engines and which are rarely found in the standard reference textbooks.

The first objective of this chapter is to make you aware of the difference between the combustion process in a real engine from either the theoretical ideal introduced in Sec. 1.5.8, or from an explosion, which is the commonly held view of engine combustion by the laity.

The second aim is to introduce you to the analysis of cylinder pressure records for the heat release characteristics which they reveal, and how such information is the most realistic available for the theoretical modeler of engine behavior. In this section, other possible theoretical models of combustion are presented and discussed.

The third objective is to make you aware of the effect of various design variables such as squish action on detonation and combustion rates and the means at hand to design cylinder heads for engines taking these factors into account.

Penultimately, the chapter refers to computer programs to assist with this design process and you are introduced to their use.

Finally, there is a section on the latest theoretical developments in the analysis of combustion processes, for it will not be many years before our understanding of engine combustion will be on a par with, say, unsteady gas dynamics or scavenging flow. The par level referred to is where the theory will design combustion chambers for two-stroke engines and the resulting engine will exhibit the designed combustion performance characteristics. It is my opinion that this level is approaching, but has not yet been achieved.

4.1 The spark-ignition process
4.1.1 Initiation of ignition

It is a well-known fact that a match thrown onto some spilled gasoline in the open atmosphere will ignite the gasoline and release a considerable quantity of heat with a significant rise in temperature. The gasoline is observed to burn as a vapor mixed with air above the remaining liquid, but rapidly vaporizing, gasoline. The procedure, for those who have witnessed it (although on safety grounds I am not recommending that the experiment be conducted), commences when the lighted match arrives at the vapor cloud above the liquid, and the ignition takes place with a "whoosh," apparently a major or minor "explosion" depending on the mass fraction of the gasoline that has been allowed to evaporate before the arrival of the match. This tends to leave us with the impression that the gasoline-air mixture within the cylinder of an IC engine will "explode" upon the application of the spark at the sparking plug. That the flammability characteristics of a petrol-air mixture are a decidedly critical phenomenon should be obvious to all who have had difficulty in starting either their lawnmower or their automobile!

What then are the requirements of an ignition process? Why does an engine fire up? When does it not fire up? The technical papers on combustion and the engineering textbooks tend to bypass such fundamental concepts, so I felt that a paragraph or two would not be amiss, especially as more difficult concepts will only be understood against a background of some fundamental understanding.

Fig. 4.1(a) depicts a two-stroke engine where a spark has ignited an air-fuel mixture and has produced a flame front burning its way through the mixture. For this to happen, like the match example before it, there had to be a gasoline vapor and air mixture, of the correct mass proportions, within the spark gap when that spark occurred. The energy in the spark provided a local rise in temperature of several thousand degrees Kelvin, which caused any gasoline vapor present to be raised above its auto-ignition temperature. The auto-ignition temperature of any hydrocarbon fuel is that temperature where the fuel now has sufficient internal energy to break its carbon-hydrogen bond structure and be oxidized to carbon dioxide and steam. In the case of gasoline the auto-ignition temperature is about 220°C. The compression process prior to the ignition point helps to vaporize the gasoline, whose maximum boiling point is about 200°C. The mass of gasoline within the spark gap, having commenced to break down in an exothermic reaction, raises the local temperature and pressure. The reaction, if it were stoichiometric, would be as given previously in Eq. 1.5.16. The actual reaction process is much more complex than this, with the gasoline molecule breaking down in stages to methane and aldehydes. Immediately after the ignition point, the initial flame front close to the spark plug has been established and heats the unburned layers of gasoline vapor-air mixture surrounding it, principally by radiation but also by convection heat transfer and by the mixture motion propelling itself into the flame front. This induces further layers of mixture to reach the auto-ignition temperature and thus the flame front moves through the combustion chamber until it arrives at the physical extremities of the chamber [4.24, 4.25, 4.26, 4.31]. The velocity of this flame front has been recorded [4.3, 1.20] in two-stroke engines between 20 and 50 m/s. It will be observed that this is hardly an "explosive" process, although it is sufficiently rapid to allow the engine to burn its fuel reasonably efficiently even at the highest engine speeds. In the example quoted by Kee [1.20], the flame speed, c_{fl}, in an engine of similar physical geometry to Fig. 4.1 was measured at 24.5 m/s at 3000 rpm. The longest flame path, x_{fl}, from the spark plug to the extremity of the combustion chamber in an engine of 85 mm bore was about 45 mm. The crankshaft rotation angle, θ_{fl}, for this flame transmission to occur is given by:

$$\theta_{fl} = \frac{6 x_{fl} \text{rpm}}{c_{fl}} \qquad (4.1.1)$$

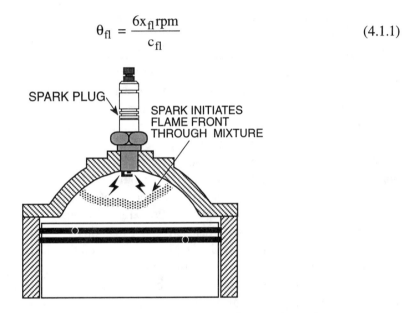

Fig. 4.1(a) Initiation of combustion in a spark-ignition two-stroke engine.

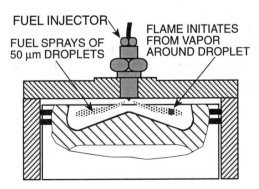

Fig. 4.1(b) Initiation of combustion in a compression-ignition two-stroke engine.

In the example quoted by Kee [1.20], the flame travel time to the chamber extremity is 33°, as from Eq. 4.1.1 the value of θ_{fl} is given by:

$$\theta_{fl} = \frac{6 \times \frac{45}{1000} \times 3000}{24.5} = 33.06°$$

That does not mean that the combustion process has been completed in 33°, but it does mean that initiation of combustion has taken place over the entire combustion space for a homogeneous charge. In Sec. 4.2.3, it will be shown that the travel time, in this case cited as 33°, coincides approximately with the maximum rate of heat release from the fuel.

4.1.2 Air-fuel mixture limits for flammability

The flammability of the initial flame kernel has a rather narrow window for success [4.25, 4.26]. The upper and lower values of the proportion by volume of gasoline vapor to air for a flame to survive are 0.08 and 0.06, respectively. As one is supplying a "cold" engine with liquid fuel, by whatever device ranging from a carburetor to a fuel injector, the vaporization rate of that gasoline due to the compression process is going to be highly dependent on the temperatures of the cylinder wall, the piston crown and the atmospheric air. Not surprisingly, in cold climatic conditions, it takes several compression processes to raise the local temperature sufficiently to provide the statistical probability of success. Equally, the provision of a high-energy spark to assist that procedure has become more conventional [4.6]. At one time, ignition systems had spark characteristics of about 8 kV with a rise time of about 25 µs. Today, with "electronic" or capacitor discharge ignition systems those characteristics are more typically at 20 kV and 4 µs, respectively. The higher voltage output and the faster spark rise time ensure that sparking will take place, even when the electrodes of the spark plug are covered in liquid gasoline. Spark duration, and even multiple sparking, also assist flame kernel growth and this is even more applicable for the ignition of a spray of liquid fuel in a stratified charging approach using direct fuel injection.

Under normal firing conditions, if the fuel vapor-air mixture becomes too lean, e.g., at the 0.06 volume ratio quoted above, then a flame is prevented from growing due to an inadequate initial release of heat. When the spark occurs in a lean mixture, the mass of fuel vapor and air ignited in the vicinity of the spark is too small to provide an adequate release of heat to raise the surrounding layer of unburned mixture to the auto-ignition temperature. Consequently, the flame does not develop and combustion does not take place. In this situation, intermittent misfire is the normal experience as unburned mixture forms the bulk of the cylinder contents during the succeeding scavenge process and will supplement the fuel supplied by it.

Under normal firing conditions, if the fuel vapor-air mixture becomes too rich, e.g., at the 0.08 volume ratio quoted above, then the flame is prevented from growing due to insufficient mass of air present at the onset of ignition. As with any flame propagation process, if an inadequate amount of heat is released at the critical inception point, the flame is snuffed out.

4.1.3 Effect of scavenging efficiency on flammability

In Sec. 3.1.5 there is mention of the scavenging efficiency variation with scavenge ratio. As the engine load, or brake mean effective pressure, bmep, is varied by altering the throttle opening thereby producing changes in scavenge ratio, so too does the scavenging efficiency, SE, change. It will be observed from Fig. 3.10, even for the best design of two-stroke engines, that the scavenging efficiency varies from 0.3 to 0.95. Interpreting this reasonably accurately as being "charge purity" for the firing engine situation, it is clear that at light loads and low engine rotational speeds there will be a throttle position where the considerable mass of exhaust gas present will not permit ignition of the gasoline vapor-air mixture. When the spark occurs, the mass of vapor and air ignited in the vicinity of the spark is too small to provide an adequate release of heat to raise the surrounding layer of unburned mixture to the auto-ignition temperature. Consequently, the flame does not develop and combustion does not take place. The effect is somewhat similar to the lean burning misfire limit discussed above.

During the next scavenging process the scavenging efficiency is raised as the "exhaust residual" during that process is composed partly of the unburned mixture from the misfire stroke and partly of exhaust gas from the stroke preceding that one. Should the new SE value prove to be over the threshold condition for flammability, then combustion does takes place. This skip firing behavior is called "four-stroking." Should it take a further scavenge process for the SE level to rise sufficiently for ignition to occur, then that would be "six-stroking," and so on. Of course, the scavenging processes during this intermittent firing behavior eject considerable quantities of unburned fuel and air into the exhaust duct, to the very considerable detriment of the specific fuel consumption of the engine and its emission of unburned hydrocarbons.

Active radical combustion

It was shown by Onishi [4.33] that it is possible under certain conditions to ignite, and ignite with great success, an air-fuel mixture in high concentrations of residual exhaust gas, i.e., at low SE values. This was confirmed by Ishibashi [4.34] and the engine conditions required are to raise the trapping pressure and temperature so that the exhaust residual remains sufficiently hot and active to provide an ignition source for the air-fuel mixture distributed through it. Hence the term "active radical," or AR, combustion. The ensuing combustion process is very stable, very efficient, and eliminates both the "four-stroking" skip firing regime and the high exhaust emissions of hydrocarbons that accompany it, as described above. It is certainly a design option open for the development of the simple carburetted engine and one which has not been sufficiently exploited. There is some evidence that several motorcycle engines of the 1930s had low speed and low load combustion processes which behaved in this manner.

The effect must not be confused with detonation, as discussed in the next section.

4.1.4 Detonation or abnormal combustion

Detonation occurs in the combustion process when the advancing flame front, which is pressurizing and heating the unburned mixture ahead of it, does so at such a rate that the unburned fuel in that zone achieves its auto-ignition temperature before the arrival of the actual flame front. The result is that unburned mixture combusts "spontaneously" and over the zone where the auto-ignition temperature has been achieved [7.57]. The apparent flame speed in this zone is many orders of magnitude faster, with the result that the local rise of pressure and temperature is significantly sharp. This produces the characteristic "knocking" or "pinking" sound and the local mechanical devastation it can produce on piston crown or cylinder head can be considerable. Actually, "knocking" is the correct terminology for what is actually a detonation behavior over a small portion of the combustion charge. A true detonation process would be one occurring over the entire compressed charge. However, as detonation in the strictly defined sense does not take place in the spark-ignition IC engine, the words "knocking" and "detonation" are interchanged without loss of meaning in the literature to describe the same effect just discussed.

The effect is related to compression ratio, because the higher the compression ratio, the smaller the clearance volume, the higher the charge density, and at equal flame speeds, the higher the heat release rate as the flame travels through the mixture. Consequently, there will be a critical level of compression ratio where any unburned mixture in the extremities of the combustion chamber will attain the auto-ignition temperature. This effect can be alleviated by various methods, such as raising the octane rating of the gasoline used, promotion of charge turbulence, squish effects, improvement of scavenging efficiency, stratifying the combustion process and, perhaps more commonly, retarding the ignition or lowering the trapped compression ratio. Some of these techniques will be discussed later.

It is clear that the combustion chamber shape has an influence on these matters. Plates 4.1 and 4.2 show the combustion chambers typical of loop-scavenged and conventionally cross-scavenged engines. The loop-scavenged engine has a compact chamber with no hot nooks or crannies to trap and heat the unburned charge, nor a hot protuberant piston crown as is the case with the cross-scavenged engine. In the cross-scavenged engine there is a high potential for fresh charge to be excessively heated in the narrow confines next to the hot deflector, in advance of the oncoming flame front.

Compounding the problem of combustion design to avoid "knocking" in the two-stroke engine is the presence of hot exhaust gas residual within the combustible charge.

In Appendix A7.1, there is a comprehensive discussion of the interrelationship between the potential for detonation, the performance characteristics, and the emissions of oxides of nitrogen.

4.1.5 Homogeneous and stratified combustion

The conventional spark-ignition two-stroke engine burns a homogeneous charge. The air-fuel mixture is supplied to the cylinder via the transfer ports with much of the fuel already vaporized during its residence in the "hot" crankcase. The remainder of the liquid fuel vaporizes during the compression process so that by the time ignition takes place, the combustion chamber is filled with a vapor-air-exhaust gas residual mixture which is evenly distributed throughout the combustion space. This is known as a *homogeneous combustion process*.

Should the fuel be supplied to the combustion space by some other means, such as direct in-cylinder fuel injection, then, because all of the vaporization process will take place during the compression process, there is a strong possibility that at the onset of ignition there will be

Chapter 4 - Combustion in Two-Stroke Engines

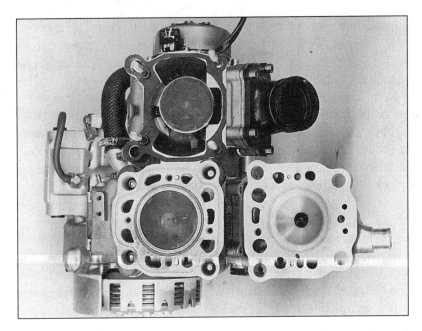

Plate 4.1 Compact combustion chamber for a loop-scavenged two-stroke engine.

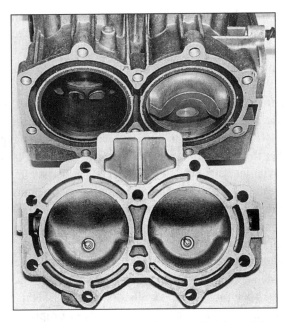

Plate 4.2 The complex combustion chamber of the conventional cross-scavenged engine.

zones in the combustion space which are at differing air-fuel ratios. This is known as a *stratified combustion process*. This stratification may be deliberately induced, for example, to permit the local efficient burning of a small mass of air and fuel of the correct proportions to overcome the problems of "four-stroking."

It is also possible to utilize charge stratification to help alleviate detonation behavior. If the extremities of the combustion chamber contain air only, or a very lean mixture, then there exists the possibility of raising engine thermal efficiency with a higher compression ratio, while lowering the potential for detonation to occur.

Further detailed discussion on this topic can be found in Chapter 7, which covers the design of engines for good fuel economy and exhaust emissions.

4.1.6 Compression ignition

This subject will be dealt with more fully in Sec. 4.3.7, but the fundamentals of the process will be discussed here to distinguish it clearly from the spark-ignition process described above. A sketch of a typical combustion chamber and fuel injector location for a direct injection (DI) diesel engine is shown in Fig. 4.1(b). In the compression-ignition process the compression ratio employed for the engine is much higher so that the air temperature by the end of compression is significantly above the auto-ignition temperature of the fuel employed. When fuel is sprayed directly into this air at this state condition and at that juncture, the fuel commences to vaporize around the droplets and, when the vapor temperature has risen to the auto-ignition temperature, the combustion commences with that vapor and the air in its immediate vicinity. Thus, in diesel combustion by complete contrast with homogeneous spark ignition, it is possible to burn one microscopic droplet of fuel in a whole "sea" of air. In such lean combustion the great advantage is that it is possible to have it occur at air-fuel ratios four or five times leaner than the stoichiometric mixture. The penalty is that efficient combustion cannot occur at air-fuel ratios which approach the stochiometric value, for now the fuel is, as it were, chasing through the combustion chamber searching for the air. In the limited time available for combustion, it is not possible to have that take place for all of the carbon component of the fuel, and thus there is a "rich limit" for diesel combustion. This rich limit is typically some 40% leaner than stoichiometric, or at an air-fuel ratio of about 20. If the air-fuel ratio is richer than this value, then the exhaust gas exhibits considerable quantities of unburned carbon particulates, or black smoke. Today, exhaust emissions legislation specifies and limits the amount of carbon particulates which may be emitted.

This effect is compounded by the heavier hydrocarbon fuel employed for diesel engines. Gasoline is typically and ideally octane, C_8H_{18}, i.e., the eighth member of the family of paraffins whose general family formula is C_nH_{2n+2}. This is much too volatile a fuel to be used in compression ignition as the ensuing pressure and temperature rates of rise would be so rapid as to cause mechanical damage to the cylinder components. The fuel employed for most automotive diesel engines is typically and ideally dodecane, $C_{12}H_{26}$, i.e., the twelfth member of the family of paraffins. As it is heavier with a more complex structure, this molecule requires more energy to separate the carbon and hydrogen atoms from each other and thus it burns more slowly, giving a less rapid rate of pressure rise during a compression-ignition process than would octane. Nevertheless, this more complex molecule does not burn to completion so readily and hence the compounding effect on black smoke production alluded to at the end of the last paragraph.

Prior to the combustion process, the fuel is injected into the air in the cylinder, which by dint of the high compression ratio is above the auto-ignition temperature of the fuel. It requires trapped compression ratios, CR_t, of about 18 to accomplish this latter effect, by comparison with the 7 or 8 normally employed for spark ignition. The fuel droplets must first be heated by the compressed air to vaporize the fuel, and then to raise that vapor temperature to the auto-ignition temperature. This, like all heat transfer processes, takes time. Naturally, there will be an upper limit of engine speed where the heating effect has not been fully accomplished and combustion will not take place efficiently, or even not at all. Thus, a diesel engine tends to have an upper speed limit for its operation in a fashion not seen for a spark-ignition power unit. The limit is imposed by the effectiveness of the heat transfer and mixing process for the particular type of diesel engine. For diesel engines in automobiles using the indirect injection (IDI) method that limit is about 4500 rpm, and for larger direct injection (DI) diesel engines for trucks or buses it is about 3000 rpm. There is a more extensive discussion of this topic in Sec. 4.3.7, in terms of the considerable differences in combustion chamber geometry for these two dissimilar approaches to the generation of combustion by compresson ignition.

4.2 Heat released by combustion
4.2.1 The combustion chamber

The combustion process is one described thermodynamically as a heat addition process in a closed system. It occurs in a chamber of varying volume proportions whose minimum value is the clearance volume, V_{cv}. This was discussed earlier in Sec. 1.4.2, and Eqs. 1.4.4 and 1.4.5 detail the values of V_{cv} in terms of the swept volume, V_{sv}, and the trapped swept volume, V_{tsv}, to attain the requisite parameters of geometric compression ratio, CR_g, and trapped compression ratio, CR_t. In Fig. 4.2, the piston is shown positioned at top dead center, tdc, so the clearance volume, V_{cv}, is seen to be composed of a bowl volume, V_b, and a squish volume, V_s. The piston has a minimum clearance distance from the cylinder head which is known as the *squish clearance,* x_s. The areas of the piston which are covered by the squish band and the bowl are A_s and A_b, respectively, and this squish action gives rise to the concept of a *squish area ratio,* C_{sq}, where:

$$C_{sq} = \frac{\text{area squished}}{\text{bore area}} = \frac{A_s}{\frac{\pi}{4} d_{bo}^2} = \frac{A_s}{A_s + A_b} \qquad (4.2.1)$$

The definitions above are equally applicable to a compression-ignition engine, including the IDI engine in Fig. 4.9.

4.2.2 Heat release prediction from cylinder pressure diagram

In Sec. 1.5.8 there is a discussion of the theoretically ideal thermodynamic engine cycle, the Otto Cycle, and the combustion process is detailed as occurring at constant volume, i.e., the imaginary explosion. The reality of the situation, in Figs. 1.14 and 1.15, is that the engine pressure diagram as measured on the QUB LS400 research engine shows a time-dependent combustion process. This is pointed out in Sec. 4.1.1, where the flame speed is detailed as having been measured at 24.5 m/s at the same test conditions.

Design and Simulation of Two-Stroke Engines

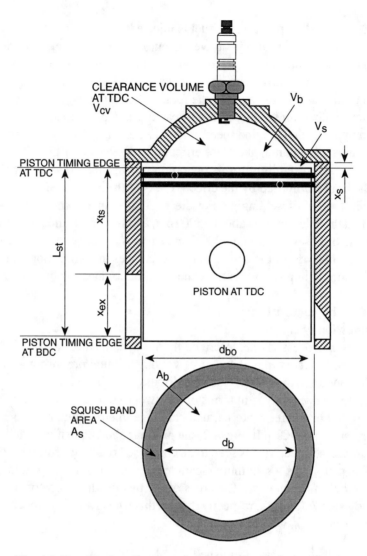

Fig. 4.2 Details of combustion chamber and cylinder geometry.

It is possible to analyze the cylinder pressure diagram and deduce from it the heat release rate for any desired period of time or crankshaft angular movement. Consider the thermodynamic system shown in Fig. 4.3, where the piston has been moved through a small crankshaft angle producing known volume changes, V_1 to V_2, the total volume above the piston, including the combustion chamber. These volumes are known from the crankshaft, connecting rod and cylinder geometry. The combustion process is in progress and a quantity of heat, δQ_R, has been released during this time step and a quantity of heat, δQ_L, has been lost through heat transfer to the cylinder walls and coolant at the same time. The cylinder pressure change has been measured as changing from p_1 to p_2 during this time step. The internal energy of the cylinder gas undergoes a change from U_1 to U_2, as does the temperature from T_1 to T_2. The

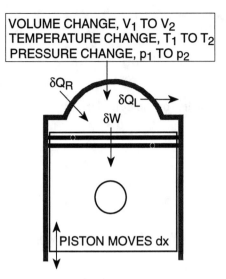

Fig. 4.3 Thermodynamic system during combustion.

work done on the piston is δW during the time interval. The First Law of Thermodynamics for this closed system states that, during this time step, such events are related by:

$$\delta Q_R - \delta Q_L = U_2 - U_1 + \delta W \tag{4.2.2}$$

Where m is the mass of gas in the cylinder and C_V is the specific heat at constant volume, the internal energy change is approximately given by:

$$U_2 - U_1 = mC_V(T_2 - T_1) \tag{4.2.3}$$

The word approximately is used above, as the value of the specific heat at constant volume is a function of temperature and also of the gas properties. During combustion both the gas properties and the temperatures are varying rapidly, such as several hundreds of degrees Kelvin per degree crankshaft, and thus C_V is not a constant as is evident from the discussion in Sec. 2.1.6. In any arithmetic application of Eq. 4.2.3, it is required to employ a value of C_V at some particular state condition such as at either T_1 or T_2 and at the gas properties pertaining at state condition 1 or 2, or at the mean of those two values and at the mean of the gas properties pertaining at state condition 1 and 2. The simplest method is to employ the properties at state condition 1, as it will always be a known condition; the properties at T_2 are usually what is being forecast and therefore are "unknowns." This caution will not be repeated throughout the thermodynamic analysis; it will be taken as read that it is included within the theory being discussed.

The pressure and temperature values at each point can be derived through the equation of state as:

$$p_2 V_2 = mRT_2 \quad \text{and} \quad p_1 V_1 = mRT_1 \tag{4.2.4}$$

where R is the gas constant.

As
$$C_V = \frac{R}{\gamma - 1} \tag{4.2.5}$$

and γ is the ratio of specific heats for the cylinder gas:

$$U_2 - U_1 = \frac{p_2 V_2 - p_1 V_1}{\gamma - 1} \tag{4.2.6}$$

The average work done on the piston during this interval is:

$$\delta W = \frac{p_1 + p_2}{2}(V_2 - V_1) \tag{4.2.7}$$

Consequently, substituting Eqs. 4.2.6 and 4.2.7 into Eq. 4.2.2:

$$\delta Q_R - \delta Q_L = \frac{p_2 V_2 - p_1 V_1}{\gamma - 1} + \frac{p_1 + p_2}{2}(V_2 - V_1) \tag{4.2.8}$$

If the combustion process had not occurred, then the compression or expansion process would have continued in a normal fashion. The polytropic compression process, seen in Figs. 1.15 and 1.16, is taking place at a relationship defined by:

$$pV^n = \text{a constant} \tag{4.2.9}$$

The index n is known from the measured pressure trace as the compression (or the expansion if post-combustion) index. Hence, p_2 would not have been achieved by combustion, but the value p_{2a} would have occurred, where:

$$p_{2a} = p_1 \left(\frac{V_1}{V_2}\right)^n \tag{4.2.10}$$

As this pseudo process is one of non-heat addition, where δQ_R is zero, the First Law of Thermodynamics shown in Eq. 4.2.8 could be rewritten to calculate the heat loss, δQ_L, by substituting p_{2a} for p_2:

$$-\delta Q_L = p_1 \left\{ \frac{V_2 \left(\frac{V_1}{V_2}\right)^n - V_1}{\gamma - 1} + \frac{(V_2 - V_1)\left(\left(\frac{V_1}{V_2}\right)^n + 1\right)}{2} \right\} \tag{4.2.11}$$

On the assumption that this heat loss behavior continues during the combustion process, substitution of this value for δQ_L into Eq. 4.2.8 yields an expression for the heat released, δQ_R:

$$\delta Q_R = \left(p_2 - p_1\left(\frac{V_1}{V_2}\right)^n\right)\left(\frac{V_2}{\gamma - 1} + \frac{V_2 - V_1}{2}\right) \qquad (4.2.12)$$

The value of the polytropic exponent, n, either n_c for compression or n_e for expansion, is known from an analysis of the measured pressure trace taken within the cylinder. A value of the ratio of specific heats for the cylinder gas, γ, could be taken as being between 1.25 and 1.32 at the elevated temperatures found during and post-combustion, and a value of 1.36 to 1.38 pre-combustion. The gas properties could be determined reasonably accurately from an exhaust gas analysis, and possibly even more accurately from a chemical analysis of a sample of the cylinder gas at the end of combustion. However, it will be seen from Sec. 2.1.6 that temperature has a much greater effect on the value of γ.

Note that the experimentally determined values of the polytropic exponents, n_c and n_e, for the compression and expansion processes for the QUB LS400 engine shown in Fig. 1.15, are 1.17 and 1.33, respectively.

Rassweiler and Withrow [4.7] presented this theory in a slightly simpler form as:

$$\delta Q_R = \left(p_2 - p_1\left(\frac{V_1}{V_2}\right)^n\right)\left(\frac{V_2}{\gamma - 1}\right) \qquad (4.2.13)$$

The error between the employment of Eqs. 4.2.12 and 4.2.13 is small, and Kee [1.20] shows it to be less than 1% for crank angle steps of 1° during the experimental pressure record analysis.

The cylinder pressure record is digitized, the values of δQ_R are determined, and the sum of all of the values of δQ_R at each crank angle step, $d\theta$, is evaluated to provide a value of the total heat released, Q_R. The total crank angle period of heat release is b°. At any angle, θ_b, after the start of heat release, the summation of δQ_R to Q_R is the value of the mass fraction of the fuel which has been burned to date; it is defined as B. This and a definition for the rate of heat release with respect to crankshaft angle, $\dot{Q}_{R\theta}$, at each and every crank angle position are to be found from:

mass fraction burned
$$B_{\theta_b} = \frac{\sum_{\theta=0}^{\theta=\theta_b} \delta Q_R}{\sum_{\theta=0}^{\theta=b} \delta Q_R}$$

heat release rate
$$\dot{Q}_{R\theta} = \frac{dQ_R}{d\theta}$$

Design and Simulation of Two-Stroke Engines

There has been a considerable volume of publications on heat release characteristics of engines, mostly four-stroke engines. To mention but four, one of the original papers related to diesel engines was presented by Lyn [4.8]; others in more recent times were given by Hayes [4.11], Martorano [4.27], and Lancaster [4.12].

4.2.3 Heat release from a two-stroke loop-scavenged engine

Secs. 1.5.8 and 1.5.9 present experimentally determined cylinder pressure diagrams in Figs. 1.14-1.16 from a QUB LS400 engine. This is a 400 cc loop-scavenged, spark-ignition, carburetted, research engine, watercooled, of 85 mm bore and 70 mm stroke. The exhaust system is untuned; in short, it is a simple muffler or silencer. Therefore, the trapping characteristics are those provided by the scavenging process and not through any dynamic assistance from a tuned exhaust pipe. The combustion chamber employed for these tests is of the "hemisphere" type shown in Fig. 4.1(a) or Fig. 4.18, with a trapped compression ratio of 6.62. It is, of course, not a true hemisphere, that is simply the jargon for all such cylinder heads with sections which are segments of a sphere. For a particular test at full throttle, the cylinder pressure records were taken by Kee [1.20] at 3000 rpm and analyzed for the values of indicated mean effective pressure, heat release rate and mass fraction burned. The flame speed is recorded during this test. It will be observed that at 24.5 m/s, or somewhat less than 55 mph, the combustion process could in no fashion be compared to an explosion!

For the record, the actual test values at this speed at wide open throttle (wot) are listed as: the imep is 6.80 bar, the bmep is 6.28 bar, the bsfc is 435 g/kWh, the air-fuel ratio is 13.0, the delivery ratio is 0.85, the exhaust carbon monoxide emission is 2.1% by volume, the exhaust hydrocarbon emission is 4750 ppm (NDIR), the exhaust oxygen level is 7.7% by volume, and the flame speed is 24.5 m/s.

The analysis of the cylinder pressure trace for the heat release rate is shown in Fig. 4.4, and that for the mass fraction burned is shown in Fig. 4.5. It is quite clear that there is an ignition delay period of 10° before the heat release becomes manifest in the combustion chamber. The heat release rate peaks at 10° atdc, 20° after the heat release commences properly at 10° btdc. In Sec. 4.1.1, where this particular example was discussed in terms of flame travel time, it was adjudged that the flame had traveled throughout the chamber in 33° crank angle. It can be seen from Fig. 4.4 that this coincides with the time from ignition to the peak of the heat release rate, and that position also coincides with the point in Fig. 4.5 where 50% of the mass fraction was burned. The position for the 50% burn point is about 10° atdc. Later, in Sec. 4.3.6 further evidence on the significance of this siting will be discussed.

Both Figs. 4.4 and 4.5 show lines declared as being "calculated." A discussion on such calculations is also contained in Sec. 4.3.6.

4.2.4 Combustion efficiency

It is possible to assess values of combustion efficiency, η_c, from such test results, where this is defined as:

$$\eta_c = \frac{\text{heat released by fuel}}{\text{heat available in fuel}} \qquad (4.2.14)$$

Chapter 4 - Combustion in Two-Stroke Engines

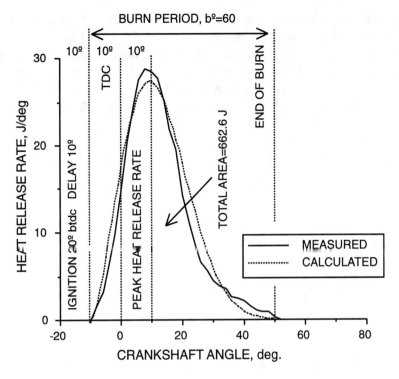

Fig. 4.4 Heat release rate for QUB LS400 at wot at 3000 rpm.

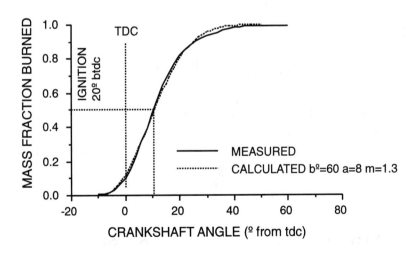

Fig. 4.5 Mass fraction burned in QUB LS400 engine at wot at 3000 rpm.

Design and Simulation of Two-Stroke Engines

This discussion is a continuation of that commenced in Sec. 1.5.7, and is best illustrated in terms of the experimental data presented already for the QUB LS400 engine. Here, the trapping efficiency was evaluated according to the procedures described in Sec. 1.6.3 and Eq. 1.6.21. The mixture was enriched at the same speed and throttle opening to ensure that the exhaust oxygen level recorded was due to scavenge losses only. A value of 7.7% oxygen was noted at an air-fuel ratio of 10.91 and this translates to a trapping efficiency of 60.2%. As the engine has a bsfc of 435 g/kWh at a bmep of 6.28 bar from an engine of 397.2 cm³ swept volume running at 50 rev/s, the cyclic consumption of fuel is given from the following sequence of equations, all of which were introduced in Secs. 1.5.7 and 1.6.1:

$$\text{Fuel supplied, kg/s} = \text{bsfc} \times \dot{W}_b = \text{bsfc} \times \text{bmep} \times V_{SV} \times \text{rps}$$

$$\text{Fuel supplied/cycle, kg} = \frac{\text{Fuel supplied/s}}{\text{rps}} = \text{bsfc} \times \text{bmep} \times V_{SV}$$

$$\text{Heat supplied/cycle, J} = C_{fl} \times \text{fuel supplied/cycle}$$

$$\text{Heat trapped/cycle, J} = \text{TE} \times \text{heat energy supplied/cycle}$$

$$\text{Heat available/cycle, J} = \text{TE} \times C_{fl} \times \text{bsfc} \times \text{bmep} \times V_{SV} \quad (4.2.15)$$

All of the symbols and notation are as used in Chapter 1. The gasoline fuel used has a lower calorific value, C_{fl}, of 43 MJ/kg. For the actual example quoted for the QUB LS400 unit, insertion of the experimental data into Eq. 4.2.15 reveals:

Heat available in fuel, J
$$= 0.602 \times 43 \times 10^6 \times \frac{0.435}{1000 \times 3600} \times 6.28 \times 10^5 \times 397.2 \times 10^{-6} = 780.23$$

The heat released from the fuel as recorded on the heat release diagram for that engine, Fig. 4.4, is the total area under the heat release curve. This is calculated from Fig. 4.4 as 662.6 J. Consequently, the combustion efficiency of this engine at that particular test point is:

$$\eta_c = \frac{662.6}{780.23} = 0.849$$

4.3 Heat availability and heat transfer during the closed cycle
4.3.1 Properties of fuels

The properties of fuels are required to assess the amount of heat available during combustion. The fuel is supplied to an engine as a liquid which must be vaporized during the compression process in a spark-ignition engine or during combustion in a diesel engine. Ideally, as declared earlier, the fuel for the spark-ignition engine is octane and dodecane for a diesel unit. The real situation is more complex than that, as Table 4.1 shows. The properties are the hydrogen to carbon molecular ratio, H/C (or n as seen in the chemical relationship CH_n), the lower calorific value, C_{fl}, the specific gravity, and the latent heat of vaporization of the fuel, h_{vap}.

Table 4.1 Properties of some fuels used in engines

FUEL	C_8H_{18}	$C_{12}H_{26}$	SU Gasoline	Auto Diesel
H/C ratio, n	2.25	2.17	1.65	1.81
C_{fl}, MJ/kg	44.79	43.5	43.0	43.3
Specific gravity	0.70	0.75	0.76	0.83
h_{vap}, kJ/kg	400	425	420	450

The tabular values show the typical properties of iso-octane, C_8H_{18}, and dodecane, $C_{12}H_{26}$. The values in the table for gasoline, labeled as SU or "super-unleaded" gasoline and for diesel fuel, labeled as "auto" diesel, are reasonably representative of the typical properties of fuels commercially available. The properties of such fuels are very dependent on the refining process which will vary from country to country, from refinery to refinery, and from the constituents and origins of the crude oil source for that particular fuel.

4.3.2 Properties of exhaust gas and combustion products

Sec. 2.1.6 shows the basic theory for the computation of the properties of a mixture of gases. The example chosen is the stoichiometric combustion of octane. This continued in the introduction given in Sec. 1.5.5. In both these examples, stoichiometric, or chemically and ideally exact, combustion was used and so all carbon burned to carbon dioxide and carbon monoxide was not formed. In all real combustion processes, dissociation takes place at elevated temperatures and pressures so that, even at stoichiometry, free carbon monoxide and free hydrogen will be created. There are two principal dissociation reactions involved and they are:

$$CO_2 \leftrightarrow CO + \tfrac{1}{2}O_2 \qquad (4.3.1)$$

$$H_2O + CO \leftrightarrow H_2 + CO_2 \qquad (4.3.2)$$

The latter equation is often known as the "water-gas" reaction. The combustion equation that includes all of these effects, and for all air-fuel ratios with a generic hydrocarbon fuel CH_n, is then completely stated as:

$$\begin{aligned} CH_n + \lambda_m(O_2 + kN_2) \\ = x_1CO + x_2CO_2 + x_3H_2O + x_4O_2 + x_5H_2 + x_6N_2 \end{aligned} \qquad (4.3.3)$$

where the value of λ_m is the molecular air-fuel ratio, and k is the nitrogen-to-oxygen molecular ratio; this is conventionally taken as 79/21 or 3.76, as seen previously in Sec. 1.5.5. The relationship between λ_m and the actual air-to-fuel ratio, AFR, is then given by:

$$\text{AFR} = \frac{\lambda_m (M_{O_2} + k M_{N_2})}{M_C + n M_H} \qquad (4.3.4)$$

where M_{O_2}, etc., are the molecular weights of the constituent gases to be found numerically in Table 2.1.1. The value of λ_m can be determined only by balancing the equation for the carbon, hydrogen, oxygen and nitrogen present in the combustion products:

carbon balance $\qquad\qquad 1 = x_1 + x_2 \qquad\qquad (4.3.5)$

hydrogen balance $\qquad\qquad n = 2x_3 + 2x_5 \qquad\qquad (4.3.6)$

oxygen balance $\qquad\qquad 2\lambda_m = x_1 + 2x_2 + x_3 + 2x_4 \qquad\qquad (4.3.7)$

nitrogen balance $\qquad\qquad 2\lambda_m k = 2x_6 \qquad\qquad (4.3.8)$

4.3.2.1 Stoichiometry and equivalence ratio

In the ideal case of stoichiometry, and hence ignoring dissociation effects, the values of x_1, x_4 and x_5 in Eq. 4.3.3 are zero, in which case the solutions of Eqs. 4.3.5 to 4.3.8 are found as:

$$x_2 = 1 \quad x_3 = \frac{n}{2} \quad \text{and} \quad \lambda_m = 1 + \frac{n}{4}$$

This reveals the stoichiometric air-fuel ratio, AFR_s, as:

$$\text{AFR}_s = \frac{\left(1 + \frac{n}{4}\right)(M_{O_2} + k M_{N_2})}{M_C + n M_H} \qquad (4.3.9)$$

To confirm this with the previous solution in Eq. 1.5.17, for iso-octane:

$$\text{AFR}_s = \frac{\left(1 + \frac{2.25}{4}\right)(31.99 + 3.76 \times 28.01)}{12.01 + 2.25 \times 1.008} = 15.02$$

Using the properties of the fuels in Table 3.1, observe that the equivalent values for AFR_s for super-unleaded gasoline is 14.2, for dodecane is 14.92, and for diesel fuel is 14.42.

The concept of *equivalence ratio*, λ, is useful in the solution of these equations and in engine technology generally. It is defined as:

$$\text{Equivalence Ratio, } \lambda = \frac{\text{AFR}}{\text{AFR}_s} \qquad (4.3.10)$$

The combination of Eqs. 4.3.4, 4.3.9 and 4.3.10 reveals:

$$\lambda = \frac{\text{AFR}}{\text{AFR}_s} = \frac{\lambda_m}{1 + \dfrac{n}{4}} \qquad (4.3.11)$$

Obviously, the equivalence ratio is unity at stoichiometry.

4.3.2.2 Rich mixture combustion

In rich mixture combustion, the value of the equivalence ratio is less than unity:

$$\lambda < 1 \qquad (4.3.12)$$

In the first instance let dissociation effects be ignored, in which case Eqs. 4.3.5 to 4.3.8 are directly soluble, hence x_5 is zero. There is insufficient oxygen to burn all of the fuel so there is no free oxygen and the value of x_4 is zero. The molecular balances become:

$$x_1 + x_2 = 1 \qquad x_3 = \frac{n}{2} \qquad 2\lambda_m = x_1 + 2x_2 + x_3 \qquad x_6 = \lambda_m k$$

Hence, for the remaining unknowns x_1 and x_2:

$$x_1 = 2 - 2\lambda_m + \frac{n}{2} \qquad x_2 = 2\lambda_m - \frac{n}{2} - 1 \qquad (4.3.13)$$

Consider a practical example where the equivalence ratio is 15% rich of stoichiometry for the combustion of super-unleaded gasoline, i.e., the value of n is 1.65 and the value of λ is 0.85. From Eq. 4.3.11 the value of λ_m is found as:

$$\lambda_m = \lambda\left(1 + \frac{n}{4}\right) = 0.85 \times \left(1 + \frac{1.65}{4}\right) = 1.20$$

Hence

$$x_1 = 2 - 2 \times 1.20 + \frac{1.65}{2} = 0.424$$

and

$$x_2 = 2 \times 1.20 - \frac{1.65}{2} - 1 = 0.576$$

also $\quad x_3 = \dfrac{1.65}{2} = 0.825 \quad x_6 = 1.20 \times 3.76 = 4.52$

The actual combustion equation now becomes:

$$CH_{1.65} + 1.20(O_2 + 3.76N_2) \\ = 0.424CO + 0.576CO_2 + 0.825H_2O + 4.52N_2 \tag{4.3.14}$$

The total moles of combustion products are 6.342 which, upon division into the moles of each gaseous component, gives the volumetric proportion of that particular gas. The exhaust gas composition is therefore 6.7% CO, 9.1% CO_2, 13.0% H_2O, and the remainder nitrogen. The emission of carbon monoxide is now significant, by comparison with the zero concentration ideally to be found at stoichiometry, the 12.5% CO_2 concentration and the 14.1% H_2O proportions by volume as seen in the equivalent analysis for octane in Sec. 2.1.6.

4.3.2.3 Lean mixture combustion
In lean mixture combustion, the value of the equivalence ratio is greater than unity:

$$\lambda > 1 \tag{4.3.15}$$

In the first instance let dissociation effects be ignored, in which case Eqs. 4.3.5 to 4.3.8 are directly soluble, hence x_5 is zero. There is excess oxygen to burn all of the fuel so there is no free carbon monoxide and the value of x_1 is zero. The molecular balances become:

$$x_2 = 1 \quad x_3 = \dfrac{n}{2} \quad 2\lambda_m = 2x_2 + x_3 + 2x_4 \quad x_6 = \lambda_m k$$

Hence, for the remaining unknown x_4:

$$x_4 = \lambda_m - 1 - \dfrac{n}{4} \tag{4.3.16}$$

Consider a practical example where the equivalence ratio is 50% lean of stoichiometry for the combustion of diesel fuel, i.e., the value of n is 1.81 and the value of λ is 1.5. From Eq. 4.3.11 the value of λ_m is found as:

$$\lambda_m = \lambda\left(1 + \dfrac{n}{4}\right) = 1.5 \times \left(1 + \dfrac{1.81}{4}\right) = 2.178$$

Hence $\quad x_4 = 2.178 - 1 - \dfrac{1.81}{4} = 0.726$

also $\quad x_3 = \dfrac{1.81}{2} = 0.905 \quad x_6 = 2.178 \times 3.76 = 8.196$

The actual combustion equation now becomes:

$$CH_{1.81} + 2.178(O_2 + 3.76N_2)$$
$$= CO_2 + 0.905H_2O + 0.726O_2 + 8.196N_2 \qquad (4.3.17)$$

The total moles of combustion products are 10.827 which, upon division into the moles of each gaseous component, gives the volumetric proportion of that particular gas. The exhaust gas composition is therefore 9.24% CO_2, 8.4% H_2O, 6.71% O_2, and the remainder nitrogen.

4.3.2.4 Effects of dissociation

In the hot, high-pressure conditions within the cylinder during combustion and expansion, gases such as carbon monoxide, carbon dioxide, steam and oxygen will associate and dissociate in chemical equilibrium. This changes the proportions of the components of exhaust gas and, as the above analysis is to be used to predict the composition of exhaust gas for use with simulation models, it is important to know the extent that dissociation effects will have on that composition.

If the above analysis is the precursor to complete flame travel models and to more detailed models of the thermodynamics of combustion as described by Douglas [4.13] and Reid [4.29-4.31] then this section is too brief and simplistic.

As the intention is to describe practical models of combustion to be included with 1D engine simulations, then the need for reasonable accuracy of exhaust gas composition assumes the higher priority. The question therefore is to what extent dissociation alters the composition of the combustion products toward the end of expansion when the cylinder state conditions would typically have a pressure of 6-8 bar and a temperature of 1400-1800 K. Consider then one of the common dissociation reactions, as in Eq. 4.3.1, but extended to cope with the reality of that situation:

$$CO + \tfrac{1}{2}O_2 \leftrightarrow xCO + (1-x)CO_2 + \tfrac{x}{2}O_2 \qquad (4.3.18)$$

The total number of moles is then N where:

$$\text{Total moles, } N = 1 + \dfrac{x}{2}$$

The partial pressures of the three components of the equilibrium mixture are, by Dalton's Laws, a function of their partial pressures in relation to the total pressure, p, in atmospheres:

$$p_{CO} = \dfrac{xp}{N} \quad p_{CO_2} = \dfrac{(1-x)p}{N} \quad p_{O_2} = \dfrac{\tfrac{x}{2}p}{N} \qquad (4.3.19)$$

The equilibrium constant for this reaction, K_p, is a function of temperature and is to be found described and tabulated in many standard texts on thermodynamics. As a fitted curve as a function of the temperature, T (K), it is as follows:

$$\log_e K_p = -68.465 + \frac{6.741}{10^2}T - \frac{2.6658}{10^5}T^2 + \frac{4.9064}{10^9}T^3 - \frac{3.3983}{10^{13}}T^4 \quad (4.3.20)$$

For the dissociation reaction it is incorporated as:

$$K_p = \frac{p_{CO}\sqrt{p_{O_2}}}{p_{CO_2}} \quad (4.3.21)$$

Substituting the values from Eq. 4.3.19 into 4.3.21 reveals:

$$x^3\left(1 - \frac{p}{K_p^2}\right) - 3x + 2 = 0 \quad (4.3.22)$$

The gas at the end of expansion is at 8 bar pressure and at 1800 K. The value for K_p for this reaction at this temperature is -8.497 defined thus:

$$\log_e K_p = -8.452 \quad \therefore \quad K_p = 2.134 \times 10^{-4}$$

The solution reveals that the value of x is 0.00224 and the volumetric proportions of the gas mixture on the right-hand side of Eq. 4.3.18 are 0.2% CO, 99.7% CO_2, and 0.1% O_2. In short, the gas composition is barely altered as a function of dissociation at these state conditions. Thus, dissociation as a factor in predicting the composition of exhaust gas is not worth real consideration. In effect, what it reveals is that at stoichiometry there is a presence of carbon monoxide and free oxygen in the exhaust gas to the extent of these proportions of the carbon dioxide. Instead of 12.5% C by volume in the exhaust gas there will be some 0.025% CO, 12.46% CO_2 and 0.0125% O_2. This corresponds with experimental test results.

However, to make the point thoroughly, if the above computation is considered again as if it were at the height of the combustion process, when the pressure would be some 40 atm and the temperature some 2500 K, then the situation is very different. The gas composition would now be 4.3% CO, 93.6% CO_2, and 2.1% O_2. Thus, dissociation should be taken into account in any detailed model of the thermodynamics of a combustion process in order to determine by calculation the heat release at any instant caused by the formation of the products.

4.3.2.5 The relationship between combustion and exhaust emissions

In Chapter 7 is a full discussion of this topic, particularly with respect to experimental measurements and their comparison with the results of theoretical simulation. Appendices

A4.1 and A4.2 present the fundamental theoretical base for the computation of exhaust emissions within an engine simulation.

It is appropriate to point out here that the analysis of Eq. 4.3.3 shows quite clearly that rich mixture combustion leads to an exhaust emission of carbon monoxide and unburned hydrocarbons. As rich mixture combustion has insufficient air present to oxidize all of the carbon and hydrogen present, the outcome is self-evident. The effects of dissociation have been shown here to lead to both CO and HC emissions, even if the mixture strength is stoichiometric.

The extent of the CO emission, either experimentally found or theoretically deduced, is shown in Sec. 7.2.1, and Fig. 7.15 gives a classic picture of the profile of CO emission decaying to a virtually non-existent level by the stoichiometric air-to-fuel ratio.

The extent of the hydrocarbon emissions is also shown in Sec. 7.2.1, but for the simple spark-ignition two-stroke engine the fuel lost, by the inefficiency of all scavenging processes, totally overshadows that emitted from the combustion process. While operating close to the stoichiometric air-fuel ratio will minimize HC emissions, and the experimental data in Fig. 7.14 shows this quite clearly, the remainder is from scavenge losses and is very considerable.

The creation of nitrogen oxides occurs as a function of temperature, so it is maximized by increasing load (i.e., bmep) levels or by combustion at the stoichiometric air-fuel ratio at any given bmep value. This is illustrated most clearly in Fig. 7.55 at full load, and in Fig. 7.59 at part load, for the same fuel-injected, two-stroke, spark-ignition engine.

For the two-stroke engine, much relief from combustion-related emissions is given by the fact that the combustion chamber in the cylinder head is normally a much simpler, more compact, and less mechanically cluttered design than that of the four-stroke engine. The "clutter" referred to are the poppet valves, and the less compact shape which ensues from their incorporation. This gives crevices into which the flame dies or is quenched, leading to incomplete combustion in those regions.

Another point of relief is the scavenging process which leaves exhaust gas residual behind in the cylinder, and this acts as a damper on the formation of nitrogen oxides. The two-stroke engine comes with this advantage "built-in," whereas the four-stroke engine requires added mechanical complexity to recirculate exhaust gas, i.e., the EGR device which is commonplace on an automobile engine to meet emissions legislation.

4.3.3 Heat availability during the closed cycle

From this discussion on dissociation it is clear that, at the high temperatures and pressures which are the reality during a combustion process, carbon dioxide cannot be produced directly at any instant during burning. Carbon monoxide is produced, but its heat of formation is much less than that yielded by a complete oxidation of the carbon to carbon dioxide. Thus the full heat potential of the fuel is not realized nor released during burning. There can be further inhibitions in this regard. The mixture can be rich so that there is insufficient oxygen to effect that process, even if it were to be ideal. There can be, and in a two-stroke engine there certainly will be, considerable quantities of exhaust gas residual present within the combustion chamber to inhibit the progress of the flame development and the efficiency of combustion. Thus the complete combustion efficiency is composed of relative sub-sets related to equivalence ratio and scavenging efficiency:

$$\eta_c = C_{burn} \eta_{af} \eta_{se} \qquad (4.3.23)$$

Experimental evidence is employed to determine these factors as the chemistry of combustion is not sufficiently advanced as to be able to provide them, although this situation is improving with the passage of time [4.29-4.31]. The factors listed in Eq. 4.3.23 are:

η_{af} The relative combustion efficiency with respect to equivalence ratio and which has a maximum value of unity, usually close to the stoichiometric value.

η_{se} The relative combustion efficiency with respect to scavenging efficiency and has a maximum value of unity at a value equal to or exceeding 90%. The term scavenging efficiency is used, when the real criterion is actually trapped charge purity. For the spark-ignition engine they are virtually the same thing, but for a diesel engine this is not the case numerically.

The coefficient C_{burn} is unashamedly a "fudge" factor to express the reality of all of the combustion effects which are virtually inexplicable by a simple analysis and are unrelated to fueling or charge purity, such as incomplete oxidation of the hydrocarbon fuel which always occurs even in optimum circumstances, incomplete flame travel into the corners of particular combustion chambers, weak or ineffective ignition systems, poor burning in crevices, and flame decay by quenching in most circumstances. However, the value of C_{burn} rarely changes by more than a few percentage points between 0.85 and 0.90, the numerical relevance of which has been seen before at the very end of Sec. 4.2.4.

Effect of trapped charge purity or scavenging efficiency

The experimentally determined relationship for the relative combustion efficiency related to charge purity or scavenging efficiency, SE, found in spark-ignition engines is set out below. For charge purities higher than 0.9, the relative combustion efficiency can be treated as unity. However,

$$\text{if } 0.51 < SE < 0.9, \eta_{se} = -12.558 + 70.108SE - 135.67SE^2 + 114.77SE^3 - 35.542SE^4$$
$$(4.3.24)$$

$$\text{if } SE < 0.51, \eta_{se} = 0.73 \qquad (4.3.25)$$

For low values of trapped charge purity, the relative combustion efficiency is modest and constant. This does not take into account the possibility of active radical (AR) combustion occurring, as discussed in Sec. 7.3.4.

For diesel engines, the charge purity should be in excess of 90% or else there will be a considerable emission of carbon particulates from combustion. Hence, for diesel engines this value should be unity or the design is in some jeopardy.

Effect of equivalence ratio for spark-ignition engines

The effect of fueling level, in terms of equivalence ratio, is best incorporated from experimental evidence. The alternative is to rely on theory, as carried out using equilibrium and dissociation theory as outlined above, by Douglas [4.13] and Reid [4.29-4.31] in an engine simulation using a flame propagation model. To execute this accurately, so as to phase the peak power point at the correct "rich" mixture level and the peak thermal efficiency at the correct "weak" mixture level, it is necessary to employ reaction kinetics within the combustion theory. This is a complex topic well beyond the scope of this design-based text and within the province of scientific research into combustion at a fundamental level.

The experimental evidence, based on many fueling loops carried out on several research engines at QUB, reveals that the relative combustion efficiency with respect to equivalence ratio, η_{af}, is measured as:

$$\eta_{af} = -1.6082 + 4.6509\lambda - 2.0746\lambda^2 \qquad (4.3.26)$$

Analysis of this function reveals that η_{af} has a maximum of unity at a λ value of about 1.12, i.e., 12% "weak" of stoichiometric, and produces a maximum total heat release at a λ value of about 0.875, i.e., at 14% "rich" of stoichiometric.

Effect of equivalence ratio for compression-ignition engines

The effect of fueling level, in terms of the trapped equivalence ratio, λ, is best incorporated from experimental evidence, as with the spark-ignition case above. The word trapped equivalence ratio is used to make the point that the exhaust residual within a diesel engine cylinder contains significant quantities of oxygen. On the other hand, for a correctly designed two-stroke diesel engine the scavenging efficiency should be above 90% and there should not be much exhaust gas residual present in any case. The experimental evidence reveals that the relative combustion efficiency with respect to equivalence ratio, η_{af}, is measured as:

$$\eta_{af} = -4.18 + 8.87\lambda - 5.14\lambda^2 + \lambda^3 \qquad (4.3.27)$$

Analysis of this function reveals that η_{af} has a maximum of unity at a λ value of 2.0, i.e., 100% "weak" of stoichiometric. Above this equivalence ratio the value is constant at unity. As a diesel engine will produce peak power at a λ value of approximately 1.25, but at impossibly high levels of black smoke emission, the equation above is sensibly applicable for equivalence ratios between unity and 2.0.

4.3.4 Heat transfer during the closed cycle

The experience at QUB is that, particularly for spark-ignition engines, the most effective and accurate method for the calculation of heat transfer from the cylinder during the closed cycle is that based on Annand's work [4.15, 4.16, 4.17]. For diesel engines, it should be added that the heat transfer equation by Woschni [4.18] is usually regarded as being equally effective for theoretical computation. Typical of the approach to the heat transfer theory proposed by Annand is his expression for the Nusselt number, **Nu**, leading to a conventional derivation for the convection heat transfer coefficient, C_h. The methodology is almost exactly that adopted for the pipe theory in Sec. 2.4. Annand recommends the following expression to connect the Reynolds and the Nusselt numbers:

$$\mathbf{Nu} = a\mathbf{Re}^{0.7} \qquad (4.3.28)$$

where the constant, a, has a value of 0.26 for a two-stroke engine and 0.49 for a four-stroke engine. The Reynolds number is calculated as:

$$\mathbf{Re} = \frac{\rho_{cy} c_p d_{cy}}{\mu_{cy}} \qquad (4.3.29)$$

The value of cylinder bore, d_{cy}, is self-explanatory. The values of density, ρ_{cy}, mean piston velocity, c_p, and viscosity, μ_{cy}, deserve more discussion.

The prevailing cylinder pressure, p_{cy}, temperature, T_{cy}, and gas properties combine to produce the instantaneous cylinder density, ρ_{cy}.

$$\rho_{cy} = \frac{p_{cy}}{R_{cy} T_{cy}}$$

During compression, the cylinder gas will be a mixture of air, rapidly vaporizing fuel and exhaust gas residual. During combustion it will be rapidly changing from the compression gas to exhaust gas, and during expansion it will be exhaust gas. Tracking the gas constant, R_{cy}, and the other gas properties listed in Eq. 4.3.29 at any instant during a computer simulation is straightforward.

The viscosity is that of the cylinder gas, μ_{cy}, at the instantaneous cylinder temperature, T_{cy}, but I have found that little loss of accuracy occurs if the expression for the viscosity of air, μ_{cy}, in Eq. 2.3.11 is employed.

The mean piston velocity is found from the dimension of the cylinder stroke, L_{st}, and the engine speed, rps:

$$c_p = 2L_{st} \text{ rps} \qquad (4.3.30)$$

Having obtained the Reynolds number, the convection heat transfer coefficient, C_h, can be extracted from the Nusselt number, as in Eq. 2.4.3:

$$C_h = \frac{C_k \mathbf{Nu}}{d_{cy}} \qquad (4.3.31)$$

where C_k is the value of the thermal conductivity of the cylinder gas and can be assumed to be identical with that of air at the instantaneous cylinder temperature, T_{cy}, and consequently may be found from Eq. 2.3.10.

Annand also considers the radiation heat transfer coefficient, C_r, to be given by:

$$C_r = 4.25 \times 10^{-9} \left(T_{cy}^4 - T_{cw}^4 \right) \qquad (4.3.32)$$

However, the value of C_r is many orders of magnitude less than C_h, to the point where it may be neglected for most two-stroke cycle engine calculations. The value of T_{cw} in the above expression is the average temperature of the cylinder wall, the piston crown and the cylinder head surfaces. The heat transfer, δQ_L, over a crankshaft angle interval, $d\theta$, and a time interval, dt, can be deduced for the mean value of that transmitted to the total surface area exposed to the cylinder gases:

as
$$dt = \frac{d\theta}{360} \times \frac{60}{rpm} \qquad (4.3.33)$$

then
$$\delta Q_L = \left(C_h\left(T_{cy} - T_{cw}\right) + C_r\right)A_{cw}dt \qquad (4.3.34)$$

The surface area of the cylinder, A_{cw}, is composed of:

$$A_{cw} = A_{cylinder\ liner} + A_{piston\ crown} + A_{cylinder\ head} \qquad (4.3.35)$$

It is straightforward to expand the heat transfer equation in Eq. 4.3.34 to deal with the individual components of it to the head or crown by assigning a wall temperature to those specific areas. It should also be noted that Eq. 4.3.34 produces a "positive" number for the "loss" of heat from the cylinder, aligning it with the sign convention assigned in Eq. 4.2.2 above.

The typical values obtained from the use of the above theory are illustrated in Table 4.2. The example employed is for a two-stroke engine of 86 mm bore, 86 mm stroke, running at 4000 rpm with a cylinder surface mean temperature, T_{cw}, of 220°C. Various timing positions throughout the cycle are selected and the potential state conditions of pressure (in atm units) and temperature (in °C units) are estimated to arrive at the tabulated values for a two-stroke engine, based on the solution of the above equations. The timing positions are in the middle of scavenging, at the point of ignition, at the peak of combustion and at exhaust port opening (release), respectively. It will be observed that the heat transfer coefficients as predicted for the radiation component, C_r, are indeed very much less than that for the convection component, C_h, and could well be neglected, or indeed incorporated by a minor change to the constant, a, in the Annand model in Eq. 4.3.27.

Table 4.2 Heat tranfer coefficients using the Annand model

Timing Position	P_{cy} (atm)	T_{cy} (°C)	Nu	Re	C_h	C_r
scavenge	1.2	250	349	29433	168	2.2
ignition	10.0	450	1058	143376	652	4.0
burning	50.0	2250	860	106688	1130	84.7
release	8.0	1200	402	36066	411	20.2

It can also be seen that the heat transfer coefficients increase dramatically during combustion, but of course that is also the position of minimum surface area and maximum gas tem-

Design and Simulation of Two-Stroke Engines

perature during the heat transfer process and will have a direct influence on the total heat transferred through Eq. 4.3.34.

4.3.5 Internal heat loss by fuel vaporization

Fuel is vaporized during the closed cycle. For the spark-ignition engine it normally occurs during compression and prior to combustion. For the compression-ignition it occurs during combustion.

Fuel vaporization for the spark-ignition engine

Let it be assumed that the cylinder mass trapped is m_t and the scavenging efficiency is SE, with an air-fuel ratio, AFR. The masses of trapped air, m_{ta}, and fuel, m_{tf}, are given by:

$$m_{ta} = m_t SE \quad \text{and} \quad m_{tf} = \frac{m_{ta}}{AFR} \qquad (4.3.36)$$

If the crankshaft interval between trapping at exhaust port closing and the ignition point is declared as θ_{vap}, and is the crankshaft interval over which fuel vaporization is assumed to occur linearly, then the rate of fuel vaporization with respect to crankshaft angle, $\dot{m}_{vap}$, is given by:

$$\dot{m}_{vap} = \frac{m_{tf}}{\theta_{vap}} \quad \text{kg/deg} \qquad (4.3.37)$$

Consequently, the loss of heat from the cylinder contents, δQ_{vap}, for any given crankshaft interval, $d\theta$, is found by the employment of the latent heat of vaporization of the fuel, h_{vap}. Numerical values of latent heat of vaporization of various fuels are to be found in Table 4.1.

$$\delta Q_{vap} = \dot{m}_{vap} h_{vap} d\theta \qquad (4.3.38)$$

It should be noted that this equation provides a "positive" number for this heat loss, in similar fashion to the application of Eq. 4.3.34.

Fuel vaporization for the compression-ignition engine

Let it be assumed that the cylinder mass trapped is m_t and the charge purity is Π, with an overall air-fuel ratio, AFR. The masses of trapped air, m_{ta}, and fuel, m_{tf}, are given by:

$$m_{ta} = \Pi m_t \quad \text{and} \quad m_{tf} = \frac{m_{ta}}{AFR} \qquad (4.4.39)$$

The crankshaft angle interval over which combustion occurs is defined as b°. Fuel vaporization is assumed to occur as each packet of fuel, $dm_{b\theta}$, is burned over a time interval and is related to the mass fraction burned at that juncture, B_{θ_b}, at a crankshaft angle, θ_b, from the onset of the combustion process. Let it be assumed that an interval of combustion is occurring over a crankshaft interval, dθ. The increment of fuel mass vaporized and burned during this time and crankshaft interval is given by dm_{vap}, thus:

$$dm_{vap} = dm_{b\theta} = \left(B_{\theta_b + d\theta} - B_{\theta_b}\right) m_{tf} h_{vap} \tag{4.3.40}$$

Consequently, the loss of heat from the cylinder contents, δQ_{vap}, for this crankshaft interval, dθ, is found by the employment of the latent heat of vaporization of the fuel, h_{vap}. Numerical values of latent heat of vaporization of various fuels are to be found in Table 4.1.

$$\delta Q_{vap} = dm_{vap} h_{vap} \tag{4.3.41}$$

It should be noted that this equation provides a "positive" number for this heat loss, in similar fashion to the application of Eq. 4.3.34, and that this is soluble only if the mass fraction burned is available as numerical information.

4.3.6 Heat release data for spark-ignition engines

Already presented and discussed is the heat release and mass fraction burned data for the QUB LS400 engine in Figs. 4.4 and 4.5. A simple model of the profile of the heat release rate curve is extracted from the experimental data and displayed in Fig. 4.6. The heat release period is b°, with a rise time of b°/3 equally distributed about tdc. The ignition delay period is 10°. It will be noted that the heat release rate profile has a "tail" of length b°/3, falling to zero from about one-sixth of the maximum value of heat release. The total area under the profile in Fig. 4.6 is the total heat released, Q_R, and is given by simple geometry, where $\dot{Q}_{R\theta}$ is the maximum rate of heat release:

$$Q_R = \frac{14 b° \dot{Q}_{R\theta}}{36} \tag{4.3.42}$$

The actual value of $\dot{Q}_{R\theta}$ in Fig. 4.4 is 28.8 J/deg and the period, b°, is 60°. The area under the model profile in Fig. 4.6 is 672 J, which corresponds well with the measured value of 662.6 J. For a theoretical total heat release of 662.6 J, one would predict from the model a maximum heat release rate, $\dot{Q}_{R\theta}$, of 28.4 J/deg.

The Vibe approach

It is possible to analyze a mass fraction burned curve and fit a mathematical expression to the experimental data. This is often referred to as the *Vibe method* [4.36]. The mathematical fit is exponential with numerical coefficients, a and m, for the mass fraction burned, B_{θ_b}, at a particular crankshaft angle, θ_b, from the onset of heat release and combustion for a total crank angle duration of b°. It is expressed thus:

$$B_{\theta_b} = 1 - e^{-a\left(\frac{\theta_b}{b°}\right)^{m+1}} \tag{4.3.43}$$

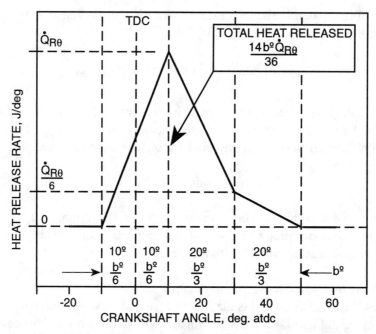

Fig. 4.6 Possible model of heat release rate for combustion simulation.

The analysis of the experimental data in Fig. 4.5 is found to be fitted with coefficients a and m of value 8 and 1.3, respectively, for a total burn period, b°, of 60° duration. This is the "calculated" data referred to in Fig. 4.5. The fit can be seen to be good and when the heat release rate is recalculated from this theoretical equation, and plotted in Fig. 4.4 as "calculated" data, the good correspondence between measurement and calculation is maintained. Thus it is possible to replace the simple line model, as shown in Fig. 4.6, with the Vibe approach.

Further data for spark-ignition engines are found in the paper by Reid [4.31] and a reprint of some is in Fig. 4.7. The data show mass fraction burned cuves for a hemispherical combustion chamber on the engine at throttle area ratio settings of 100%, 25% and 10% in Figs. 4.7(a)-(c), respectively. These data come from an engine of similar size and type to the QUB LS400, but with a bore-stroke ratio of 1.39. The engine speed is 3000 rpm and the scavenging efficiency for the cylinder charge in the three data sets in (a) to (c) are approximately 0.8, 0.75 and 0.65, respectively; the scavenge ratio was measured at 0.753, 0.428 and 0.241, respectively. It is interesting to note the increasing advance of the ignition timing with decreasing cylinder charge purity and the lengthening ignition delay which accompanies it. Nevertheless, the common factor that prevails for these mass fraction burned curves (and the comment is equally applicable to Fig. 4.5) is that the position of 50% mass fraction burned is almost universally phased between 5° and 10° atdc. In other words, optimization of ignition timing means that, taking into account the ignition delay, the burn process is phased to provide an optimized pressure curve on the piston crown and that is given by having 50% of the fuel burned by about 7.5° atdc. The 50% value for the mass fraction burned, B, usually coincides with the peak heat release rate, $\dot{Q}_{R\theta}$.

Chapter 4 - Combustion in Two-Stroke Engines

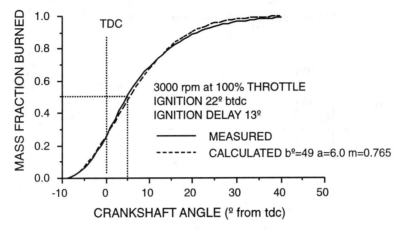

Fig. 4.7(a) Mass fraction burned at 100% throttle area ratio.

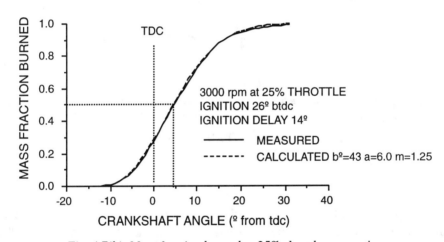

Fig. 4.7(b) Mass fraction burned at 25% throttle area ratio.

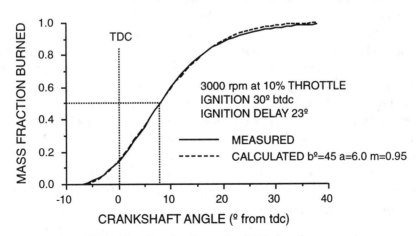

Fig. 4.7(c) Mass fraction burned at 10% throttle area ratio.

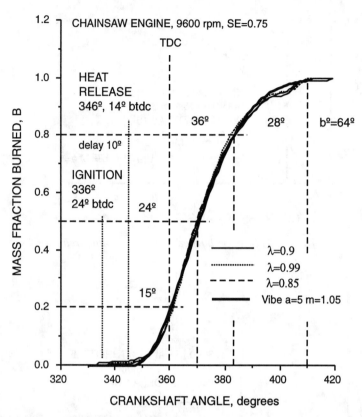

Fig. 4.7(d) Mass fraction burned characteristics of a chainsaw engine.

It can be observed that the data from Reid [4.31] are fitted with a common "a" coefficient of 6.0 and the "m" coefficient is 0.765, 1.25 and 0.95, with burn durations of 49°, 43° and 45°, in Figs. 4.7(a) to (c), respectively.

More information on combustion characteristics in two-stroke engines is in Fig. 4.7(d) for a chainsaw engine running at 9600 rpm. The data are for full throttle operation, but in such a unit the scavenging efficiency is modest at 0.75, for the delivery ratio is low so that the bmep produced approaches 4 bar at a high trapping efficiency, giving good fuel consumption and low hydrocarbon exhaust emissions. It should not be thought that the scavenging of this engine is inferior; indeed the very opposite is the case as this is the actual engine defined as "loopsaw" in Figs. 3.12, 3.13 and 3.19. The mass fraction burned characteristics are for three fueling levels, with λ ranging from near stoichiometric to 0.85. This has little effect on the measured B characteristic. The measured curves are seen to be modeled by Vibe coefficients, a and m, of 5 and 1.05, respectively. The burn period, b°, at the high engine speed of 9600 rpm, and a scavenging efficiency of 0.75, is long at 64°. The 50% burn position is at 10° atdc; the periods of 0-20%, 0-50%, and 0-80% burn are recorded as 15°, 24° and 36°, respectively.

Further information on combustion characteristics in two-stroke engines is in Fig. 4.7(e) for a Grand Prix racing motorcycle engine running at 10,350 rpm where the bmep is 9 bar.

Chapter 4 - Combustion in Two-Stroke Engines

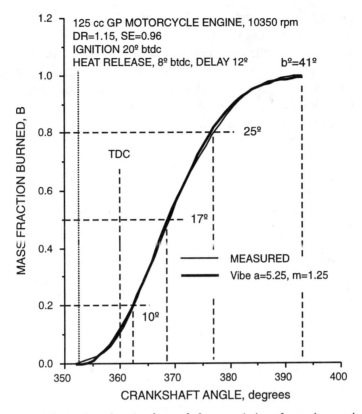

Fig. 4.7(e) Mass fraction burned characteristics of a racing engine.

The data are recorded by Cartwright [4.35] in the course of the preparation of that reference. The measured data can be seen to be modeled by Vibe coefficients, a and m, of 5.25 and 1.25, respectively. The burn period, $b°$, at an engine speed of 10,350 rpm in the middle of the power band, and a scavenging efficiency of 0.95, is quite brief at 41°. Note that the fuel being consumed is aviation gasoline with a motor octane number (MON) of 100, and a H/C molecular ratio of 2.1; it is colloquially referred to as "avgas." The 50% burn position is at 9° atdc; the periods of 0-20%, 0-50%, and 0-80% burn are recorded as 10°, 17° and 25°, respectively.

There are many common factors in these diagrams of mass fraction burned. One which is almost universal is the siting of the 50% point for the mass fraction burned, B. The optimized burn profile rarely has this point before 5° atdc nor later than 10° atdc. It can also be observed, from Fig. 4.5 and Figs. 4.7(a), (b), (d), and (e) that the ignition delay for engines, with scavenging efficiencies at 0.75 and above, are quite commonly between 10° and 14°, whereas at the lower scavenging efficiency, to be found in Fig. 4.7(c), this rises to 23° crankshaft angle. However, I would not wish this to become the universal "law of ignition delay," for the evidence is too flimsy for that conclusion. The modeler should note that increasing the value of the coefficient, a, and decreasing the value of the coefficient, m, advances the rate of combustion.

4.3.7 Heat release data for compression-ignition engines

In Sec. 4.1.6 it was pointed out that combustion by compression ignition requires the rapid heating of the droplets in the fuel spray, initially to vaporize the liquid, and then to promote the rise of that vapor temperature to the auto-ignition temperature in hotter air. The smaller the droplet, but more importantly the higher the relative velocity between fuel droplet and air, the more rapid will be the transfer of heat from air to fuel to accomplish this effect. There are two ways this can be accomplished, either fast-moving fuel and slow-moving air, or fast-moving air and slow-moving fuel.

The direct injection (DI) engine uses the fast-moving fuel approach by employing high-pressure injection into a bowl formed in the piston crown (see Fig. 4.1(b)) to give the highest speed for the fuel droplets. Recent design trends are for ever-higher injection line pressures, up to 1000 bar or more, to give more rapid motion to the fuel droplets. As the air and fuel relative velocities are virtually fixed at any engine speed, this tends to give a low-speed limit for the engine, typically 3000 rpm in automotive applications; there are exceptions to this statement.

The indirect injection (IDI) engine uses the fast-moving air approach by employing compression-created rapid swirl in a side combustion chamber into which low-pressure fuel injection can be employed to provide only a moderate speed of movement for the fuel droplets. The fuel injection line pressures are rarely higher than 250 bar, which also implies both a cheaper and a quieter injection system well suited to automobile applications. As the swirling air flow (see Fig. 4.9) is compression created, the relative velocity of air to fuel tends to rise with engine speed and so the speed limit for this engine is higher than the DI engine. The limiting speed tends to be that of the (mechanical/hydraulic) fuel injection equipment rather than the combustion process.

4.3.7.1 The direct injection (DI) diesel engine

A sketch of a typical combustion chamber for such an engine is shown in Fig. 4.1(b). This shape is often referred to as a "Mexican hat," for obvious reasons. In practice, many other DI designs are employed from the simple "bowl in piston" as shown in Fig. 4.18 or Fig. 3.41, to "squish lip" designs, "wall wetting," etc. [4.32].

Combustion in DI diesel engines is characterized by rapid burning around the tdc position and a sketch of a typical heat release rate curve is shown in Fig. 4.8(a). Apart from the high compression ratio, which in itself provides high thermal efficiency as seen in Eq. 1.5.22, the rapid burn around tdc approaches the ideal for combustion which is a constant volume process. However, the rapid rates of pressure rise make it extremely noisy, i.e., the typical DI diesel "rattle."

In Fig. 4.8(a) the sharp spike of heat release close to the tdc position is known as "pre-mixed" combustion and the remainder as "diffusion" burning. The position of the peak of pre-mixed burning and the profile of the diffusion burning period tends to be influenced by engine speed, as forecast in Sec. 4.1.6. For automotive engines, or so-called "high-speed" diesel engines used in automotive applications, some simple information can be dispensed on these factors for engine speeds between 1000 and 2600 rpm. Above or below these speeds, the maximum or minimum values of 2600 or 1000 should be used in the theory below.

Chapter 4 - Combustion in Two-Stroke Engines

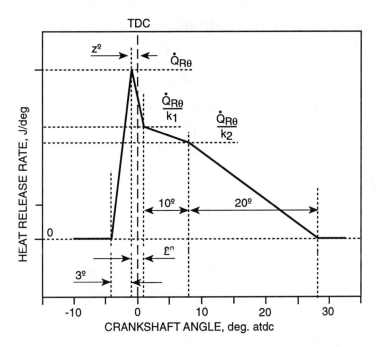

Fig. 4.8(a) Possible model of heat release rate for DI diesel combustion.

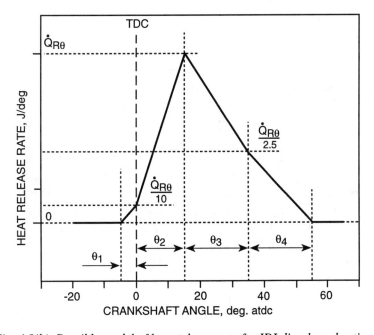

Fig. 4.8(b) Possible model of heat release rate for IDI diesel combustion.

The position of the peak of pre-mixed burning is at $z°$ btdc, where:

$$z° = 4.1206 - \frac{1.251}{10^3} \text{rpm} \tag{4.3.44}$$

The value of the coefficient, k_1, is recommended as:

$$k_1 = 0.81255 + \frac{1.0618}{10^4} \text{rpm} - \frac{6.667}{10^8} \text{rpm}^2 \tag{4.3.45}$$

The value of the coefficient, k_2, is recommended as:

$$k_2 = 1.6377 - \frac{8.5053}{10^4} \text{rpm} + \frac{1.6167}{10^7} \text{rpm}^2 \tag{4.3.46}$$

It will be observed from Fig. 4.8(a) that the intervals for each period are fixed at 3°, 2°, 10° and 20°, i.e., a total burn period, b°, of just 35°, which is rapid indeed. In other words, the phasing of the entire diagram is controlled by the value of $z°$ with respect to the tdc position.

The combustion in DI engines is heavily influenced by swirl of the air with respect to the injected fuel spray and droplets. In opposed piston uniflow-scavenged engines, such swirl is easily arranged; indeed, too easily arranged and often overdone. In loop-scavenged units, it cannot, for this would deteriorate the quality of the scavenging process. However, when the fuel injection rate is optimized for the particular air flow pattern within a particular combustion chamber geometry, the optimized heat release rate curve tends to be very similar to that shown in Fig. 4.8(a), after the theory in Eqs. 4.3.44-46 has been applied.

4.3.7.2 The indirect injection (IDI) diesel engine
A sketch of a typical combustion chamber for such an engine is shown in Fig. 4.9. This shape, and particularly the cut-out on the piston crown, is often referred to as a Ricardo Comet design. In practice, many other IDI designs are employed [4.32].

Combustion in IDI diesel engines is characterized by less rapid burning around the tdc position, by comparison with the DI design, and a sketch of a typical heat release rate curve is shown in Fig. 4.8(b). As both Figs. 4.8(a) and (b) are drawn to scale, the slower combustion in the IDI engine can be seen clearly. By definition, this deviation away from constant volume combustion is less efficient, in practice by some 10% of the thermal efficiency of the DI unit. It is also less noisy as the rates of pressure rise are reduced. It is also less efficient because of the air pumping loss engendered by the piston motion into, and by higher pressure burned charge post combustion from, the combustion chamber where the fuel is injected. Due to the pressure drop, and hence temperature drop, into the combustion chamber during compression, it is necessary to employ a glow plug (heater) to assist in starting the engine.

You may well ask the purpose of this engine, if its thermal efficiency is so compromised. The answer lies in the high-speed swirl created through the throat of area, A_t, into the side chamber. The rotational speed of swirl, N_{sw}, is designed to be some 20 to 25 times the rotational speed of the engine and is ensured by designing the area of the throat, A_t, to be some 1.0

Chapter 4 - Combustion in Two-Stroke Engines

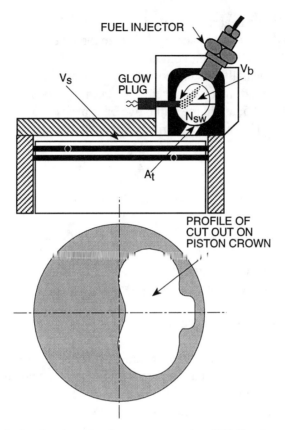

Fig. 4.9 Combustion chamber geometry of an IDI diesel engine.

to 1.5% of the piston area, A_p. The disposition of the clearance volume, V_{cv}, into that which is within the bowl, V_b, and that which is above the piston, V_s, is about equal; that is, V_s roughly equals V_b. The compression ratio employed is higher than the DI engine due to the temperature drop through the throat into the combustion bowl. A four-stroke cycle IDI diesel engine has a geometric compression ratio, CR_g, between 20 and 22 and for the two-stroke engine that is roughly equivalent to a trapped compression ratio, CR_t, of between 17 and 19, depending on the exhaust port timing, the supercharge pressure level, and the extent of the exhaust tuning. The purpose of the shape of the cut-out on the piston crown is to assist the mixing of the burning charge emanating from the combustion bowl with the rest of the air above the piston; in effect, combustion in the bowl takes place at air-fuel ratios approaching the stoichiometric and then "torches" into the main chamber to continue as "lean" combustion.

The result is a high-speed engine with a high specific output, providing less noise and black smoke than the DI unit, and which can run in automobile applications to engine speeds in excess of 4500 rpm. It is ideal for a loop-scavenged, two-stroke engine as it does not require in-cylinder swirling air flow to permit high-speed combustion, as is preferable for the DI design.

The typical profile of a heat release rate diagram for an IDI diesel engine is shown in Fig. 4.8(b). It is slower and takes place later after tdc than the DI unit. At the richer fueling levels, i.e., for λ between 1.5 and 2.0, the combustion period, $b°$, can be as long as 60°. As the combustion process is virtually at stoichiometric conditions in the side chamber, the greater the quantity of fuel injected then the longer is the burning process. The k_1 and k_2 ratios are virtually fixed at 10 and 2.5, as are the initial two burn periods, θ_1 and θ_2. They are typically 5° and 15°, respectively. The latter two periods, θ_3 and θ_4, are affected by fueling levels, thus:

for $\lambda < 2$ $\qquad\qquad\qquad\qquad \theta_3 = \theta_4 = 20 \qquad\qquad\qquad\qquad$ (4.3.47)

for $\lambda \geq 2$ $\qquad\qquad\qquad\qquad \theta_3 = \theta_4 = 30 - 5\lambda \qquad\qquad\qquad$ (4.3.48)

The peak of the heat release rate diagram is at 15° atdc, much later than a DI diesel, or even spark-ignition combustion with a retarded ignition timing. However, this leads to a lesser pressure and temperature rise around the tdc point, providing much lower NO_x emissions than either the DI diesel engine or most spark-ignition units.

4.4 Modeling the closed cycle theoretically

For the engine designer who wishes to predict the performance characteristics of the engine, it is clear that the theory already given in Chapter 2 on unsteady gas flow and in Chapter 3 on scavenging permits the preparation of a computer model that will predict the thermodynamic and gas-dynamic behavior of an engine at any throttle opening and any rotational speed. All of this theory pertains to the open cycle, from exhaust port opening to exhaust port closing. The remaining part of the computer analysis which is required must cover the closed cycle period when the combustion phase takes place. However, it is pertinent in this chapter on combustion to point out the various possibilities that exist for formulating such a model. There are three main methods possible and each of these will be discussed in turn:
  (a) a heat release or mass fraction burned model
  (b) a one-dimensional flame propagation model for spark-ignition engines
  (c) a three-dimensional combustion model for spark-ignition engines

4.4.1 A simple closed cycle model within engine simulations

One of the simplest models of engine combustion is to use the heat release analysis presented in Sec. 4.2.2, in reverse. In particular, Eq. 4.2.12 would become a vehicle for the prediction of the pressure, p_2, at the conclusion of any incremental step in crankshaft angle, if all of the other parameters were known as input data. This relies on the assumption of data values for such factors as the combustion efficiency and the polytropic exponents during compression and expansion. At the same time, a profile has to be assumed for the heat release during the combustion phase as well as a delay period before that heat release commences.

At The Queen's University of Belfast, heat release models have been used for many years, since their original publication on the subject [4.9] until some ten years later [4.10]. That it is found to be effective is evident from the accuracy of correlation between measurement and calculation of engine performance characteristics described in those publications.

The heat release characteristics of homogeneously charged, spark-ignition, two-stroke engines have been found to be remarkably similar, which is not too surprising as virtually all of the engines have rather similar combustion chambers. Such would not be the case for four-stroke cycle engines, where the combustion chamber shape is dictated by the poppet valve mechanisms involved. A uniflow two-stroke engine of the type illustrated in Fig. 1.5 would conform more to a four-stroke combustion model than one for a two-stroke engine.

One of the simplifying assumptions in the use of a simple heat release model for combustion is that the prediction of the heat loss to the cylinder walls and coolant is encapsulated in the selection of the polytropic exponents for the compression and expansion processes. This will be regarded by some as an unnecessary and potentially inaccurate simplification. However, as all models of heat transfer are more or less based on empirical forms, the use of experimentally determined polytropic exponents could actually be regarded as a more realistic assumption. This is particularly true for the two-stroke engine, where so many engine types are similar in construction. The key for success is to have a complete map of the polytropic indices of compression and expansion to cover the entire speed and load range of any engine. Unfortunately, such information does not exist, and much of that which is published is very contradictory.

The modeler who wishes to use this simple approach, in the absence of better experimental evidence from a particular engine, could have some confidence in using polytropic indices of 1.25 and 1.35 for the compression and expansion phases, respectively, and 1.34 for the numerical value of γ in Eq. 4.2.12, together with a value between 0.8-0.9 for the overall combustion efficiency, η_c.

4.4.2 A closed cycle model within engine simulations

A slightly more complex approach, but one which is much more complete and of greater accuracy, is to employ all of the theory provided above on heat transfer, fuel vaporization, heat release rates or mass fraction burned behavior, and solve the First Law of Thermodynamics as expressed earlier in Eq. 4.2.8 at every step in a computation, but extended to include vaporization of fuel.

$$\delta Q_R - \delta Q_L - \delta Q_{vap} = \frac{p_2 V_2 - p_1 V_1}{\gamma - 1} + \frac{p_1 + p_2}{2}(V_2 - V_1)$$

All of the physical geometry of the engine will be known, as will physical parameters such as all surface areas and their temperature. The open cycle model employed, such as provided in Chapter 2, must give the initial masses, purities and state conditions of the cylinder contents at the onset of the closed cycle. The air-fuel ratio will be known, so that either the mass of fuel to be injected, or to be vaporized during compression, can be computed. A heat release rate, or mass fraction burned, profile must be assumed so that at at any juncture during combustion the heat to be released into the combustion chamber can be determined. The heat transfer to or from the cylinder walls can be found at any juncture using the Annand model of Sec. 4.3.4.

Using the theory of Sec. 2.1.6, at any point thereafter in the closed cycle, will yield the properties of the cylinder gas at any instant, during compression and before combustion, during combustion, and in the final expansion process prior to release at exhaust port or valve opening.

The computation time step is dt, corresponding to a crankshaft angle, $d\theta$. The notation as used in Sec. 4.2.2 is re-employed, and Fig. 4.3 is still applicable. The crankshaft is at θ_1 degrees and moves to θ_2. All properties and values at position 1 are known, and the new cylinder volume, V_2, is known geometrically. It is assumed that the gas properties at position 1 will persist for the small time step, dt, as a function of the temperature, T_1, and purity, Π_1, at the beginning of the time step, using the theory of Sec. 2.1.6. These will be updated at the end of that time step as a function of the temperature, T_2, and purity, Π_2, to be those at the commencement of another.

Let us deal with each phase of the closed cycle in turn in terms of the acquisition of each of the numerical values of the terms of Eq. 4.2.8. The basic solution of the Eq. 4.2.8 is for the new pressure, p_2:

as from Sec. 2.1.3
$$G_6 = \frac{\gamma + 1}{\gamma - 1}$$

then
$$p_2 = \frac{2(\delta Q_R - \delta Q_L - \delta Q_{vap}) + p_1(G_6 V_1 - V_2)}{G_6 V_2 - V_1} \qquad (4.4.1)$$

Consequently, from the state equation:

$$T_2 = \frac{p_2 V_2}{m_2 R} \qquad (4.4.2)$$

and from mass continuity:

$$m_2 = m_1 + dm \qquad (4.4.3)$$

The heat transfer term, δQ_L, is determined from Sec. 4.3.4.

Compression in spark-ignition engines (Eqs. 4.3.36-38)

$$dm = \dot{m}_{vap} d\theta \qquad \delta Q_{vap} = \dot{m}_{vap} h_{vap} d\theta \qquad \delta Q_R = 0$$

Compression in compression-ignition engines

$$dm = 0 \qquad \delta Q_{vap} = 0 \qquad \delta Q_R = 0$$

Combustion in spark-ignition engines (Eqs. 4.3.23-27)

$$dm = 0 \qquad \delta Q_{vap} = 0 \qquad \delta Q_R = \eta_c \frac{\dot{Q}_{R\theta_1} + \dot{Q}_{R\theta_2}}{2} d\theta$$

Combustion in compression-ignition engines (Eqs. 4.3.23-41)

$$dm = dm_{b\theta} = dm_{vap} \quad \delta Q_{vap} = dm_{vap} h_{vap}$$

$$\delta Q_R = \eta_c \frac{\dot{Q}_{R\theta_1} + \dot{Q}_{R\theta_2}}{2} d\theta$$

Expansion in spark-ignition and compression-ignition engines

$$dm = 0 \quad \delta Q_{vap} = 0 \quad \delta Q_R = 0$$

In the above equations where the heat release rates at the beginning and end of the time steps, $\dot{Q}_{R\theta_1}$ and $\dot{Q}_{R\theta_2}$, are employed for the acquisition of the heat release term, this may be replaced by the equivalent expression using the mass fraction burned approach for the fuel quantity consumed in the time period. Using the symbolism employed previously, it is:

$$\delta Q_R = \eta_c \left(B_{\theta_b + d\theta} - B_{\theta_b} \right) m_{tf} C_{fl} \qquad (4.4.4)$$

Gas purity throughout the closed cycle

It is essential to trace the gas purity because, together with temperature, this provides the essential information on the gas properties at any instant. At the commencement of the closed cycle, the scavenging efficiency is SE_t, the purity is Π_t, and is composed of the purity of the air, Π_a, and the exhaust gas, Π_{ex}. It should be re-emphasized that the purity of exhaust gas is not necessarily zero for, if it is lean combustion in spark-ignition engines or all combustion in diesels, the exhaust gas will contain oxygen. The discussion is in Sec. 4.3.2. Therefore the purity of exhaust gas is given by:

$$\text{if } \lambda < 1 \quad \Pi_{ex} = 0 \qquad (4.4.5)$$

$$\text{and if } \lambda \geq 1 \quad \Pi_{ex} = \frac{\lambda - 1}{\lambda} \qquad (4.4.6)$$

The actual constituents of the gas mixture are dictated by the theory in Sec. 4.3.2 and are fed to the fundamental theory in Sec. 2.1.6 for the calculation of the actual gas properties of the ensuing mixture. During compression the gas purity is constant and is dictated by:

$$\Pi_t = SE_t \Pi_a + (1 - SE_t) \Pi_{ex} \qquad (4.4.7)$$

During combustion the gas purity at an interval, θ_b, from the onset of burn is dictated by:

$$\Pi_{\theta_b} = \left(1 - B_{\theta_b}\right) \Pi_t + B_{\theta_b} \Pi_{ex} \qquad (4.4.8)$$

During expansion the gas purity is constant and is Π_{ex}.

Design and Simulation of Two-Stroke Engines

A more accurate combustion model in two zones

The theory presented here shows a single zone combustion model. A simple extension to burning in two zones is given in Appendix A4.2. Arguably, what is presented there is merely a more accurate single zone model. This is not accidental, as the computation of any combustion process based on heat release data (from a Rassweiler and Withrow analysis), or on a mass fraction burned curve (in the Vibe fashion), must theoretically replay that approach in precisely the same manner as the data were experimentally gathered. Those experimental data are referred to, and analyzed with reference to, a single zone, i.e., the entire combustion chamber. Thermodynamically replay it back into a computer simulation in any other way and the end result is totally inaccurate; perhaps "theoretically meaningless" is a better choice of words to describe a lack of mathematical logic.

4.4.3 A one-dimensional model of flame propagation in spark-ignition engines

One of the simplest models of this type was proposed by Blizard [4.2] and is of the eddy entrainment type. The model was used by Douglas [4.13] at QUB and has been expanded greatly by Reid [4.29-4.31]. In essence, the procedure is to predict the mass fraction burned curves as seen in Fig. 4.7 and then apply equilibrium and dissociation thermodynamics to the in-cylinder process.

The model is based on the propagation of a flame as shown in Fig. 4.1, and as already discussed in Sec. 4.1.1. The model assumes that the flame front entrains the cylinder mass at a velocity which is controlled by the in-cylinder turbulence. The mass is entrained at a rate controlled by the flame speed, c_{fl}, which is a function of both the laminar flame speed, c_{lf}, and the turbulence velocity, c_{trb}.

The assumptions made in this model are:
(a) The flame velocity is the sum of the laminar and turbulence velocities.
(b) The flame forms a portion of a sphere centered on the spark plug.
(c) The thermodynamic state of the unburned mass which has been entrained is identical to that fresh charge which is not yet entrained.
(d) The heat loss from the combustion chamber is to be predicted by convection and radiation heat transfer equations based on the relative surface areas and thermodynamic states of burned and unburned gases. There is no heat transfer between the two zones.
(e) The mass fraction of entrained gas which is burned at any given time after its entrainment is to be estimated by an exponential relationship.

Clearly, a principal contributor to the turbulence present is squish velocity, of which there will be further discussion in Sec. 4.5. The theoretical procedure progresses by the use of complex empirical equations for the various values of laminar and turbulent flame speed, all of which are determined from fundamental experiments in engines or combustion bombs [4.1, 4.4, 4.5].

It is clear from this brief description of a turbulent flame propagation model that it is much more complex than the heat release model posed in Secs. 4.4.1 and 4.4.2. As the physical geometry of the clearance volume must be specified precisely, and all of the chemistry of the reaction process followed, the calculation requires more computer time. By using this

Chapter 4 - Combustion in Two-Stroke Engines

theoretical approach, the use of empirically determined coefficients, particularly for factors relating to turbulence, has increased greatly over the earlier proposal of a simple heat release model to simulate the combustion process. It is somewhat questionable if the overall accuracy of the calculation has been greatly improved, although the results presented by Reid [4.29-4.31] are impressive. There is no doubt that valuable understanding is gained, in that the user obtains data from the computer calculation on such important factors as exhaust gas emissions and the flame duration. However, this type of calculation is probably more logical when applied in three-dimensional form and allied to a more general CFD calculation for the gas behavior throughout the cylinder leading up to the point of ignition. This is discussed briefly in the next section.

4.4.4 Three-dimensional combustion model for spark-ignition engines

From the previous comments it is clearly necessary that reliance on empirically determined factors for heat transfer and turbulence behavior, which refer to the combustion chamber as a whole, will have to be exchanged for a more microscopic examination of the entire system if calculation accuracy is to be enhanced. This is possible by the use of a combustion model in conjunction with a computational fluid dynamics model of the gas flow behavior within the chamber. Computational fluid dynamics, or CFD, was introduced in Chapter 3, where it was shown to be a powerful tool to illuminate the understanding of scavenge flow within the cylinder.

That the technology is moving toward providing the microscopic in-cylinder gas-dynamic and thermodynamic information is seen in the paper by Ahmadi-Befrui *et al.* [4.21]. Fig. 4.10 is taken directly from that paper and it shows the in-cylinder velocities, but the calculation holds all of the thermodynamic properties of the charge as well, at a point just before ignition. This means that the prediction of heat transfer effects at each time step in the calculation will take place at the individual calculation mesh level, rather than by empiricism for the chamber as a whole, as was the case in the preceding sections. For example, should any one surface or side of the combustion bowl be hotter than another, the calculation will predict the heat transfer in this microscopic manner giving new values and directions for the motion of the cylinder charge. This will affect the resulting combustion behavior.

This calculation can be extended to include the chemistry of the subsequent combustion process. Examples of this have been published by Amsden *et al.* [4.20] and Fig. 4.11 is an example of their theoretical predictions for a direct injection, stratified charge, spark-ignition engine. Fig. 4.11 illustrates a section through the combustion bowl, the flat-topped piston and cylinder head. Reading across from top to bottom, at 28° btdc, the figure shows the spray droplets, gas particle velocity vectors, isotherms, turbulent kinetic energy contours, equivalence ratio contours, and the octane mass fraction contours. The paper [4.20] goes on to show the ensuing combustion of the charge. This form of combustion calculation is preferred over any of those models previously discussed, as the combustion process is now being theoretically examined at the correct level. However much computer time such calculations require, they will become the normal design practice in the future, for computers are becoming ever more powerful, ever more compact, faster and less costly with the passage of time.

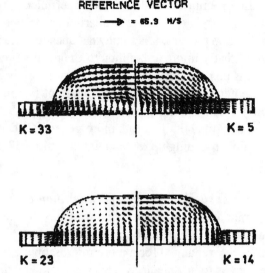

Fig. 4.10 CFD calculations of velocities in the combustion chamber prior to ignition (from Ref. [4.21]).

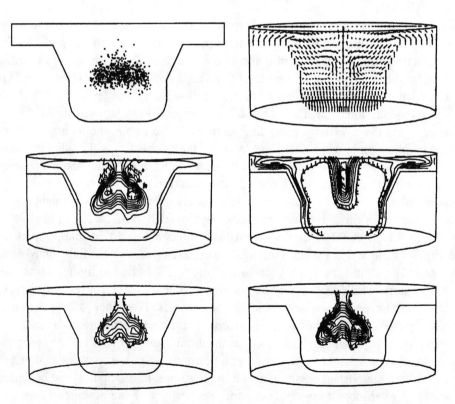

Fig. 4.11 Detailed calculation of gasoline combustion in a fuel-injected SI engine (from Ref. [4.20]).

4.5 Squish behavior in two-stroke engines

It has already been stated in Sec. 4.1.4 that detonation is an undesirable feature of spark-ignition engine combustion, particularly when the designer attempts to operate the engine at a compression ratio which is too high. As described in Sec. 1.5.8, the highest thermal efficiency is to be obtained at the highest possible compression ratio. Any technique which will assist the engine to run reliably at a high compression ratio on a given fuel must be studied thoroughly and, if possible, designed into the engine and experimentally tested. Squish action, easily obtained in the uncluttered cylinder head area of a conventional two-stroke engine, is one such technique. A photograph of squish action, visually enhanced by smoke in a motored QUB-type deflector piston engine, is shown in Plate 4.3.

4.5.1 A simple theoretical analysis of squish velocity

From technical papers such as that given by Ahmadi-Befrui *et al.* [4.21], it is clear that the use of a CFD model permits the accurate calculation of squish velocity characteristics within the cylinder. An example of this is seen in Fig. 4.10, a CFD calculation result for in-cylinder velocities at the point of ignition at 20° btdc. This insight into the squish action is excellent, but it is a calculation method unlikely to be in common use by the majority of designers, at least for some time to come. Because a CFD calculation uses much time on even the fastest of super-computers, the designer needs a more basic program for everyday use. Indeed, to save both super-computer and designer time, the CFD user will always need some simpler guidance tool to narrow down to a final design before employing the fewest possible runs of the CFD package. An analytical approach is presented below which is intended to

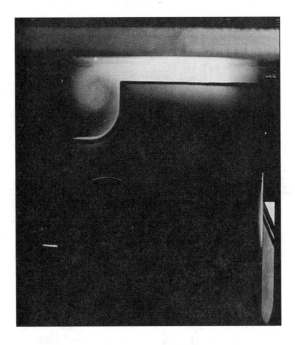

Plate 4.3 The vigorous squish action in a QUB-type cross-scavenged engine at the end of the compression stroke.

fulfill that need. This is particularly helpful, principally because the conventional two-stroke engine has a simple cylinder head, so the designer can conceive of an almost infinite variety of combustion chamber shapes. This is obvious from an examination of some of the more basic shapes shown in Fig. 4.13.

There have been several attempts to produce a simple analysis of squish velocity, often with theory more empirically based than fundamentally thermodynamic. One of the useful papers which has been widely quoted in this area is that by Fitzgeorge [4.22]. Experimental measurements of such phenomena are becoming more authoritative with the advent of instrumentation such as laser doppler anemometry, and the paper by Fansler [4.23] is an excellent example of what is possible by this accurate and non-intrusive measurement technique. However, the following theoretical procedure is one which is quite justifiable in thermodynamic terms, yet is remarkably simple.

Figs. 4.2 and 4.12 represent a compression process inducing a squished flow between the piston and the cylinder head. The process commences at trapping, i.e., exhaust port closure. From Sec. 1.5.6, the value of trapped mass, m_t, is known and is based on an assumed value for the trapped charge pressure and temperature, p_t and T_t. At this juncture, the mass will be evenly distributed between the volume subtended by the squish band, V_{st}, and the volume subtended by the bowl, V_{bt}. The actual values of V_{s1} and V_{b1} at any particular piston position, as shown in Fig. 4.12, are a matter of geometry based on the parameters illustrated in Fig. 4.2. From such input parameters as squish area ratio, c_{sq}, the values of squish area, A_s, bowl area, A_b, and bowl diameter, d_b, are calculated from Eq. 4.2.1.

For a particular piston position, as shown in Fig. 4.12, the values of thermodynamic state and volumes are known before an incremental piston movement takes place.

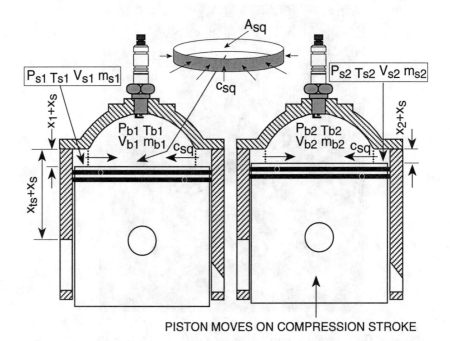

Figure 4.12 Simple theoretical model of squish behavior.

Chapter 4 - Combustion in Two-Stroke Engines

The fundamental thesis behind the calculation is that all state conditions equalize during any incremental piston movement. Thus, the values of cylinder pressure and temperature are p_{c1} and T_{c1}, respectively, at the commencement of the time step, dt, which will give an incremental piston movement, dx, an incremental crank angle movement, $d\theta$, and a change of cylinder volume, dV.

However, if equalization has taken place from the previous time step, then:

$$p_{c1} = p_{s1} = p_{b1} \quad T_{c1} = T_{s1} = T_{b1} \quad \rho_{c1} = \rho_{s1} = \rho_{b1}$$

and

$$V_{c1} = V_{s1} = V_{b1} \quad m_{c1} = m_t = m_{s1} + m_{b1}$$

For the next time step the compression process will occur in a polytropic manner with an exponent, n, as discussed in Sec. 4.2.2. If the process is considered ideal, or isentropic, that exponent is γ, the ratio of specific heats for the cylinder gas. Because of the multiplicity of trapping conditions and polytropic compression behavior, it is more logical for any calculation to consider engine cylinders to be analyzed on a basis of equality. Therefore, the compression process is assumed to be isentropic, with air as the working fluid, and the trapping conditions are assumed to be at the reference pressure and temperature of 1 atm and 20°C. In this case the value of γ is 1.4. The initial cylinder density is given by:

$$\rho_t = \frac{p_t}{RT_t} = \frac{101325}{287 \times 293} = 1.205 \quad kg/m^3$$

The value of the individual compression behavior in the squish and bowl volumes, as well as the overall macroscopic values, follows:

$$\frac{p_{s2}}{p_{s1}} = \left(\frac{V_{s1}}{V_{s2}}\right)^\gamma \quad \frac{p_{b2}}{p_{b1}} = \left(\frac{V_{b1}}{V_{b2}}\right)^\gamma \quad \frac{p_{c2}}{p_{c1}} = \left(\frac{V_{c1}}{V_{c2}}\right)^\gamma$$

$$\frac{\rho_{c2}}{\rho_{c1}} = \frac{V_{c1}}{V_{c2}}$$

If a squish action takes place, then the value of squish pressure, p_{s2}, is greater than either p_{c2} or p_{b2}, but:

$$p_{s2} > p_{c2} > p_{b2}$$

The squish pressure ratio, P_{sq}, causing gas flow to take place, is found from:

$$P_{sq} = \frac{p_{s2}}{p_{b2}} \tag{4.5.1}$$

At this point, consider that a gas flow process takes place within the time step so that the pressures equalize in the squish band and the bowl, equal to the average cylinder pressure. This implies movement of mass from the squish band to the bowl, so that the mass distributions at the end of the time step are proportional to the volumes, as follows:

$$m_{s2} = m_t \frac{V_{s2}}{V_{c2}} \qquad (4.5.2)$$

During the course of the compression analysis, the mass in the squish band was considered to be the original (and equalized in the manner above) value of m_{s1}, so the incremental mass squished, dm_{sq}, is given by:

$$dm_{sq} = m_{s1} - m_{s2} = m_t \left(\frac{V_{s1}}{V_{c1}} - \frac{V_{s2}}{V_{c2}} \right) \qquad (4.5.3)$$

This incremental movement of mass, occurring in time, dt, is further evaluated by the continuity equation to determine the squish velocity, c_{sq}, by:

$$dm_{sq} = \dot{m}_{sq} dt \qquad (4.5.4)$$

where
$$\dot{m}_{sq} = \rho_{c2} A_{sq} c_{sq} \qquad (4.5.5)$$

and as
$$dt = \frac{d\theta}{360 \times rps} \qquad (4.5.6)$$

then combining Eqs. 4.5.3-4.5.6 to resolve the value of squish velocity:

$$c_{sq} = \frac{V_{c2}}{A_{sq}} \left(\frac{V_{s1}}{V_{c1}} - \frac{V_{s2}}{V_{c2}} \right) \frac{360 \, rps}{d\theta} \qquad (4.5.7)$$

The value of the squish flow area, A_{sq}, requires definition and is shown in Fig. 4.12. It is the annular area between the bowl and the squish band formed as a "curtain," the length of which is the line separating the squish band from the bowl, and the height which is the mean of the piston movement from x_1+x_s to x_2+x_s during the time interval of the calculation.

From Eq. 4.5.7 it can be seen that the value of squish velocity is independent of the gas properties selected for the computation, but the value of squish pressure ratio will be dependent on them.

The remaining factor to be evaluated is the value of A_{sq}, the area through which the incremental mass, dm_{sq}, will pass during the time step, dt. This will have various values depending on the type of cylinder head employed, irrespective of the parameters previously mentioned. For example, consider the cases of the basic cylinder head types (a)-(d), as shown in Fig. 4.13.

Chapter 4 - Combustion in Two-Stroke Engines

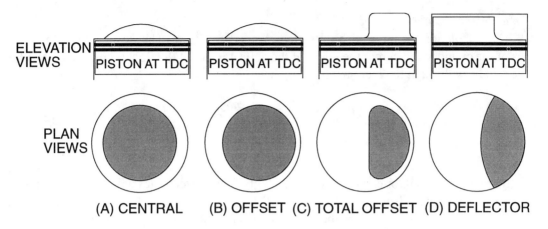

Fig. 4.13 Combustion chamber types analyzed in Prog.4.1, SQUISH VELOCITY.

The central chamber, type (a)

The following is the analysis for the squish flow area, A_{sq}. The flow will be radial through the circumference of the squish band, and if the flow is considered to take place midway through the incremental step, then:

$$A_{sq} = \pi d_b \left(x_s + \frac{x_1 + x_2}{2} \right) \qquad (4.5.8)$$

The offset chamber, type (b)

The following analysis for the squish flow area, A_{sq}, will apply. If the chamber edge approximately coincides with the cylinder bore, then about one-quarter of the potential flow path is blocked by comparison with a central chamber. The flow will still be radial through a segment of the cylinder, and if the flow is considered to take place midway through the incremental step, then:

$$A_{sq} = 0.75 \times \pi d_b \left(x_s + \frac{x_1 + x_2}{2} \right) \qquad (4.5.9)$$

The total offset chamber, type (c)

The following is the analysis for the squish flow area, A_{sq}. The flow will be approximately across a cylinder diameter, and if the flow is considered to take place midway through the incremental step, then:

$$A_{sq} = d_{bo} \left(x_s + \frac{x_1 + x_2}{2} \right) \qquad (4.5.10)$$

The deflector chamber, type (d)

The calculation for the squish flow area, A_{sq}, is as follows. The flow will be along the axis of the cylinder and across the width of the deflector, and if the flow is considered to take place midway through the incremental step, then:

$$A_{sq} = x_d \left(x_s + \frac{x_1 + x_2}{2} \right) \quad (4.5.11)$$

where x_d is the deflector width.

It is clear that even for equal values of c_{sq}, higher squish velocities will be given for those chambers with more offset.

It is also possible to determine the turbulence energy induced by this flow, on the assumption that it is related to the kinetic energy created. The incremental kinetic energy value at each time step is dKE_{sq}, where:

$$dKE_{sq} = dm_{sq} \frac{c_{sq}^2}{2} \quad (4.5.12)$$

The total value of turbulence kinetic energy squished, KE_{sq}, is then summed for all of the calculation increments over the compression process from trapping to tdc.

4.5.2 Evaluation of squish velocity by computer

The equations in Sec. 4.5.1 are programmed in Prog.4.1, SQUISH VELOCITY, available from SAE. All of the combustion chamber types shown in Fig. 4.13 can be handled by this program. You are prompted to type in the chamber type by name, either "central," "offset," "total offset," or "deflector." Of course, there will appear a design for a combustion chamber which is not central, yet is not sufficiently offset as to be described by category (b) in Fig. 4.13. Such a chamber would be one where there is still a significant radial clearance from the bore edge so that the squish flow can proceed in a radial fashion around the periphery of the bowl. In that case, and it is one requiring user judgment, the "chamber type?" prompt from the computer program should be answered with "central."

A typical output from the program is shown in Fig. 4.14, where all of the relevant data input values are observed, with the output values for maximum squish velocity and maximum squish pressure ratio at the crank angle position determined for these maxima, and the total kinetic energy which has been squished. The screen picture is dynamic, i.e., the piston moves from trapping to tdc, and the squish velocity graph is also dynamically created. By such means the operator obtains an enhanced design feel for the effect of squish action, for it is upon such insight that real design experience is built.

The sketch of the engine, cylinder, and cylinder head which appears on the screen is drawn to scale, although the head is drawn as a "bathtub" type purely for ease of presentation. Thus, the physical changes incurred by altering bore-stroke ratio, squish area ratio, squish clearance, exhaust port timing and trapped compression ratio are immediately obvious to the designer at the same time as the squish velocity characteristics.

Chapter 4 - Combustion in Two-Stroke Engines

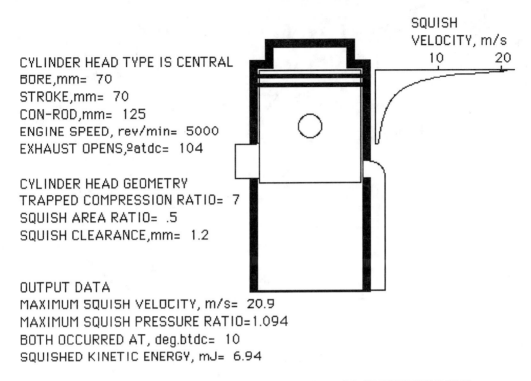

Fig. 4.14 *Typical screen and printer output from Prog.4.1, SQUISH VELOCITY.*

4.5.3 Design of combustion chambers to include squish effects

To emphasize the points already made regarding squish action, the computer program was used to predict the squish effects in various combustion chamber configurations when mounted on a common engine block. The engine, selected at random, has bore, stroke and rod dimensions of 70, 70 and 125 mm, respectively. It runs at 5000 rpm and has an exhaust closing point at 104° btdc. The chamber types (a), (b) and (c) have C_{sq} values of 0.5, while the QUB deflector type (d) has a C_{sq} value of 0.7 and a deflector width of 67 mm. Various squish clearances are now inserted as data in the computer program, ranging from a mechanically possible narrow clearance of 1.2 mm up to 3 mm. The results are plotted in Figs. 4.15 and 4.16 and the common data are printed on the figures.

There is considerable disparity between the values of squish velocity for the various types of chamber, even for those of equal C_{sq} and squish clearance. The total offset and deflector chambers can have squish velocity values at least double that of a central chamber. Naturally, as the kinetic energy values are based on the square of the velocity, the KE_{sq} graph enhances the level of disparity among the several chambers. It is interesting to compare the order of magnitude of the highest value of KE_{sq}, shown in Fig. 4.16 as 56 mJ, with that given by Amsden [4.20] and shown here as Fig. 4.10. The highest contour value of turbulent kinetic energy given in Fig. 4.10 is 263 mJ and the lowest value is 32 mJ. However, that engine also had considerable swirl motion to the in-cylinder gas flow.

In Eq. 4.3.1, it is reported that the total flame velocity is the sum of the laminar flame velocity and the turbulence velocity. If that statement has any authenticity then the flame

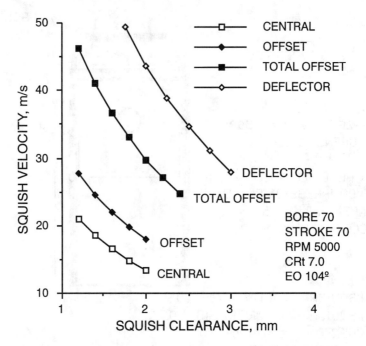

Fig. 4.15 Squish velocity in various combustion chamber types.

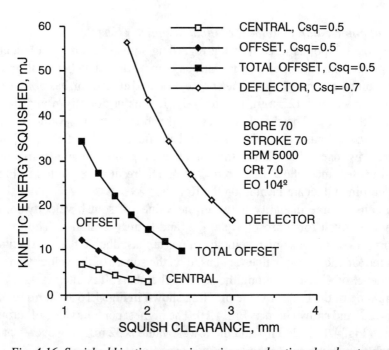

Fig. 4.16 Squished kinetic energy in various combustion chamber types.

velocities observed in various combustion chambers must bear some correspondence with squish velocity. Kee [1.20] reports flame speeds measured in a loop-scavenged engine under the same test conditions as already quoted in Sec. 4.2.3. The chamber was central and the measured flame speed was 24.5 m/s. With the data inserted for the combustion chamber involved, Prog.4.1 predicted maximum squish velocity and squished kinetic energy values of 11.6 m/s and 2.61 mJ, respectively.

In further tests using a QUB-type deflector piston engine of the same bore and stroke as the loop-scavenged engine above, Kee [1.20] reports that flame speeds were measured at the same speed and load conditions at 47 m/s within the chamber and 53 m/s in the squish band. The average flame velocity was 50 m/s. With the data inserted for the deflector combustion chamber involved, Prog.4.1 predicted maximum squish velocity and squished kinetic energy values of 38.0 m/s and 50.2 mJ, respectively. In other words, the disparity in squish velocity as calculated by the theoretical solution is seen to correspond to the measured differences in flame speed from the two experimental engines. That the squish action in a QUB-type cross-scavenged engine is quite vigorous can be observed in Plate 4.3.

If the calculated squish velocity, c_{sq}, is equated to the turbulence velocity, c_{trb}, of Eq. 4.3.1 and subtracted from the measured flame velocity, c_{fl}, in each of Kee's experimental examples, the laminar flame velocity, c_{lf}, is predicted as 12.9 and 12.0 m/s for the loop-scavenged and QUB deflector piston cases, respectively. That the correspondence for c_{lf} is so close, as it should be for similar test conditions, reinforces the view that the squish velocity has a very pronounced effect on the rate of burning and heat release in two-stroke engines. It should be added, however, that any calculation of laminar flame velocity [4.2, 4.5] for the QUB 400 type engine would reveal values somewhat less than 3 m/s. Nevertheless, flame speed values measured at 12 m/s would not be unusual in engines with quiescent combustion chambers, where the only turbulence present is that from the past history of the scavenge flow and from the motion of the piston crown in the compression stroke.

The design message from this information is that high squish velocities lead to rapid burning characteristics and that rapid burning approaches the thermodynamic ideal of constant volume combustion. There is a price to be paid for this, evidenced by more rapid rates of pressure rise which can lead to an engine with more vibration and noise emanating from the combustion process. Further, if the burning is too rapid, too early, this can lead to (a) high rates of NO_x formation (see Appendix A4.2) and (b) slow and inefficient burning in the latter stages of combustion [1.20, 4.29]. Nevertheless, the designer has available a theoretical tool, in the form of Prog.4.1, to tailor this effect to the best possible advantage for any particular design of two-stroke engine.

One of the beneficial side effects of squish action is the possible reduction of detonation effects. The squish effect gives high turbulence characteristics in the end zones and, by inducing locally high squish velocities in the squish band, increases the convection coefficients for heat transfer. Should the cylinder walls be colder than the squished charge, the end zone gas temperature can be reduced to the point where detonation is avoided, even under high bmep and compression ratio conditions. For high-performance engines, such as those used for racing, the design of squish action must be carried out by a judicious combination of theory and experimentation. A useful design starting point for gasoline-fueled, loop-scavenged engines with central combustion chambers is to keep the maximum squish velocity between 15 and 20 m/s at the peak power engine speed. If the value is higher than that, the mass trapped in the end zones of the squish band may be sufficiently large and, with the faster flame front velocities engendered by a too-rapid squish action, may still induce detonation. However, if natural gas were the fuel for the engine, then squish velocities higher than 30 m/s would be advantageous to assist with the combustion of a fuel which is notoriously slow burning. As with most design procedures, a compromise is required, and that compromise is different depending on the performance requirements of the engine, and its fuel, over the entire speed and load range.

Design of squish effects for diesel engines

The use of squish action to enhance diesel combustion is common practice, particularly for DI diesel combustion. Of course, as shown in Fig. 4.9, the case of IDI diesel combustion is a special case and could be argued as one of exceptional squish behavior, in that some 50% of the entire trapped mass is squeezed into a side combustion chamber by a squish area ratio exceeding 98%!

However, for DI diesel combustion as shown in Fig. 4.1(b) where the bowl has a more "normal" squish area ratio of some 50 or 60% with a high trapped compression ratio typically valued at 17 or 18, then an effective squish action has to be incorporated by design. Fig. 4.17 shows the results of a calculation using Prog.4.6, with a squish area ratio held constant at 60% with a squish clearance at 1.0 mm, for an engine with a 90 mm bore and stroke and an exhaust closing at 113° btdc. While this is a spurious design for low compression ratio, gasoline-burning engines, nevertheless the computations are conducted and the graph in Fig. 4.17 is drawn for trapped compression ratios from 7 to 21. The design message to be drawn is that the higher the compression ratio the more severely must the squish effect be applied to acquire the requisite values of squish velocity and squish kinetic energy. What is an extreme and unlikely design prospect for the combustion of gasoline at a trapped compression ratio of 7 becomes an effective and logical set of values for DI diesel combustion at a trapped compression ratio of 20. Designers and developers of combustion chambers of diesel engines should note this effect and be guided by the computation and not by custom and practice as observed for combustion chamber design as applied successfully to two-stroke engines for the burning of gasoline.

4.6 Design of combustion chambers with the required clearance volume

Perhaps the most important factor in geometrical design, as far as a combustion chamber is concerned, is to ensure that the subsequent manufactured component has precisely the correct value of clearance volume. Further, it is important to know what effect machining tolerances will have on the volume variations of the manufactured combustion space. One of the significant causes of failure of engines in production is the variation of compression ratio from cylinder to cylinder. This is a particular problem in multi-cylinder engines, particularly if they are high specific performance units, in that one of the cylinders may have a too-high compression ratio which is close to the detonation limit but is driven on by the remaining cylinders which have the designed, or perhaps lower than the designed, value of compression ratio. In a single- or twin-cylinder engine the distress caused by knocking is audibly evident to the user, but is much less obvious acoustically for a six- or eight-cylinder outboard engine.

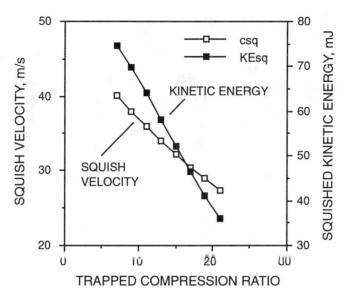

Fig. 4.17 Effect of compression ratio on squish behavior.

So that the designer may assess all of these factors, as well as be able to produce the relevant data for the prediction of squish velocity by Prog.4.1, there are six design programs included in the Appendix Listing of Computer Programs to cover the main possibilities usually presented to the designer as combustion chamber options. The physical geometry of these six combustion chambers is shown in Fig. 4.18, and all but one are intended primarily for use in loop-scavenged engines. Clearly Prog.4.7 is specifically for use for QUB-type deflector piston engines. The names appended to each type of combustion chamber, in most cases being the jargon applied to that particular shape, are also the names of the computer programs involved. To be pedantic, the name of computer program, Prog.4.4, is BATHTUB CHAMBER. Actually, the "bowl in piston" combustion system would also be found in uniflow-scavenged engines, as well as loop-scavenged engines, particularly if they are diesel power units, or even gasoline-fueled but with direct in-cylinder fuel injection.

An example of the calculation is given in Fig. 4.19 for the hemisphere combustion chamber, a type which is probably the most common for loop-scavenged two-stroke engines. A photograph of this type of chamber appears in Plate 4.1, where the bowl and the squish band are evident. A useful feature of this calculation is the data insertion of separate spherical radii for the piston crown and the squish band. A blending radius between the bowl and the squish band is also included as data. The output reflects the tapered nature of the squish band, but it is within the designer's control to taper the squish band or have it as a parallel passage. In the calculation for squish velocity, a tapered squish clearance would be represented by the mean of the two data output values; in the case shown it would be the mean of 1.6 and 2.1 mm, or 1.85 mm. Similarly, the designer can tailor the geometry of any of the central types of chamber to have no squish band, i.e., a zero value for the squish area ratio, should that quiescent format be desirable in some particular circumstance.

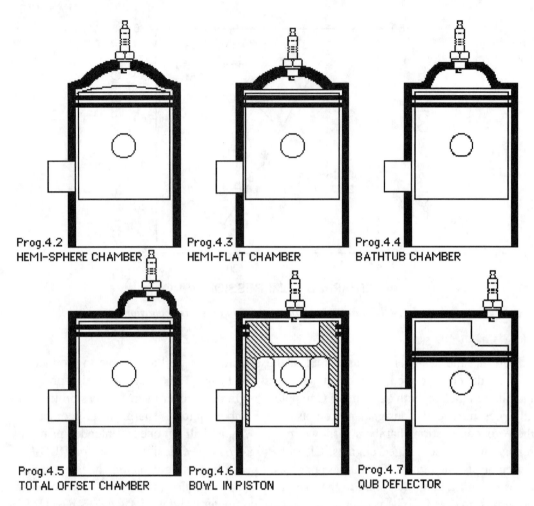

Fig. 4.18 Combustion chamber types which can be designed using the programs.

As with most of the programs associated with this book, the sketch on the computer screen is drawn to scale so that you may estimate by eye, as well as by number, the progress to a final solution. In the operation of any of these six programs, the computer screen will produce an actual example upon start-up, just like that in Fig. 4.19, and invite you to change any data value until the required geometry is ultimately produced. Finally, the printer output is precisely what you see on the screen, apart from the title you typed in.

4.7 Some general views on combustion chambers for particular applications

The various figures in this chapter, such as Figs. 4.13, 4.18 and 4.19 and Plates 4.1-4.3, show most of the combustion chambers employed in spark-ignition two-stroke engines. This represents a plethora of choices for the designer and not all are universally applicable for both homogeneous and stratified charge combustion. The following views should help to narrow that choice for the designer in a particular design situation.

Chapter 4 - Combustion in Two-Stroke Engines

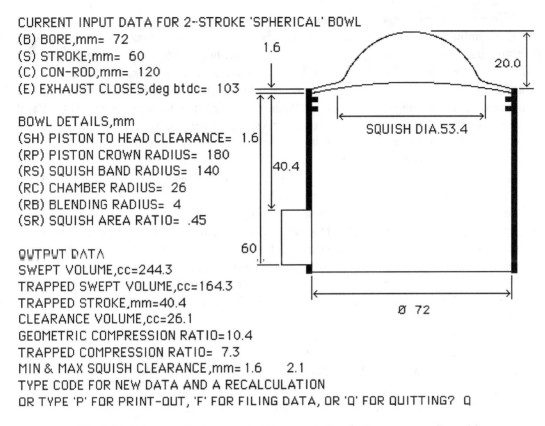

```
CURRENT INPUT DATA FOR 2-STROKE 'SPHERICAL' BOWL
(B) BORE,mm= 72
(S) STROKE,mm= 60
(C) CON-ROD,mm= 120
(E) EXHAUST CLOSES,deg btdc= 103

BOWL DETAILS,mm
(SH) PISTON TO HEAD CLEARANCE= 1.6
(RP) PISTON CROWN RADIUS= 180
(RS) SQUISH BAND RADIUS= 140
(RC) CHAMBER RADIUS= 26
(RB) BLENDING RADIUS= 4
(SR) SQUISH AREA RATIO= .45

OUTPUT DATA
SWEPT VOLUME,cc=244.3
TRAPPED SWEPT VOLUME,cc=164.3
TRAPPED STROKE,mm=40.4
CLEARANCE VOLUME,cc=26.1
GEOMETRIC COMPRESSION RATIO=10.4
TRAPPED COMPRESSION RATIO= 7.3
MIN & MAX SQUISH CLEARANCE,mm= 1.6    2.1
TYPE CODE FOR NEW DATA AND A RECALCULATION
OR TYPE 'P' FOR PRINT-OUT, 'F' FOR FILING DATA, OR 'Q' FOR QUITTING? Q
```

Fig. 4.19 Screen and printer output from computer design program, Prog.4.2.

4.7.1 Stratified charge combustion

There will be further discussion in Chapter 7 on stratified charging of the cylinder, as distinct from the combustion of a stratified charge. A stratified charge is one where the air-fuel ratio is not common throughout the chamber at the point of combustion. This can be achieved by, for example, direct injection of fuel into the cylinder at a point sufficiently close to ignition that the vaporization process takes place as combustion commences or proceeds. By definition, diesel combustion is stratified combustion. Also by definition, the combustion of a directly injected charge of gasoline in a spark-ignition engine is potentially a stratified process. All such processes require a vigorous in-cylinder air motion to assist the mixing and vaporization of air with the fuel. Therefore, for loop- and cross-scavenged engines, the use of a chamber with a high squish velocity capability is essential, which means that all of the offset, total offset or deflector designs in Fig. 4.13, and the bowl in piston design in Fig. 4.18, should be examined as design possibilities. In the case of the uniflow-scavenged engine, the bowl in piston design, as in Fig. 4.1(b) is particularly effective because the swirling scavenge flow can be made to spin even faster in the bowl toward the end of compression. This characteristic makes it as attractive for two-stroke diesel power units as for direct injection four-stroke diesel engines. It is also a strong candidate for consideration for the two-stroke spark-ignition engine with the direct injection of gasoline, irrespective of the method of scavenging.

4.7.2 Homogeneous charge combustion

The combustion of a homogeneous charge is in many ways an easier process to control than stratified charge combustion. On the other hand, the homogeneous charge burn is always fraught with the potential of detonation, causing both damage and loss of efficiency. If the stratified charge burning is one where the corners of the chamber contain air only, or a very lean mixture, then the engine can be run in "lean-burn, high compression" mode without real concern for detonation problems. The homogeneous charge engine will always, under equality of test conditions, produce the highest specific power output because virtually all of the air in the cylinder can be burned with the fuel. This is not the case with any of the stratified charge burning mechanisms, and the aim of the designer is to raise the air utilization value to as high a level as possible, with 90% regarded as a good target value for that criterion.

For homogeneous charge combustion of gasoline, where there is not the need for excessive air motion due to squish, the use of the central, bathtub, or offset chambers is the designer's normal choice. This means that the designer of the QUB deflector chamber has to take special care to control the squish action in this very active combustion system. Nevertheless, for high specific output engines, particularly those operating at high rotational speeds for racing, the designer can take advantage of higher flame speeds by using the dual ignition illustrated in Fig. 4.20(a).

For many conventional applications, the central squish system shown in Fig. 4.20(b) is one which is rarely sufficiently explored by designers. The designer is not being urged to raise the squish velocity value, but to direct the squished charge into the middle of the chamber. As shown in the sketch, the designer should dish the piston inside the squish band and proportion the remainder of the clearance volume at roughly 20% in the piston and 80% in the head. By this means, the squished flow does not attach to the piston crown, but can raise the combustion efficiency by directly entering the majority of the clearance volume. Should such

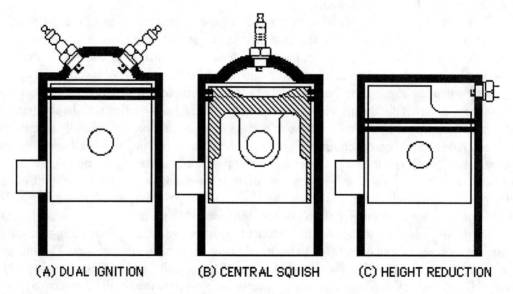

Fig. 4.20 Alternative combustion designs.

a chamber be tried experimentally and fail to provide an instant improvement to the engine performance characteristics, the designer should always remember that the scavenging behavior of the engine is also being altered by this modification.

For combustion of other fuels, such as kerosene or natural gas, which are not noted for having naturally high flame speed capabilities, the creation of turbulence by squish action will speed up the combustion process. In this context, earlier remarks in the discussion on stratified charge combustion become applicable in this homogeneous charge context.

Combustion chambers for cross-scavenged engines

For conventional cross-scavenged engines, it is regretted that little direct advice can be given on this topic, for the simple truth is that the complex shapes possible from the deflector design are such that universal recommendations are almost impossible. Perhaps the only common thread of information which has appeared experimentally over the years is that combustion chambers appear to perform most efficiently when placed over the center of the cylinder, with just a hint of bias toward the exhaust side, and with very little squish action designed to come from either the scavenge side or the exhaust side of the piston. Indeed, some of the best designs have been those which are almost quiescent in this regard.

For unconventional cross-scavenged engines, as discussed in Sec. 3.5.3, the associated computer program, Prog.3.3(a), includes a segment for the design of a combustion chamber which is central over the flat top of the piston and squishes from the deflector areas. It is also possible to squish on the flat top of the piston over the edges of the deflector, although no experimental data exist to confirm that would be successful.

References for Chapter 4

4.1 B. Lewis, G. Von Elbe, <u>Combustion, Flames and Explosions of Gases</u>, Academic Press, 1961.

4.2 N.C. Blizard, J.C. Keck, "Experimental and Theoretical Investigation of Turbulent Burning Model for Internal Combustion Engines," SAE Paper No. 740191, Society of Automotive Engineers, Warrendale, Pa., 1974.

4.3 T. Obokata, N. Hanada, T. Kurabayashi, "Velocity and Turbulence Measurements in a Combustion Chamber of SI Engine under Motored and Firing Conditions by LDA with Fibre-Optic Pick-up," SAE Paper No. 870166, Society of Automotive Engineers, Warrendale, Pa., 1987.

4.4 W.G. Agnew, "Fifty Years of Combustion Research at General Motors," *Prog.Energy Combust.Sci.*, Vol 4, pp115-155, 1978.

4.5 R.J. Tabaczynski, "Turbulence and Turbulent Combustion in Spark-Ignition Engines," *Prog.Energy Combust.Sci.*, Vol 2, p143, 1977.

4.6 M.S. Hancock, D.J. Buckingham, M.R. Belmont, "The Influence of Arc Parameters on Combustion in a Spark-Ignition Engine," SAE Paper No. 860321, Society of Automotive Engineers, Warrendale, Pa., 1986.

4.7 G.M. Rassweiler, L. Withrow, "Motion Pictures of Engine Flames Correlated with Pressure Cards," SAE Paper No. 800131, Society of Automotive Engineers, Warrendale, Pa., 1980.

4.8 W.T. Lyn, "Calculations of the Effect of Rate of Heat Release on the Shape of Cylinder Pressure Diagram and Cycle Efficiency," *Proc.I. Mech.E.*, No 1, 1960-61, pp34-37.

4.9 G.P. Blair, "Prediction of Two-Cycle Engine Performance Characteristics," SAE Paper No. 760645, Society of Automotive Engineers, Warrendale, Pa., 1976.

4.10 R. Fleck, R.A.R. Houston, G.P. Blair, "Predicting the Performance Characteristics of Twin Cylinder Two-Stroke Engines for Outboard Motor Applications," SAE Paper No. 881266, Society of Automotive Engineers, Warrendale, Pa., 1988.

4.11 T.K. Hayes, L.D. Savage, S.C. Sorensen, "Cylinder Pressure Data Acquisition and Heat Release Analysis on a Personal Computer," SAE Paper No. 860029, Society of Automotive Engineers, Warrendale, Pa., 1986.

4.12 D.R. Lancaster, R.B. Krieger, J.H. Lienisch, "Measurement and Analysis of Engine Pressure Data," SAE Paper No. 750026, Society of Automotive Engineers, Warrendale, Pa., 1975.

4.13 R. Douglas, "Closed Cycle Studies of a Two-Stroke Cycle Engine," Doctoral Thesis, The Queen's University of Belfast, May, 1981.

4.14 R.J. Tabaczynski, S.D. Hires, J.M. Novak, "The Prediction of Ignition Delay and Combustion Intervals for a Homogeneous Charge, Spark Ignition Engine," SAE Paper No. 780232, Society of Automotive Engineers, Warrendale, Pa., 1978.

4.15 W.J.D. Annand, "Heat Transfer in the Cylinders of Reciprocating Internal Combustion Engines," *Proc.I. Mech.E.*, Vol 177, p973, 1963.

4.16 W.J.D. Annand, T.H. Ma, "Instantaneous Heat Transfer Rates to the Cylinder Heat Surface of a Small Compression Ignition Engine," *Proc.I. Mech.E.*, Vol 185, p976, 1970-71.

4.17 W.J.D. Annand, D. Pinfold, "Heat Transfer in the Cylinder of a Motored Reciprocating Engine," SAE Paper No. 800457, Society of Automotive Engineers, Warrendale, Pa., 1980.

4.18 G. Woschni, "A Universally Applicable Equation for the Instantaneous Heat Transfer Coefficient in the Internal Combustion Engine," SAE Paper No. 670931, Society of Automotive Engineers, Warrendale, Pa., 1967.

4.19 A.A. Amsden, J.D. Ramshaw, P.J. O'Rourke, J.K. Dukowicz, "KIVA—A Computer Program for Two- and Three-Dimensional Fluid Flows with Chemical Reactions and Fuel Sprays," Los Alamos National Laboratory Report, LA-102045-MS, 1985.

4.20 A.A. Amsden, T.D. Butler, P.J. O'Rourke, J.D. Ramshaw, "KIVA—A Comprehensive Model for 2-D and 3-D Engine Simulations," SAE Paper No. 850554, Society of Automotive Engineers, Warrendale, Pa., 1985.

4.21 B. Ahmadi-Befrui, W. Brandstatter, H. Kratochwill, "Multidimensional Calculation of the Flow Processes in a Loop-Scavenged Two-Stroke Cycle Engine," SAE Paper No. 890841, Society of Automotive Engineers, Warrendale, Pa., 1989.

4.22 D. Fitzgeorge, J.L. Allison, "Air Swirl in a Road-Vehicle Diesel Engine," *Proc.I. Mech.E.*, Vol 4, p151, 1962-63.

4.23 T.D. Fansler, "Laser Velocimetry Measurements of Swirl and Squish Flows in an Engine with a Cylindrical Piston Bowl," SAE Paper No. 850124, Society of Automotive Engineers, Warrendale, Pa., 1985.

4.24 G.F.W. Zeigler, A. Zettlitz, P. Meinhardt, R. Herweg, R. Maly, W. Pfister, "Cycle Resolved Two-Dimensional Flame Visualization in a Spark-Ignition Engine," SAE Paper No. 881634, Society of Automotive Engineers, Warrendale, Pa., 1988.

4.25 P.O. Witze, M.J. Hall, J.S. Wallace, "Fiber-Optic Instrumented Spark Plug for Measuring Early Flame Development in Spark Ignition Engines," SAE Paper No. 881638, Society of Automotive Engineers, Warrendale, Pa., 1988.

4.26 R. Herweg, Ph. Begleris, A. Zettlitz, G.F.W. Zeigler, "Flow Field Effects on Flame Kernel Formation in a Spark-Ignition Engine," SAE Paper No. 881639, Society of Automotive Engineers, Warrendale, Pa., 1988.

4.27 L. Martorano, G. Chiantini, P. Nesti, "Heat Release Analysis for a Two-Spark Ignition Engine," International Conference on the Small Internal Combustion Engine, Paper C372/026, Institution of Mechanical Engineers, London, 4-5 April, 1989.

4.28 J.B. Heywood, Internal Combustion Engine Fundamentals, McGraw-Hill, New York, ISBN 0-07-100499-8, 1989.

4.29 M.G.O. Reid, "Combustion Modelling for Two-Stroke Cycle Engines," Doctoral Thesis, The Queen's University of Belfast, May, 1993.

4.30 M.G.O. Reid, R. Douglas "A Closed Cycle Model with Particular Reference to Two-Stroke Cycle Engines," SAE Paper No. 911847, Society of Automotive Engineers, Warrendale, Pa., 1991.

4.31 M.G.O. Reid, R. Douglas "Quasi-Dimensional Modelling of Combustion in a Two-Stroke Cycle Spark Ignition Engine," SAE Paper No. 941680, Society of Automotive Engineers, Warrendale, Pa., 1994.

4.32 L.R.C. Lilly (Ed.), Diesel Engine Reference Book, Butterworths, London, ISBN 0-408-00443-6, 1984.

4.33 S. Onishi, S.H. Jo, K. Shoda, P.D. Jo, S. Kato, "Active Thermo-Atmosphere Combustion (ATAC)—A New Combustion Process for Internal Combustion Engines," SAE Paper No. 790501, Society of Automotive Engineers, Warrendale, Pa., 1979.

4.34 Y. Ishibashi, Y. Tsushima, "A Trial for Stabilising Combustion in Two-Stroke Engines at Part Throttle Operation," International Seminar on 'A New Generation of Two-Stroke Engines for the Future,' Institut Français du Pétrole, Paris, November 1993.

4.35 A. Cartwright, R. Fleck, "A Detailed Investigation of Exhaust System Design in High Performance Two-Stroke Engines," SAE Paper No. 942515, Society of Automotive Engineers, Warrendale, Pa., 1994, pp131-147.

4.36 I.I. Vibe, "Brennverlauf und Kreisprozeb von Verbrennungs-motoren," VEB Technik Berlin, 1970.

4.37 F.M. Coppersmith, R.F. Jastrozebski, D.V. Giovanni, S. Hersh, "A Comprehensive Evaluation of Stationary Gas Turbine Emissions Levels," Con Edison Gas Turbine Test Program, Air Pollution Control Association Paper 74-12, 1974.

4.38 R.W. Schefer, R.D. Matthews N.P. Ceransky, R.F. Swayer, "Measurement of NO and NO_2 in Combustion Systems," Paper 73-31 presented at the Fall meeting, Western States Section of the Combustion Institute, El Segundo, California, October 1973.

4.39 Ya.B. Zeldovitch, P. Ya. Sadovnikov, D.A. Frank-Kamenetskii, "Oxidation of Nitrogen in Combustion," (transl. by M Shelef) Academy of Sciences of USSR, Institute of Chemical Physics, Moscow-Leningrad, 1947.

4.40 National Standards Reference Data System, Table of Recommended Rate Constants for Chemical Reactions Occuring in Combustion, QD502 /27.

4.41 G.A. Lavoie, J.B. Heywood, J.C. Keck, *Combustion Science Technology*, 1, 313, 1970.

4.42 G. De Soete, *Rev. Institut Français du Pétrole*, 27, 913, 1972.

4.43 V.S. Engleman, W. Bartok, J.P. Longwell, R.B. Edleman, Fourteeth Symposium (International) on Combustion, The Combustion Institute, 1973, p755.

Appendix A4.1 Exhaust emissions

The combustion process

The combination of Eqs. 4.3.3 and 4.3.4 permits the determination of the molecular composition of the products of combustion for any hydrocarbon fuel, CH_n, at any given air-to-fuel ratio, AFR.

The mass ratio of any given component gas "G" within the total, ε_G, is found with respect to the total molecular weight of the combustion products, M_c:

$$M_c = \frac{x_1 M_{CO} + x_2 M_{CO_2} + x_3 M_{H_2O} + x_4 M_{O_2} + x_5 M_{CH_4} + x_6 M_{N_2}}{x_1 + x_2 + x_3 + x_4 + x_5 + x_6} \times \quad (A4.1.1)$$

$$\varepsilon_G = \frac{x_G M_G}{M_c} \quad (A4.1.2)$$

Hence if the engine power output is $\dot{W}$ (in kW units), the delivery ratio is DR, the trapping efficiency is TE, and the engine speed is in rpm, many of the brake specific pollutant emission figures can be determined from an engine simulation, or from design estimations, from combustion in the following manner. The total mass of air and fuel trapped each cycle within the engine, m_{cy}, is given by, using information from Eq. 1.5.2:

$$m_{cy} = TE \times DR \times m_{dref} \times \left(1 + \frac{1}{AFR}\right) \quad (A4.1.3)$$

The mass of gas pollutant "G" produced per hour is therefore:

$$\dot{m}_G = 60 \times rpm \times m_{cy} \times \varepsilon_G \quad kg/h \quad (A4.1.4)$$

and the brake specific pollutant rate for gas "G," bsG, is found as:

$$bsG = \frac{\dot{m}_G}{\dot{W}} \quad kg/kWh \quad (A4.1.5)$$

Carbon monoxide emissions

This is obtained only from the combustion source, and is found as:

$$bsCO = \frac{\dot{m}_{CO}}{\dot{W}} \quad kg/kWh \quad (A4.1.6)$$

Combustion-derived hydrocarbon emissions

These are found as:

$$bsHC_{comb} = \frac{\dot{m}_{CH_4}}{\dot{W}} \quad kg/kWh \quad (A4.1.7)$$

For the simple two-stroke engine they are but a minor contributor by comparison with those from scavenge losses, if scavenging is indeed being conducted by air containing fuel, and in Chapter 7 there is considerable discussion of this topic. It is also a very difficult subject theoretically and the chemistry is not only complex but highly dependent on the mechanism of flame propagation and its decay, quenching or otherwise at the walls or in the crevices of the chamber.

Scavenge-derived hydrocarbon emissions

The mass of charge lost per hour through the inefficiency of scavenging is found as:

$$\dot{m}_{HCscav} = 60 \times \text{rpm} \times \frac{DR \times (1 - TE) \times m_{dref}}{AFR} \quad \text{kg/h} \quad (A4.1.8)$$

Consequently these devolve to a brake specific pollutant rate as:

$$bsHC_{scav} = \frac{\dot{m}_{HC_{scav}}}{\dot{W}} \quad \text{kg/kWh} \quad (A4.1.9)$$

Total hydrocarbon emissions

The total hydrocarbon emission rate is then given by the sum of that in Eq. A4.1.7 and Eq. A4.1.9 as bsHC:

$$bsHC = bsHC_{comb} + bsHC_{scav} \quad (A4.1.10)$$

Emission of oxides of nitrogen

Extensive field tests have shown that nitric oxide, NO, is the predominant nitrogen oxide emitted by combustion devices in recent investigations by Coppersmith [4.37], Schefer [4.38] and Zeldovitch [4.39].

The two principal sources of NO in the combustion of conventional fuels are oxidation of atmospheric (molecular N_2) nitrogen and to a lesser extent oxidation of nitrogen containing compounds in the fuel (fuel nitrogen).

The mechanism of NO formation from atmospheric nitrogen has been extensively studied by several prominent researchers. It is generally accepted that in combustion of lean and near stoichiometric air-fuel mixtures the principal reactions governing formation of NO from molecular nitrogen are those proposed by Zeldovitch [4.39].

$$O + N_2 \leftrightarrow NO + N \quad (A4.1.11)$$
$$N + O_2 \leftrightarrow NO + O \quad (A4.1.12)$$

The forward and reverse rate constants for these reactions have been measured in numerous experimental studies, and kinetic details for this model have been sourced from NSRDS [4.40].

The researchers, Lavoie *et al.* [4.41], have suggested that the reaction described by,

$$O + OH \leftrightarrow O_2 + H \tag{A4.1.13}$$

should also be included. This statement should not be disregarded as the argument for its inclusion with the Zelovitch equations from above is that during rich and near stoichiometric air-fuel ratios this third, rate-limiting condition prevails. If this model is used to predict the formation rate of NO in the cylinder of a compression-ignition engine, where the richest trapped air-fuel ratio will approximately be a value of 20, then the NO rate model may be able to exclude this third, rate-limiting equation.

The NO formation rate is much slower than the combustion rate and most of the NO is formed after the completion of the combustion due to the high temperatures present in the combustion zone. Therefore Eqs. A4.1.11 and A4.1.12 are decoupled from the combustion model.

It was reported by researchers [4.41,4.42] that measurements of the NO formed in the post-flame-front zone was greater than that predicted by the reaction kinetics. Several models have been formed to deal with this scenario, which is due to "prompt NO" formation.

The reaction kinetics which are an integral part of the model are generally formulated under ideal laboratory conditions in which the combustion occurs in a shock tube, but clearly this is somewhat remote from the closed cycle combustion taking place in the spark-ignition or the compression-ignition engine.

The temperature calculated in the cylinder as a product of the increase in pressure corresponding to the period of burn is the average in-cylinder temperature. This temperature is the product of two distinct zones in the cylinder, namely the burn zone comprising the products of combustion and the unburned zone composed of the remaining air and exhaust gas residual. From researcher De Soete [4.42], it was proposed that a greater rate of NO formation was recorded in the post-flame zone whose temperature has been raised by the passing flame. To emulate this, the NO_x model uses the average temperature, T_b, in the burn zone. As described previously, the flame packet contains air, fuel and exhaust gas residual. This packet mass varies with time step and its position during the heat release process. At the conclusion of the burning of each packet in the computation time step, normally about 1° crankshaft, the burn zone has new values of mass, volume, temperature, and oxygen and nitrogen mass concentrations.

The NO rate formation model may now be generally described as follows:

$$\frac{dNO}{dt} = k \times f\left[m_{b\,O_2}, m_{b\,N_2}, V_b, T_b\right] \tag{A4.1.14}$$

where V_b is the volume of the burn zone, T_b is the temperature in the burn zone and k is a rate-limiting constant. The symbols, $m_{b\,O_2}$ and $m_{b\,N_2}$ refer to the mass of oxygen and nitrogen, respectively, within the burn zone. Once the formation of NO is determined as a function of time, its formation in any given time-step of an engine simulation is determined, and the summation of that mass increment over the combustion period gives the total mass formation of the oxides of nitrogen.

In practice, the execution within a computer simulation is not quite as straightforward, because it is necessary to solve the equilibrium and dissociation behavior within the burn zone. The amount of free oxygen within the burn zone is a function of the local temperature and pressure, as has been discussed in Sec. 4.3.2. Two equilibrium reactions must be followed closely, the first being that for the carbon monoxide given in Eq. 4.3.18 and the second for the so-called "water-gas" reaction:

$$H_2O + CO \leftrightarrow H_2 + CO_2 \qquad (A4.1.15)$$

It is clear that this can be solved only by the availability of a two-zone burning model and this is described in Appendix A4.2. There the results of a computer simulation of an engine which predicts NO formation using the above theory are presented.

General

The use of these theories to predict pollutant gas emission rates is illustrated in Chapter 7 and compared with typical measurements from two-stroke cycle engines. In particular, in Appendix A7.1, this theory is employed to show the effect of the change of compression ratio on both the performance characteristics and on the formation of nitric oxide.

Appendix A4.2 A simple two-zone combustion model
The combustion process

Sec. 4.4.2 details a single-zone combustion model. The physical parameters for the theoretical solution of a combustion process in two zones are summed up in the sketch in Fig. A4.1. The assumption pervading this approach is that the pressure in the unburned and burned zones are equal at the beginning and end of a time step in a computation, which is represented by a time interval, dt, or a crankshaft interval, $d\theta$. The piston movement produces volume variations from V_1 to V_2 during this period, and so the mean cylinder conditions of pressure and temperature change from P_1 to P_2, and T_1 to T_2, respectively. The total cylinder mass, m_c, is constant but the masses in the unburned and burned zones, m_b and m_u, change with respect to the increment of mass fraction burned, dB, during this time interval, thus:

$$dB = B_2 - B_1 = B_{\theta_1 + d\theta} - B_{\theta_1} \qquad (A4.2.1)$$

and
$$m_{b2} = m_{b1} + dB \times m_c \qquad (A4.2.2)$$

$$m_{u2} = m_{u1} - dB \times m_c \qquad (A4.2.3)$$

also
$$m_c = m_{b1} + m_{u1} \quad \text{and} \quad m_c = m_{u2} + m_{b2} \qquad (A4.2.4)$$

The purities in both the burned and unburned zones are known. That in the burned zone is zero and at the initial temperature, T_{b1}, the theory of Sec. 2.2.6 can be deployed to determine the gas properties at that temperature, with respect to the air-to-fuel ratio and the particular hydrocarbon fuel being used, to find the numerical values of gas constant, R_b, specific heat at constant volume, C_{vb}, and the ratio of specific heats, γ_b. The purity in the unburned zone is that at the trapping condition, Π_t. In the unburned zone, at the temperature, T_{u1}, the properties

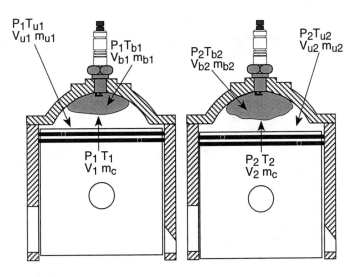

Fig. A4.1 Simple theoretical model for two-zone combustion.

of gas constant, R_u, specific heat at constant volume, C_{vu}, and the ratio of specific heats, γ_u, may also be determined from Sec. 2.1.6. The average properties of the entire cylinder space at the commencement of the time step, and which are assumed to prevail during it, may be found as:

$$R_1 = B_1 R_b - (1 - B_1) R_u \tag{A4.2.5}$$

$$C_{v1} = B_1 C_{vb} - (1 - B_1) C_{vu} \tag{A4.2.6}$$

$$\gamma_1 = \frac{C_{p1}}{C_{v1}} = \frac{R_1 + C_{v1}}{C_{v1}} \tag{A4.2.7}$$

Using heat release data, or a mass fraction burned approach as described in Sec. 4.4.2, the overall behavior for the entire cylinder space may be found as before, using Eq. 4.4.1, or even more simply below if the time interval is sufficiently short. A sufficiently short time interval is defined as 1° crankshaft.

$$T_2 = \frac{\delta Q_R - \delta Q_L + m_c C_{v1} T_1 - p_1 (V_2 - V_1)}{m_c C_{v1}} \tag{A4.2.8}$$

and

$$p_2 = \frac{m_c R_1 T_2}{V_2} \tag{A4.2.9}$$

For the properties within the two zones, a simple solution is possible only if some assumption is made regarding inter-zone heat transfer. Without such an assumption, it is not possible to determine the individual volumes within each zone, and hence the determination of individual zone temperatures cannot be conducted, as the thermodynamic equation of state must be satisfied for each of them, thus:

$$T_{u2} = \frac{p_2 V_{u2}}{m_{u2} R_u} \quad \text{and} \quad T_{b2} = \frac{p_2 V_{b2}}{m_{b2} R_b} \tag{A4.2.10}$$

Some researchers [4.29-4.31] have employed the assumption that there is zero heat transfer between the zones during combustion. This is clearly unrealistic, indeed it is inter-zonal heat transfer which induces detonation. The first simple assumption which can be made to instill some realism into the solution is that the process in the unburned zone is adiabatic. At first sight this also appears too naive, but this is not the case as the alternative restatement of the adiabatic assumption is that the unburned zone is gaining as much heat from the burned zone as it is losing to the surfaces of the cylinder and the piston crown. The more this assumption is examined, in terms of the relative masses, volumes, surface areas and temperatures of the two zones, the more logical it becomes. With it, the solution is straightforward:

$$\frac{p_2}{p_1} = \left(\frac{\rho_2}{\rho_1}\right)^{\gamma_u} = \left(\frac{m_{u2}}{m_{u1}} \times \frac{V_{u1}}{V_{u2}}\right)^{\gamma_u} \qquad (A4.2.11)$$

The volume of the unburned zone, V_{u2}, is the only unknown in the above equation. Consequently, the volume of the burned zone is:

$$V_{b2} = V_2 - V_{u2} \qquad (A4.2.12)$$

and the temperatures in the two zones may be found using Eq. A4.2.10.

This is arguably a more accurate solution for the single-zone theory described in Sec. 4.4.2. There, the average gas properties for the entire cylinder space are determined as in Eqs. A4.2.5-A4.2.7, but the individual properties of the trapped air and the burned gas are determined less realistically using the average cylinder temperature, T_1, instead of the zone temperatures, T_{u1} and T_{b1}, employed here. In the example given below, it is clear that this will induce errors; while not negligible, they are small.

The use of this theory within an engine simulation

In Chapters 4-7, a chainsaw engine is frequently employed as a design example for a computer simulation. The data used here for the computation are given in Sec. 5.5.1 for the standard engine. The fuel employed in the simulation is unleaded gasoline with a stoichiometric air-to-fuel ratio of 14.3. The results of a computer simulation with air-to-fuel ratios of 12, 13, 14, 15 and 16, are shown in Fig. A4.2, on which are displayed the cylinder temperatures as predicted by this two-zone combustion model. The mean cylinder temperature, and the temperatures in the burned and unburned zones, are indicated on this figure around the tdc period. The rapid rise of temperature in the burned zone to nearly 2400°C is clearly visible, and this peaks at about 10° atdc. The mean cylinder temperature peaks at about 2000°C, but this occurs at some 30° atdc. The temperature in the unburned zone has a real peak at about 15° atdc, whereas the spike at 50° atdc indicates the engulfment of, and disappearance of, the unburned zone by the combustion process. The maximum burn zone temperatures recorded during the simulations are shown in Fig. A4.3, with the highest value shown to be at an AFR of 13, and not closer to the stoichiometric point, i.e., where λ is unity.

The formation of nitric oxide (NO)

The formation of oxides of nitrogen is very, indeed exponentially, dependent on temperature. In Fig. A4.3, as captured from Fig. A4.2, the results of a simulation incorporating the two-zone burn model show the peak temperatures in both the burned and the unburned zones with respect to the same fueling changes. The burn zone temperature reaches a maximum before the stoichiometric air-to-fuel ratio. The unburned zone encounters its highest peak temperature at the richest air-fuel ratio.

The Zeldovitch [4.39] approach to the computation of NO formation, as in Eq. A4.1.14, takes all of these factors into account. The profile of the calculated NO formation, shown in Fig. A4.4 with respect to air-fuel ratio, is quite conventional and correlates well with the

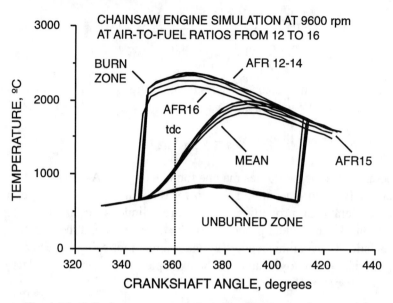

Fig. A4.2 Cylinder temperatures in a simple two-zone burn model.

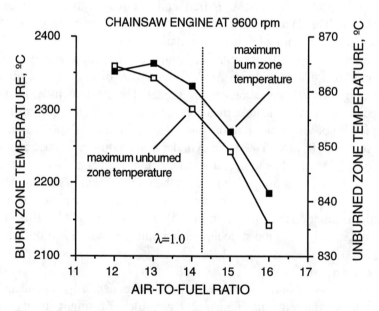

Fig. A4.3 The effect of fueling on peak temperatures in the two zones.

measured data shown in the same figure. That the bsNO peaks after the AFR location of maximum burn zone temperature may be surprising. The reason is, while the maximum burn zone temperature is decreasing toward the stoichiometric air-to-fuel ratio, the oxygen concentration therein is rising; see Fig. A4.8.

The measured data are recorded at the same engine speed and burn the same gasoline employed in the simulation; further correlation of this simulation with measured performance

Chapter 4 - Combustion in Two-Stroke Engines

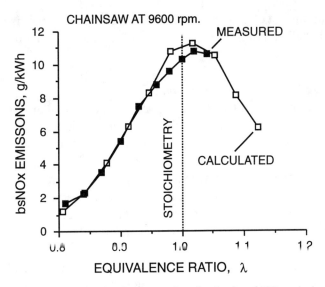

Fig. A4.4 Comparison of measured and calculated NO emissions.

characteristics is found in Sec. 5.5.1; there also is a description of this engine, its physical construction and its geometrical data. Further examples of measured bsNOx data for a quite different engine, which have similar profiles with respect to fueling, are to be found in Figs. 7.55 and 7.59.

The formation of nitric oxide, NO, is time related, as evidenced by Eq. A4.1.14. This is shown clearly by the results of the simulation of the chainsaw engine in Fig. A4.5. The highest rates of formation are at the air-to-fuel ratios of 14 and 15, which bracket the stoichiomet-

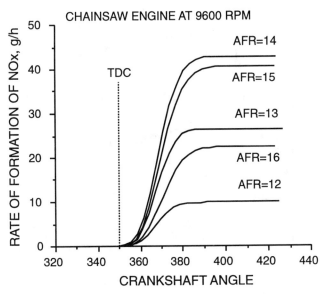

Fig. A4.5 Nitrogen oxide emission growth rate with respect to fueling.

ric value. Although it is not clearly visible on the printed figures, there is a small decrease in NO formation at the tail end of most of the line graphs, indicating that the rates of formation operate in both directions.

The time-related equilibrium and dissociation behavior

As discussed in Appendix A4.1, it is necesssary to solve simultaneously the dissociation reactions for CO/CO_2 and the "water-gas" reaction. The results of such a computation have great relevance, not only for the NO formation model discussed above, but also for the exhaust gas properties presented in Sec. 2.1.6. The time-related cylinder gas properties expand on the information which emanates from the employment of the relatively simple chemistry found at the start of Sec. 4.3.2. Exhaust gas is ultimately the very same gas that occupies the burn zone at the point of exhaust opening, for by that time the combustion period is over and the "burn zone" covers the entire volume of the cylinder at that juncture.

The same simulation of the chainsaw, which has provided the data for Figs. A4.2 to A4.5, is queried further to yield the time-histories of the mass concentrations of some of the gases within the burn zone. These are extracted from the simulation of the chainsaw at 9600 rpm and plotted in Figs. A4.6 to A4.9. The simulation, identical in every respect to that employed earlier in this Appendix, is conducted for the "standard" chainsaw input data at air-to-fuel ratios from 12 to 16 on unleaded gasoline.

Oxygen and its effect on NO formation

From the specific standpoint of the discussion above on the formation of nitric oxide, Fig. A4.8 is the most relevant. It is the mass concentration of oxygen, together with the nitrogen, which yields the nitric oxide. At equal burn zone temperatures and total mass, the rate of

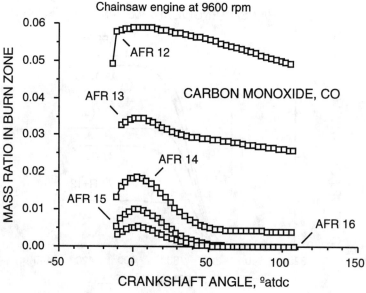

Fig. A4.6 *Mass ratio of carbon monoxide in cylinder burn zone.*

Chapter 4 - Combustion in Two-Stroke Engines

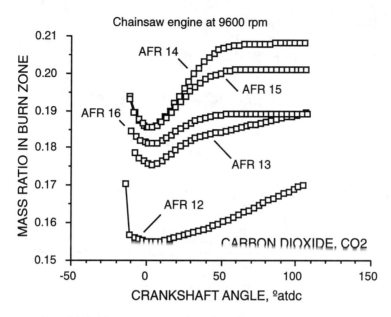

Fig. A4.7 Mass ratio of carbon dioxide in cylinder burn zone.

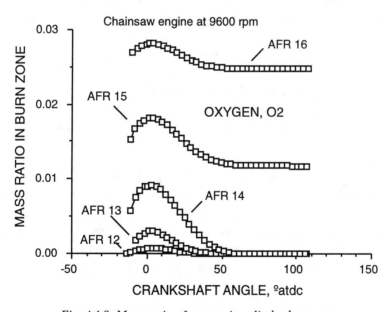

Fig. A4.8 Mass ratio of oxygen in cylinder burn zone.

formation of nitric oxide will be almost directly proportional to the mass concentration of oxygen. It is interesting to note that the mass concentrations of oxygen at air-to-fuel ratios of 12 and 14 are 0.0007 and 0.009, respectively, which are a factor of 13 different. The NO created at the same fueling levels is 8.2 and 35.6 g/h, respectively, which is only a factor of

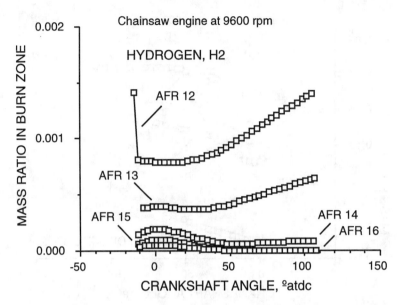

Fig. A4.9 Mass ratio of hydrogen in cylinder burn zone.

4.3 different. The exponential effect of burn zone temperature is now clarified, for the peak burn zone temperatures at the AFR values of 12 and 14 are 2352°C and 2333°C, respectively, which are almost identical. From a more general viewpoint, Fig. A4.8 shows that the dissociation of oxygen is dependent on temperature. However, at any given fueling level, by 50° atdc, which is 40° crankshaft before the exhaust port opens, the mass concentration of oxygen has stabilized at an appropriate level. The leaner the fueling, the greater the oxygen content of the exhaust gas, although it is only significantly high for air-to-fuel ratios leaner than stoichiometric. The simple chemistry of Eq. 4.3.3 already supplies that information.

Carbon monoxide

Simple chemistry dictates that the richer the fueling the greater the CO content. Fig. A4.6 confirms this. Dissociation at high temperature temporarily increases this during combustion. At air-to-fuel ratios close to the stoichiometric value, the behavior is not unlike that for oxygen, i.e., stabilization has occurred well in advance of the exhaust port opening at 107° atdc. However, at rich mixture combustion, the dissociation behavior continues right up to exhaust opening, although the variation in real terms is quite small.

Carbon dioxide

The time-related behavior for air-to-fuel ratios from 12 to 16 is shown in Fig. A4.7. The process is the reverse of that for the carbon monoxide and the oxygen, in that carbon dioxide is reformed by association as the temperature falls with time as illustrated in Fig. A4.2. In short, the higher the temperature, the greater the dissociation of the carbon dioxide. Again, as with the carbon monoxide, the richest mixture combustion is the slowest to reform with the passage of time, and is barely complete as the exhaust port opens and freezes the process.

Note that the closer the AFR during the combustion is to the stoichiometric, the higher the carbon dioxide mass concentration, which means that the most efficient extraction of heat has been obtained from the fuel. This reinforces the empirical relationship, based on measurements, given in Eq. 4.3.26 for the relative combustion efficiency with respect to fueling.

Hydrogen

This is probably the most complex reaction behavior of the series, for it can be seen in Fig. A4.9 that it is dissimilar on either side of the stoichiometric fueling level, which is at an AFR of 14.3 for this particular (unleaded gasoline) fuel. Close to, or richer than the stoichiometric value, where λ is unity, the hydrogen is more pronouncedly dissociated than at lean mixture burning.

At rich mixtures, and at higher temperatures in the burn zone, more hydrogen is created leaving more free oxygen. As the cylinder contents cool (if 1800°C can be described as "cooler") the oxygen is able to combine with the hydrogen to form steam, and reduce the free hydrogen content of the embryo exhaust gas. At an AFR of 12 the process is barely complete by exhaust port opening at 107° atdc. The proportions involved are tiny, as a mere 0.15% of the cylinder mass is free hydrogen, even at the richest mixture combustion. Nevertheless, it is upon such complete chemistry that the computation depends for its accurate deduction of the oxygen mass concentration, without which level of accuracy the prediction of nitric oxide (NO) formation would become hopelessly incorrect.

General

As discussed in Appendix A4.1, it is evident that this combustion model, using both equilibrium chemistry and reaction kinetics, is absolutely necessary to provide temperature data, and mass concentration data for all of the relevant gas species, in the burned zone for the prediction of the mass emissions of oxides of nitrogen and, ultimately, the properties of exhaust gas.

Chapter 5

Computer Modeling of Engines

5.0 Introduction

In the first four chapters of this book, you are introduced to the two-stroke engine, to the unsteady nature of the gas flow in, through and out of it, to the nature of the scavenging process and its potential for retention of the fresh charge supplied for each cycle, and to the combustion process where the characteristics of burning of the air and fuel in a two-stroke engine are explained. In each case, be it cylinder geometry, unsteady gas flow, scavenging or combustion, theoretical models are described which can be solved on a digital computer. The purpose of this chapter is to bring together those separate models to illustrate the effectiveness of a complete model of the two-stroke engine. This simulation is able to predict all of the pressure, temperature and volume variations with time within an engine of specified geometry, and calculate the resulting performance characteristics of power, torque, fuel consumption and air flow.

Because of the (almost) infinite number of conceivable combinations of intake and exhaust geometry which could be allied to various selections of engine geometry, it is not possible to demonstrate a computer solution that will cope with every eventuality.

Within the chapter, it is proposed to use engine simulation programs to illustrate many of the points made in earlier chapters regarding the design of various types of engines and to highlight those further principles not previously raised regarding the design of two-stroke engines.

Perhaps one of the most useful aspects of engine modeling is that the simulation allows the designer to imagine the unimaginable, by being able to see on a computer screen the temporal variations of pressure, volume and gas flow rate that take place during the engine cycle. Many of the parametric changes observed in this visual manner, together with their net effect on power output and fuel consumption, provide the designer with much food for design thought. Computer output is traditionally in the form of numbers, and graphs of the analysis are normally created after the computation process. In the era when such analysis was conducted on mainframe computers, the designer acquired the output as a package of printout and (sometimes) pictorial information, and then pondered over it for its significance. This post-analytical approach, regarded at the time as a miracle of modern technology, was much less effective as a design aid than the desktop microcomputer presenting the same information in the same time span, but allowing the designer to watch the engine "run" on the computer screen in a slow-motion mode. It is from such physical insights that a designer conceives of future improvements in, say, power or fuel consumption or exhaust emissions. Let it

Design and Simulation of Two-Stroke Engines

not be forgotten that it is the human brain that synthesizes while the computer merely analyzes. The pace of development of stand-alone computer systems, in terms of performance for cost, is such that the comments made above are virtually out of date as they are being typed into just such a computer!

In this chapter I propose to illustrate, by using computer models (available from SAE), the operational behavior of spark-ignition engines in very fine detail, often by predicting data on thermodynamic properties throughout the engine which, at this point in history, are not capable of being measured. For instance, the measurement of temperature as a function of time has always posed a serious instrumentation problem, yet a computer model has no difficulty in calculating this state condition anywhere within the engine at intervals of 1° crankshaft while the engine is turning at 10,000 rpm. As it cannot be measured, there is no means of telling the accuracy of this prediction. On the other hand, the measurement of pressure as a function of time is a practical instrumentation exercise, although it must be conducted carefully so that comparison with the calculations can be carried out effectively.

Through the illustrations from the output data obtained by running the engine simulations, you should be able to comprehend the ramifications of changing design variables on an engine, such as port timings and areas, or compression ratios, or the dimensions of the exhaust or intake ducting.

The types of engine simulated are selected to show the effects of exhaust tuning, or the lack of it, on the ensuing performance characteristics. A complete discussion on this subject would occupy a book by itself, so the simulations described here will be restricted to spark-ignition engines. Should some of the discussion also be applicable to diesel engines, appropriate statements will be inserted into the text.

The first engine to be simulated is a spark-ignition engine which does not use exhaust system tuning to achieve its performance characteristics. The engine is a chainsaw with simple intake and exhaust silencers in the form of what can only be described as "boxes" or plenums and, as it is sold as a product where compactness and light weight are at a premium, very small "boxes" are used as intake and exhaust silencers. The engine is small, it has a swept volume less than 100 cm^3, and produces a target bmep of some 4 bar at up to 11,000 rpm.

The second engine simulated is also a spark-ignition unit, but which uses all of the tuning effect it can acquire from the ducting attached to the cylinder. It is a reed-valved engine used for Grand Prix motorcycle racing and here the target bmep is 11 bar at engine speeds up to 13,000 rpm.

The third engine simulated is the multi-cylinder, externally scavenged, spark-ignition design, with direct fuel injection, where the optimization of exhaust tuning from a compact exhaust manifold is paramount to producing a high-performance product. In short, the engine uses a blower and is colloquially referred to as being supercharged.

5.1 Structure of a computer model

The key elements required of any computer program to model a two-stroke engine, as indicated first in Sec. 2.18.10, are:
- (i) The physical geometry of the engine so that all of the port areas, the cylinder volume and the crankcase volume are known at any crankshaft angle during the rotation of the engine for several revolutions at a desired engine speed.

Chapter 5 - Computer Modeling of Engines

(ii) A model of the unsteady gas flow in the inlet, transfer and exhaust ducts of the engine.
(iii) A model of the unsteady gas flow at the ends of the inlet, transfer and exhaust ducts of the engine where they encounter cylinders, crankcases, reed blocks, and the atmosphere; or branches, expansions, contractions, and restrictions within that ducting.
(iv) A model of the thermodynamic and gas-dynamic behavior within the cylinder and in the crankcase of the engine while the ports are open, i.e., the open cycle period.
(v) A model of the thermodynamic behavior within the cylinder of the engine while the ports are closed, i.e., the closed cycle period.
(vi) A model of the scavenge process so that the proportion of fresh charge retained within the cylinder can be predicted.

Virtually all of the material related to the above topics has already been described and discussed in the earlier chapters, and here they are brought together as major programs to describe a complete engine. These programs are lengthy, running to some five thousand lines of computer code for even the simplest of single-cylinder engines. Just three items remain to be discussed, prior to introducing the considerable array of design information which comes from such computer simulations, and they are, (i) the acquisition of the physical geometry of an engine required to be employed as input data for such a simulation, (ii) heat transfer in the crankcase of a crankcase compression engine, and (iii) the friction characteristics to be assigned to a particular engine geometry so that the computations detailing the indicated performance characteristics may be translated into the brake performance behavior required of any design, as first debated in Secs. 1.5 and 1.6.

5.2 Physical geometry required for an engine model

A detailed description of engine geometry was carried out in Secs. 1.4.3-1.4.6, which formed the subject of three simple computer programs, Progs.1.1-1.3. Chapter 3, on scavenging flow, presented more computer programs that provide geometrical information on the areas of ports arising from various types of engine scavenging. The geometrical information required for a simulation program is even more detailed, requiring all that was needed for those earlier programs as well as complete information regarding the geometry of all of the ducting attached to the engine, and of the areas of the porting where the ducts are attached to a cylinder of the engine. The number of possible combinations of engine ducting which can be attached to an engine cylinder or crankcase are too great to be included in this text so a representative set of generic types will be shown.

5.2.1 The porting of the cylinder controlled by the piston motion

The simulation model needs to know the area of any port in the cylinder through which gas can pass to or from the ducting connected to it. Most ports in a simple two-stroke engine are controlled by the piston, as discussed initially in Sec. 1.1. The simulation also needs to know the precise area of any port being opened or closed by the piston by crank rotation. If the area is established at any crankshaft position, then after the application of coefficients of discharge as discussed in Appendix A2.3, the thermodynamic and gas-dynamic simulation can proceed as described in detail in Sec. 2.18.

The variations for the geometry of engine ports as conceived by practicing designers can be quite diverse, but they fall into two categories, the simple or the complex. The simple port layout is shown in Fig. 5.1(a) and the complex is illustrated in Fig. 5.1(b).

The simple port layout with piston crown control

This is sketched in Fig. 5.1(a) and typically applies to the exhaust and scavenge ports. The crankshaft has turned an angle, θ, from the top dead center, tdc, position and the piston has moved a length, x_θ, from the tdc point, having uncovered a port at an angle, θ_1, from tdc, and will fully uncover it after turning an angle, θ_2; the piston travel lengths are $x_{\theta 1}$ and $x_{\theta 2}$, respectively. For exhaust and scavenge (transfer) ports, almost universally, the value of θ_2 is 180° or at bdc, in which case the value of $x_{\theta 2}$ is the stroke length, L_{st}.

The area of the port at any juncture is shown in the figure as A_θ for a rectangular port of width, x_p, and top and bottom corner radii, r_t and r_b, respectively.

The width across the port, $x_{p\theta}$, at any crankshaft angle, θ, will be determined by the piston position being in the middle of the port, or across either of the corner radii, and is found from simple geometrical considerations, thus:

$$\text{if } x_\theta < r_t + x_{\theta 1} \text{ then } x_{p\theta} = x_p - 2r_t + 2\sqrt{r_t^2 - (r_t - x_\theta + x_{\theta 1})^2} \qquad (5.2.1)$$

$$\text{if } x_\theta \geq r_t + x_{\theta 1} \text{ and } x_\theta < x_{\theta 2} - r_b \text{ then } x_{p\theta} = x_p \qquad (5.2.2)$$

$$\text{if } x_\theta > x_{\theta 2} - r_b \text{ then } x_{p\theta} = x_p - 2r_b + 2\sqrt{r_b^2 - (r_b - x_{\theta 2} + x_\theta)^2} \qquad (5.2.3)$$

The area of the port is then given by, where an incremental crankshaft movement, $d\theta$, is accompanied by a corresponding incremental piston movement, dx:

$$A_\theta = \int_{x=x_{\theta 1}}^{x=x_\theta} x_{p\theta} dx = \int_{\theta=\theta_1}^{\theta=\theta} x_{p\theta} \frac{dx}{d\theta} d\theta \qquad (5.2.4)$$

This is a relatively complex piece of calculus, and is much more readily solved on a computer by conducting the calculation in even increments of angle, thereby conducting the integral as a simple summation process for any port. The information is loaded into a data file, and by linearly interpolating from that data file, the value of port area can be determined at any given point within the computation. It is seen thus:

$$A_\theta = \sum_{\theta=\theta_1}^{\theta=\theta} \frac{x_{p\theta} + x_{p\theta+\Delta\theta}}{2} (x_{\theta+\Delta\theta} - x_\theta) \qquad (5.2.5)$$

The summation can be conducted equally effectively by employing even increments of angle, $\Delta\theta$, of 1° crankshaft, or even increments of piston movement, Δx, of 0.1 mm.

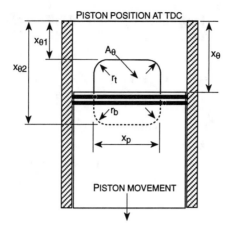

Fig. 5.1(a) Area of a regularly shaped port in a cylinder wall.

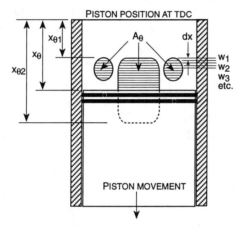

Fig. 5.1(b) Area of an irregularly shaped port in a cylinder wall.

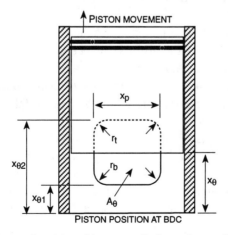

Fig. 5.1(c) Area of a piston skirt controlled port in a cylinder wall.

Design and Simulation of Two-Stroke Engines

The maximum area of the port, A_{max}, is calculated relatively easily from the fact that it is rectangular in shape and has corner radii.

$$A_{max} = x_w(x_{\theta 2} - x_{\theta 1}) - \left(2 - \frac{\pi}{2}\right)\left(r_t^2 + r_b^2\right) \tag{5.2.6}$$

Virtually all scavenge ports are rectangular in projected area terms and are normally fully open precisely at bdc. Most exhaust ports are also rectangular in profile, although here the variation in shape, as seen in practice, is greater, and are normally fully open precisely at bdc; but the lower edge in some configurations does not extend completely to the bdc position.

The more complex port layout with piston crown control

This is shown in Fig. 5.1(b). The majority of the cases that have to be treated in this manner fall into two categories. The first is where the top edge of the port is distorted for some particular reason. Perhaps the center section is not horizontal but is curved so as to give the piston ring an easier passage from its expansion into the port to being cajoled into following the profile of the cylinder bore. Perhaps a vertical slot has been profiled into the top edge to give a longer, more gentle, blowdown process and reduce the exhaust noise [8.14]. Perhaps, as is sketched in Fig. 5.1(b), two extra exhaust ports have been added in the blowdown region of the exhaust port to make that process more rapid; this is commonly seen in racing engine design. For whatever reason, the profile becomes too complex for simple algebraic analysis and so the approach seen through Eq. 5.2.5 is extended to expedite the computation of the port area.

The particular geometry is analyzed to determine the widths, w_1, w_2, w_3, etc., of the port at even increments of height, Δx, from the top to the bottom of the port. An input data file of this information is presented, to be analyzed using Eq. 5.2.5, and a second data file is generated giving the port area, A_θ, with respect to piston position, x_θ, at various crankshaft angles, θ, from the tdc position. Throughout the engine simulation this file is indexed and linear interpolation is employed to determine the precise area at the particular crankshaft position at that instant during the analysis.

Hence, during an engine simulation using the theory of Secs. 2.16 and 2.17, the effective throat area of a port controlled by the piston crown at any juncture (see Eq. 2.16.4) is defined by A_t, and the numerical value of A_t is that given by A_θ in Eqs. 5.2.4 or 5.2.5. To carry out this exercise during simulation, be it for a simple regularly shaped port or a more complex shape, the data listed in Figs. 5.1(a) or 5.1(b) are required as input data to it.

The simple port layout with piston skirt control

This is shown in Fig. 5.1(c) and typically applies to the intake ports. The crankshaft is at an angle, θ, before the tdc position and the piston has moved a length, x_θ, from the bdc point, having uncovered a port at an angle, θ_1, before tdc, and will fully uncover it after turning an angle, θ_2, also before tdc. It is normal design practice to fully uncover the intake ports at or before tdc. The piston travel lengths are $x_{\theta 1}$ and $x_{\theta 2}$, respectively, but are measured from bdc. For intake ports, almost universally, the value of θ_2 is $0°$ or at tdc, in which case the value of $x_{\theta 2}$ is the stroke length, L_{st}.

Chapter 5 - Computer Modeling of Engines

The area of the port at this juncture is shown in the figure as A_θ for a rectangular port of width, x_p, and top and bottom corner radii, r_t and r_b, respectively. Notice that the position of the top and bottom port radii are in the same physical locations for the port, in a cylinder which is considered to be sitting vertically. Thus the analysis for the instantaneous port area, A_θ, presented in Eqs. 5.2.1 to 5.2.5, is still applicable except that the values for the port radii must be juxtaposed so as to take this nomenclature into account. The relationship for the maximum port area, A_{max}, in Eq. 5.2.6 is correct as it stands.

Hence, during an engine simulation using the theory of Secs. 2.16 and 2.17, the effective throat area of a port controlled by the piston crown at any juncture (see Eq. 2.16.4) is defined by A_t, and the numerical value of A_t is that given by A_θ in Eqs. 5.2.4 or 5.2.5, taking into account the above discussion. To carry out this exercise during simulation, be it for a simple regularly shaped port, or a more complex shape, all of the data listed in Fig. 5.1(c) are required as input data to it.

5.2.2 The porting of the cylinder controlled externally

The use of valves of various types to control the flow through the ports of an engine was introduced in Sec. 1.3. For the intake system, reed valves, disc valves and poppet valves are commonly employed. For the exhaust system, poppet valves have been commonly used, particularly for diesel engines. As a further refinement to the piston control of ports leading to the exhaust system, timing edge control valves have been utilized, particularly for the high-performance engines found in racing motorcycles. The areas of the ports or apertures leading into the cylinders of the engine must be accounted for in the execution of a simulation model of an engine.

The use of poppet valves

Poppet valves are not normally used in simple two-stroke engines, but are to be found quite conventionally in uniflow-scavenged diesel engines used for marine applications or in trucks and buses, such as those shown in Plates 1.4 and 1.7. The area for flow through a poppet valve, at any juncture of its lift, is described in detail in Appendix A5.1.

Hence, during an engine simulation using the theory of Secs. 2.16 and 2.17, the geometric throat area of a port controlled by the piston crown at any juncture (see Eq. 2.16.4) is defined by A_t, and the numerical value of A_θ is that given by A_t in Appendix A5.1 in Eqs. A5.4 and A5.5. To carry out this exercise during simulation, all of the data shown in Fig. A5.1 are required as input data to it.

The use of a control valve for the port timing edge

The common practice to date has been to apply timing control valves to the exhaust ports of the engine. A typical arrangement for an exhaust port timing control valve is sketched in Fig. 5.2. In Fig. 5.2(a) the valve is fully retracted so that the timing control edge of the valve coincides with the top edge of the exhaust port. The engine has a stroke dimension of L_{st} and a trapped stroke of length x_{ts}. When the valve is rotated clockwise through a small angle and held at that position, as in Fig. 5.2(b), normally at a lower point in the engine speed range, the trapped stroke is effectively changed to x_{ets} and the total area of the port is reduced. Clearly, the blowdown area and the blowdown timing interval are also reduced. While such a valve

Design and Simulation of Two-Stroke Engines

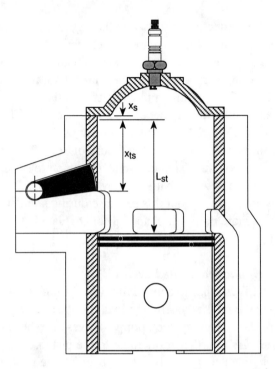

Fig. 5.2(a) Exhaust control valve in fully retracted position.

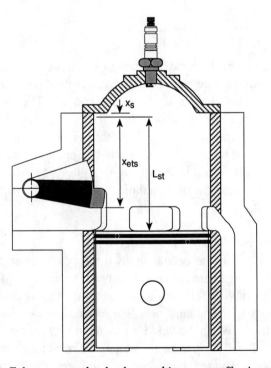

Fig. 5.2(b) Exhaust control valve lowered increases effective trapped stroke.

Chapter 5 - Computer Modeling of Engines

can never seal the port to the extent that a piston can accomplish within the cylinder bore, the design of the valve can be such that it closely follows the piston profile over the width of the exhaust port(s). There are many practical designs for such valves, ranging from the cylindrical in section [7.4] to guillotine designs and the lever type shown in Fig. 5.2.

If the simulation model is to incorporate such a timing edge control valve, then several factors must be taken into account. The first is the altered port profile which can be incorporated into the execution of the analytic process described above. Even the "leakage" profile past the valve can be accounted for by assigning an appropriate width to the port to simulate the area of the leakage path. However, if such a valve is found to seal the cylinder at a timing point other than the fully open position, the simulation model must take into account the change of trapped compression ratio, CR_t. This is the consequence of the fact that the clearance volume of the engine will have been deduced from the trapped compression ratio at the full height of the exhaust port. Hence the modified trapped compression ratio with the timing valve lowered is:

$$CR_{t\ modified} = (CR_t - 1)\frac{X_{ets}}{X_{ts}} + 1 \tag{5.2.7}$$

The use of a disc valve for the intake system

This has been a very popular intake system for many years, particularly for single-cylinder engines, and especially in motorcycles. It is very difficult to incorporate into a multi-cylinder design. A sketch of the significant dimensions of such a valve is shown in Fig. 5.3 and a photograph of one fitted to an engine is shown in Plate 1.8. Further pertinent information and discussion on disc valve design is seen in Figs. 6.28 and 6.29 and in Sec. 6.4, where the nomenclature is common with this section and the sketches provide further aid to understanding of the analysis below. The total opening period of such a valve, ϕ_{max}, is readily seen as the combined angles subtended by the disc and the port:

$$\phi_{max} = \phi_p + \phi_d \tag{5.2.8}$$

Thus, the total opening period is distributed around the tdc position in an asymmetrical manner, as presented before in Fig. 1.8. The maximum area of the port can be shown to be a segment of an annulus between two circles of radius r_{max} and r_{min}, less the corner radii, r_p.

$$A_{max} = \pi\frac{\phi_d}{360}\left(r_{max}^2 - r_{min}^2\right) - r_p^2(4 - \pi) \tag{5.2.9}$$

The instantaneous area, A_θ, is found by a similar theoretical approach to that seen in Eqs. 5.2.1-5.2.4, except that the term for the movement of the disc with respect to the angular crankshaft movement is now more simply deduced in that it is linear. For example, if the disc has fully uncovered the corner radii at opening, the instantaneous area at a juncture of $\theta°$ from the opening position is given by:

$$A_\theta = \pi\frac{\theta}{360}\left(r_{max}^2 - r_{min}^2\right) - r_p^2\left(2 - \frac{\pi}{2}\right) \tag{5.2.10}$$

Design and Simulation of Two-Stroke Engines

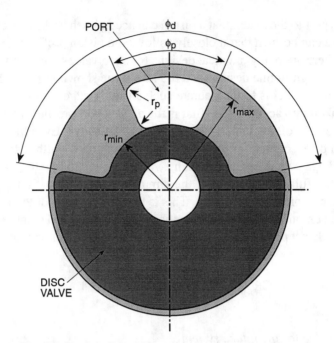

Fig. 5.3 Design dimensions of an intake system disc valve.

As the port corner radii are being uncovered by the edge of the disc, the precise solution to the geometrical problem is more complex but can be solved with little loss of accuracy by the iterative approach given above for rectangular ports in Eq. 5.2.1. If x_p is considered to be the port height, after $\theta°$ from the opening position:

$$x_p = r_{max} - r_{min} \tag{5.2.11}$$

then

$$x_{p\theta} = x_p - 2r_p + 2\sqrt{r_p^2 - h^2} \tag{5.2.12}$$

if

$$h = \pi \frac{\theta}{360}(r_{max} + r_{min}) \tag{5.2.13}$$

where h is the length along the port centerline and is less than the corner radius, r_p. The area of the port during this early stage can be found by the same method given in Eq. 5.2.4, but with the term for $dx/d\theta$ replaced by the following equation, where the angle is expressed in degrees:

$$\frac{dx}{d\theta} = \frac{\pi(r_{max} + r_{min})}{360} \tag{5.2.14}$$

A similar iterative approach is employed for the later stages of port opening or closing when the relevant corner radii are being uncovered or covered, respectively.

Chapter 5 - Computer Modeling of Engines

Hence, during an engine simulation using the theory of Secs. 2.16 and 2.17, the effective throat area of the port posed by the disc valve system at any juncture (see Eq. 2.16.4) is defined by A_t, and the numerical value of A_t is that given by A_θ in Eq. 5.2.10. To carry out this exercise during simulation, all of the data shown in Fig. 5.3 are required as input data to it.

The use of a reed valve for the intake system

The use of an automatic valve for the control of intake flow is quite common in pulsating air-breathing devices, not only for motorcycles and outboards [1.12, 1.13] but for air and refrigerant compressors [6.6] and in pulsejets [5.22], such as the German V1 "doodlebug." The format of the design can assume many mechanical configurations, but the V block shape shown in Fig. 5.4, Fig. 6.27, or Plate 6.1, has become commonplace in high-performance motorcycles. In outboards, and in other uses where space is at a premium or the specific performance requirement is not so great, a simpler flat plate design is often found.

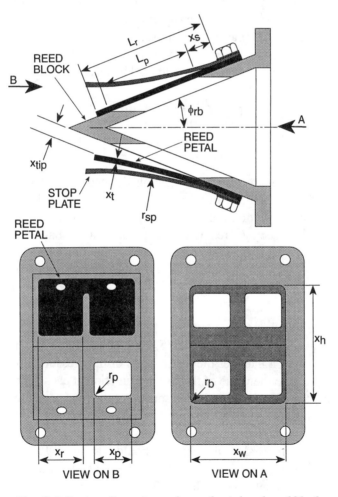

Fig. 5.4 Design dimensions of a reed petal and reed block.

The reed block is placed in the crankcase, or in the scavenge ducts, of a naturally aspirated spark-ignition engine, and permits air flow into the crankcase when the crankcase pressure falls below the atmospheric pressure, and shuts again when it exceeds it. When the reed lifts, and both pressure drop and particle flow take place through the reed block port and across the reed into the crankcase, a complex pattern of pressure is applied to the reed surfaces on both sides. It is the summation of all of these pressures over the entire surface area of the reed petal which gives rise to the force that causes the reed to open, and a combination of these forces and the dynamic spring characteristics of the reed petal which makes it reseat itself. Further pertinent discussion relating to reed valve operation and design is found in Sec. 6.3.

At QUB, much research has been carried out on this subject [5.9, 5.14-5.19] and the following is a precis of those findings as applied to their incorporation into a computer model of an engine fitted with a reed valve.

Hinds [5.18] showed that the reed valve can be treated as a pressure-loaded cantilevered beam clamped at one end and forced into various modes of vibration by the crankcase pressure acting on one side of the beam, and the superposition pressure at that position in the inlet tract acting upon the other. The basic solution devolves to the determination of the natural frequency in cycles per second, f_j, or ω_j in radians per second, of the first few modes of vibration, j, from the basic theory of pressure loaded beams, thus:

$$f_j = \frac{\omega_j}{2\pi} = \frac{(\beta_j L_r)^2}{2\pi} \sqrt{\frac{YI}{\rho A L_r^4}} = \frac{(\beta_j L_r)^2}{2\pi} \sqrt{\frac{Y x_t^2}{12 \rho L_r^4}} \qquad (5.2.15)$$

In the above equation, apart from those symbols defined in Fig. 5.4, the value of Y is the Young's Modulus of the reed petal material, ρ is its density, and I and A are the second moment of area and cross-section area in the plane of bending, respectively. The second moment of area and cross-section area of a rectangular reed petal are given by:

$$I = \frac{x_r x_t^3}{12} \qquad A = x_r x_t \qquad (5.2.16)$$

The values of the function related to mode, $\beta_j L_r$, in Eq. 5.2.15 for the first five modes of vibration are determined as 1.875, 4.694, 7.855, 10.996, and 14.137, respectively.

The theoretical solution includes the distribution of the pressure along the length of the reed, on both sides, to give the value of the forcing function at each segment along the reed, usually in steps of 1 mm. The pressure on the surface of the reed petal facing the crankcase is assumed to be the crankcase pressure. The pressure on the surface of the reed petal facing the reed block is assumed to be the superposition pressure at that end of the inlet tract [5.16]. Bounce off the reed stop must also be included, should it be in the form shown in Fig. 5.4 or be it a simple tip stop as is often found in practice. The final outcome is a computation of the lift of the reed at each element along its length from the clamping point to the tip for each of the modes of vibration it is following. By correlating theory with experiment, Hinds [5.18] showed that it was quite adequate to consider only the first two modes of vibration and that the higher orders could safely be neglected.

For a fuller description of the theoretical model of reed valve motion the thesis by Hinds [5.18] or by Houston [5.19] should be consulted. Houston [5.19] extends the work of Hinds to the movement of reed petals which are other than rectangular in plan profile.

The reed block poses two potential restrictions to the flow into the engine, the area at entrance to the reed ports in the block, and the area posed by the lifted reed petals. From a gas-dynamic standpoint, the area of the intake duct at the reed block end is denoted by A_{IP}, and is the area defined from Fig. 5.4 as:

$$A_{IP} = x_w x_h - (4 - \pi) r_b^2 \qquad (5.2.17)$$

The effective area posed by the reed ports, A_{rp}, is given by, where n_{rp} is the number of ports:

$$A_{rp} = n_{rp} \left(L_p x_p - r_p^2 (4 - \pi) \right) \sin \phi_{rb} \qquad (5.2.18)$$

However, the effective area of flow past the reed petals is a much more debatable issue. For example, flow into the crankcase from the side of the reeds is possible, but this is unlikely between closely spaced adjacent reeds, and possibly not if the reed block is tightly packed into the mouth of the crankcase. The only certain flow direction is tangentially past the reed tip, which at some particular instant has a tip lift of dimension, x_{tip}. This maximum flow area at that juncture can be defined as A_{rd}, where n_r is the number of reed petals, thus:

$$A_{rd} = n_r x_{tip} x_p \qquad (5.2.19)$$

Actually, this statement is still too simplistic, for the throat area of the flow past the petal is more realistically given by the reed lift, x_{act}, at a point on the reed at the end of the port, i.e., the combination of the distances L_p and x_s. The more correct solution for the throat of the reed flow area is given by the insertion of x_{act} into Eq. 5.2.19 as a replacement for x_{tip}. It may seem to the reader that the reed throat area term in Eq. 5.2.19 should be multiplied by the cosine of the reed block half angle, for the same reason as the port area is modified by the sine of that angle in Eq. 5.2.18. The fact is that the discharge coefficients, determined using the approach of Appendix A2.3, employ the geometric area as given in Eq. 5.2.19, thereby automatically taking into account any effects caused by the reed block angle. As a supporting argument in this same context, it will be observed that the reed port area for a flat plate reed is always computed correctly; when ϕ_b as 90° is applied in Eq. 5.2.18, the sine of that value is unity.

Hence, during an engine simulation using the theory of Secs. 2.16 and 2.17, the effective throat area of the port posed by the reed system at any juncture (see Eq. 2.16.4) is defined by A_t, and the numerical value of A_t is that given by either A_{rp} or A_{rd}, whichever is the lesser. To carry out this exercise during simulation, all of the data shown in Fig. 5.4, with the exception of x_{tip} and x_{act} which are computed, are required as input data to it, together with the Young's Modulus and the density of the reed petal material.

Design and Simulation of Two-Stroke Engines

To assist with the analysis of computed data, and to compare it with that measured, it is useful to declare a *reed tip lift ratio*, C_{rdt}, defined at any instant with respect to the reed length, as:

$$C_{rdt} = \frac{x_{tip}}{L_r} \qquad (5.2.20)$$

The empirical solution for reed valve design, given in Sec. 6.3, may help to further elucidate you about the application of some of the theory shown here.

5.2.3 The intake ducting

The intake ducting must be designed to suit the type of induction system, i.e., be it by reed valve, or be it piston skirt controlled, or through the use of a disc valve, all of which have been discussed initially in Sec. 1.3. The modeling of the port has been discussed above.

A typical intake duct is shown in Fig. 5.5. It commences at the inlet port, marked as A_{IP}, and extends to the atmosphere at an inlet silencer box of volume, V_{IB}, and breathes through an air filter of effective diameter, d_F. The area of the inlet port, A_{IP}, is normally the area seen in Sec. 5.2 described as the maximum area at the cylinder or crankcase port, A_{max}, but is not necessarily so for it is not uncommon in practice that there is a step change at that position. The point is best made by re-examining Figs. 2.16 and 2.18, as the area at the end of the intake system denoted here by A_{IP} is actually the area denoted by A_2 in Figs. 2.16 and 2.18. It is good design practice to ensure that A_2, i.e., A_{IP}, precisely equals the maximum port area; it is not always seen in production engines. Should the engine have a reed valve intake system, then it is normal for A_{IP} to equal the value defined as such in Eq. 5.2.17. Should the engine have a disc valve intake system, then it is normal for A_{IP} to equal the value defined as A_{max} in Eq. 5.2.9.

Somewhere within this intake duct there will be a throttle, or combined throttle and venturi if included with a carburetor, of effective diameter, d_{tv}. The throttle and venturi are modeled using the restricted pipe theory from Sec. 2.12. All of the other pipe sections, of lengths L_1, L_2, and L_4, with diameters ranging from d_0 which is equivalent to area A_{IP}, to d_1, d_2, d_3, and d_4, are modeled using the tapered pipe theory found in Sec. 2.15.

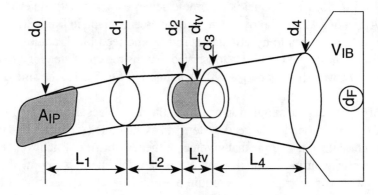

Fig. 5.5 Dimensions of an intake system including the carburetor.

In this connection the *throttle area ratio*, C_{thr}, is defined as the area of the venturi or the area set by the throttle plate, whichever is the lesser, with respect to the downstream section in Fig. 5.5, as:

$$C_{thr} = \frac{A_{tv}}{A_2} = \frac{d_{tv}^2}{d_2^2} \qquad (5.2.21)$$

At the end of the intake system, adjacent to the airbox, or to the atmosphere if unsilenced as in many racing engines, the diameter is marked as dimension d_4. The input data system to any simulation must be made aware if the geometry there provides a bellmouth end, or is a plain-ended pipe, for Secs. 2.8.2 and 2.8.3 make the point that the pressure wave reflection regime is very dependent on the type of pipe end employed. In other words, the coefficients of discharge of bellmouth and plain-ended pipes are significantly different [5.25].

5.2.4 The exhaust ducting

The exhaust ducting of an engine has a physical geometry that depends on whether the system is tuned to give high specific power output or is simply to provide silencing of the exhaust pressure waves to meet noise and environmental regulations. Even simple systems can be tuned and silenced and, as will be evident in the discussion, the two-stroke engine has the inestimable advantage over its four-stroke counterpart in that the exhaust system should be "choked" at a particular location to provide that tuning to yield a high power output.

Compact untuned exhaust systems for industrial engines

Many industrial engines such as those employed in chainsaws, weed trimmers, or generating sets have space limitations for the entire package, including the exhaust and intake ducting, yet must be well silenced. A typical system is shown sketched in Fig. 5.6. The system has a box silencer, typically some ten or more cylinder volumes in capacity, depending on the aforementioned space limitations. The pipe leading from the exhaust port is usually parallel and has a diameter, d_1, representing area, A_2, in Fig. 2.16, which normally provides a 15-20% increase over the maximum exhaust port area, A_{max}, defined by Eq. 5.2.6. The dimension, d_0, corresponds to the maximum port area, A_{max}. The distribution of box volumes is dictated by silencing and performance considerations, as are the dimensions of the other pipes, dimensioned by lengths and diameter as L_2 and d_2, and L_3 and d_3, respectively.

Tuned exhaust systems for high-performance, single-cylinder engines

Many racing engines, such as those found in motorcycles, snowmobiles and skijets, use the tuned expansion chamber exhaust shown in Fig. 5.7. The dimension, d_0, corresponds to the maximum port area, A_{max}. The pipe leading to dimension, d_1, may be tapered or it can be parallel, depending on the whim of the designer. The first few sections leading to the maximum diameter, d_4, are tapered to give maximum reflective behavior to induce expansion waves, and the remainder of the pipe contracts to reflect the "plugging" pulsations essential for high power output. The empirical design of such a pipe is given in Chapter 6, Sec. 6.2.5. The dimension, d_4, is normally some three times larger than d_1.

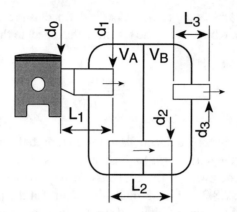

Fig. 5.6 Dimensions of a chainsaw exhaust system.

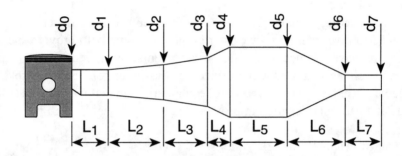

Fig. 5.7 Dimensions of a tuned exhaust system.

The tail-pipe, normally parallel of diameter, d_7, is usually about one-half the diameter of that at d_1. The tail-pipe can lead directly to the atmosphere, but it is extremely noisy as such, vide Plate 5.1. The regulations for motorcycle racing specify a silencer, which is typically a short, straight-through absorption device wrapped around the tail-pipe as seen in Fig. 2.6, the inclusion of which in a simulation is barely noticeable on the ensuing gas dynamics. A discussion on the design of such silencers is found in Chapter 8.

Tuned exhaust systems for high-performance, multi-cylinder engines

Sketches of a typical arrangement of the exhaust manifold and exhaust system of a multi-cylinder two-stroke engine are shown in Figs. 5.8(a) and (b). They are drawn in the context of a three-cylinder engine, but the logical extension of the arrangement to twin-cylinder units and to four or more cylinders is quite evident. It will become clear in the ensuing discussion that the three-cylinder engine has distinct tuning possibilities for the creation of high specific power characteristics which are denied the twin-cylinder and the four-cylinder engine, particularly if they are gasoline-fueled and spark-ignited and have exhaust port timings which open at 100° atdc or earlier. It will also become clear that, if the exhaust port timings are low, as may be the case for a well-designed supercharged engine, then a four-cylinder layout could be the optimum design. The close coupled exhaust manifold of the two-stroke engine will be

Chapter 5 - Computer Modeling of Engines

Plate 5.1 The QUB 500 single-cylinder 68 bhp engine with the expansion chamber exhaust slung underneath the motorcycle (photo by Rowland White).

demonstrated to contain a significant potential for tuning to provide high-performance characteristics not possible in a similar arrangement for a four-stroke engine.

The sketches in Fig. 5.8 show the lengths and diameters necessary as input data for a simulation to be conducted. The systems contain an exhaust box which, in the case of the outboard engine, is also the transmission housing leading to the propeller and can be seen in the photograph of the OMC V8 engine in Plate 5.2. That same type of engine, sketched in Fig. 5.8(a), has an exhaust manifold referred to as a "log" type in the jargon of such designs, and contains branches which are effectively T junctions. The manifolds may contain splitters which assist the flow in turning toward the exhaust box and, if effective in that regard, the appropriate branch angles can be inserted into the input data file as is required for the solution of the non-isentropic theory set out in Sec. 2.14. Should the bend at cylinder number 1, i.e., that at the left-hand end of the bank of cylinders sketched, be considered to be tight enough to warrant an appropriate insertion of loss for the gas flow going around it, then the theory of Sec. 2.3.1 can be employed.

The design shown in Fig. 5.8(b) is more appropriate to that used for an automotive engine, be it for an automobile or a motorcycle, or for a diesel-engined vehicle, where space is not at such a premium as it is for an outboard engine. The gas flow leaving the manifold toward the exhaust box, which may contain a catalyst or be a silencer, does so by way of a four-way branch. The theory for this flow regime requires an extension to that presented in Sec. 2.14. The sketch shows the junction at a mutual 90° for each pipe, but in practice the pipes have been observed to join at steeper angles, such as 60°, to the final exit pipe. This may not always be a design optimum, for the closer coupling of the cylinders normally takes precedence over the branch angle as a tuning criterion, as what may appear to provide an easier flow for the gas to exit the manifold may reduce the inter-cylinder cross-charging

Design and Simulation of Two-Stroke Engines

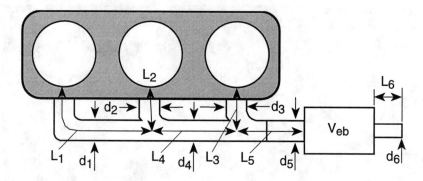

Fig. 5.8(a) *Geometry of the exhaust system of a three-cylinder outboard engine.*

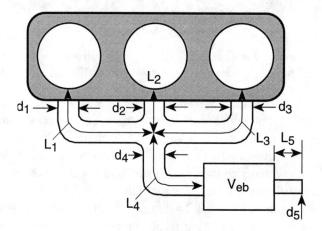

Fig. 5.8(b) *Geometry of the exhaust system of a three-cylinder automotive engine.*

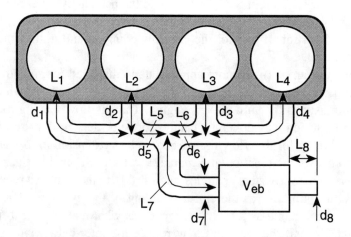

Fig. 5.8(c) *Geometry of the exhaust system of a four-cylinder automotive engine.*

374

Chapter 5 - Computer Modeling of Engines

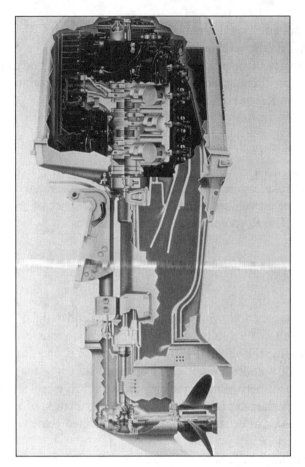

Plate 5.2 A cut-away view of a 300 hp V8 outboard motor (courtesy of Outboard Marine Corporation).

characteristics. This subtlety of design will be discussed below during the presentation of data acquired by simulation.

Should the design approach given in Fig. 5.8(b) be extended to a four-cylinder engine, then the manifold exit toward the exhaust box reverts to a three-way branch, normally mutually at 90°, as that exit is now located in the middle of the bank of cylinders, i.e., between cylinders numbered 2 and 3 of a four-cylinder unit. This is shown in Fig. 5.8(c). This is particularly true for an automotive diesel or spark-ignition engine, but less correct to be stated categorically if it is for an outboard design, as in Plate 5.2, where the branch lengths L_1, L_2, L_3, etc., are now of some considerable length in a design optimized for power and also where space limitations may prevent full optimization of its potential.

5.3 Heat transfer within the crankcase

The heat transfer characteristics in the crankcase of a crankcase compression two-stroke engine is not a topic found in the technical literature. Therefore, I present a logical approach to the analysis without any experimental proof that it is accurate. That research will be conducted, but as yet has not been done.

The approach is virtually identical to that used in Sec. 4.3.4, and as proposed by Annand [4.15-4.17], for the heat transfer within the cylinder. Annand recommends the following expression to connect the Reynolds and the Nusselt numbers:

$$\mathbf{Nu} = a\mathbf{Re}^{0.7} \qquad (4.3.28)$$

where the constant, a, has a value of 0.26 for in-cylinder heat transfer in a two-stroke engine and 0.49 for a four-stroke engine. It seems logical to take the value of "a" as being 0.26, as this is the crankcase of a two-stroke engine, and in the absence of any experimental data to confirm its selection. The Reynolds number is calculated as:

$$\mathbf{Re} = \frac{\rho_{cc} c_f d_f}{\mu_{air}} \qquad (5.3.1)$$

Within this equation the following terms are selected and defined. The crankcase is rarely filled with other than air and partially vaporized fuel and therefore the properties can be taken to be as for air. Actually, the simulation tracks the precise properties at any instant, but the analysis is more readily understood if air is assumed as being the resident gas. The heat transfer takes place in air between the spinning crankshaft flywheel and crankcase wall interface, so the crankshaft diameter is employed in the determination of Reynolds number and is defined as d_f. The values of density, ρ_{cc}, crankshaft surface velocity, c_f, and viscosity, μ_{air}, require further discussion.

The prevailing crankcase pressure, p_{cc}, temperature, T_{cc}, and gas properties combine to produce the instantaneous crankcase air density, ρ_{cc}.

$$\rho_{cc} = \frac{p_{cc}}{R_{air} T_{cc}} \qquad (5.3.2)$$

The viscosity is that of air, μ_{air}, at the instantaneous crankcase temperature, T_{cc}, and the expression for the viscosity of air as a function of temperature in Eq. 2.3.11 is employed.

The maximum crankshaft surface velocity is found from the dimension of the flywheel diameter, d_f, and the engine speed, rps:

$$c_f = \pi d_f \text{rps} \qquad (5.3.3)$$

Having obtained the Reynolds number, the convection heat transfer coefficient, C_h, can be extracted from the Nusselt number, as in Eq. 2.4.3 or 4.3.31:

$$C_h = \frac{C_k \mathbf{Nu}}{d_f} \qquad (5.3.4)$$

where C_k is the value of the thermal conductivity of air at the instantaneous crankcase temperature, T_{cc}, and consequently may be found from Eq. 2.3.10.

The value of T_{ccw} is defined as the average temperature of the crankcase wall surface, the exposed crankshaft and connecting rod surfaces, the surface of the underside of the piston, and the exposed cylinder liner surface. The heat transfer, δQ_{cc}, over a crankshaft angle interval, $d\theta$, and a time interval, dt, can be deduced for the mean value of that transmitted to the total surface area exposed to the cylinder gases:

as
$$dt = \frac{d\theta}{360} \times \frac{60}{\text{rpm}} \qquad (5.3.5)$$

then
$$\delta Q_{cc} = C_h A_{ccw}(T_{cc} - T_{ccw})dt \qquad (5.3.6)$$

The total surface area within the crankcase, A_{ccw}, is composed of:

$$A_{ccw} = A_{crankcase} + A_{piston} + A_{crankshaft} \qquad (5.3.7)$$

It is straightforward to expand the heat transfer equation in Eq. 5.3.6 to deal with the individual components of it, by assigning a wall temperature to each specific area noted in Eq. 5.3.7. It should also be noted that Eq. 5.3.6 produces a "positive" number for the "loss" of heat from the cylinder, aligning it with the sign convention assigned in Sec. 4.3.4.

The typical values obtained from the use of the above theory are illustrated in Table 5.1. The example employed is for a two-stroke engine of 86 mm bore, 86 mm stroke, running at 4000 rpm with a flywheel diameter of 150 mm. Various state conditions for the crankcase gas throughout the cycle are selected and the potential state conditions of pressure (in atm units) and temperature (in °C units) are estimated to arrive at the tabulated values for a two-stroke engine, based on the solution of the above equations. The three conditions selected are, in table order: at standard atmospheric conditions, at the height of crankcase compression, and at the peak of the suction process.

Table 5.1 Crankcase heat transfer using the Annand model

P_{cc} (atm)	T_{cc} (°C)	Nu	Re	C_h (W/m²K)
1.0	20	1770	299,037	316
1.2	120	1448	224,483	321
0.6	60	1078	146,493	210

It can be seen that the convection heat transfer coefficient does not vary greatly over the range of operational conditions within the crankcase, but the variation is sufficiently significant as to warrant the inclusion of the analytical technique within an engine simulation. The significance is such as to justify an experimental program to provide a more accurate value of the constant, a, in the Annand model than that which is assumed here as applicable to a two-stroke engine crankcase. The values determined for the convection heat transfer coefficient, C_h, lying beteen 200 and 320 W/m²K, are much as one would determine from an even simpler analysis, or could be culled from the literature describing somewhat similar physical situations in other fields of engineering.

5.4 Mechanical friction losses of two-stroke engines

As the designs of two-stroke engines contain details of physical construction which are relevant variables as far as the friction losses are concerned, it is not possible to determine a fundamental theoretical approach to this topic. The best I can offer is a series of empirical relationships which have been shown to correlate quite well with experimental observations on various types of two-stroke engines. The situation is more complex for the two-stroke engine in comparison to the four-stroke unit in that two-stroke engines are manufactured with both "frictionless" bearings, i.e., rolling element bearings using balls, rollers or needle rollers, and also with hydrodynamic bearings, i.e., the oil pressure-fed plain bearings as seen in virtually all four-stroke cycle engines. The friction characteristics of any engine are also related to the type of lubrication, but virtually all engines which have crankshafts supported by rolling element bearings employ a total loss oiling system into the crankcase of the engine. This is normally in the form of (i) pre-mixed lubricating oil and fuel supplied via the carburetor and hence to the various bearings and friction surfaces, (ii) a pump supplying a metered quantity of oil directly to the bearings and friction surfaces, or (iii) a pump supplying a metered quantity of oil into the inlet tract to be broken up by the air flow and distributed to the bearings and friction surfaces. As a historical note, the British word for pre-mixed oil and gasoline (petrol) was "petroil."

The friction characteristics of two-stroke engines can be divided into several classifications, i.e., those with rolling element bearings or those with plain bearings, and for those units which employ spark ignition or compression ignition. Virtually all compression-ignition, two-stroke engines use plain bearings. All of the input or output data in the empirical equations quoted below are in strict SI units; the friction mean effective pressure, fmep, is in Pascals; the engine stroke, L_{st}, is in meters; the engine speed, N, is in revolutions per minute.

It will be observed that the equations below relating friction mean effective pressure to the variables listed above are of a straight-line format:

$$fmep = a + bL_{st}N$$

where a and b are constants. It is interesting to note, and it would be supported by more fundamental theory on lubrication and friction, that the value of the constant, a, is zero for rolling element bearings. The term in these equations, which combines stroke and engine speed, is piston speed in all but name.

Spark-ignition engines with rolling element bearings

This classification covers virtually all small engines from industrial units such as chainsaws and weed trimmers to the outboard, the motorcycle and the snowmobile. From the experimental evidence, there appear to be somewhat proportionately higher friction characteristics for small industrial engines, i.e., of cylinder capacity less than 100 cm^3, and I have never satisfactorily resolved whether or not the bank of information which resulted in the provision of Eq. 5.4.1 also incorporates the energy loss associated with the cooling fan normally employed on such engines.

industrial engines $\qquad$ fmep = 150L$_{st}$N $\qquad$ (5.4.1)

motorcycles, etc. $\qquad$ fmep = 105L$_{st}$N $\qquad$ (5.4.2)

Spark-ignition engines with plain bearings
This set of engines is normally found in prototype automobiles using direct injection of fuel into the cylinder and with the air supply to the engine provided by a supercharger.

$$fmep = 25,000 + 125L_{st}N \qquad (5.4.3)$$

Compression-ignition engines with plain bearings
This group of engines is normally found in prototype automobiles, and in trucks, buses and generating sets, using direct injection of fuel into the cylinder and with the air supply to the engine provided by a supercharger or a turbocharger. Here, there appear to be two classifications, with somewhat lesser friction characteristics appearing in the automobile set where the engine is more lightly constructed and runs to a higher piston speed. The breakdown point appears to lie between engines with stroke lengths above or below 100 mm. However, diesel engines have higher compression and combustion pressure loadings than spark-ignition engines, and as the bearing and piston ring designs must cope with this loading, they are proportionately greater in size or number. This gives a higher friction content for this type of engine.

automobiles $\qquad$ fmep = 34,400 + 175L$_{st}$N $\qquad$ (5.4.4)

trucks $\qquad$ fmep = 61,000 + 200L$_{st}$N $\qquad$ (5.4.5)

5.5 The thermodynamic and gas-dynamic engine simulation

Virtually everything written up to this point within this text has been oriented toward this section of this chapter. The theory of unsteady gas flow in Chapter 2, the theory of scavenging behavior in Chapter 3, and the theory of combustion and heat transfer in the engine cylinder in Chapter 4, are all brought together into a single computational format and linked together to simulate an engine.

For those who have studied the publications in Refs. [2.31-2.35, 2.40-2.41, 5.20-5.21] you will realize that much has been learned, researched, and published on engine modeling since I presented a simple engine simulation program based on Benson's publications on the method of characteristics, together with its computer coding [3.34].

The applications selected here, to illustrate the extent of the design information that comes from a more advanced simulation technique, are a chainsaw, a racing engine and a multi-cylinder supercharged unit with direct in-cylinder fuel injection for automotive use. The simulation is assumed to reach equilibrium after some nine or ten cycles of computation and the data for that cycle are collected and stored as being representative of the engine performance characteristics based on the input data file being employed.

5.5.1 The simulation of a chainsaw

The chainsaw engine employed for the simulation is a production unit with known measured performance characteristics. It is of 65 cm^3 swept volume, with a bore of 48 mm and a stroke of 36 mm. The trapped compression ratio is 7 and the crankcase compression ratio is 1.5.

The induction system is as sketched in Fig. 5.5 and uses a diaphragm carburetor of 22 mm bore. The combination of the venturi and the relative positioning of the throttle and choke butterflies provide a maximum throttle area ratio, C_{thr}, of 0.61.

The intake system access to the crankcase is controlled by the piston skirt, while access to the cylinder is controlled by the piston crown. It has exhaust, transfer, and inlet port opening timings of 108° atdc, 121° atdc, and 75° btdc, respectively. The exhaust, transfer and intake ports are all of the "regular" type as sketched in Fig. 5.1(a) and (c). The single exhaust port is 28 mm wide, the four transfer ports have a total effective width of 47 mm and the single inlet port is 28 mm wide. All ports have top and bottom corner radii.

The engine is air cooled and has an exhaust box silencer exactly as sketched in Fig. 5.6, with a total volume of 560 cm^3. The final outlet pipe, d_3, from the silencer is 12 mm diameter and the first pipe has a diameter, d_1, of 25 mm and a length, L_1, of 40 mm. The compact nature of the exhaust silencer is evident.

During the simulation, it is necessary to assume values for the mean wall temperatures of the various elements of the ducting and the engine. The values selected are, in °C: cylinder surfaces, 200; crankcase surfaces, 100; inlet duct wall, 50; transfer duct wall, 190; exhaust duct walls, 200; inlet box wall, 40; exhaust box walls, 200.

The combustion model employed is exactly as shown in Fig. 4.7(d) for a chainsaw, with an ignition delay of 10°, a combustion duration, b°, of 64°, and Vibe constants, a and m, of 5 and 1.05, respectively. It uses unleaded gasoline at an equivalence ratio, λ, of 0.9, and is spark-ignited with an ignition timing of 24° btdc. The burn coefficient, C_{burn}, is 0.85.

The scavenge model used in the simulation is as given in Sec. 3.3.1 and is characterized by the κ_0, κ_1 and κ_2 coefficients numerically detailed for a "loopsaw" in Fig. 3.16. Within the simulation these data are applied through the theory given in Secs. 3.3.1 to 3.3.3.

The friction characteristics assumed during the simulation are as described above in Eq. 5.4.1.

Correlation of simulation with measurements, Figs. 5.9-5.13

Fig. 5.9 shows the close correlation of the measured and simulated power and torque (as bmep) of the engine over the speed range from 5400 to 10,800 rpm. The attained bmep is quite modest at some 4 bar, but when the measured and simulated characteristics of delivery ratio and trapping efficiency are examined in Fig. 5.10, it can be seen that the engine attains this torque level with a peak delivery ratio, DR, of only 0.53 but with a high trapping efficiency of over 70%. This too is not so surprising, as it can be seen from Figs. 3.12, 3.13 and 3.19 that the "loopsaw" scavenging characteristics were indeed quite excellent and virtually up to a "uniflow" standard.

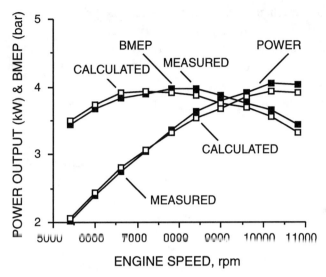

Fig. 5.9 Measured and computed power and bmep characteristics of a chainsaw.

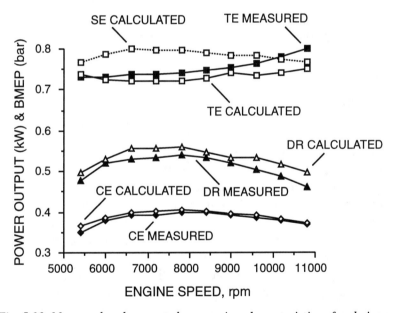

Fig. 5.10 Measured and computed scavenging characteristics of a chainsaw.

There is close correlation of the measured and simulated characteristics of delivery ratio and trapping efficiency in Fig. 5.10. As they are so close to the experimental values, they provide the correct information for the computation of charging efficiency, CE, and, with the subsequent employment of a comprehensive closed cycle model, give the requisite correlation with the measured power and torque. Also in Fig. 5.10 is the computation of the scavenging efficiency, SE, for which no measurement is available for comparison purposes; the computation predicts that it has a peak value of 0.8 at 6600 rpm.

With the simulation closely predicting air flow and power, the potential for accurately simulating the measured brake specific fuel consumption, bsfc, and the emissions of hydrocarbons, bsHC, is realized in Fig. 5.11.

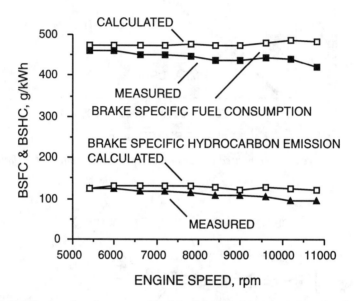

Fig. 5.11 Measured and computed bsfc and bsHC characteristics of a chainsaw.

The closed cycle simulation, relying on the combustion and heat transfer theory of Chapter 4, is seen to give a more than adequate representation of that behavior in Figs. 5.12 and 5.13. In Fig. 5.12 are the measured and computed cylinder pressure diagrams at 9600 rpm, and while the error on peak pressure is relatively small, the computation of the angular position of peak pressure is completely accurate. In Fig. 5.13 at the same speed is the comparison of measurement and calculation of the cylinder pressures during compression and expansion. The Annand model of heat transfer in Sec. 4.3.4, and the fuel vaporization model in Sec. 4.3.5, can be seen to provide very accurate simulation of the compression and expansion processes. The measured data are averaged over 100 engine cycles.

Design data available from the simulation, Figs. 5.14-5.27

The computation provides extensive information for the designer, much of which can be measured only with great difficulty, or even not at all due to the lack of, or the non-existence of, the necessary instrumentation. Much of this information is required so that the designer can comprehend the internal gas-dynamic and thermodynamic behavior of the engine and the influence that changes to engine geometry have on the ensuing performance characteristics. The following is a sample of the range and extent of the design information which an accurate

Chapter 5 - Computer Modeling of Engines

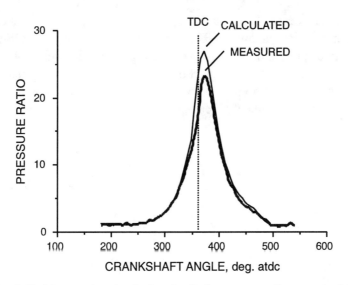

Fig. 5.12 Measured and calculated cylinder pressure diagrams at 9600 rpm.

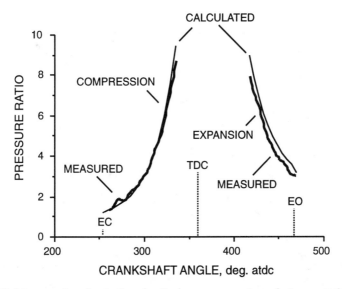

Fig. 5.13 Measured and calculated cylinder pressures in a chainsaw at 9600 rpm.

simulation provides. If the accuracy of the simulation is not adequate, the designer cannot rely on any further information that the simulation produces. Manifestly, from Figs. 5.9 to 5.13, the simulation of the chainsaw falls into the "sufficiently accurate" category.

Mechanical losses, Figs. 5.14 and 5.15

Figs. 5.14 and 5.15 show the interrelationship between friction, pumping losses and mechanical efficiency. In Fig. 5.14 are the calculated friction and pumping mean effective pressures, fmep and pmep, respectively. For a two-stroke engine employing crankcase compression, the value of pumping loss is low at 0.3 bar, and it does not increase significantly, even at the much higher delivery ratios observed for the racing engine in Fig. 5.29. Friction is the greater of the two parasitic losses. The impact of the combination of friction and pumping to reduce the indicated work to the brake related value (see Eq. 1.6.9) is found in Fig. 5.15. Over the engine speed range the mechanical efficiency falls by about 6% to a low of some 80%. By

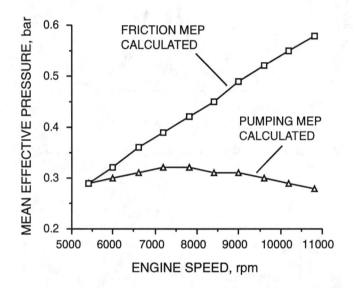

Fig. 5.14 *Measured and computed friction and pumping characteristics of a chainsaw.*

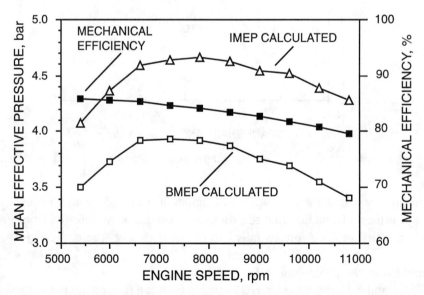

Fig. 5.15 *Mechanical efficiency and mep relationships for a chainsaw.*

four-stroke cycle engine standards, and particularly for such a small engine, that would be a very high value. It is clear that if the imep could be raised without jeopardizing other factors, then the mechanical efficiency could be improved and the specific fuel consumption reduced.

Gas flow through the cylinder, Figs. 5.16 to 5.18

The flow of gas in any direction is caused by pressure difference. Fig. 5.16 shows the pressure behavior within the chainsaw engine at 9600 rpm. The high-pressure events during combustion, shown in Fig. 5.12 at the same speed, are deleted for reasons of clarity. The atmosphere, from which the engine breathes air, and exhausts to, has a pressure ratio of unity. Apart from minor periods of backflow at 140° and 220° atdc, the cylinder is always emptying its contents into the exhaust pipe. It can also be seen that, during the entire open cycle period, the pressure at the exhaust port is uniformly above 1.15 atm, thus the exhaust system provides a dynamic "back-pressure" on the cylinder which decays slowly until the exhaust port opens again.

As the transfer of air from the crankcase to the cylinder for the scavenge process is conducted by pressure difference, it can be seen that this process lasts only from about 130° to 190° atdc. In other words, the scavenge period is concluded just after bdc, and the cylinder proceeds to spill some of its contents, either into the exhaust port or back into the crankcase. This is hardly an optimum design procedure, but the consequential design question is what to do about it.

The pressure at the cylinder end of the transfer duct exhibits a backflow during the first 10° of transfer port opening, as the cylinder pressure exceeds the crankcase pressure at that juncture during blowdown. The consequences of the backflow on the local purity and temperature can be found in Figs. 5.17 and 5.18. In Fig. 5.17, the purity at the cylinder end of the

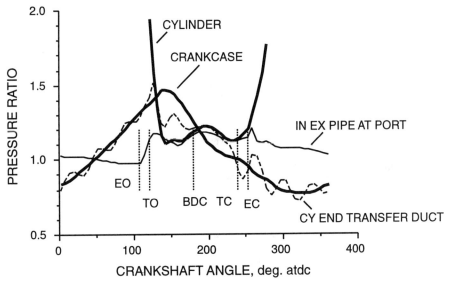

Fig. 5.16 Pressure within a chainsaw at 9600 rpm.

transfer duct drops temporarily from 0.95 to below 0.90 and, in Fig. 5.18, the temperature rises from 180°C to 280°C. When scavenging actually commences, it is with gas which is both hotter and less pure than it might otherwise be.

The scavenging process, Figs. 5.17 to 5.21

Fig. 5.19 shows the dynamic behavior of scavenge ratio by volume, SRv, trapping efficiency by volume, TEv, scavenging efficiency SE, and charging efficiency, CE, over the cycle

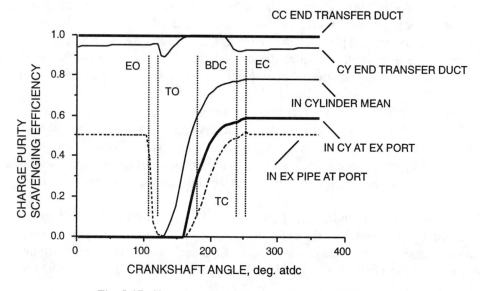

Fig. 5.17 Charge purity within a chainsaw at 9600 rpm.

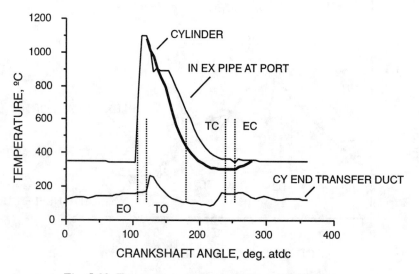

Fig. 5.18 Temperature within a chainsaw at 9600 rpm.

of events in the chainsaw at 9600 rpm. The scavenging efficiency rises smoothly to 0.8 as the trapping efficiency falls to a value just below that.

The discussion above, on pressure differentials, is reinforced by the graph in Fig. 5.20 which draws the pressure diagrams for the scavenge period in finer detail than Fig. 5.16. Scavenging flow takes place from the transfer duct from about 132° atdc until 190° atdc, i.e., only 10° abdc. As the scavenge port does not shut for another 45° or so, and the cylinder pressure is higher than either the scavenge duct or the exhaust pipe at the port, spillage of the cylinder contents is inevitable, as discussed above. The deleterious effect on charging effi-

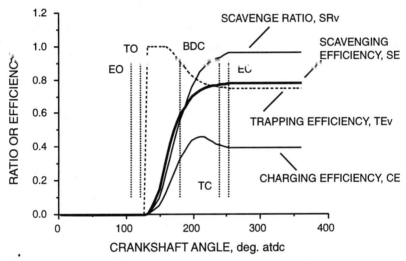

Fig. 5.19 SE, SR, TE, and CE within a chainsaw at 9600 rpm.

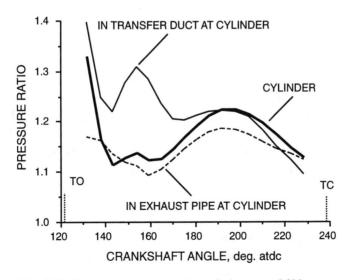

Fig. 5.20 Scavenging pressures in a chainsaw at 9600 rpm.

ciency is seen here, for the charging efficiency drops from a peak of 0.47 to 0.40, which will deteriorate the ensuing power and torque, bsfc and bsHC, by at least that amount, i.e., 15%. The design question is, as written above: What can be done about it?

In Fig. 5.17 it is discovered where this spilled charge goes. After 190° atdc, the purity at the cylinder end of the scavenge duct decreases from unity to a value of about 0.90, and, apart from some localized mixing within the transfer duct, remains at that value when the port shuts until the next time it is opened, at which point the aforementioned blowdown backflow takes place and reduces it even further. At the exhaust port, and in the exhaust pipe, spillage causes the purity there to rise from 190° atdc until exhaust closure. You should note the subtlety of the point being made: scavenging has ceased but spillage takes over. The other apparent contradiction is that, during this same period, the cylinder scavenging efficiency, i.e., purity, continues to rise. The explanation is that the spilled charge, i.e., that at the exhaust port, is of lesser purity than the cylinder mean and while mass is lost from the cylinder, the average purity of the cylinder charge is improved.

The scavenge model from the theory given Sec. 3.3.2, and shown graphically in Fig. 3.19 for the "loopsaw" test cylinder, is seen in operation in Fig. 5.17, in that the purity at the exhaust port tracks the cylinder purity and lagging behind it as dictated by Eq. 3.3.12.

The temperatures of the gas during scavenging, in various locations beside or within the cylinder, are shown in Figs. 5.18 and 5.21. In Fig. 5.18, at the cylinder end of the scavenge duct, the blowdown backflow causes an initial rise to 240°C, and then flow from the crankcase drops it down to about 100°C by bdc, whence cylinder spillage, with gas at a lesser purity and at some 300°C, increases it to nearly 200°C by transfer port closure. After that, with mixing caused by internal pressure wave motion in the transfer duct, and by expansion during the induction phase, the temperature decays to about 130°C by the port opening point. The cylinder temperature shows more dramatic behavior, falling from over 1000°C at the onset of scavenging, to a minimum at about 250°C with the influx of colder air from the crankcase, and then it rises toward the beginning of compression.

In Fig. 5.18, the temperature in the exhaust pipe at the exhaust port has an interesting profile. It tracks cylinder temperature but is always higher, as the gas which is leaving the cylinder has a lesser purity and therefore a higher temperature than the mean of the cylinder contents. This follows the discussion in Sec. 3.3.3 regarding the temperature differential model for air and exhaust gas at this very juncture. The behavior of this model is shown in Fig. 5.21, and the apparently inexplicable flat in the exhaust gas temperature profile in Fig. 5.18 is now clarified. At the beginning of scavenging, the entering air is rapidly heated to nearly 600°C, but falls as more enters and cools the entire cylinder contents. On the other hand, the exhaust gas, during this early period of "perfect displacement" scavenging up to 160° atdc, is leaving the cylinder without benefit of this cooling through mixing, which explains the "flat" on the profile. The temperature differential model, as defined in Eqs. 3.3.13 to 3.3.18, is seen to control the average in-cylinder air and exhaust gas profiles in a logical manner. The word logical is employed as there is no means of ever confirming this statement with experimental data. The apparently instantaneous equalization of temperatures at the trapping point is merely the computer program informing the cylinder contents that they will be theoretically treated as a homogeneous gas from that point onward in the compression process.

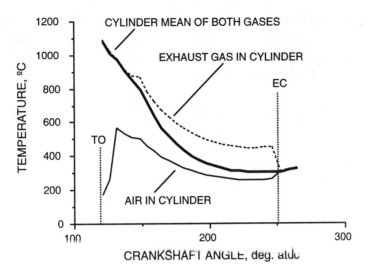

Fig. 5.21 Temperatures within scavenge model for a chainsaw at 9600 rpm.

The exhaust system behavior, Figs. 5.22 and 5.23

The silencer (muffler) is a compact two-box design with considerable restriction to flow so as to achieve an adequate silencing effect. In Fig. 5.22, the pressures at the cylinder outlet, in boxes A and B, and in the final outlet pipe are plotted over the engine cycle at 9600 rpm. As debated above, it is seen that the basic behavior is, during the entire open cycle, to raise the "back-pressure" on the cylinder to above 1.1 atm, and to restrict the scavenge flow. The addition to that debate is that this graph shows the origin of that resistance; it is the final outlet pipe backing up first box B and then box A. Noise, as is discussed more fully in Chapter 8 in terms of this exhaust system geometry on this engine, is caused by a combination of mass

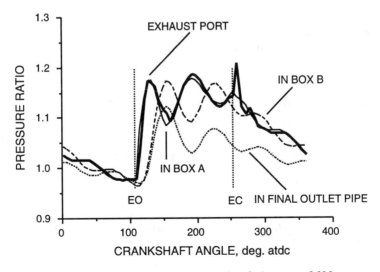

Fig. 5.22 Pressures in exhaust system of a chainsaw at 9600 rpm.

flow rate, pressure amplitude, number of pulsations and the sharpness of their pressure rise, of those pulsations which enter the atmosphere. The sharpness of the pressure rise on the major pulsation is seen to be progressively reduced from that at the exhaust pipe at the port, to that in box A, to that in box B, and finally in the outlet pipe as it progresses toward the atmosphere.

A similar progression of change takes place for the gas temperature, shown in Fig. 5.23. The dramatic events in the first pipe are evened out in box A, even more so in box B, and exit as a relatively constant value at about 320°C. For a chainsaw this is an important design issue, as legislation exists regarding the maximum value of the exit gas temperature, on the quite logical grounds that it must not auto-ignite the sawdust within the vicinity of the device!

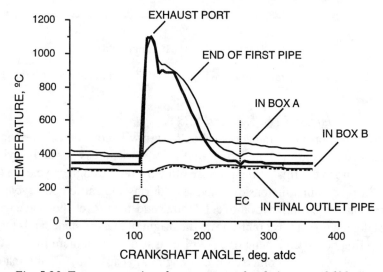

Fig. 5.23 Temperatures in exhaust system of a chainsaw at 9600 rpm.

The intake system behavior, Figs. 5.24 and 5.25

The intake system is controlled by the piston skirt opening the port at 75° btdc, and provides a relatively constant delivery ratio of 0.5 over most of the speed range. In Fig. 5.24, the delivery ratio, the pressures in the crankcase, and at both ends of the intake ducting, are shown over one cycle at an engine speed of 9600 rpm. The apparently curious abrupt transition of the DR curve, from 0.5 to 0.0 at bdc, is caused by the computer program resetting the counter for cumulative air flow back to zero, in advance of the next induction process. A similar abrupt transition due to counter resetting is seen in Fig. 5.19, in this case for trapping efficiency.

The system works well, in that the delivery ratio increases progressively from zero to its maximum, with no backflow, aided by a modest ramming behavior in the intake duct prior to port closure. The short intake duct provides those pressure oscillations from the reflection of the initial suction wave, some five or so, and so the fundamental frequency of the sound propagated into the atmosphere can be seen to be some four to five times the natural frequency of the engine speed, i.e., between 640 and 800 Hz.

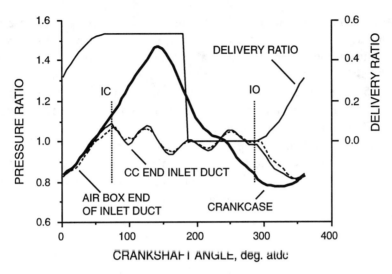

Fig. 5.24 Pressure related to air flow in a chainsaw at 9600 rpm.

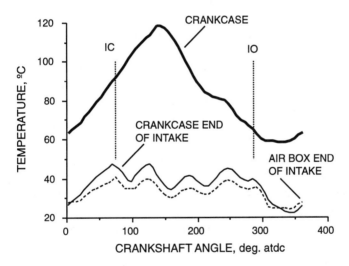

Fig. 5.25 Temperatures in the intake system of a chainsaw at 9600 rpm.

The crankcase pressure has already dropped to 0.8 atm by the time the inlet port opens, sending a sharp, i.e., noisy, pulsation as an intake wave into the inlet duct, which peaks at about the tdc position. The ensuing crankcase pumping action raises its pressure to about 1.5 atm, aided at that juncture by the higher pressure cylinder backflow into the scavenge ducts, as discussed above.

The temperatures throughout the intake process are shown in Fig. 5.25 for the crankcase air and at both ends of the intake duct. Heat transfer in the crankcase and inlet tract is responsible for the dichotomy which exists. The air in the crankcase never drops below 60°C, whereas most of the air in the inlet duct oscillates around 40°C. During induction into the crankcase,

Design and Simulation of Two-Stroke Engines

the combination of incoming air at 40°C flowing into an expanding crankcase volume containing air already at about 65°C, decreases the crankcase temperature somewhat to about 60°C. This seems curiously insufficient as a drop in temperature; however, in terms of the mass of air already within the crankcase, the entering quantity is quite small. Put crudely, the DR value is 0.5, or about 32.5 cm^3 in volume terms for a 65 cm^3 engine. The crankcase compression ratio is 1.5, so its maximum volume is 195 cm^3, therefore the entering quantity is only some 17% of the total in residence.

As the air is gulped into the intake duct from the airbox muffler, the temperature around the tdc point briefly drops down close to the atmospheric value. The peak temperature of the crankcase air rises to 120°C and, as shown in Fig. 5.21, can be heated within the cylinder up to 600°C during the early stages of scavenging. Not surprisingly, the majority of fuel vaporization occurs within the cylinder.

The cylinder pressure and temperature, Figs. 5.26 and 5.27

The design objective for the engine, in terms of the torque and power it will produce, or the NO_x emission it may create, is summed up in Fig. 5.26. It is, after all, cylinder pressure which pushes on the piston area to create that torque and power. It is the level of peak cylinder temperature that influences directly the amount of the emissions of oxides of nitrogen it will produce; this statement, although true, is too naive and requires the amplification supplied in Appendices A4.1 and A4.2. The open cycle period is indicated in the figure and the pressure events during it are barely visible on a diagram drawn to this scale. They would appear to have no influence on the behavioral outcome of the engine in terms of its performance characteristics. From the discussion above that is known to be incorrect, for it is the state conditions of the cylinder at trapping, dictated by the gas-dynamic behavior of the breathing and scavenging system, which results in the pressure and temperature created within the cylinder during the closed cycle period. The temperature profile during the open cycle period is demonstrably much more dramatic during the open cycle period.

The point being made is better illustrated in Fig. 5.27. In the next section, the discussion will focus on high-performance engines, with a 125 cm^3 motorcycle racing engine used as the design and simulation example. In Fig. 5.27, there is drawn the calculated cylinder pressure diagrams of the chainsaw engine and the racing motorcycle engine. They are both high-speed engines, but the disparity in the attained cylinder pressures and the attained bmep and specific power output could not be greater. One outperforms the other by a factor approaching three. That means that the specific trapped cylinder mass of air and fuel must be three or more times greater for the motorcycle engine than the chainsaw engine. This disparity of cylinder filling and emptying must occur during the events of the open cycle. On this diagram, drawn to this scale, there is little hint from the two pressure traces during the open cycle that anything untoward, or even different, is occurring.

The next section explains how it is possible, and how a simulation procedure by computer can incorporate it by design.

Further simulations involving this chainsaw engine

In Chapter 6, where the focus is on design assistance for the selection of engine dimensions for either simulation or experimental development, the chainsaw engine discussed here

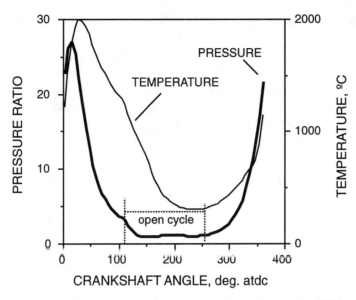

Fig. 5.26 Cylinder pressure and temperature in a chainsaw at 9600 rpm.

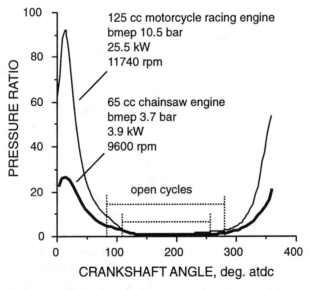

Fig. 5.27 Comparison of cylinder pressures in a chainsaw and a racing engine.

is employed as a working example. The effects on the performance characteristics of relatively minor changes of exhaust port timing, or for transfer port timing, are shown in Figs. 6.8-6.11, and in Figs. 6.12-6.15, respectively.

In Chapter 7, where the focus is on fuel economy and emissions, the chainsaw engine is again employed as one of the working examples. The results of the simulation, shown here to give good correlation with measured exhaust hydrocarbon emissions, is repeated to illustrate that, as well as HC emissions, the emissions of carbon monoxide and exhaust oxygen content may also be calculated as a function of air-fuel ratio or throttle opening; this is shown in Figs. 7.13-7.15 and Figs. 7.16-7.18, respectively. The emissions of nitrogen oxides is a complex topic, but is covered in Appendices A4.1 and A4.2, using this engine and these input data as the working example. Perhaps of even greater interest, in Figs. 7.19 and 7.20, the scavenging characteristics are changed from the LOOPSAW quality used above to four other types, some better and some worse, and the various, differing performance characteristics are derived by simulation. Lastly, the basic physical dimensions of the engine are used to design a low-emissions, simple two-stroke engine and the reduction of the bsHC emissions from the 120 g/kWh level here to 25 g/kWh is a matter of great interest for those involved in R&D in this area.

Finally, in Chapter 8, the chainsaw engine data are used as the basis for elaboration on the principles and the practice of designing intake and exhaust silencers.

As a consequence, these exercises in simulation, most of which are shown to correlate well with typical measured data, provide the necessary insight into the behavior of an engine which is too complex for the human mind to comprehend. This reinforces the point made frequently in the discussion above, that the interrelationships between the design parameters of an engine are so complicated that only an accurate simulation will unravel them. Once computed, the conclusion seems obvious. Prior to that, they are inexplicable. At that point, the human mind starts to design.

5.5.2 The simulation of a racing motorcycle engine

The engine employed for the simulation is a production unit with known measured performance characteristics. It is of 125 cm^3 swept volume, with a bore of 56 mm, a stroke of 50.6 mm, and a connecting rod length of 110 mm. The trapped compression ratio is 9 and the crankcase compression ratio is 1.35. It is the engine as described by Cartwright [4.35] and the exhaust system being used here in the simulation is the pipe he describes as "A2" in that publication. The discharge coefficients for the exhaust port of this engine are shown in Figs. A2.4 to A2.7, but are discussed more thoroughly elsewhere [5.25].

The induction system is as sketched in Fig. 5.5 and uses a slide carburetor of 38 mm bore. There is no real venturi in a racing carburetor and the thin throttle slide provides a maximum throttle area ratio, C_{thr}, of unity.

The intake system access to the crankcase is controlled by a reed valve, as sketched in Fig. 5.4. There are six "glass-fiber" petals, each 38 mm long, 22.7 mm wide, and 0.42 mm thick. The reed ports in the block are 32 mm long, 19.6 mm wide, have 1.0 mm corner radii, and start 4 mm from the clamp point. The reed block half angle is 23.5°, and the entry area to the block is 38 mm high, 38 mm wide with 19 mm corner radii. In this design there is no stopplate *per se*, but a tip movement limit stop, some 13 mm above the reed, is evident. Further general discussion of the properties of reed petals, including the glass-fiber material used here, is found in Sec. 6.3.2.

The piston controls the exhaust and transfer ports. It has exhaust and transfer port opening timings of 81° atdc and 115° atdc, respectively. The transfer ports are of the "regular" type as sketched in Fig. 5.1(a). There are six transfer ports giving a total effective width of 80.4 mm, and each has 1.0 mm corner radii. The exhaust port is of the "irregular" profile, almost exactly as shown in Fig. 5.1(b), with the top section varying from 40 to 53 to 40 mm effective width over some 10 mm, and then remaining virtually parallel at 40 mm down to the bdc position. A further aspect of the exhaust port timing is that it has an exhaust control valve, as sketched in Fig. 5.2. If the exhaust control valve perfectly sealed the cylinder at closing timings of 85, 90, 95, 100, 105 and 110° btdc, then the trapped compression ratio would be raised from 9.0 to 9.6, 10.3, 11.1, 11.8, 12.4 and 13.1, respectively. The valve does not seal the port in this ideal manner, but does so quite effectively, and in its fully lowered position closes the port at 95° btdc.

The engine is liquid cooled and has a tuned exhaust system as sketched in Fig. 5.7. The lengths L_1 to L_7 are 83, 189, 209, 65, 78, 205 and 250 mm, respectively. The diameters d_1 to d_7 are 37.5, 48.5, 100, 116, 116, 21 and 21 mm, respectively.

During the simulation, it is necessary to assume values for the mean wall temperatures of the various elements of the ducting and the engine. The values selected are, in °C: cylinder surfaces, 200; crankcase surfaces (which in this engine receives some coolant), 80; inlet duct wall, 30; transfer duct wall, 100; exhaust duct walls, 350.

The combustion model employed is exactly as shown in Fig. 4.7(e) for a racing engine, with an ignition delay of 12°, a combustion duration, b°, of 41°, and Vibe constants, a and m, of 5.25 and 1.25, respectively. The actual engine uses aviation gasoline, the properties of which are given in Sec. 4.3.6, at an air-to-fuel ratio of 11.5. It is spark-ignited with an ignition timing of 20° btdc up to 11,500 rpm, when the system in practice retards the spark linearly until it is at 14° btdc at 12,300 rpm. The simulation incorporates the experimental ignition timing curve. The burn coefficient, C_{burn}, is 0.85.

The scavenge model used in the simulation is as given in Sec. 3.3.1 and is characterized by the κ_0, κ_1 and κ_2 coefficients numerically detailed for the "YAM14" cylinder in Fig. 3.16. The particular racing engine cylinder has not been scavenge tested, so its precise behavior is unknown. Therefore, the scavenging characteristics of a multiple port, loop-scavenged, motorcycle engine, with relatively good quality scavenging, has been assumed. Within the simulation these data are applied through the theory given in Secs. 3.3.1 to 3.3.3.

The friction characteristics assumed during the simulation are as described above in Eq. 5.4.1.

Correlation of simulation with measurements, Figs. 5.28-5.33

The measured performance characteristics of power and torque (as bmep) are shown in Fig. 5.28 and compared to those computed by the simulation over the speed range of the engine. The correlation for power and bmep is good. In Fig. 5.29, the measured and computed behavior for delivery ratio, trapping efficiency and charging efficiency are shown; here, too, the correlation is good both in amplitude and profile. The high values of delivery ratio and charging efficiency, compared to the equivalent diagram for the chainsaw in Fig. 5.10, explains the disparity in the bmep and power attained in each case. It does not explain how they are achieved.

The simulation of the exhaust gas temperature, not just as a bulk mean value recorded in the middle of the pipe system, but everywhere throughout the pipe at every instant of time, is a vital issue if the simulation is to accurately phase the dynamic events within the long tuned exhaust pipe. It is not possible to record temperature-time histories with the same accuracy as

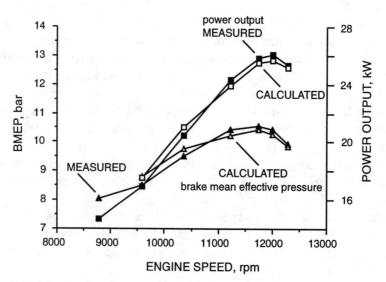

Fig. 5.28 Measured and computed performance characteristics of a racing engine.

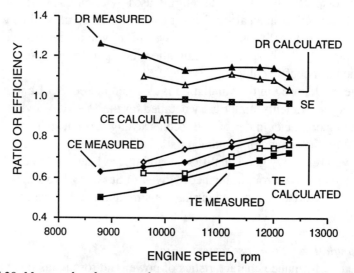

Fig. 5.29 Measured and computed scavenging characteristics of a racing engine.

pressure-time histories, so the simulation records its cumulative assessment of the mean temperature in the center of the pipe system which is then compared with a corresponding measurement by a thermocouple. Such a comparison is given in Fig. 5.30, and it can be seen that the mean value recorded by the thermodynamic and gas-dynamic simulation compares well with the measurement taken at the same physical location.

The effect of the exhaust timing control valve

It will be observed that the simulated data in Figs. 5.28 to 5.30 do not contain a computation point at 8800 rpm. The engine has an exhaust timing control valve which is still lowered

Chapter 5 - Computer Modeling of Engines

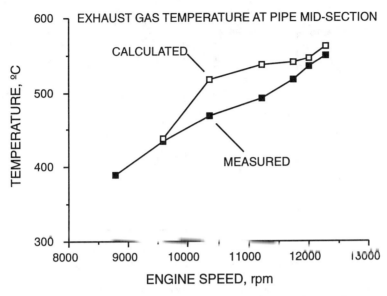

Fig. 5.30 Mean exhaust gas temperature for a racing engine.

at this speed, but while its actual position is not known, it is known to be fully lifted by 9500 rpm, as sketched in Fig. 5.2(a). In Fig. 5.31, the results of a simulation at 8500 rpm of the effects of changing the timing value of the exhaust control valve, including the consequential change of trapped compression ratio given above, reveals that the optimum value of that timing is about 100° btdc for its closing condition. The simulation indicates a value of 8 bar bmep, which is seen to be close to the measured value at 8780 rpm in Fig. 5.28. In practice, the fully lowered position of the valve on this engine corresponds to a closing point about 95° btdc. The simulated cylinder and exhaust pipe pressures are shown in Fig. 6.21, where a discussion on tuning related to speed focuses on this issue.

The exhaust pressure diagrams

Fig. 5.32 shows the comparisons of the measured and simulated pressure diagrams at the physical location of the pressure transducer in the exhaust pipe. That location is at 100 mm from the plane of the exhaust port. The same figure shows the pressure-time diagram computed as being at the exhaust port. It is a measure of the rapidly changing pressure dynamics in such highly tuned pipes that such a major shift of pressure-time history exists over such a short distance. The correspondence between measurement and calculation is quite good, both in amplitude, phasing and profile. This is especially so as the simulation treats the first 83 mm pipe leading from the exhaust port as a parallel duct of 37.5 mm diameter. The actual duct does not really resemble that description and it would be difficult to model it accurately with a 1D simulation.

The closed cycle model

As with the chainsaw engine discussed earlier, it is instructive to note that the identical Annand model is employed here for heat transfer within the racing engine. It should be pointed

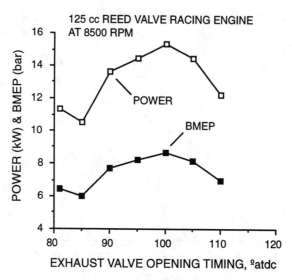

Fig. 5.31 Simulation of exhaust control valve timing on engine performance.

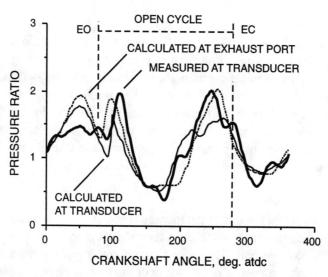

Fig. 5.32 Measured and calculated exhaust pipe pressure at 11,740 rpm.

out that all of the thermodynamic and gas-dynamic theory, heat transfer, etc., is identical within the software employed to simulate both these engines. In Fig. 5.33, as with the earlier Fig. 5.13, a comparison of measurement and simulation is made for the cylinder pressure diagrams during expansion and compression. As with that earlier figure, there is good correlation with profile, suggesting that the heat transfer coefficients predicted by the Annand model, and the inclusion of the fuel vaporization model during compression, are functioning in a most satisfactory manner. In the simulation, the ignition timing may be too advanced, or

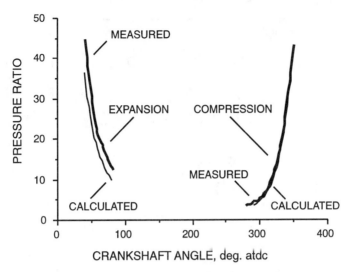

Fig. 5.33 Measured and calculated cylinder pressures in a racing engine at 11,740 rpm.

the burn period too short, for the cylinder pressure decay line lies earlier than the measurement. This viewpoint is confirmed by the position of the value of the simulated peak cylinder pressure and its magnitude (see Fig. 5.27) for it is 1° earlier and is 5 bar higher than the measured values.

The origins of high charging efficiency, Fig. 5.34

Large pressure oscillations are observed in the measured and computed values for the in-cylinder and exhaust pipe pressure transducer locations in Fig. 5.32. This greatly enhances the already considerable rates of mass flow through the engine by ports which are very large by comparison with the chainsaw discussed earlier; see also Figs. 6.2 and 6.7 to observe the physical differences. Nevertheless, as shown by the chainsaw in Fig. 5.19, when scavenging stops, the charging efficiency, which is virtually a linear function of bmep, is deteriorated by spillage by an amount approaching 20% of the CE value. The opposite happens in an engine with a tuned exhaust system, as can be seen in Fig. 5.34.

In Fig. 5.34 are plotted the simulated pressures within the engine cylinder and in the exhaust pipe at the exhaust port, together with the charging efficiency, over one engine cycle at 11,740 rpm. The fundamental behavior of the tuned pipe is clear. The exhaust pulse is large due to the high charging efficiency providing a greater cylinder mass and, when burned and expanded, a high pressure at the release position. The peak of the exhaust pulse is nearly 2 atm in amplitude. The reflection of that pulse at the diffuser provides a strong expansion wave which returns to the exhaust port and pulls the cylinder pressure at bdc down to about 0.7 atm. The suction pulse amplitude is about 0.5 atm. At that point in time, the crankcase is at 1.3 atm (see Fig. 5.35) which gives an overall pressure ratio for flow from the crankcase to the cylinder of about 1.85. Such a pressure ratio is almost certain to give sonic particle flow through the transfer ports. Not surprisingly, with this high pressure differential forcing rapid air flow, the charging efficiency curve goes from 0.1 to about 0.7 in the short time frame from

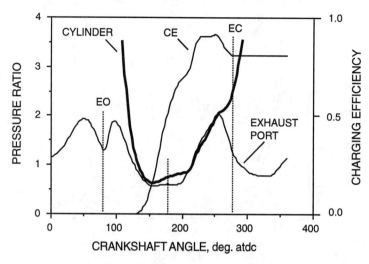

Fig. 5.34 The charging of a racing engine at 11,740 rpm.

150° atdc to 200° atdc. That can be contrasted with the chainsaw, which rose to a peak CE value of 0.5 in the same time period (see Fig. 5.19), at which point onward, spillage dropped the peak value of CE back to 0.4. However, in the racing engine, at this very juncture, the onward transmittal of the exhaust pulse to the rear cone of the tuned pipe has provided a compression wave reflection of some 2.0 atm in amplitude which returns to the exhaust port. The cylinder pressure is raised over the next 70° crankshaft from 1.0 atm to over 2.0 atm and, as density is related to pressure through the state equation, the charging efficiency goes from 0.7 to over 0.9. In the last critical 10° or 15° before exhaust closure, spillage again occurs and the charging efficiency drops back to just over 0.8. In short, the outcome is twice the value of CE which is trapped by the chainsaw.

The tuners of racing two-stroke engines should note this result carefully, for this engine and this exhaust pipe are not far removed from the "state-of-the-art, c. 1994," so a better exhaust pipe design could recover the lost 10% of charging efficiency and potentially raise the (after transmission) bmep level from the present 10.5 bar to 11.8 bar!

The tuners should also note the presence of "resonance" in the design of exhaust systems for such high specific output engines. If you examine the exhaust pressure trace in Fig. 5.34 you will note that there is an "exhaust pulse" prior to the appearance of the actual blowdown pulse at the release point, EO. This is not, of course, leakage from the cylinder. It is the second return of the plugging pulse from the previous cycle which has "echoed" off the rapidly closing exhaust port acting as an increasingly closed end. This precursor to blowdown is a large pulse of 2.0 atm but it too is "echoing" off the closed ended exhaust pipe at the port in a superposition manner; thus, in reality it has a magnitude of some 1.5 atm. It effectively strengthens and broadens the ensuing exhaust pulse when it appears some 50° later. The result is the broad, deep suction pulse around the bdc position and the enhanced amplitude plugging pulse seen at 250° atdc. The effective harnessing of this second-order phenomenon is called "resonance" and is yet another vital factor in the design of high specific output engines.

Chapter 5 - Computer Modeling of Engines

Nevertheless, the explanation of the mechanism of the achievement of high specific performance characteristics in the simple, ported two-stroke engine is quite straightforward. The fundamental principles are (i) provide a "plugging" pulse prior to trapping to seal in the cylinder charge and prevent spillage, (ii) provide a "suction" pulse around the bdc position to assist the scavenging flow in reaching a high charging efficiency prior to sealing, (iii) provide a "suction" pulse around the bdc position to extract air from the crankcase and lower its pressure so as to commence induction vigorously, and (iv) harness the exhaust pulse resonance effects to enhance the basic mechanisms of (i), (ii) and (iii), if that is possible within the design configuration. This process is significantly related to engine speed, as the further illustrations from this simulation of cylinder and exhaust pipe pressure show in Figs. 6.18 to 6.20 for engine speeds of 9600, 11,200 and 12,300 rpm, respectively. They are near the bottom, near the peak, and at the upper end of the useful power band of this highly tuned engine. The basic tuning and charging action seen in Fig. 5.32 is observed to appear in Figs. 6.18-6.20 to a greater or lesser degree. Further discussion occurs in Sec. 6.2.5 with respect to the design of the expansion chambers for two-stroke engines and their speed-related tuning characteristics. A discussion on speed-related tuning is also in Sec. 6.2.5.

The behavior of the reed valve induction system, Fig. 5.35

Fig. 5.35 shows the results of computer simulation of the reed valve motion, the delivery ratio, and the forcing pressures to create those effects, namely within the crankcase and the superposition pressure at the entrance to the reed block which is at the crankcase end of the inlet tract. The reed valve motion is plotted as the tip lift ratio, C_{rdt}.

It can be seen that the reed is open for most of the cycle, but it is impelled into action by a strong pressure ratio across it which lasts from about 190° atdc to 240° atdc. The pressure-created impulse is significant, for it coincides with (i) the crankcase pressure having been lowered below atmospheric pressure by the gas-dynamic tuning action emanating from the

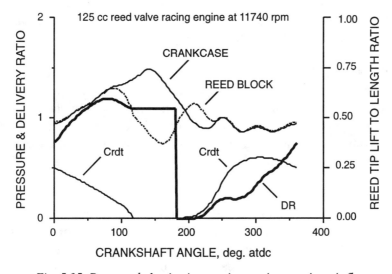

Fig. 5.35 Pressure behavior in a racing engine causing air flow.

exhaust pipe, and (ii) an intake system pressure pulsation which is above atmospheric pressure. At any other period the overall pressure differential across the reed petals is relatively small. It will be seen that it is this initial "jab" of force on the reed petal, from 190° atdc to 240° atdc, which provides the energy to swing the reed off its seat to a tip lift ratio of 0.25. The delivery ratio at that point, with the reed lifted to 80% of its maximum lift, is miniscule at 0.15 and is even falling slightly. The momentum given to the reed petal continues the lifting process, with some minor pressure assistance, until the peak lift point occurs at a C_{rdt} value of 0.3 at about 310° atdc. For a 38-mm-long reed, that means that the tip has lifted some 11.4 mm. At that point, with some further minor pressure differential continuing to flow air past it and acting against its closure, the "spring" forces in the petal material dominate and the reed petal reseats itself. During the period from maximum reed lift, the pressure difference across it may be small, but it lifts the delivery ratio, DR, from 0.3 at 60° btdc to 1.2 at 75° atdc, a period of 135°. The intake system provides a modest ramming action, up to 1.25 atm by 90° atdc, but eventually the increasing crankcase compression pressure supersedes it and backflow occurs, taking some 10% off the delivery ratio. The reed finally reseats itself at 115° atdc. As it had opened at 200° atdc, the total opening period of the reed is 275°.

The designer should note most carefully that the reed action is controlled by the joint action of the exhaust system and the intake system dynamics. To attain a high delivery ratio, the intake system must provide two ramming actions, the conventional one which prevents significant backflow as the reed is closing and another which assists in impelling the reed rapidly off its seat as early as possible.

This information can be compared to that given in Chapter 1 in Figs. 1.8(c) and (d), and the discussion will supplement that found later.

Further simulation involving this racing motorcycle engine

In Chapter 6, where the focus is on empirical design assistance for the physical geometry of the engine and its ducting, this reed-valved engine and its tuned exhaust system is used to illustrate the discussion. In Sec. 6.1.2, the dimensions of the cylinder porting are shown to fit into the same basic design rules regarding specific time areas as for the chainsaw. In Sec. 6.2.5, where simple design rules for the tuned expansion chamber are expressed, there are further simulation results for the pressure-time histories in the exhaust system of the racing engine at 8500, 9600, 11,200 and 12,300 rpm, shown in Figs. 6.18-6.21, respectively. That in Fig. 6.21 is particularly interesting, for it graphs the pressure behavior in the exhaust pipe at 8500 rpm, when the exhaust timing valve is lowered to open at 100° atdc, and provides the necessary insight into the bmep-speed relationship seen here in Fig. 5.31. The other diagrams have the exhaust timing valve raised, in Figs. 6.18-6.20, so they illustrate the exhaust pressure wave control of a bmep curve which starts at 9600 rpm and concludes at 12,300 rpm.

5.5.3 The simulation of a multi-cylinder engine

This discussion concerns a multi-cylinder, spark-ignition engine with direct in-cylinder injection of gasoline. Such engines have been built in recent times as prototypes for automobiles [5.23] and simulation has been attempted for such units using the homentropic method of characteristics theory [5.24]. The simulation here uses the non-isentropic theory presented in Chapter 2, together with the scavenging and combustion theory given in Chapters 3 and 4.

A three-cylinder engine

The illustration is for the engine designed by Thornhill [5.24] and it is a V6 3.2-liter engine with a 90 mm bore and a 82 mm stroke. One bank of three cylinders is simulated, each one firing at 120° intervals, with a 1-3-2 firing order. The supercharger included within the simulation is a Roots blower, geared at 1.5 times off the engine crankshaft, and the full map of its characteristics for mass flow rate, air temperature rise, and isentropic efficiency, as a function of pressure ratio and rotational speed, is incorporated into the simulation. The physical geometry of the engine is given in some detail by Thornhill [5.24] and the salient points of that geometry are repeated here so as to give meaning to the discussion.

The exhaust port opens at 95° atdc and the scavenge ports at 120° atdc. The exhaust port is simulated as a 55-mm-wide port with 16 mm corner radii. The scavenge ports are simulated as four ports with a total effective width of 74 mm and with 4 mm corner radii. The trapped compression ratio is 7.0. The exhaust system is as sketched in Fig. 5.8(b) with the lengths, L_1 and L_3, being 200 mm, and length, L_2, being 100 mm, the diameters, d_1, d_2, and d_3, are each 50 mm. The downpipe from the manifold, d_4, is 55 mm in diameter and the length, L_4, is 1 m before the 10-liter expansion box, i.e., volume, V_{eb}.

An intercooler is employed between the blower and the engine. The scavenging model is assumed to be that appropriate to cylinder YAM1, as shown in Fig. 3.16. The ignition timing is at 25° btdc, with an air-fuel ratio of 14.5 on unleaded gasoline, for a simulation conducted at 3500 rpm to show the effects of the compact manifold on the charging characteristics of the engine. The summary of the overall simulation for the 3.2-liter V6 engine at 3500 rpm shows the performance characteristics to be: power 136 kW, bmep 7.47 bar, bsfc 245 g/kWh, DR 0.80, SE 0.877, TE 0.77, CE 0.615, η_m 0.88, air flow 626 kg/h, fuel flow 33.4 kg/h, fuel injected per shot 26.5 mg, peak cylinder pressure 46 bar, peak cylinder temperature 2560 K, position of peak pressure 18° atdc, mean blower supply pressure 1.23 bar, and mean air supply temperature 38°C. Although they are not the primary objective of the discussion, the high specific output and the low specific fuel consumption of the engine are quite evident. The discussion in Chapter 7 focuses more clearly on the benefits of direct in-cylinder fuel injection to reduce hydrocarbon emissions and fuel consumption in two-stroke engines.

The principal objective is to show the influence of the compact manifold on the charging characteristics of the engine, and some of the fundamental design requirements of a blower-scavenged engine by comparison with that generated by a crankcase compression pump. This is summarized in Figs. 5.36(a)-(c) for the pressures in the cylinder during the open cycle, the pressures at the cylinder in the exhaust and scavenge ports, and the charging efficiency which this produces. The three sub-figures show these effects for cylinders 1-3, respectively.

The effect of blower scavenging

The principal difference between blower scavenging and crankcase compression scavenging is that blower scavenging normally lasts for the majority of the duration that the scavenge ports are open. This can seen clearly in Fig. 5.36(a), where the scavenge port pressure is virtually constant and almost always above the cylinder pressure. The charging efficiency rises to a plateau by 210° atdc, which is then held to the trapping point. This should be contrasted with the chainsaw in Fig. 5.20, where the sharper scavenging process reaches its ceiling at this point and then spills nearly 10% of its contents out of the cylinder.

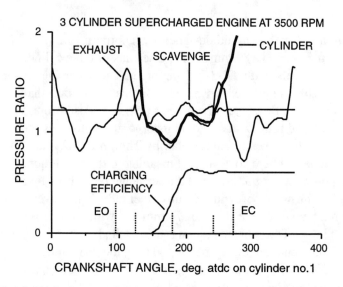

Fig. 5.36(a) Open cycle pressures and charging for cylinder No. 1.

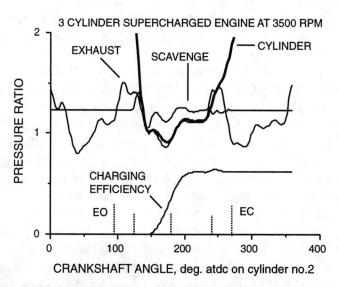

Fig. 5.36(b) Open cycle pressures and charging for cylinder No. 2.

The peak pressure supplied by the blower, to flow a delivery ratio of 0.8, is just 1.20 atm. The crankcase compression engine reached a peak pressure in the crankcase of 1.5 atm to flow a delivery ratio of 0.5. However, the lower peak pressure in the blower-scavenged engine has one drawback: It is essential to arrange for a longer blowdown period so that the cylinder pressure can drain off to a level approaching that in the supply plenum. Even though the blowdown period is long for what is a relatively low-speed engine, it is 25° crankshaft, and the exhaust port is apparently of an adequate width, the cylinder and scavenge line pressures do not equalize until 30° after the scavenge ports have opened. This is seen by the

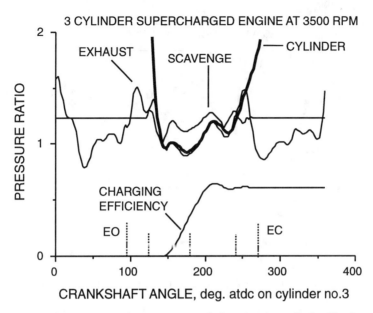

Fig. 5.36(c) Open cycle pressures and charging for cylinder No. 3.

charging efficiency curve commencing at 150° atdc. Further information on this topic is given in Fig. 5.37, for the purity and temperature in the scavenge ports at cylinder 1, an end cylinder. The blowback into the exit from the scavenge ports is considerable, with the purity dipping to 0.4 and the temperature rising to 600°C. It is almost bottom dead center before the situation has been recouped and the purity has risen again to unity. It is a much more extreme characteristic than that seen for a crankcase compression engine in Figs. 5.17 and 5.18. It also illustrates a common failing of supercharged designs, in that they are rarely endowed with adequate blowdown characteristics, for the cylinder backflow into the scavenge ports backs up the compressor, increasing the pumping losses significantly. Designers of such engines should re-examine Figs. 3.40 and 3.41, for it is possible you may have initially considered the exhaust port width shown there as excessive. In the light of the evidence given here, you may wish to reconsider that opinion.

The exhaust tuning of a three-cylinder engine

The pressure characteristics within the exhaust pipe at the exhaust port are also shown in Figs. 5.36(a)-(c) for each of the three cylinders. Just prior to the closing of the exhaust port, there is a "plugging" pulse reminiscent of that seen for the racing engine in Fig. 5.34. While this engine does not have a tuned expansion chamber, the effect of employing the compact manifold is almost the same. The "plugging" pulse effect in this case is caused by another cylinder blowing down into the manifold. In the case of Fig. 5.36(a) for cylinder 1, the "plugging" pulse emanates from the opening of the exhaust port of cylinder 3. In Fig. 5.32(b) it is from cylinder 1; in Fig. 5.32(c) it is from cylinder 2. This information is gleaned from the firing order being 1-3-2. The effect is to hold the charging efficiency at its plateau level, in this case about 0.61. This behavior in the multi-cylinder engine is known as "cross-charging."

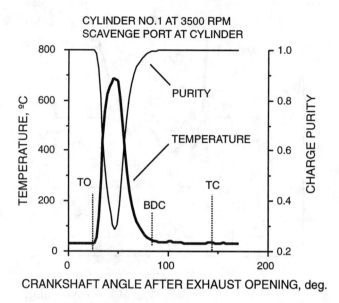

Fig. 5.37 Temperature and purity in scavenge port for blown engine.

It is also observable that the pressure diagrams for the three cylinders are not dissimilar, but they are not identical. In this case, the design point is "good," in that the net effect is the provision of almost identical charging efficiencies for all three cylinders. It is not always so, and the designer must bear this point in mind because, in the case of a fuel injection system dispensing equal quantities of fuel per shot to each cylinder, a poorly optimized manifold design can wreak havoc with the inter-cylinder air-fuel ratios.

The exhaust port timing employed in this supercharged spark-ignition design example is typical of that used in three-cylinder outboard marine units, where they are normally aspirated by crankcase compression. Nevertheless, it is clear from Fig. 5.36 that it requires that length of exhaust timing to accommodate the passage of the cross-charging pulse to the other end of a compact manifold, to assist the charging of that cylinder. It could be argued that the manifold should be made more compact and employ a shorter exhaust period but with a wider exhaust port. Consider the practicalities of that statement in light of the physical dimensions of the engine being discussed here. The manifold in the present design example is 400 mm from the exit of cylinder 1 to the entrance of cylinder 3; i.e., length L_1 plus L_3 as shown in Fig. 5.8(b), each given as 200 mm above. With a cylinder bore of 90 mm, and a minimum inter-cylinder spacing of 110 mm, the absolute minimum end-to-end manifold length (L_1 plus L_3) value becomes 220 mm. However, by the time a 55-mm-diameter exhaust pipe is turned through 90° from its exit direction and toward the other end cylinder, and it is difficult to accomplish this in less than 70 mm of pipe length, so a further 140 mm is added to the absolute minimum of 220 mm. This raises it to 360 mm, compared to the 400 mm length employed here. Hence, there is not much room for design maneuver in that context. Thus, while long exhaust port periods do provide good trapping and charging in three-cylinder engines, the period length is actually extended beyond the optimum of that required for good breathing in most two-stroke engines, simply to acquire the performance enhancement from exhaust tuning, with the more peculiar needs of high-speed racing engines being the exception to this comment.

There comes a point in the design process for any multi-cylinder engines when consideration should always be given to a four-cylinder engine, with firing intervals of 90°, and employing lower exhaust port timings with a compact manifold, as shown in Fig. 5.8(c). Its possible adoption depends on many factors, and while for many engines the three-cylinder option remains the optimum, no designer can afford to blindly accept "custom and practice" in one field as being universally applicable in all circumstances.

A four-cylinder engine

To illustrate the points made above regarding the effectiveness of a four-cylinder engine optimally designed with the correct port timings, port widths and areas, together with a compact manifold, a four cylinder engine with the same bore and stroke, is simulated at the same engine speed point, 3500 rpm. The bore and stroke are 90 and 82 mm, respectively, but the engine capacity is now reduced to 2100 cm^3. The same Roots supercharger is employed within the simulation, but downsized by 33%. The exhaust and inlet opening timings are at 108° and 132° atdc, respectively, and there are twelve ports around the cylinder, as in Fig. 3.41, four of which are exhaust ports and eight of which are scavenge ports. The piston rings are unpegged. The four exhaust ports are each of 19.8 mm effective width and the eight transfer ports are each of 16.2 mm effective width. The scavenging is simulated as being of YAM1 quality (see Fig. 3.16) as in the three-cylinder engine simulation. The very considerable increase in the effective width of the exhaust and scavenge ports, over that in the three-cylinder engine example, is clearly evident. The exhaust manifold corresponds to the sketch in Fig. 5.8(c). The lengths of the various limbs are denoted as L_1 to L_6 and they are 200, 100, 100, 200, 50 and 50 mm, respectively. The diameters of the various limbs are denoted as d_1 to d_7 and they are each set at 48 mm.

All other factors for this engine are as for the three-cylinder engine simulation, i.e., trapped compression ratio, air-to-fuel ratio, intercooler or combustion characteristics, or engine and duct surface temperatures, etc.

While the main point of the simulation is to illustrate the effectiveness of the trapping and charging characteristics of a four-cylinder engine, a summary of the ensuing performance characteristics is very relevant for direct comparison with the three-cylinder engine simulation. The bmep is 7.9 bar, bsfc is 268 g/kWh, DR is 0.839, SE is 0.813, TE is 0.843, CE is 0.712, and η_m is 0.87. By comparison with the three-cylinder engine, it will be observed that the attained bmep and supplied air flow rate are higher, and the trapping and charging efficiencies are superior, while the specific fuel consumption and mechanical efficiency are somewhat inferior. The latter effects are due to the greater pumping loss from a blower which is operating at a higher pressure level to deliver the same air flow rate, as can be seen in Figs. 5.38 and 5.39.

The exhaust tuning of the four-cylinder engine

The cylinder firing order is 1-3-4-2. This means that the exhaust pulse from cylinder number 3 is that which cross-charges cylinder number 1, and that from cylinder number 1

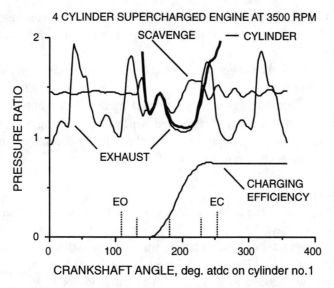

Fig. 5.38 Open cycle pressures and charging for cylinder No. 1.

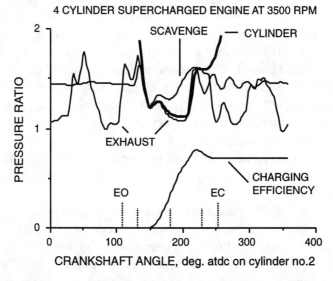

Fig. 5.39 Open cycle pressures and charging for cylinder No. 2.

then cross-charges cylinder number 2. Figs. 5.38 and 5.39 show the charging efficiency characteristics of cylinders 1 and 2, an end and a middle cylinder in the four-cylinder bank. Also displayed are the pressures in the cylinder, and in the exhaust and scavenge ducts at their respective ports. The charging efficiency behavior is observed to be slightly different for the two cylinders. In cylinder number 1 it is seen to rise to 0.744 before falling back to 0.729. In cylinder number 2 it is seen to rise to 0.779 before falling back to 0.726. The phasing and amplitude of the arriving cross-charging pressure pulsations at the exhaust ports are seen to be slightly different for these two cylinders, much as observed for the end and middle cylinders of the three-cylinder engine in Figs. 5.36(a) and (b). Nevertheless the basic cross-charging action of the four-cylinder engine is readily observed in either Fig. 5.38 or 5.39. Naturally, the pressure record in the exhaust pipe shows the appearance of four exhaust pulsations, whereas in the three-cylinder engine there are obviously three such pressure waves.

The scavenge plenum "mean" pressure is seen to be 1.45 bar, by comparison with 1.23 bar for the three-cylinder engine, and this leads to the higher pumping loss for the four-cylinder engine, commented on above. The port open periods of the four-cylinder engine are considerably narrower than those in the three-cylinder engine, which can be observed from the locations of the port timing bars in Fig. 5.38 by comparison with those in Fig. 5.36.

The tuning of the four-cylinder engine with a compact manifold is just as effective as that of a three-cylinder unit, provided that the exhaust port timings and the manifold are designed correctly. There is a further advantage in the four-cylinder layout of an engine designed for application in an automobile. It is inherent in the port timings which can be employed for an externally scavenged unit destined to be used in conjunction with direct in-cylinder fuel injection. In the optimized four-cylinder engine the exhaust port opened at 108° atdc, by comparison with 95° for the three-cylinder engine discussed earlier. In the three-cylinder engine, to attain good stratified combustion at low speeds and light load down to the idle condition, it was necessary to employ an exhaust timing control valve of the type shown in Fig. 5.2 which could be lowered to close the port at timings approaching 120° btdc. In the four-cylinder design, with the port already trapping naturally by 108° btdc, it may be possible to eliminate this timing control device entirely, or to replace it with a simpler, and more durable, butterfly type valve at the manifold exit. An exhaust timing control valve occupies much of the front face of the cylinder above the exhaust port and seriously deteriorates the cooling there, which leads to local distortion of the cylinder bore (see Fig. 5.2). To be able to eliminate this feature would greatly enhance the durability and longevity of the design. This fact, in conjunction with the employment of unpegged rings on the piston, offers the prospect of an externally scavenged, four-cylinder engine with improved durability over a three-cylinder design, without any sacrifice of the engine performance characteristics. This discussion is recalled in Chapter 7, where the design of engines to meet exhaust emission and fuel consumption targets is described.

The above discussion on multi-cylinder, spark-ignition engines has equal relevance for a similar debate on the design of multi-cylinder, compression-ignition engines.

5.6 Concluding remarks

The simulation of engines has been greatly enhanced by the use of the theoretical methods outlined in Chapters 2-4. This much is evidenced by the examples employed here, not solely because of the accuracy of correlation with experiments, but rather more by the extent of reliable information on gas properties and state conditions throughout the engine. This greatly assists the designer in comprehending the effect on performance characteristics which are due to alterations of the physical geometry of the engine and its ducting or its scavenging.

References for Chapter 5

5.1 F. Nagao, Y. Shimamoto, "Effect of the Crankcase Volume and the Inlet System on the Delivery Ratio of Two-Stroke Cycle Engines," SAE Paper No. 670030, Society of Automotive Engineers, Warrendale, Pa., 1967.

5.2 G.P. Blair, "Gas Flow Techniques Applied to a Sleeve-Valve Engine Design," *Automotive Design Engineering*, Vol 9, April 1970.

5.3 G.P. Blair, M.B. Johnston, "The Development of a High Performance Motor Cycle Engine," *Proc.I. Mech.E.*, 1970-1971, Vol 185, 20/71, pp273-283.

5.4 G.P. Blair, W.L. Cahoon, "The Design and Initial Development of a High Specific Output, 500cc Single Cylinder, Two-Stroke, Racing Motorcycle Engine," SAE Paper No. 710082, Society of Automotive Engineers, Warrendale, Pa., 1971.

5.5 G.P. Blair, "Further Development of a 500cc Single Cylinder Two-Cycle Engine for Motorcycle Racing and Motocross Applications," SAE Paper No. 740745, Society of Automotive Engineers, Warrendale, Pa., 1974.

5.6 R. Mellde, J. Eklund, K. Knutsson, "Use of the Two-Stroke Cycle Engine as a Power Unit for a Passenger Car," SAE Paper No. 650008, Society of Automotive Engineers, Warrendale, Pa., 1965.

5.7 V. Sathe, P.S. Myers, O.A. Uyehara, "Parametric Studies using a Two-Stroke Engine Cycle Simulation," SAE Paper No. 700124, Society of Automotive Engineers, Warrendale, Pa., 1970.

5.8 R.B. Krieger, R.R. Booy, P.S. Myers, O.A. Uyehara, "Simulation of a Crankcase Scavenged Two-Stroke, SI Engine and Comparison with Experimental Data," SAE Paper No. 690135, Society of Automotive Engineers, Warrendale, Pa., 1969.

5.9 G.P. Blair, E.T. Hinds, R. Fleck, "Predicting the Performance Characteristics of Two-Cycle Engines Fitted with Reed Induction Valves," SAE Paper No. 790842, Society of Automotive Engineers, Warrendale, Pa., 1979.

5.10 D.S. Sanborn, "Paper Powerplants Promote Performance Progress," SAE Paper No. 750016, Society of Automotive Engineers, Warrendale, Pa., 1975.

5.11 G.J. Van Wylen, R.E. Sonntag, Fundamentals of Classical Thermodynamics, 2nd Edition, SI Version, Wiley, New York, 1976.

5.12 K.J. Patton, R.G. Nitschke, J.B. Heywood, "Development and Evaluation of a Friction Model for Spark-Ignition Engines," SAE Paper No. 890836, Society of Automotive Engineers, Warrendale, Pa., 1989.

5.13 H. Naji, R. Said, R.P. Borghi, "Towards a Turbulent Combustion Model for Spark-Ignition Engines," SAE Paper No. 890672, Society of Automotive Engineers, Warrendale, Pa., 1989.

5.14 K. Landfahrer, D. Plohberger, H. Alten, L.A. Mikulic, "Thermodynamic Analysis and Optimization of Two-Stroke Gasoline Engines," SAE Paper No. 890415, Society of Automotive Engineers, Warrendale, Pa., 1989.

5.15 E.T.Hinds, G.P.Blair, "Unsteady Gas Flow Through Reed Valve Induction Systems," SAE Paper No. 780766, Society of Automotive Engineers, Warrendale, Pa., 1978, p10.

5.16 R.Fleck, G.P.Blair, R.A.R.Houston, "An Improved Model for Predicting Reed Valve Behaviour in Two-Stroke Cycle Engines," SAE Paper No. 871654, Society of Automotive Engineers, Warrendale, Pa., 1987, p21.

5.17 R.Fleck, R.A.R.Houston, G.P.Blair, "Predicting the Performance Characteristics Of Twin Cylinder Two-Stroke Engines for Outboard Motor Applications," SAE Paper No. 881266, Society of Automotive Engineers, Warrendale, Pa., 1988, p25.

5.18 E.T. Hinds, "Intake Flow Characteristics of a Two-Stroke Cycle Engine fitted with Reed Valves," PhD Thesis, The Queen's University of Belfast, August 1978.

5.19 R.A.R. Houston, "Performance Modelling of Multi-Cylinder Reed Valve, Two-Stroke Cycle Engines," PhD Thesis, The Queen's University of Belfast, May 1988.

5.20 G.P. Blair, S.J. Kirkpatrick, R. Fleck, "Experimental Evaluation of a 1D Modelling Code for a Pipe containing Gas of Varying Properties," SAE Paper No. 950275, Society of Automotive Engineers, Warrendale, Pa., 1995, p14.

5.21 G.P. Blair, S.J. Kirkpatrick, D.O. Mackey, R. Fleck, "Experimental Evaluation of a 1D Modelling Code for a Pipe System containing Area Discontinuities," SAE Paper No. 950276, Society of Automotive Engineers, Warrendale, Pa., 1995, p16.

5.22 J.S.Richardson, D.W.Artt, G.P.Blair, "A Computer Model of a Pulsejet Engine," SAE Paper No. 820953, Society of Automotive Engineers, Warrendale, Pa., 1982, p28.

5.23 D. Thornhill, R. Fleck, "Design of a Blower-Scavenged, Piston Ported, V6, Two-stroke Automotive Engine," SAE Paper No. 930980, Society of Automotive Engineers, Warrendale, Pa., 1993.

5.24 D. Thornhill, R. Fleck, "A Generic Engine Simulation Program Applied to the Development of a V6 Automotive Two-Stroke Engine," SAE Paper No. 940396, Society of Automotive Engineers, Warrendale, Pa., 1994.

5.25 G.P. Blair, H.B. Lau, B.D. Ragunathan, A. Cartwright, D.O. Mackey, "Coefficients of Discharge at the Apertures of Engines," SAE Paper No. 952138, Society of Automotive Engineers, Warrendale, Pa., 1995.

Appendix A5.1 The flow areas through poppet valves

The physical geometry of the poppet valve and its location are shown in Fig. A5.1. It is characterized by a lift, L, above a seat at an angle, ϕ, which has inner and outer diameters, d_{is} and d_{os}, respectively. The valve stem diameter, d_{st}, partially obscures the aperture.

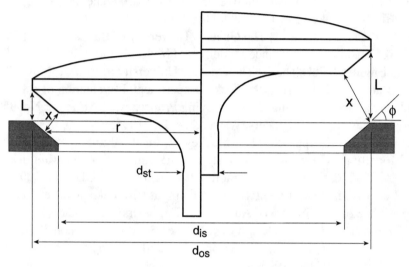

Fig. A5.1 Valve curtain areas at two lift positions.

The port-to-pipe-area ratio, k, for this particular geometry is often simplistically expressed as:

$$\text{Area ratio, } k = \frac{\text{valve curtain area}}{\text{pipe area}} = \frac{A_t}{A_p} = \frac{\pi d_{is} L}{\frac{\pi}{4} d_p^2} \tag{A5.1}$$

For accuracy of incorporation of poppet valve flow into an engine simulation, it is clear from the above analysis that it is vital to calculate correctly the geometrical throat area of the restriction, A_t. For ports in two-stroke engine cylinders, plain pipe ends, throttles, nozzles and venturis, it is relatively easy to deduce the geometrical throat area of any particular geometry in terms of its projected area which is normal to the gas particle flow direction (see Sec. 5.2). For poppet valves this requires more careful examination than that provided by the simple statement seen in Eq. A5.1. At QUB this was first deduced by McConnell [2.7].

Observe from Fig. A5.1 that the valve curtain area at the throat, when the valve lift is L, is that which is represented by the frustum of a cone defined by the side length dimension, x, the valve seat angle, ϕ, and the inner or outer seat diameters, d_{is} and d_{os}; or also of dimension, r, depending on the amount of valve lift, L. This is a distinctly different situation from that predicted in Eq. A5.1 where the valve curtain area is considered simply as the surface area of a cylinder of diameter, d_{is}, and height, L.

The surface area of a frustum of a cone is given by:

$$\text{surface area} = \pi x \frac{d_{major} + d_{minor}}{2}$$

where x is the length of the sloping side and d_{minor} and d_{major} are its top and bottom diameters.

The effective area of the seat of the valve through which the gas flows to or from the port is given by the seat area less the valve stem area, thus:

$$A_{seat\ eff} = \frac{\pi}{4}\left(d_{is}^2 - d_{st}^2\right) \tag{A5.2}$$

There is no point in designing a poppet valve that lifts to an extent whereby the effective valve curtain area exceeds this value.

The dimension, x, through which the gas flows has two values which are sketched in Fig. A5.1. On the left, the lift is sufficiently small that the value, x, is at right angles to the valve seat and, on the right, the valve has lifted beyond a lift limit, L_{lim}, where the value, x, is no longer normal to the valve seat angle, ϕ. By simple geometry, this limiting value of lift is given by:

$$L_{lim} = \frac{d_{os} - d_{is}}{2\sin\phi\cos\phi} = \frac{d_{os} - d_{is}}{\sin 2\phi} \tag{A5.3}$$

For the first stage of poppet valve lift where:

$$L \leq L_{lim}$$

then the valve curtain area, A_t, is given from the values of x and r as:

$$x = L\cos\phi$$

$$r = \frac{d_{is}}{2} + x\sin\phi$$

whence
$$A_t = \pi L \cos\phi \left(d_{is} + L\sin\varphi\cos\phi\right) \tag{A5.4}$$

For the second stage of poppet valve lift where:

$$L > L_{lim}$$

then the valve curtain area, A_t, is given from the value of x as:

$$x = \sqrt{\left(L - \frac{d_{os} - d_{is}}{2}\tan\phi\right)^2 + \left(\frac{d_{os} - d_{is}}{2}\right)^2}$$

whence

$$A_t = \pi\left(\frac{d_{os} + d_{is}}{2}\right)\sqrt{\left(L - \frac{d_{os} - d_{is}}{2}\tan\phi\right)^2 + \left(\frac{d_{os} - d_{is}}{2}\right)^2} \qquad (A5.5)$$

The simple criterion for maximum valve lift, ignoring the effect of the valve stem diameter, can be found from Eq. A5.1 by inserting the area ratio, k, as a maximized value of unity, which produces the information that it is 25% of the inner seat diameter. From the above theory this is manifestly much too simple a conclusion. The effect of a (conventional) valve seat angle of 45°, and typical values for valve stem diameters, is to empirically raise that ratio from 0.25 to more nearly 0.3 in practice. Empiricism is quite inadequate in this context, as the above theory must be employed by the designer to ensure that the flow areas of the poppet valve and its port are carefully matched at the design stage. The mathematical reality for the assessment of maximum valve lift is that the discharge coefficient must be incorporated into the debate as:

$$C_{da}A_t = A_{\text{seat eff}} \qquad (A5.6)$$

and the maximum lift is then found from the curtain area term, A_t, by iteration for valve lift until an equality for Eq. A5.6 is obtained.

To ensure accuracy regarding the prediction of mass flow rates and of the magnitude of pressure wave formation, this analysis of valve curtain area must be incorporated into both the reduction of measured data for the determination of the actual coefficient of discharge, C_{da}, measured in steady flow experiments and also into the replay of such data within any engine simulation incorporating unsteady gas flow. For C_{da} determination this is described in Appendix A2.3.

Chapter 6

Empirical Assistance for the Designer

6.0 Introduction

The first five chapters of this book introduce the two-stroke engine, scavenging, unsteady gas flow, combustion, and then the combination of all of the preceding topics into computer modeling of a complete engine. Having just finished reading Chapter 5, you can be forgiven for feeling somewhat bemused. In that chapter, it is assured that modeling reveals the "inside story" on engine behavior, and hopefully that is how it is presented. At the same time you will be conscious that, for the engine computer simulations presented in that chapter to be of practical use in a design mode, the designer will to have to produce real input data in the form of banks of numbers as input data files for that software to function.

Some of you may feel that it will be as simple as opening up a computer program such as those listed in the Appendix Listing of Computer Programs, inserting the numbers for the engine and geometry as the mood takes you, and letting the computer inform you of the outcome regarding the pressure wave action and the performance characteristics. Of course that is an option open to the designer, but I, and possibly also the designer's employer, would not consider it as the most rapid approach to optimize a design to meet a particular need. Depending on the complexity of the geometry of an engine, there could be up to one hundred different data numbers or labels which can be changed in such engine simulations for a single calculation, and the statistical probability of achieving an optimized engine design by an inspired selection of those numbers is about on a par with successfully betting on racehorses. Others may feel so overwhelmed at the range of choices available for the data value of any one parameter that you could spend more time trying to estimate a sensible data value than using the engine simulation for the predictive task.

What is required is what is presented in this chapter, namely some empirical guidance for the data values to be used in engine simulation in particular and engine design in general. There are the academic purists who will react adversely to the revelation that this textbook, written by a university Professor and which can be read by undergraduate students, will contain empirical guidance and advice. The contrary view is that the design and creation of engineering artifacts involves the making of decisions which are based just as much on past experience as they are on theoretical analyses. For those whose past experience is limited, such as undergraduate engineering students, they require the advice and guidance, both in lectures and from textbooks, from those with that know-how.

The empirical guidance to be given will range from the simplistic, such as the maximum port width that the designer can select without encountering piston ring breakage, to the complex, such as the design of the entire geometry of an expansion chamber to suit a given engine. Actually, what will be regarded as simplistic by some will be considered by others as inconceivable. To quote the apparently simple example above, the experienced engineer in industry will have little problem in deciding that the maximum width of an exhaust port could be as large as 70% of the cylinder bore dimension, provided that suitably large corner and top edge radii are used to compress the piston ring inside the timing edges. On the other hand, the archetypical undergraduate university student will not have that past experience to fall back on as empirical guidance and has been known to spend inordinate amounts of time peering at a computer screen which is awaiting an input, unable to make a decision.

There exists the danger that the empirical guidance to be given will be regarded as the final design, the "quick-fix formula" so beloved by a human race ever desirous of cutting corners to get to a solution of a problem. Let a word of caution be sounded in that case, for no simplistic approach will ever totally provide an exact answer to a design question. A good example of that is the use of the heat release model of combustion within the engine modeling programs presented in Chapter 5. While the quality of the answer is acceptable in design terms today, the thinking engineer will realize the limitations imposed by such an approach, and will strive to incorporate ever more sophisticated models of combustion and heat transfer into those engine design programs, particularly as CFD theory becomes more practical and desk-top computers calculate ever faster.

The empirical advice will incorporate matters relating to the engine simulations shown in Chapter 5. The following is a list of the topics to be covered:

(i) The estimation of the data for porting characteristics to meet a design requirement for a given engine cylinder at a given engine speed to produce a particular power output.

(ii) The translation of the data in (i) into port timings, widths, heights and areas for a particular engine geometry.

(iii) The estimation of the data for the geometry of an expansion chamber exhaust system for a high specific output engine so that it is tuned at a desired engine speed and matched to the flow requirements of the engine.

(iv) The preliminary design of reed valve and disc valve induction systems for two-stroke engines.

6.1 Design of engine porting to meet a given performance characteristic

The opening part of this section will discuss the relationship between port areas and the ensuing gas flow, leading to the concept of *specific time area* as a means of empirically assessing the potential for that porting to produce specific performance characteristics from an engine. The second part will deal with the specific time area analysis of ports in engines of a specified geometry and the predictive information to be gleaned from that study. The third part will examine the results of that study by its analysis through an engine modeling program.

Chapter 6 - Empirical Assistance for the Designer

6.1.1 Specific time areas of ports in two-stroke engines

Fig. 6.1 is a graphical representation of the port areas as a function of crank angle in a two-stroke engine. The actual data presented is for a chainsaw engine design somewhat similar to that discussed in Chapter 5 and with the geometrical data given in Fig. 6.2. On the grounds that engineers visualize geometry, the cylinder is presented in Fig. 6.2, as seen by a computer-generated sketch based on the numerical data. Therefore, it is the movement of the piston past those ports, be they exhaust, transfer or inlet, as the crankshaft turns which produces the area opening and closing relationships presented in Fig. 6.1.

Some of the numbers generated are of general interest. Note that the maximum area of the inlet port and the exhaust port are about equal. The transfer port total area is clearly less than the inlet and exhaust areas. The area reached by the exhaust port at the point of transfer port opening, i.e., at the official end of the blowdown period, is about one-quarter of the maximum area attained by the exhaust port. The flat top to the inlet area profile is due to that particular design being fully uncovered for a period of 52° around tdc. The symmetry of the port area diagrams around the tdc or bdc positions is created by the piston control of all gas flow access to the cylinder; this would not be the case for ports controlled by a disc valve, reed valve or poppet valve mechanism (see Sec. 1.3).

It has already been demonstrated that the performance characteristics of an engine are related to the mass flow rates of gas traveling through the ports of the engine. For example, in Figs. 5.9 and 5.10, a chainsaw engine is shown to produce bmep which is directly related to the work produced per cycle, virtually as a direct function of the mass of air which it breathes per cycle, i.e., its delivery ratio. The following simplified theory for gas flow through a port or valve is fundamental to a discussion of that cylinder filling process. Let A_θ be the instantaneous area of an open port at any crank angle θ after its opening point, and let c_θ be the

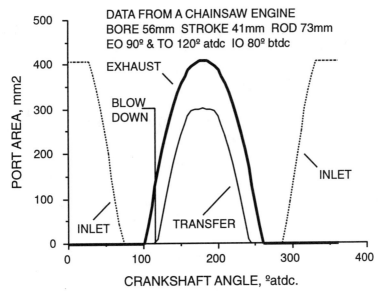

Fig. 6.1 Port areas by piston control in a two-stroke engine.

Design and Simulation of Two-Stroke Engines

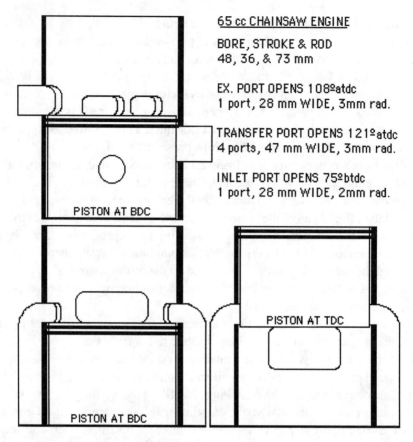

Fig. 6.2 *Computer-generated picture of chainsaw engine cylinder.*

particle velocity through it, and ρ_θ be the density of the gas. The continuity equation (see Sec. 2.1.4) dictates that the mass rate of flow, $\dot{m}_\theta$, is:

$$\dot{m}_\theta = \rho_\theta A_\theta c_\theta \qquad (6.1.1)$$

If this situation persists for a small time interval, dt, the increment of flow to pass through would be:

$$dm = \dot{m}_\theta dt \qquad (6.1.2)$$

The total gas flow, m_p, to pass through the port is the integral of all flow increments for the entire duration of the port opening, either t_p in time units or θ_p in crankshaft angle units. This is written mathematically as:

$$m_p = \int_{t=0}^{t=t_p} \rho_\theta A_\theta c_\theta dt = \int_{\theta=0}^{\theta=\theta_p} \rho_\theta A_\theta c_\theta \frac{dt}{d\theta} d\theta \qquad (6.1.3)$$

Chapter 6 - Empirical Assistance for the Designer

This equation applies to all ports and to all gas flow through the engine. Within the engine and gas-dynamic models of Chapters 2 and 5, this summation exercise is carried out at each calculation step so that the values of delivery ratio, scavenge ratio, etc., are ultimately determined. The term, A_θ, which is the instantaneous area of the port at any instant, employed in Eq. 6.1.3, is determined using the theoretical approach detailed in Sec. 5.2.

As most two-stroke engines have the symmetrical porting geometry shown in Fig. 6.1, and as the variations of density and gas velocity with time are somewhat similar from engine to engine, there might exist some mathematical function linking the remaining terms to the gas flow transmitted. Perhaps more importantly, as the gas flow which is transmitted is related to that which is trapped and burned with fuel to produce power, the relationship could extend to a more direct linkage to engine performance.

For example, the application of Eq. 6.1.3 to the inlet port within an engine model predicts the total air mass ingested into the engine on each cycle. The point has already been made above that this must have a direct relationship to the work output from that cycle when that air is burned with fuel in the correct proportions. As already pointed out, this can be seen in Figs. 5.9 and 5.10 from the close correspondence between the delivery ratio and bmep with engine speed for the chainsaw engine.

The net work output per cycle, from Fig. 1.16, is given by:

$$\text{Work/cycle} = V_{sv}\, \text{bmep} \tag{6.1.4}$$

If the relationship is linear, then the division of air mass ingested by work output should be "a constant":

$$\frac{m_{\text{air ingested}}}{V_{sv}\text{bmep}} = \frac{\int_{t=0}^{t=t_p} \rho_\theta A_\theta c_\theta\, dt}{V_{sv}\text{bmep}} = \text{a constant} \tag{6.1.5}$$

On the assumption that the temporal variations of density, ρ_θ, and particle velocity, c_θ, are somewhat similar, even for different engines, incorporating these so-called constants into the right-hand side of Eq. 6.1.5 reduces it to:

$$\frac{\int_{t=0}^{t=t_p} A_\theta\, dt}{V_{sv}} \propto \text{bmep} \tag{6.1.6}$$

This can also be written as:

$$\frac{\int_{\theta=0}^{\theta=\theta_p} A_\theta\, \frac{dt}{d\theta}\, d\theta}{V_{sv}} \propto \text{bmep} \tag{6.1.7}$$

The left-hand side of Eq. 6.1.6 or Eq. 6.1.7 is known as the "time-integral of area per unit swept volume" for the port of a two-stroke engine. This expression is often shortened colloquially to *specific time area*, A_{sv}. The units of "time-integral of area per unit swept volume," or specific time area, can be seen to be:

$$\frac{m^2 s}{m^3} = s/m$$

The evaluation of the area term in Eq. 6.1.6 or 6.1.7 has been discussed in Sec. 5.2.1. The summation of the term $A_\theta d\theta$ is the port area shown in Fig. 6.1 for each of the ports on that diagram. The term $dt/d\theta$ is a fixed relationship between crankshaft angle and time given by:

$$\frac{dt}{d\theta} = \frac{\text{time per revolution}}{\text{degrees per revolution}} = \frac{60 \div \text{rpm}}{360} = \frac{1}{6 \times \text{rpm}} \qquad (6.1.8)$$

On the assumption that the mass of gas which flows into an engine on each cycle must also flow through it and ultimately out of it, the evaluation of specific time area, A_{sv}, for inlet ports, transfer ports and exhaust ports can be carried out and an attempt made to relate that data to the bmep produced by an engine of a known swept volume, V_{sv}, at a known engine speed, rpm.

Another important flow parameter can be evaluted by calculating the A_{sv} for the blowdown period between the exhaust port opening and the transfer port opening. For the cylinder pressure to fall from the release point to a value approaching that in the scavenge ducts in order to avoid excessive backflow into the transfer ports, a certain proportion of the cylinder mass must pass out of the exhaust port during that period. The higher the bmep in a given engine, the greater will be the trapped charge mass and pressure, and therefore, so will be the need for a larger specific time area to let the cylinder pressure fall appropriately during the blowdown phase of the exhaust process. The blowdown value of A_{sv} is indicated in Fig. 6.1.

At QUB over the years a considerable volume of data has been collected from engine design and testing, not only from in-house designs but from production engines as part of R&D programs. A small fraction of this "expert system" data is illustrated in Figs. 6.3-6.6 for the specific time areas related to brake mean effective pressure. On those diagrams, the engine cylinder sizes range from 50 cm³ to 500 cm³, the speeds at which the performance data were recorded range from 3000 to 12,000 rpm, and the bmep spread is from 3.5 bar to 11 bar. The graphical scatter is, not unexpectedly, considerable. The reasons are fairly self-evident, ranging from engines which have over- or under-designed ports, to engines with exhaust systems attached in varying degrees of tuning level from the Grand Prix racing type to that for an outboard motor or a chainsaw. Further complications arise, such as the use of differing crankcase compression ratio, or the employment of supercharging; these factors do not emerge in any simplistic analysis of the specific time area of engines. Despite such logic, in each of these figures, trends for a bmep-A_{sv} relationship can be seen to exist and a straight line opinion as to the best fit is drawn on each diagram; in the case of the transfer ports a band of a

Chapter 6 - Empirical Assistance for the Designer

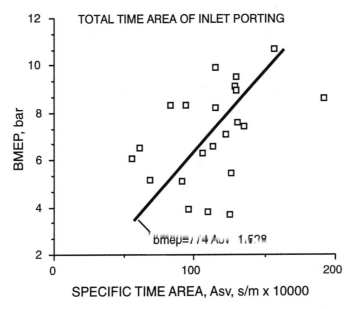

Fig. 6.3 *Specific time areas for inlet ports from measured data.*

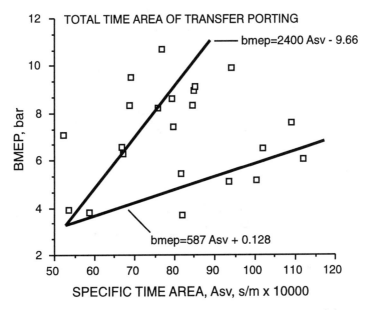

Fig. 6.4 *Specific time areas for transfer ports from measured data.*

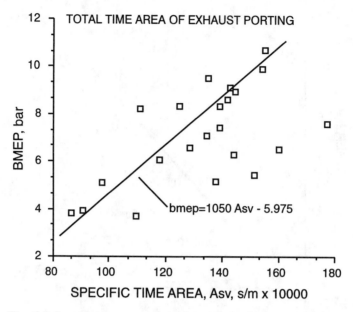

Fig. 6.5 Specific time areas for exhaust ports from measured data.

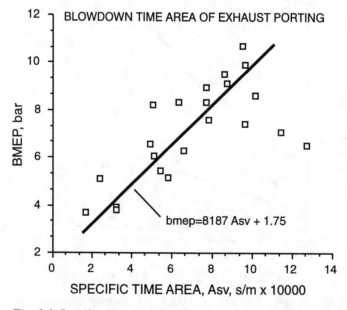

Fig. 6.6 Specific time areas for blowdown from measured data.

Chapter 6 - Empirical Assistance for the Designer

higher and a lower level is shown as the most informed view of a more complex relationship coming from the "logic" above.

The arithmetic relationships from those diagrams are rewritten below. In each equation the unit of bmep is bar and the specific time area is in units of s/m.

For inlet ports, the specific time area is labeled as A_{svi}.

$$bmep = 774A_{svi} - 1.528 \tag{6.1.9}$$

For transfer ports, the upper band of specific time area is labeled as A_{svt1}.

$$bmep = 2400A_{svt1} - 9.66 \tag{6.1.10}$$

For transfer ports, the lower band of specific time area is labeled as A_{svt2}.

$$bmep = 587A_{svt2} + 0.128 \tag{6.1.11}$$

For exhaust ports, the specific time area is labeled as A_{sve}.

$$bmep = 1050A_{sve} - 5.975 \tag{6.1.12}$$

For exhaust blowdown, the specific time area is labeled as A_{svb}.

$$bmep = 8187A_{svb} + 1.75 \tag{6.1.13}$$

However, while these functions are easily solved using a pocket calculator, a simple computer program is referred to in the Appendix Listing of Computer Programs as Prog.6.1, TIMEAREA TARGETS, and is available from SAE. For those not very familiar with programming in Basic this straightforward computer program will serve as another useful example of data insertion, simple calculation, and data presentation on both the computer screen and the printer.

As an example of the use of this program and of the analysis represented by Eqs. 6.1.7-6.1.11, consider the case of two engines similar to those studied in Chapter 5, a chainsaw engine and a racing engine. As observed there, these engines are at opposite ends of the performance spectrum. Imagine that a design is to be formulated for two new engines to satisfy the very criteria which are known to be attainable by virtue of both the measurements and the engine modeling carried out in Chapter 5 on such engines. The design brief might read as follows:

(a) A chainsaw engine is needed to produce 3.9 kW at 9600 rpm, with a bmep of 3.75 bar as a potentially obtainable target. That this would necessitate an engine of 65 cm^3 swept volume is found from Eq. 1.6.6 in Sec. 1.6.1. All of the porting is to be piston controlled.

(b) A high-performance 125 cm^3 engine is needed for racing, where 26.5 kW is required to be competitive. For mechanical reasons, it is decided to try to produce this power

Design and Simulation of Two-Stroke Engines

at 11,500 rpm. Using Eq. 1.6.6, this translates into the production of 11 bar bmep. All of the cylinder porting is to be piston controlled and the induction by a reed valve directly into the crankcase.

Upon inserting the information on the bmep level for each engine into Eqs. 6.1.9-6.1.13, or using Prog.6.1, the following information on specific time area is obtained for each engine. All units are s/m.

Table 6.1 Specific time areas for a chainsaw and a racing engine

Specific time area, s/m	A_{sve}	A_{svb}	A_{svt1}	A_{svt2}	A_{svi}
65 cm³ chainsaw	0.0095	0.00027	0.0057	0.0066	0.0071
125 cm³ racer	0.0162	0.00113	0.0086	0.0185	0.0162

Note that the specific time area requirements for the porting of the racing engine are much larger than for the chainsaw engine. Although the cylinder sizes are only 25% different, the larger cylinder of the racing engine is expected to produce about three times the torque at an engine speed that has only 70% of the time available for filling it. To assist with the visualization of what that may imply in terms of porting characteristics, a computer-generated sketch of a cylinder of a 125 racing engine of "square" dimensions, complete with a piston-controlled induction system of the requisite size, is shown in Fig. 6.7. A comparison with the similar sketch for the chainsaw engine in Fig. 6.2 reveals the considerable physical differences in both port timings and area. It is clear that the large ports needed for a Grand Prix engine are open much longer than they are closed.

The next step is to analyze the porting characteristics of the two engines discussed in Chapter 5, and to determine if the criteria noted above have any relevance to those which are known to exist.

6.1.2 The determination of specific time area of engine porting

Discussion of the calculation of port areas in a two-stroke engine has been a recurring theme throughout this book, most recently in Sec. 5.2 with respect to engine simulation. The calculation procedure for specific time area is encapsulated within any program that will compute the area of any port in an engine as a function of crankshaft angle, θ, or time. The value required is for A_θ, i.e., the areas in Fig. 6.1 for inlet, transfer, exhaust or blowdown. As it is unlikely that the A_{sv} will be determined by a direct mathematical solution due to the complexity of the relationship between the instantaneous value of A as a function of θ, the computer solution by summation at crankshaft intervals of one or two degrees, i.e., $\Sigma A_\theta d\theta$, will provide adequate numerical accuracy. When the $\Sigma A_\theta d\theta$ is determined, that value is inserted into the appropriately combined Eqs. 6.1.7 and 6.1.8 as follows:

$$\text{Specific time area, } A_{sv} = \frac{\sum_{\theta=0}^{\theta=\theta_p} A_\theta d\theta}{6V_{sv}\text{rpm}} \text{ s/m} \qquad (6.1.14)$$

Chapter 6 - Empirical Assistance for the Designer

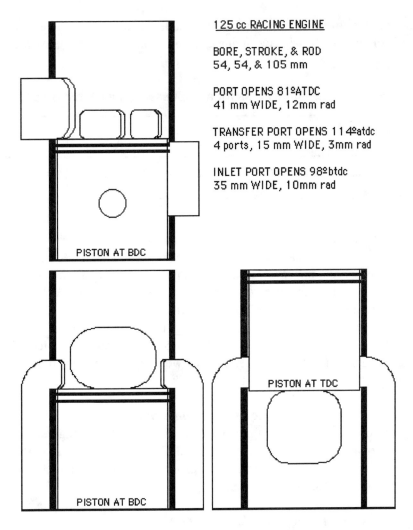

Fig. 6.7 Computer-generated picture of 125 Grand Prix engine cylinder.

Those who wish to use this equation should note that the units are strictly SI, and that the area, A, is in m² and V_{sv} in m³. By sheer coincidence, you can enter the value of A in mm² and V_{sv} in cm³ and the units are self-correcting to s/m.

The data required to perform such a calculation are the physical geometry of the piston, connecting rod and crankshaft and the timings, widths and corner radii of the porting for the engine, as in Fig. 5.1(a). Those data are sufficient to produce Fig. 6.1, Fig. 6.2 and Fig. 6.7 by a computer program, which is demonstrably in command of all of the necessary information. If the porting is more complex, as in Fig. 5.1(b), then the calculation can be conducted in stages for each port segment to sum up to the total values. To assist with the deduction of specific time area values for proposed or existing engines, a computer program is referred to in the Appendix Listing of Computer Programs as Prog.6.3, TIME-AREAS. The data insertion format is similar in style to others in the series. The program output is in the form of screen or line-printer output without graphics, so Table 6.2 is a precis of the use of Prog.6.3 for the analysis of the actual chainsaw engine and the Grand Prix racing geometries discussed in Chapter 5.

Table 6.2 Measured time areas for a chainsaw and a GP racing engine

Specific time area, s/m	A_{sve}	A_{svb}	A_{svt}	A_{svi}
65 cm³ chainsaw	0.0071	0.00015	0.0067	0.00110
125 cm³ racer	0.0164	0.00124	0.0096	reed valve

It can be seen that the specific time area values for the actual engines are very similar to those from Table 6.1, predicted as being necessary to produce the required performance characteristics. In other words, the specific time areas predicted as requirements, and found as examples, are sufficiently close to give some confidence in the predictive tool as an initial design step. There are some differences which will be debated below, together with the designer's options and methodologies when tackling the initial steps in the design process.

The topic of specific time area is not mentioned very frequently in technical papers. However, a useful reference in this context, which would agree with much of the above discussion, is that by Naitoh and Nomura [6.5].

6.1.3 The effect of changes of specific time area on a chainsaw

Consider the 65 cm³ chainsaw engine as the example and let an engine simulation of it provide the performance characteristics for the debate. The raw dimensions of the porting are shown in Fig. 6.2 and the remaining data are as given in Sec. 5.5.1. The "standard" exhaust port timing is an opening timing of 108° atdc.

A change of exhaust port timing with others constant

Let all other data values in the simulation stay constant, but change the exhaust port timing to 104°, 106° and 110° and compare the ensuing performance characteristics with each other and the "target" specific time area values seen in Table 6.1. The A_{sv} values that emanate from these timing changes are summarized in Table 6.3.

Table 6.3 Time areas for various exhaust timings for a chainsaw

Exhaust timings	A_{sve}	A_{svb}	A_{svt}
104° atdc	0.00827	0.00027	0.0067
106° atdc	0.00765	0.00021	0.0067
108° atdc (std)	0.00705	0.00015	0.0067
110° atdc	0.00648	0.00011	0.0067
target values	0.0095	0.00027	0.0066

Chapter 6 - Empirical Assistance for the Designer

Observe that the time area values at an exhaust port timing of 108° are the same as those given in Table 6.2 as measured for the chainsaw. It can be seen that the transfer time area is virtually on its "target" value, that for blowdown it is at a 104° atdc timing, and the total time area "target" for the exhaust port is somewhat higher than any to be examined.

The results of the simulation are presented in Figs. 6.8 to 6.11, for bmep, air flow, specific fuel consumption and specific hydrocarbon emissions, respectively. In each figure the performance parameter is shown with respect to the exhaust and blowdown time areas and the exhaust port timing is also indicated. As far as power and torque are concerned, in Fig. 6.8 it would appear that the target time area values are very relevant, for the trend is to peak at the 104° port timing value. The air flow, plotted as delivery ratio in Fig. 6.9, climbs toward the 104° exhaust port timing, but the scale is narrow and it is virtually constant. The specific fuel consumption, shown in Fig. 6.10, would appear to reach a plateau at the 104 and 106° exhaust port timings, but superior to that achieved at the standard 108° timing value. However, when the hydrocarbon emissions are examined in Fig. 6.11, the basic trend is for their deterioration with a longer exhaust period. The best HC emission value is seen to be at a 106° exhaust timing. In modern times, with hydrocarbon emissions now controlled by legislation in many countries, the designer can no longer merely optimize the design for power alone.

From this analysis, the standard port timing and area can be seen to be a compromise between the acquisition of power and the reduction of HC emissions. It is also clear that a designer using a time area analysis would have arrived into this area of engine geometry more rapidly than by merely guessing port timings and widths, and subsequently indulging in an experimental R&D program. The use of the engine simulation ensures that the design process is even more rapid and more accurate.

A change of transfer port timing with others constant

Let the exhaust port timing remain at the "standard" value of 108° atdc and change the transfer port timing from 117° atdc to 125° atdc in 2° steps. This will have the effect of altering the pumping behavior of the crankcase and of changing the blowdown time area as well as that of the transfer ports. The results of these geometrical alterations are summarized

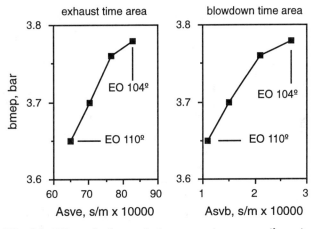

Fig. 6.8 Effect of exhaust timing on engine torque (bmep).

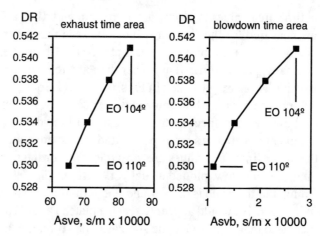

Fig. 6.9 *Effect of exhaust timing on engine air flow (DR).*

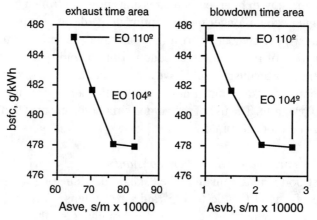

Fig. 6.10 *Effect of exhaust timing on specific fuel consumption.*

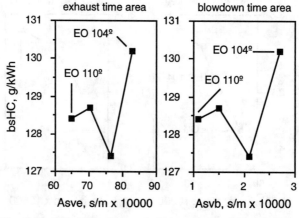

Fig. 6.11 *Effect of exhaust timing on hydrocarbon emissions.*

in Table 6.4. The standard transfer port timing is on the "target" value, so the timing variations are such as to investigate greater and lesser values of specific time area. It can be seen that the "target" value of blowdown specific time area is obtained by a transfer port timing of 125° atdc. Increasing the transfer time area reduces that for blowdown, and vice versa.

Table 6.4 Time areas for various transfer timings for a chainsaw

Transfer timings	A_{sve}	A_{svb}	A_{svt}
117° atdc	0.00705	0.00007	0.00813
119° atdc	0.00705	0.00011	0.00739
121° atdc (std)	0.00705	0.00015	0.00670
123° atdc	0.00705	0.00020	0.00603
125° atdc	0.00705	0.00026	0.00542
target values	0.0095	0.00027	0.00660

The results of the simulation are presented in Figs. 6.12 to 6.15, for bmep, air flow, specific fuel consumption and specific hydrocarbon emissions, respectively. In each figure the performance parameter is shown with respect to the transfer port and blowdown time areas and the transfer port timing is also indicated.

More transfer port area should give more air flow, but only if the rest of the design is optimized to permit it. At a 117° timing there may be more transfer time area but the blowdown is inadequate. The backflow reduces the air flow, as can be seen in Fig. 6.13. The air flow peaks at transfer port timings of 121° and 123° atdc. The effect on bmep and power at this speed is to give a peak at a transfer port timing of 123° atdc. When the equally important specific fuel consumption is investigated in Fig. 6.14, and the even more important specific hydrocarbon emissions in Fig. 6.15, the optimization at a transfer port timing of 123° atdc is now evident. Much more importantly, the optimization is at specific time area values of trans-

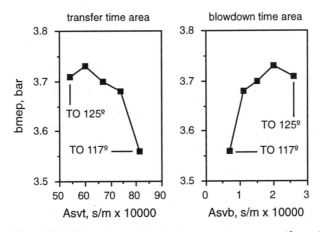

Fig. 6.12 Effect of transfer timing on engine torque (bmep).

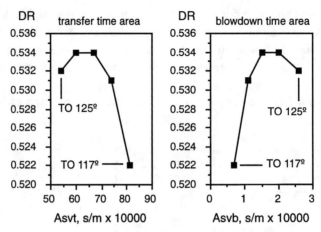

Fig. 6.13 Effect of transfer timing on engine air flow (DR).

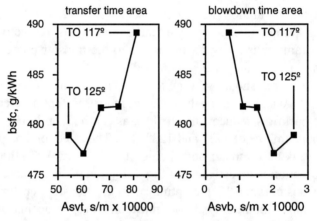

Fig. 6.14 Effect of transfer timing on specific fuel consumption.

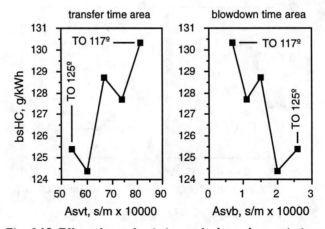

Fig. 6.15 Effect of transfer timing on hydrocarbon emissions.

Chapter 6 - Empirical Assistance for the Designer

fer A_{svt} about 0.006, and blowdown A_{svb} about 0.0002; they are not significantly different from the target values.

By changing the transfer port timings, and holding the exhaust port timing and area constant, the interaction of the pumping action, the scavenging and the blowdown process are now revealed.

A time area analysis is, as with any empirical tool, never totally accurate. It is, however, remarkably effective in the production of data on port timings and widths as an initial data set for the application of engine simulation methods. The demonstrations given in this section show, within a relatively narrow optimization zone, the effect of minor changes to port timings on the engine performance characteristics. Without using empirical methods, much human and computer time is wasted before reaching such an optimization zone in which the more accurate simulation methods can be applied.

6.2 Some practical considerations in the design process

The design of a new engine is always the moment for a fundamental examination of the selection of the basic engine dimensions. The reason for the new engine may be market driven, i.e., a new device needs to be powered; it may be legislation driven, i.e., to meet new regulations regarding noise, exhaust emissions or fuel quality; it may be durability driven, i.e., the product may be required to last longer at higher loading levels. In any event the opportunity for an optimized selection of the basic engine dimensions and its construction should always be taken. For the student reader, this section should provide some basic factual information to assist with the design process. For the more experienced designer, there will be both new information and challenges to previously held convictions.

6.2.1 The acquisition of the basic engine dimensions

The process of selection of engine size, its bore and stroke and number of cylinders has already been made above on the basis of meeting a performance target. The discussion in Sec. 1.7 also laid down some controlling ground rules in this regard.

Within these ground rules, the designer attempts to optimize this geometry. The designer of the racing engine and the chainsaw, featured extensively in this and earlier chapters, will opt for an "oversquare" design for differing reasons. Both engines are "high-speed" units, so the lower piston speed assists better lubrication and resistance to seizure. The racing design will go to a bore-stroke ratio of about 1.2 on the grounds that any higher value may lead to combustion problems from detonation at the 11 bar bmep levels seen in practice. Very oversquare racing designs have been made and raced, but they are often replaced later with less radical bore-stroke ratios as the unequal struggle with the loss of expansion stroke, high surface-to-volume ratio, etc., becomes evident. The chainsaw designer will go to even higher levels as detonation is less of a problem at 4 bar bmep and the prime requirement is to meet legislation on noise and exhaust emissions and machine vibration.

For the chainsaw engine, the oversquare layout of 48 mm bore and 36 mm stroke is selected because this is a handheld power tool and the minimum mechanical vibration level is not just desirable, it is a legislative requirement in many countries. Mechanical vibration is a function of the out-of-balance forces in an engine and are related to the inertial forces. Where

m_b is the out-of-balance mass of the piston, connecting rod, etc., the maximum value of such forces, F, is given by:

$$F \approx m_b L_{st} \, rpm^2$$

It can be seen that this is apparently a linear function of the engine stroke, L_{st}, and so the lowest possible stroke length is used to minimize the vibration felt by the operator. Thus bore-stroke ratios of up to 1.4 are quite common in the design of handheld power tools, even though the potential bmep attainment of the engine is compromised by the reduced port areas available for any given set of port timings. As an alleviating factor, the highest crankcase compression ratio will always be produced by the largest bore-stroke ratio, and the scavenging of a simple loop-scavenged engine is relatively unaffected by bore-stroke ratio, both of which are vital factors in the design of this type of motor.

Where there are no overriding considerations of the type discussed above, the two-stroke engine designer should opt for a bore-stroke ratio between 1.0 and 1.2. However, for diesel engines, where combustion is totally important, bore-stroke ratios are typically optimized at 0.9 for high-speed engines and can be as low as 0.5 for low-speed marine engines. In the latter special case, the values of bore and stroke are gross by comparison with the chainsaw engine, as the bore is about 900 mm and the stroke is some 1800 mm!

6.2.2 The width criteria for the porting

The designers of high-performance two-stroke engines tend to favor the use of single exhaust and inlet ports, if at all possible. Although a divided port for exhaust or inlet is seen in many designs, it is always a highly stressed feature and the lubrication of the hot, narrow exhaust bridge is never an easy proposition at high piston speeds. Furthermore, as the need for large ports with long timings is self-evident for a high specific output engine, from Fig. 6.16 and the required specific time area criterion, every mm of bore circumference around the bdc position is needed for ports and not port bridges, so that the port opening periods can be kept as short as possible.

Exhaust ports

The designer of the racing engine will opt for a single exhaust port whose width is 75% of the bore dimension, and insert large corner radii to compress the piston ring into its groove before it passes the timing edge on the compression stroke. The net physical effect of the data chosen is seen in Fig. 6.16. The top edge itself can also be radiused to enhance this effect. To further enhance the blowdown effect without compromising the port width criteria, extra ports are added in the blowdown region, as shown in Fig. 5.1(b).

In the case of the chainsaw engine of Fig. 6.2, and other such motors of a more modest bmep and power requirement, where the designer is not under the same pressure for effective port width at the bdc position, the typical criterion for the effective exhaust port width is to employ 55% of the bore dimension. For many such engines, a long and reliable engine life is an important feature of the design, so the ports are made to ensure a smooth passage of the rings over the port timing edges. (See Chapter 8 for further comments on the design of this port for noise considerations [8.14].)

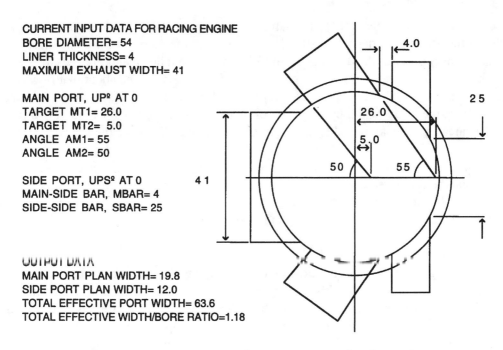

Fig. 6.16 Scavenge port layout for a piston ported racing engine.

Inlet ports

The inlet port width of a piston ported racing engine is deduced in much the same manner, with the port width to bore criterion for a single port being about 65% of the bore dimension, even though the piston rings may not intrude into the port. As all ports have to fit within the bore circumference, it is clearly necessary to distribute that available for the exhaust, transfer and the inlet ports. Often the inlet port width is conveniently made to coincide with the carburetor flow diameter, so the designer allows that dimension to be used, permitting a straight gas flow path for two sides of the intake duct. For a racing engine design with a wide single inlet port, the corner radii of the port need to be very generous around the bdc period by the upper timing edge of that port if a piston ring is not to be trapped, assuming that the rings actually so intrude. As for the exhaust port [8.14], Chapter 8 includes further comment on noise generation from the induction system with respect to this port design.

Transfer ports

The transfer port width for calculation purposes can be decided relatively easily. The geometrical layout can be effected by using Prog.3.4, although the experienced designer will know that it is mechanically possible to pack in a total effective transfer port width of between 1.2 and 1.35 times the bore dimension in most loop-scavenged designs. The designer, experienced or not, will wish to determine the port width criteria in some detail, and for this tedious exercise Prog.3.4 is invaluable.

Fig. 6.16 shows the result of a few moments of time spent at the computer screen for a piston ported racing engine of 54 mm bore and stroke, confirming that the total port width selected at 60 mm is mechanically acceptable and that a potential width of 68 mm is available with a little attention to design detail. Also observe that the rear transfer ducts must curve around the inlet port, for the back bar is only 25 mm wide and the inlet port needs to be some 35 mm across at the widest point to accept a carburetor which will be about that flow diameter. Because the widths of the ports are proportioned in the ratio of 5:3 in favor of the main transfer port, the actual width of the main port is relatively narrow and cannot trap a piston ring; the area can be maximized by inserting small corner radii, hence the selection of the 3 mm corner radii in the sketch drawn as Fig. 6.7. During the operation of Prog.3.4, always bear in mind the advice given in Sec. 3.5.4.1 regarding the orientation of the main transfer ports. It is clear that the use of reed or disc valve induction will free up more bore circumference for exhaust and transfer ports; however, much further complexity may be added to the overall design package, especially if it is a multi-cylinder unit.

Unpegged rings

If the design issue is durability, then the use of unpegged piston rings becomes a vital topic, as discussed in Sec. 3.5.2. To employ unpegged rings the port should subtend no more than 25° at the cylinder center. This is particularly important for diesel engines, for good piston sealing during compression is a vital feature of its operation. Here the blown loop-scavenged engine, as shown in Fig. 3.40, can be readily designed to incorporate the relevant port width criteria. For the spark-ignition engine, the cross-scavenged engine, as shown in Fig. 3.32, is easily designed with unpegged rings. It is more difficult for the simple loop-scavenged engine to incorporate unpegged rings as the ports are normally wider and fewer in number.

6.2.3 The port timing criteria for the engine

Having determined the port widths available for the various ports, the designer will then index Prog.6.3 to calculate the specific time areas of those ports. Naturally, you have the option of writing a personal numerical solution to the theory presented above for the derivation of specific time area. The values for port timings are estimated and simply inserted as data into Prog.6.3, then subsequently modified until the values for the deduced specific time area match those predicted as being required by Prog.6.1, or Figs. 6.3-6.6 to provide the requisite performance characteristics. The program also automatically produces the port height and area information for all ports. This is also helpful, for the inlet port maximum area is a very useful guide to carburetor size, i.e., its flow diameter. If one matches ports in this manner, the carburetor flow area, for an engine with the inlet port controlled by the piston skirt, can be set to between 85% and 95% of the inlet port flow area, as an approximate guide. For engines with induction through reed valves or disc valves, a separate discussion on this topic is found in Secs. 6.3.2 and 6.4.

6.2.4 Empiricism in general

Always remember that empirical design criteria are based on past history and experience. By definition, the answers emanating from empiricism are not always correct and the result of any design decision should be checked by an overall engine modeling program and, ulti-

Chapter 6 - Empirical Assistance for the Designer

mately, the acid test of dynamometer testing of a firing engine. At the same time, it is an unwise move to totally ignore recommendations which are based on the accumulation of past history and experience. In that regard, examination of that which already exists in production always provides a guide to the possible, and a study of the compendia of design data on two-stroke engines from the Technical University of Graz [3.32] can assist in the design process.

6.2.5 The selection of the exhaust system dimensions

The subject of exhaust systems for two-stroke engines is one of considerable complexity, not simply because the unsteady gas dynamics are quite difficult to trace within the human mind, but also because the optimum answer in any given application is arrived at from a combination of the desirable gas dynamics and the necessity of physical constraints caused by the market requirements. A good example is the chainsaw engine discussed as a design example within this book. The conclusion in Chapter 5 is that the compact exhaust system necessary for a handheld tool is a serious limitation on the engine power output. The designer of that unit is well aware of the problem, but is unable to silence the engine adequately within the constraint of the given package volume without impairing performance.

As is evident from the above statements and from the discussion in Chapter 5, the complexity of the pressure wave reflections in a tuned system are such that it is difficult to keep track of all of the possible combinations of the pulsations within the mind. However, as Chapter 5 demonstrates, the computer will carry out that task quite well, but only after the numbers for the lengths and diameters have been inserted into it for subsequent analysis. That puts the problem of selecting those numbers right back into the human domain, which implies some form of basic understanding of the process. This applies equally to the design of the compact chainsaw system as it does to the expansion chamber for the racing engine.

The exhaust system for an untuned engine

The first objective is to allow the creation of the exhaust pulse without restriction, thus the downpipe diameter should be at least as large in area as the total exhaust port effective area. If the system is untuned then there is almost certainly a silencer to be incorporated within it. As exhaust noise is a function of exhaust pulse amplitude and also of the rate of rise of that exhaust pulse, there is little point in having a downpipe from the engine which is the minimum criteria diameter as stated above. Thus the downpipe can be more comfortably sized between 1.2 and 1.4 times the port area. The silencer box, if it has a volume at least 10 times the volume of the cylinder swept volume and does not have an outlet pipe which is unduly restrictive, can be situated at an appropriate distance from the engine so that a suction reflection, of whatever sub-atmospheric amplitude, can assist the engine to dispose of the exhaust gas and promote a stronger flow of fresh charge through the motor. The length of the downpipe from the piston face to the entrance to the silencer box is labeled in Fig. 5.6 as L_1, and employed in mm units for convenience in this empirical calculation. This pipe length can be easily assessed from an empirical knowledge of the following: the total exhaust period in degrees, θ_{ep}; the probable mean exhaust gas temperature in °C, T_{exc}; the engine speed of rotation in rev/min, rpm.

Design and Simulation of Two-Stroke Engines

The front of all pressure waves travels at the local speed of sound, a_0, which from Eq. 2.1.1 is:

$$a_0 = \sqrt{\gamma R(T_{exc} + 273)} \quad \text{m/s} \tag{2.1.1}$$

For engines below 5 bar bmep the value of T_{exc} will be in the region of 350°-450°C, and R and γ can be estimated to be as for air at atmospheric conditions, i.e., 287 J/kgK and 1.4, respectively.

Although the exhaust pulse peak will propagate at supersonic velocity, the suction reflection will travel at subsonic speed and the average can be approximated to the local speed of sound (see Sec. 2.1.3). If you assume that the exhaust pulse peak occurs at 15° after exhaust port opening, you would wish the peak of any suction reflection to return at or about the bdc piston position. This gives a reflection period in degrees crankshaft of θ_{rp}:

$$\theta_{rp} = \frac{\theta_{ep}}{2} - 15 \tag{6.2.1}$$

The time, t_{rp}, taken for this double length travel along the pipe, L_1, must equal the reflection period, where:

$$t_{rp} = \frac{\theta_{rp}}{360} \times \frac{60}{\text{rpm}} = \frac{2L_1}{1000 a_0} \tag{6.2.2}$$

Incorporating Eq. 6.2.1 and rearranging Eq. 6.2.2 to obtain the pipe length, L_1, which is in mm units:

$$L_1 = \frac{41.66 a_0 (\theta_{ep} - 30)}{\text{rpm}} \tag{6.2.3}$$

If the appropriate numbers are inserted into this equation for a chainsaw at 8000 rpm, with an assumed exhaust gas mean temperature of 400°C, and an exhaust period of 160° crankshaft, then L_1 is derived as 352 mm. As there is clearly no possibility of installing a 352-mm-long downpipe on a compact chainsaw, the designer is left to dimension a silencer box which will merely have the least possible restrictive effect on gas flow from the engine, while still silencing it to the legislative requirement. This effect is abundantly evident from the discussion in Chapter 5. Fortunately the two aims are not mutually incompatible, as the largest possible box can be ingeniously designed into the available space so as to give the best power output and the best silencing effect. See Chapter 8 for an expanded discussion on noise and silencing.

The basic rule of thumb for simple exhaust silencer boxes, be they a single box, or the double box design as seen in Fig. 5.6, is to set the total box volume to at least ten swept volumes and the final outlet pipe to about 50% of the area of the exhaust port.

Chapter 6 - Empirical Assistance for the Designer

For engines not so constrained by package volume, such as a generator set, or even a brush cutter if one uses the handle in an innovative fashion, the designer can obtain some power advantage by using Eq. 6.2.3.

The exhaust system of the high-performance engine

Having been through the fundamentals of unsteady gas flow in Chapter 2 and the results of computer engine modeling of high-performance engines in Chapter 5, you will find this empirical study of flow in an expansion chamber a lighter mathematical affair. Useful references for further perusal are the technical papers by Naitoh *et al.* [6.4, 6.5], Cartwright [4.35], or Blair [2.18]. As stated in the introduction, it is the designer who has to provide the numbers for the physical dimensions in Fig. 5.7 and so the fundamental understanding gained in Chapters 2 and 5 must be translated into the ability to deduce appropriate data for the calculation. To aid with the empirical design of an expansion chamber, a computer program, Prog.6.2v2, is included in the Appendix Listing of Computer Programs.

Fig. 6.17 shows the basic action taking place in an optimized system, such as in the racing engine at 11,740 rpm in Fig. 5.34. The sketch is drawn with time on the vertical axis and distance on the horizontal axis. The port timings appropriate to that time scale for that particular speed are marked on the time axis as EO, TO, BDC, TC and EC, for the porting events from exhaust and transfer opening through bdc to transfer and exhaust closing, respectively. The important factor in this discussion is that the scale on the vertical axis is time, not crank-

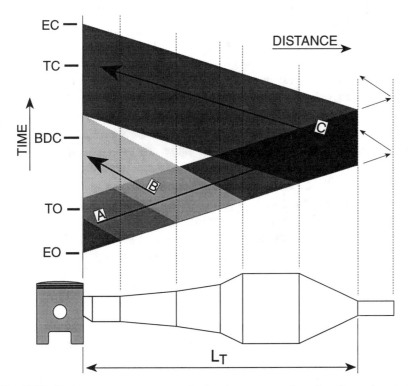

Fig. 6.17 Optimum pressure wave phasing in an expansion chamber exhaust pipe.

shaft angle. The exhaust pulse (A) is seen to leave the exhaust port and to propagate rightward toward the atmosphere, primarily sending suction reflections (B) back toward the engine from the diffuser, and compression wave reflections (C) from the tail nozzle. The density of shading in Fig. 6.17 is intended to represent superposition pressure, where the more dense the shading, the higher the pressure. The exhaust pulse will propagate in a parallel pipe with no reflections, but on arrival at a change in section area, be it gradual or abrupt, reflections will occur in the manner described in Chapter 2. As debated in Chapter 5, in the optimized design this primary reflection behavior ensures that (i) the peak of the suction reflection returns to the exhaust port around the bdc period to aid the scavenging and induction process and (ii) the plugging pulse returns to the exhaust port in time to hold in the fresh cylinder charge, and enhance the charging efficiency, before the exhaust port actually closes. The exhaust pressure-time history of the racing engine in Fig. 5.34, aiding both the scavenging and charging processes, can be regarded as a classic example of that statement.

This picture of time and distance is one which will remain ostensibly similar at all engine speeds, therefore it is the crankshaft timing points which move on the vertical axis as a linear function of the engine rotation rate. At low engine speeds, the time span from EO to EC will increase, moving the marker for EC and the others "farther up" the time axis allowing the plugging pulse to interfere with the scavenging process and contribute nothing to the trapping of charge in the cylinder as the port is being closed by the piston. At high engine speeds beyond that for an optimized behavior, the time span from EO to EC will be shortened. The result at a high engine speed beyond the optimum is that the plugging pulse appears at the exhaust port after it is already closed so that the trapping process receives no benefit from its arrival.

The behavior of tuning as related to engine speed is illustrated by further information from the simulation of the same racing engine utilized in Chapter 5. Figs. 6.18 to 6.20 show the cylinder and exhaust pipe pressures at the exhaust port for engine speeds of 9600, 11,200 and 12,300 rpm, respectively. They represent the bottom, the middle and the top end of the speed range where high torque is available. In Fig. 6.18 at 9600 rpm, the "plugging" pulse can be seen to arrive just in time to give effective trapping, but some spillage back into the exhaust pipe is observed by the falling cylinder pressure just prior to trapping at exhaust closure. The presence of a suction pulse at exhaust closure further exacerbates the spillage action and, at all lower speeds, will do so with increasing significance. By 11,200 rpm, in Fig. 6.19, the trapping by exhaust tuning shows only a hint of that spillage. Finally, at the top end of the speed range, in Fig. 6.20 at 12,300 rpm, then trapping is "perfect," but the delivery ratio has dropped and friction loss has increased, so the power curve seen in Fig. 5.28 is beginning to fall. Nevertheless, the effective speed range at a high bmep is nearly 3000 rpm. The extension of good torque to the highest speed point is aided by a retarded ignition timing giving higher exhaust temperatures and assisting with some "re-tuning" of the speed of pressure wave propagation within the exhaust pipe.

One of the methods available to extend that band of speed downward, over which the plugging pulse would provide a significant gas trapping action, is for the pipe to remain the same length dimension physically but for the exhaust opening period, θ_{ep}, to be reduced at low engine speeds by a valving mechanism. Such a mechanism to control the exhaust port timing and duration is described in Secs. 5.2.2 and 7.3, and shown in Fig. 5.2. The effect on

Chapter 6 - Empirical Assistance for the Designer

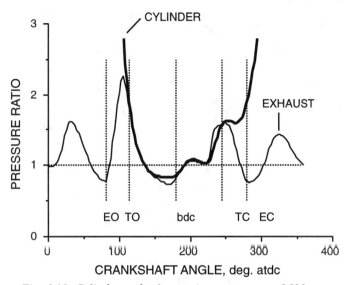

Fig. 6.18 Cylinder and exhaust pipe pressures at 9600 rpm.

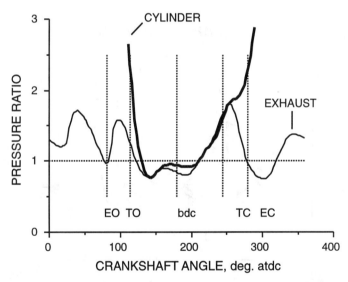

Fig. 6.19 Cylinder and exhaust pipe pressures at 11,200 rpm.

power and torque has already been discussed for a racing engine in Sec. 5.5.2 and shown graphically in Fig. 5.31. The optimized position of the exhaust timing control valve is found by simulation to open at 100° atdc for an engine speed of 8500 rpm. Fig. 6.21 shows the cylinder and exhaust pipe pressures at the exhaust port at 8500 rpm. The narrower exhaust period now relegates the suction wave at the exhaust port at the trapping point to be in the same relative position as it is at 9600 rpm, in Fig. 6.18. The result is effective trapping and a near equality of bmep attainment at 8500 and 9600 rpm.

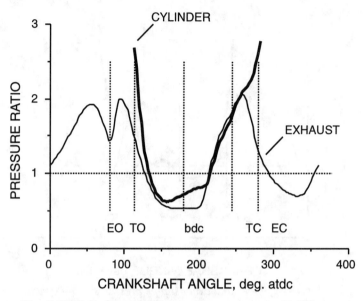

Fig. 6.20 Cylinder and exhaust pipe pressures at 12,300 rpm.

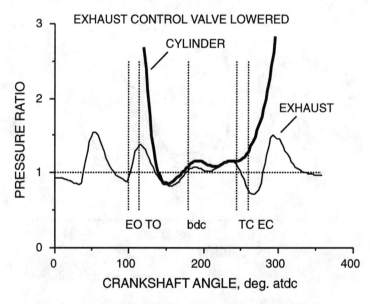

Fig. 6.21 Cylinder and exhaust pipe pressures at 8500 rpm.

An alternative design approach is to be able to vary the tuned length, L_T, as a function of engine speed and this has been done in some go-kart racing machines with trombone-like sliding sections within a tuned exhaust pipe. Yet another method is to inject water into the exhaust pipe at the lower engine speeds of the power band to lower its temperature [6.7]. All of these approaches to broadening the effective power band of the engine rely on some linear relationship between the exhaust period, θ_{ep}, the tuned length, L_T, and the reference speed for pressure wave propagation, a_0. That this is the case, empirically speaking only, is seen from Eq. 6.2.3 and Fig. 6.17.

The principal aim of the empirical design process is to phase the plugging reflection correctly for the engine speed desired for peak power, which means calculating the tuning length from piston face to tail-pipe entry, L_T. The next important criterion is to proportion the tail-pipe exit area as a function of the exhaust port area so that the pipe will empty in a satisfactory manner before the arrival of the next exhaust pulse. The third factor is to locate the end of the diffuser and the beginning of the tail nozzle so that the spread of the suction and plugging effects are correctly phased. The fourth function is to locate the second half of the diffuser so that the reverse reflection of the plugging pulse from it, and the primary reflection of it from the exhaust port, can recombine with the next outgoing exhaust pulse, thereby producing the resonance effect demonstrated in Fig. 5.32 or Fig. 6.20. This will happen only in racing engines with very long exhaust port periods, typically in excess of 190°.

A multi-stage diffuser is an efficient method of reflecting the exhaust pulse and the shallower first stages reduce the potential for energy losses from shock formation at the entrance to that diffuser. The shocks form as high-amplitude compression exhaust waves meet strong suction waves which are progressing back to the engine from the diffuser. As seen in Sec. 2.15, this raises the local gas particle velocity, and in the diffuser entrance area this can attempt to exceed a Mach number of unity. The result is a shock which effectively "destroys" the wave energy and, although the energy reappears as heat, this does nothing for the retention of the strength of the all-important pressure wave action. The slower rate of area expansion of the first stage of the diffuser (see also Sec. 2.15) smears the reflection process more widely and efficiently over the entire length of the diffuser.

The entirety of the empirical design process is founded in very simple equations which have been tried and tested for many years in the theoretical and experimental development of expansion chambers for racing engines at QUB. As remarked earlier, remember that this is an empirical approach, not a precise calculation, and is the starting point for optimization by simulation, as seen in Chapter 5, preferably in combination with the experimental testing of the firing engine [4.35].

As this is an empirical calculation for the lengths and diameters shown in Fig. 5.7, and are conventionally discussed in mm units, the data are inserted and produced by the calculation in those units. The data required for the calculation are in blocks, the first of which is: engine bore, stroke, connecting rod length, exhaust port total opening period, θ_{ep}, in units of crankshaft degrees, number of exhaust ports, maximum width of each exhaust port, and the radii in the corners of each exhaust port. If the exhaust port shape is complex, as in Fig. 5.1(b), some preliminary arithmetic must be carried out prior to computation to find the "mean" exhaust port width, so that the data are inserted accurately into the calculations in the above format.

The above set of data permits the calculation of the exact exhaust port area. Incidentally, it is presumed that the designer has followed some design process to this point and the data being used are matched to the required performance of the engine. With this knowledge of the exhaust port flow area, equivalent to a diameter, d_0, the empirical calculation describes this as the ideal exhaust pipe initial diameter. In a racing pipe, where maximizing the plugging pressure gives the highest trapping efficiency, there seems little point in having the first downpipe diameter, d_1, much more than 1.05 times larger than d_0, for in unsteady gas flow, the larger the pipe, the lower the pressure wave amplitude for the same mass rate of gas flow (see Sec. 2.1.4). Knowing the ideal value of d_1, the calculation ascribes values for the tail-pipe diameter so that the pipe will provide the requisite plugging reflection, yet still empty the system during each cycle. Using the geometrical nomenclature of Fig. 5.7, depending on the expected bmep level of the engine, the tail-pipe diameter, d_7, is found as,

for a (track racing) engine at about 11 bar bmep: $\quad d_7 = 0.6 d_0$

for a (motocross) engine at about 9 bar bmep: $\quad d_7 = 0.65 d_0$

for an (enduro) engine at about 8 bar bmep: $\quad d_7 = 0.7 d_0$

The latter will have a comprehensive silencer following the tuned section. Motocross and road-racing engines have rudimentary silencers, and the design philosophy being presented is not at all affected by the attachment of a simple absorption silencer section in conjunction with the tail-pipe. That, as with all such debate on silencers, is to be discussed in Chapter 8.

The next set of data is required to calculate lengths, all of which will be proportioned from the tuned length, L_T, to be calculated: engine speed for peak power, rpm; exhaust gas temperature, T_{exc}, as estimated in the pipe mid-section, in °C.

For those who may not have immediate access to temperature data for expansion chambers, the potential values have a loose relationship with bmep. Therefore, if you expect to have the design approach a Grand Prix racing engine power level, then an exhaust temperature of 600°C can be anticipated. If the engine is a lesser rated device, such as an enduro motorcycle, then a temperature of 500°C could be safely used in the calculation. As the speed of sound is the real criterion, then as it is proportional to the square root of the Kelvin temperature, the selection of exhaust pipe temperature is not quite as critical as it might appear to be. The speed of sound, a_0, is given by Eq. 2.1.1:

$$a_0 = \sqrt{\gamma R (T_{exc} + 273)} \quad \text{m/s} \tag{2.1.1}$$

The calculation of L_T is based on the return of the plugging pulse within the exhaust period. The most effective calculation criterion, based on many years of empirical observation, is the simplest. The reflection period is decreed to be the exhaust period and the speed of wave propagation is the local speed of sound in the pipe mid-section. Much more complex solutions have been used from time to time for this empirical calculation, but this approach gives the most reliable answers in terms of experimental verification from engine testing.

As in Eq. 6.2.2, the reflection time, t_{rp}, taken for this double length travel along the tuned length, L_T, must equate to the reflection period which, in this particular case, has been found to equal the exhaust period:

$$t_{rp} = \frac{\theta_{rp}}{360} \times \frac{60}{\text{rpm}} = \frac{\theta_{ep}}{6\,\text{rpm}} = \frac{2 L_T}{1000 a_0}$$

Chapter 6 - Empirical Assistance for the Designer

Rearranging for the pipe length, L_T, which is in mm units:

$$L_T = \frac{83.3 a_0 \theta_{ep}}{rpm} \qquad (6.2.4)$$

The remainder of the empiricism presented is straightforward, following the nomenclature of Fig. 5.7. The end of the diffuser is positioned as in Fig. 6.17, about two-thirds of the distance between the exhaust port and the tail-pipe entry. The pipe segment lengths are proportioned as follows:

first pipe and diffuser:

$$L_1 = 0.11 L_T \quad L_2 = 0.275 L_T \quad L_3 = 0.183 L_T \quad L_4 = 0.092 L_T \qquad (6.2.5)$$

mid-section, rear cone and tail-pipe:

$$L_5 = 0.11 L_T \quad L_6 = 0.24 L_T \quad L_7 = L_6 \qquad (6.2.6)$$

For the diameters the following advice is offered. First, for the initial pipe diameter, d_1:

$$d_1 = k_1 d_0 \qquad (6.2.7)$$

The value for the constant, k_1, ranges from 1.125 for broadly tuned, enduro type engines to 1.05 for road-racing engines of the highest specific power output.

$$d_4 = k_2 d_0 \qquad (6.2.8)$$

The value for the constant, k_2, ranges from 2.125 for broadly tuned, enduro type engines to 3.25 for road-racing engines of the highest specific power output. The normal design criterion is for the value of d_5 to be the same as d_4, i.e., the center section is of constant area.

The remaining diameters to be determined are that for d_2 and d_3 and these are calculated by the following expressions:

$$d_2 = d_1 e^{x_{12}} \quad \text{and} \quad d_3 = d_1 e^{x_{13}} \qquad (6.2.9)$$

where

$$x_{12} = \left(\frac{L_2}{L_2 + L_3 + L_4}\right)^{k_h} \times \log_e\left(\frac{d_4}{d_1}\right) \qquad (6.2.10)$$

and

$$x_{13} = \left(\frac{L_2 + L_3}{L_2 + L_3 + L_4}\right)^{k_h} \times \log_e\left(\frac{d_4}{d_1}\right) \qquad (6.2.11)$$

and

$$1.25 < k_h < 2.0 \qquad (6.2.12)$$

Design and Simulation of Two-Stroke Engines

The diffuser is designed as a "horn" and the exponent, k_h, is called the "horn coefficient."

It is self-evident from any of the discussion on gas flow, that all diameters are the internal diameters of the exhaust pipe, and that all lengths are distances along the centerline of a pipe which may be curved for some segments.

Some years ago, I presented a paper concerning the empirical design of expansion chambers [2.18]. It is instructive to compare the above discussion with that paper and to note that the empiricism has actually become simpler with the passage of both time and experience.

For those who wish to have these calculations performed automatically, the Appendix Listing of Computer Programs includes Prog.6.2v2, EXPANSION CHAMBER. This carries out the useful dual function of deducing all of the above dimensions and data and the computer screen and line printer will show a dimensioned sketch of the exhaust system. An example of that is shown for a prediction of an expansion chamber for the 125 racing motorcycle engine used as one of the working examples in Chapter 5. The screen/line-print output is shown in Fig. 6.22. If the data predicted by the empirical design program are compared with that given in Sec. 5.5.2, a considerable level of close correspondence will be observed. Recall that the engine modeling program predicted in Fig. 5.28 a high specific power performance for an engine with a very similar exhaust system. This should give you some confidence that this simple program, Prog.6.2v2, albeit overlaid with some considerable practical experience, enables the rapid and efficient preparation of the geometry of an expansion chamber for further complete analysis by an engine simulation. Prog.6.2v2 permits you to vary the diffuser profile by changing the horn coefficient. In Fig. 6.22, the values of d_2 and d_3 are displayed as 66 and 99 mm, respectively, for a horn coefficient of 1.5. If the horn coefficient

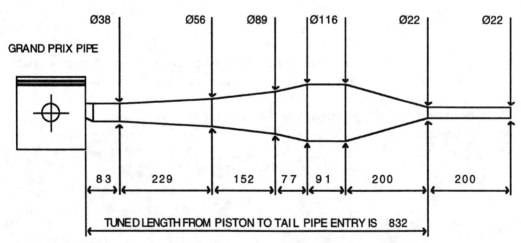

Fig. 6.22 Expansion chamber design for a racing engine using Prog.6.2v2.

Chapter 6 - Empirical Assistance for the Designer

is changed to 1.25 and 2, i.e., the limit values in Eq. 6.2.12, then the values of d_2 and d_3 become 61 and 92 mm, and 50 and 82 mm, respectively. The higher the value of the horn coefficient, then the more "extreme" becomes the area profile of the diffuser with respect to length along it.

Another useful feature of Prog.6.2v2 is that there is available a second page of line-print output. This gives the manufacturing data for the various cones of the expansion chamber so that they may be marked out, cut out, and rolled from flat sheet metal, then subsequently seam welded into the several cones. The cones and parallel sections are butt welded to form a pipe for either competition or dynamometer testing. An example of the actual line-printer output of the manufacturing data for the racing engine design of Fig. 6.22 is shown in Fig. 6.23. In that figure, RX and RY are the radii from the center of a circle marked on the metal sheet, and φ is the angle of the flat sheet segment. This information alone is quite sufficient to permit the complete marking of the sheet metal. The extra information for the chords, CHX and CHY, allows an alternative method of marking out the flat sheet metal segment, or provides a cross-check on the accuracy of setting out the angle φ.

6.2.6 Concluding remarks on data selection

From the methods discussed above, information is selected for a data bank for the main engine geometry and porting. Those who pursue this empirical approach by using the several computer packages for the creation of the data bank necessary for the operation of an engine

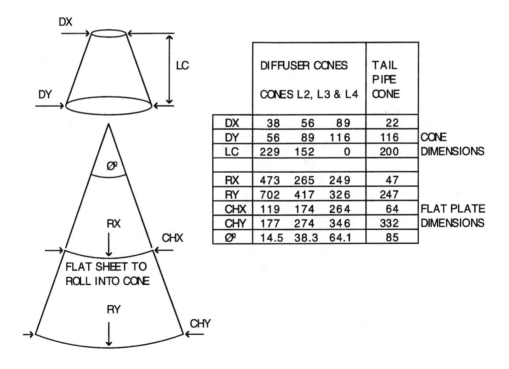

	DIFFUSER CONES CONES L2, L3 & L4			TAIL PIPE CONE	
DX	38	56	89	22	
DY	56	89	116	116	CONE
LC	229	152	0	200	DIMENSIONS
RX	473	265	249	47	
RY	702	417	326	247	
CHX	119	174	264	64	FLAT PLATE
CHY	177	274	346	332	DIMENSIONS
φ	14.5	38.3	64.1	85	

Fig. 6.23 Manufacturing data for the expansion chamber given by Prog.6.2v2.

modeling program will discover that with a little practice the data bank can be assembled in a remarkably short time. This approach reduces the number of times the engine simulations need to be run to create an optimized design to meet the perceived requirement for the performance characteristics. As the engine modeling programs take the longest time to run on the computer, indexing in the first instance the virtually instantaneous response of the empirical design programs is an effective use of the designer's time. Perhaps more important, it tends to produce porting and exhaust systems which are at least matched in the first instance. An example of that is seen in the above discussion of the design for the chainsaw, where the exhaust and transfer ports could have been better matched by an initial application of the empirical approach.

There will be the natural tendency, about which you were cautioned earlier, to regard the empirical deductions as a final design for either the porting or the exhaust system. That warning is repeated here, for I made that mistake sufficiently often to realize its validity!

6.3 Empirical design of reed valves for two-stroke engines
The alternatives, reed valves and disc valves

Sec. 1.3 introduces the use of mechanical devices which permit asymmetrical timing of the exhaust or induction process. The three types discussed are poppet valves, disc valves and reed valves. Needless to add, the inventors of this world have produced other ingenious devices for the same purpose, but the three listed have withstood the test of time and application. It is proposed to discuss the design of disc valves and reed valves in this section, but not poppet valves, for they are covered in the literature of four-stroke cycle engines to an extent which would make repetition here just that. As the design of reed valves and disc valves is not covered extensively in the literature, it is more logical that the appropriate space within this book should be devoted to the "unknown." Some preliminary discussion and illustration of the use of reed valves, and the geometry of disc valves, has been given in Chapter 5.

The use of reed valves for the induction system has always been common for outboard motors and increasingly so for motorcycle and other types of two-stroke engines. The incorporation of a reed valve into an engine is shown in Fig. 1.7(b) and the details of a reed block assembly are sketched in Fig. 5.4, with a photograph giving further illustration in Plate 6.1. The technology of the design has improved greatly in recent times with the use of new materials and theoretical design procedures. The new materials, such as plastics reinforced by either glass-fiber or carbon-fiber, are effective replacements for the conventional use of spring steel for the reed petal. In particular, any failure of a plastic reed petal in service does not damage the engine internally, whereas it would in the case of a steel reed.

The use of disc valves for the induction system has been confined mostly to high-performance racing engines although there are some production examples of lesser specific power output. The initial discussion in Chapter 1 gives a sketch of the installation in Fig. 1.7(a) and Plate 1.8. A more detailed drawing of the disc and the inlet port which it uncovers is shown in Fig. 5.5. There is not a large body of technical information available on the design and development of disc valves, but the papers by Naitoh *et al.* [6.4, 6.5] and the book by Bossaglia [6.3] contain valuable insights to the design and development process.

The incorporation of these induction systems into an engine modeling program is straightforward in the case of the disc valve and complex for a reed valve design. That it has been

Chapter 6 - Empirical Assistance for the Designer

Plate 6.1 A reed valve block with six steel reeds and stop-plate. The block is rubber coated to reduce both noise and damage to bouncing reeds.

successfully achieved for the reed valve case is evident from the technical papers presented [1.13, 4.10, 5.9, 5.16-5.19]. Actually there is a considerable body of literature on reed valves when used in air and refrigeration compressors as automatic valves and the technical paper by MacClaren [6.6] will open the door to further references from that source. The modeling of the disc valve case within an engine design program is regarded as requiring a simple extension to the simulation to accommodate the inlet port area geometry as uncovered by the disc. The discussion in Sec. 6.4 clarifies this statement.

However, just as the parameters for the geometry of the piston-controlled porting of the engine have to be assembled in some logical manner, it is necessary to be able to arrive at an initial decision on the dimensions of the reed or disc valve in advance of either the use of a computer modeling package or the experimental testing of prototype devices. The following discussion, and that in Sec. 6.4 for disc valves, should be of practical assistance within that context.

6.3.1 The empirical design of reed valve induction systems

The reed valve induction system is installed in the engine between the atmosphere and the crankcase of the engine, as shown in Figs. 1.7(b) and 5.4. If the carburetor is the fueling device, it is placed between the reed valve and the atmosphere. If low-pressure fuel injection is used, the injector nozzle can be situated in the same position as a carburetor, but it is also possible to inject the fuel directly into the crankcase. A photograph of a cylinder of a motor-cycle engine with a reed induction system appears in Plates 1.9 and 4.1 and a closer view of the block, petals and stop-plate is shown in Plate 6.1. The typical opening and closing characteristics of this automatic valve are illustrated in Fig. 1.8(c). To expand this information from

Design and Simulation of Two-Stroke Engines

that calculated by a simulation of a racing engine in Fig. 5.35, and to illustrate the point made above regarding the incorporation of a reed valve model into an engine simulation program, a few of the results from the experimental and theoretical paper by Fleck *et al.* [1.13] are presented here as Figs. 6.24-6.26. The engine used as the research tool in this paper is the high-performance YPVS RD350LC twin-cylinder Yamaha motorcycle engine, with a peak bmep of nearly 8 bar at 9000 rpm. Each cylinder has a block holding four reed petals and ports. In this case, designated as RV1, they are steel reed petals of 0.20 mm thickness, i.e., from Fig. 5.4 the dimension x_t is 0.20 mm.

The upper portion of Fig. 6.24 shows the crankcase and inlet port pressure at a "low" engine speed of 5430 rpm. Those pressure diagrams are predicted by the engine simulation program and you can see that they are very similar in profile to those observed for the piston ported engine reported in Chapter 5. In short, there is nothing particularly unusual about the pressure difference across the reed valve by comparison with that observed for a piston-controlled induction system. The solid line in the lower half of that figure shows the measured reed tip lift in mm, and that predicted by the reed valve simulation within the computer program is the dashed line. The close correspondence between the calculation and measure-

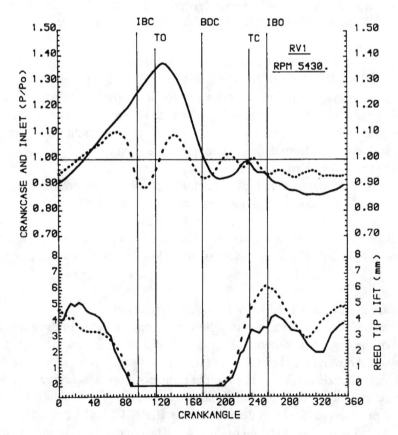

Fig. 6.24 The crankcase and inlet port pressure and reed lift behavior at a low engine speed.

Chapter 6 - Empirical Assistance for the Designer

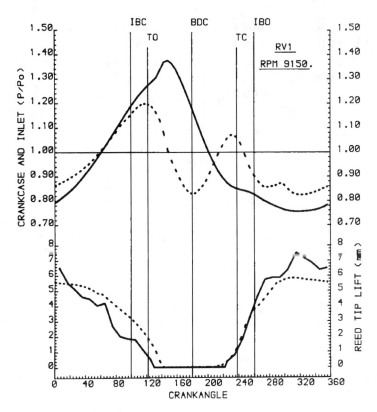

Fig. 6.25 The crankcase and inlet port pressure and reed lift behavior at a high engine speed.

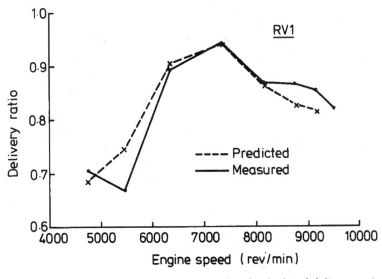

Fig. 6.26 The comparison between measured and calculated delivery ratio from an engine model incorporating a reed valve simulation.

ment of reed lift is evident, as is the resulting calculation and measurement of delivery ratio over the entire speed range of the engine, to be seen in Fig. 6.26. The timing of opening and closing of the reed petal at the low speed of 5430 rpm shows the reed opening at 160° btdc and closing at 86° atdc, which is an asymmetrical characteristic about tdc.

At the "high" engine speed of 9150 rpm, the reed petal opens at 140° btdc and closes at 122° atdc. This confirms the initial view, expressed in Sec. 1.3.4 and Figs. 18(c) and (d), that the reed petal times the inlet flow behavior like a disc valve at low engine speeds and as a piston-controlled intake port at high engine speeds. In other words, it is asymmetrically timed at low speeds and symmetrically timed at high speeds.

Within this same paper by Fleck *et al.* [1.13] there is given a considerable body of experimental and theoretical evidence on reed valve characteristics for steel, glass-fiber- and carbon-fiber-reinforced composites when used as reed petal materials. The paper [1.13] gives the dimensions of the engine with its tuned expansion chamber exhaust system and associated measured power data, which you will find instructive as another working example for the comparison of Prog.6.2v2 with experiment.

6.3.2 *The use of specific time area information in reed valve design*

From the experimental and theoretical work at QUB on the behavior of reed valves, it has been found possible to model in a satisfactory fashion the reed valve in conjunction with an engine modeling program. This implies that all of the data listed as parameters for the reed valve block and petal in Fig. 5.4 have to be assembled as input data to run an engine simulation. The input geometrical data set for a reed valve is even more extensive than that for a piston-controlled intake port, adding to the complexity of the data selection task by the designer before running an engine model. This places further emphasis on the use of some empirical design approach to obtain a first estimate of the design parameters for the reed block and petals, before the insertion of that data set into an engine modeling calculation. I propose to pass on to you my data selection experience in the form of an empirical design for reed valves as a pre-modeling exercise.

In Sec. 6.1 there is a discussion on specific time area and its relevance for exhaust, transfer and intake systems. The flow through a reed valve has to conform to the same logical approach. In particular, the value of specific time area for the reed petal and reed port during its period of opening must provide that same numerical value if the flow of air through that aperture is to be sufficient to provide equality of delivery ratio with a piston-controlled intake port. The aperture through the reed valve assembly is seen in Fig. 5.4 and, from the discussion in Sec. 5.2, is to be composed of two segments: the effective reed port area in the flow direction as if the reed petal is not present; and the effective flow area past the reed petal when it has lifted to its maximum, caused either by the gas flow or as permitted by the stop-plate. Clearly, there is little point in not having these two areas matched. For example, if the design incorporates a large reed port area but a stiff reed which will barely lift under the pressure differential from the intake side to the crankcase, very little fresh charge will enter the engine. Equally, the design could have a large flexible reed which lifted easily but exposed only a small reed port, in which case that too would produce an inadequate delivery ratio characteristic. In a matched design, the effective reed port area should be larger than the effective flow

area past the petal, but not by a gross margin. Therefore, the empirical design process is made up of the following elements:
  (i)   Ensure that the effective reed port area has the requisite specific time area, on the assumption that the reed petal will lift at an estimated rate for an estimated period.
  (ii) Ensure that the reed will lift to an appropriate level based on its stiffness characteristics and the forcing pressure ratio from the crankcase.
  (iii) Ensure that the natural frequency of vibration of the reed petal is not within the operating speed range of the engine, thereby causing interference with criterion (ii) or mechanical damage to the reed petal by the inevitable fatigue failure.

The data required for such a calculation are composed of the data sketched in Fig. 5.4 and for the physical properties of Young's Modulus and density of the reed petal. Of general interest, Fleck *et al.* [1.13] report the physical properties of both composites and steel when employed as reed petal materials, as recorded in a three-point bending test. They show that a glass-fiber-reinforced composite material has a Young's Modulus, Y, of 21.5 GN/m^2 and a density of 1850 kg/m^3. The equivalent data for carbon-fiber-reinforced composite and steel is measured in the same manner and by the same apparatus. The value of Young's Modulus for steel is 207 GN/m^2 and its density is 7800 kg/m^3. The value of Young's Modulus for a carbon-fiber-reinforced composite material is 20.8 GN/m^2 and its density is 1380 kg/m^3. Within the paper there are more extensive descriptions of the specifications of the glass-fiber and carbon-fiber composite materials actually used as the reed petals.

The opening assumption in the calculation is that the specific time area required is the same as that targeted in Eq. 6.1.9 as A_{svi} related to bmep.

$$A_{svi} = \frac{bmep + 1.528}{774}$$

The theoretical relationship for specific time area for inflow, Eq. 6.1.14, is repeated below:

$$\text{Specific time area, } A_{svi}, \text{ s/m} = \frac{\int_{\theta=0}^{\theta=\theta_p} A d\theta}{6 V_{sv} \text{rpm}}$$

A standard opening period, θ_p, of 200° crankshaft is assumed with a lift-to-crank angle relationship of isosceles triangle form. The maximum lift of the reed is discussed below so that the area, A, may be calculated. The time for each degree of crank rotation is given by $dt/d\theta$. From Eq. 6.1.14, the $Ad\theta$ term is evaluated with swept volume, V_{sv}, inserted as m^3, and if the area, A, is in m^2 units, the units for the time-area, A_{svi}, remain as s/m.

$$\int A d\theta = 6 A_{svi} V_{sv} \text{rpm} \quad \text{m}^2\text{deg} \tag{6.3.1}$$

Design and Simulation of Two-Stroke Engines

The required reed flow area past the petals, A_{rd}, is given by determining the area of an isosceles triangle of opening which spans an assumed period, θ_p, of 200°:

$$A_{rd} = \frac{2\int A d\theta}{\theta_p} = \frac{12 A_{svi} V_{sv} \text{rpm}}{\theta_p} \quad m^2 \qquad (6.3.2)$$

At this point the designer has to estimate values for the reed port dimensions and determine a flow area which will match that required from the time-area analysis. That step is to assign numbers to the reed port dimensions shown in Fig. 5.4, and is found from Eq. 5.2.18. The notation for the data is also given in Fig. 6.27. The effective reed port area is declared as A_{rp}:

$$A_{rp} = n_{rp}\left(L_p x_p - r_p^2(4 - \pi)\right)\sin\phi_{rb} \qquad (6.3.3)$$

The required reed flow area, A_{rp}, and the port area, A_{rd}, are compared and the port dimensions adjusted until the two values match from Eqs. 6.3.2 and 6.3.3. If anything, you should always err slightly on the generous side in apportioning reed port area. Without carrying out this form of design calculation, one tends to err on the restrictive side because the eye, viewing it on a drawing board or a CAD screen, tends to see the projected plan area and not the all-important effective area in the flow direction.

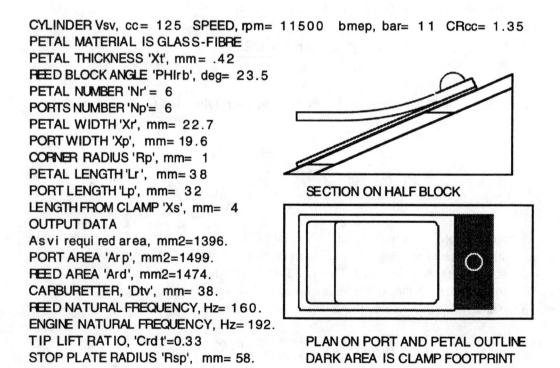

```
CYLINDER Vsv, cc= 125  SPEED, rpm= 11500  bmep, bar= 11  CRcc= 1.35
PETAL MATERIAL IS GLASS-FIBRE
PETAL THICKNESS 'Xt', mm= .42
REED BLOCK ANGLE 'PHIrb', deg= 23.5
PETAL NUMBER 'Nr' = 6
PORTS NUMBER 'Np'= 6
PETAL WIDTH 'Xr', mm= 22.7
PORT WIDTH 'Xp', mm= 19.6
CORNER RADIUS 'Rp', mm= 1
PETAL LENGTH 'Lr', mm= 38
PORT LENGTH 'Lp', mm= 32
LENGTH FROM CLAMP 'Xs', mm= 4
OUTPUT DATA
Asvi required area, mm2=1396.
PORT AREA 'Arp', mm2=1499.
REED AREA 'Ard', mm2=1474.
CARBURETTER, 'Dtv', mm= 38.
REED NATURAL FREQUENCY, Hz= 160.
ENGINE NATURAL FREQUENCY, Hz= 192.
TIP LIFT RATIO, 'Crd t'=0.33
STOP PLATE RADIUS 'Rsp', mm= 58.
```

SECTION ON HALF BLOCK

PLAN ON PORT AND PETAL OUTLINE
DARK AREA IS CLAMP FOOTPRINT

Fig. 6.27 Computer screen output from Prog.6.4, REED VALVE DESIGN.

The carburetor flow area can be estimated from this required area in the same manner as is effected for a piston-controlled intake port, although with a larger restriction factor, C_c, from the reed port area to the carburetor flow area. This is to create a greater pressure differential across the reed, assisting it to respond more rapidly to the fluctuating crankcase pressure. The carburetor flow diameter is notated as d_{tv} in Fig. 5.5, and is deduced by:

$$d_{tv} = \sqrt{C_c \frac{4A_{rd}}{\pi}} \qquad (6.3.4)$$

where the restriction factor, C_c, varies from application to application. The value is bmep dependent, so the value of C_c will range from 0.65 to 0.85 for engines with bmep aspirations from 4 bar to 11 bar, respectively.

The next requirement is for calculations to satisfy the vibration and amplitude criteria (ii) and (iii) above. These are connected to theory found in texts on "vibrations" or on "strength of materials" such as that by Morrison and Crossland [6.1] or Tse et al. [6.2], and vibration has already been discussed in Sec. 5.2.2 and theory given in Eq. 5.2.15. The natural frequency of the forcing function on the reed is created by the pressure loading across the reed from intake pipe to crankcase, and the first fundamental frequency is equal to the engine speed in cycles per second, i.e., rps. The first-order natural frequency is the one to which some attention should be paid, for it is my experience that it tends to be quite close to the maximum speed of engine operation of a well-matched design. As long as the reed natural frequency is 20% higher than the engine natural frequency, there is little danger of accelerated reed petal damage through fatigue.

The final step is to ensure that the reed petal will lift sufficiently during the pumping action of the crankcase. This is precisely what the engine modeling program and reed simulation in Chapter 5, or of Fleck et al. [1.13], is formulated to predict, and provides in Figs. 6.24-6.26. Therefore, an empirical approach is aimed at getting a first estimate of that lift behavior. Fleck et al. [1.13] show reed tip lift to length ratios, C_{rdt}, in the region of 0.15-0.3. That will be used as the lift criterion for a reed petal undergoing design scrutiny when it is treated as a uniformly loaded beam by a proportion of the pressure differential estimated to be created by the crankcase compression ratio, CR_{cc}.

The mean pressure differential, Δp, assumed to act across the reed is found by estimating that this is a linear function of CR_{cc} in atmosphere units, where 18% is regarded as the mean effective value of the maximum:

$$\Delta p = 0.18(CR_{cc} - 1) \times 101325 \quad N/m^2 \qquad (6.3.5)$$

From Morrison and Crossland [6.1], the deflection, x_{tip}, at the end of the uniformly loaded beam representing the reed petal, is given by:

$$x_{tip} = \frac{\Delta p \, x_r \, L_r^4}{8 Y I} \qquad (6.3.6)$$

where the parameter, I, is the second moment of area given in Eq. 5.2.16. The maximum tip lift ratio, C_{rdt}, for this empirical calculation is then:

$$C_{rdt} = \frac{x_{tip}}{L_r} \tag{6.3.7}$$

Consequently, for implementation in Eq. 6.3.1, the time area of the reeds, for a number of reeds, n_r, is given by:

$$\int A d\theta = \frac{n_r x_{tip} x_p \theta_p}{2} \quad m^2 deg \tag{6.3.8}$$

where θ_p can been assigned a value of 200° for the isosceles triangle of lift to x_{tip}.

The above equations, referring to a combination of time areas and the mechanics of reed lift, i.e., Eqs. 6.3.1-6.3.8, should be strictly carried out in SI units, or arithmetic inaccuracies will be the inevitable consequence.

The last geometrical dimension to be calculated is the stop-plate radius which should not permit the tip of the reed to move past a C_{rdt} value of 0.3, but should allow the reed petal to be tangential to it should that lift actually occur. That a lift limit for C_{rdt} of 0.3 is realistic is seen in Fig. 5.35 for a racing engine. This relationship is represented by the following trigonometrical analysis, where the normal limit criterion is for a tip lift ratio of 0.3:

$$r_{sp} = \frac{(1 - C_{rdt}^2) L_r}{2 C_{rdt}} \tag{6.3.9}$$

6.3.3 The design process programmed into a package, Prog.6.4

This calculation is not as convenient to carry out with an electronic calculator as that for the expansion chamber, as a considerable number of cycles of estimation and recalculation are required before a matched design emerges. Therefore, a computer program has been added to those presented and available from SAE, Prog.6.4, REED VALVE DESIGN. This interactive program has a screen output which shows a plan and elevation view to scale of the reed block and reed petal under design consideration. A typical example is that illustrated in Fig. 6.27, which is also a design of a reed block and petals for a 125 racing engine which has already been discussed in Chapter 5 as a design example. The top part of the sketch on the computer screen shows an elevation section through one-half of the reed block, although it could just as well represent a complete block if that were the design goal. The lower half of the sketch shows a projected view looking normally onto the reed port and the petal as if it were transparent, with the darkened area showing the clamping of the reed by the stop-plate. The scaled dimension to note here is the left-hand edge of the dark area, for that is positioned accurately, while some artistic license has been used to portray the extent of the clamped area rightward of that position!

Chapter 6 - Empirical Assistance for the Designer

On the left are all of the input and output data values of the calculation, any one of which can be readily changed, whereupon the computer screen refreshes itself, virtually instantaneously, with the numerical answers and the reed block image. Note that the values of Young's Modulus and reed petal material density are missing from those columns. The data values for Young's Modulus and material density are held within the program as permanent data and all that is required is to inform the program, when it asks, if the petal material is steel, glass-fiber, or carbon-fiber. The information is recognized as a character string within the program and the appropriate properties of the reed material are indexed from the program memory.

The reed valve design program, Prog.6.4, is used to empirically design reed blocks and petals in advance of using the data in engine modeling programs. This program is useful to engineers, as it is presumed that they are like me, designers both by "eye" and from the numerical facts. The effectiveness of such modeling programs [1.13] has already been demonstrated in Chapter 5 and in Figs. 6.8-6.15, so it is instructive to examine the empirical design for the 125 Grand Prix engine shown in Fig. 6.27 and compare it with the data declared for the actual engine in Chapter 5, Sec. 5.5.2.

The glass-fiber reed has a natural frequency of 160 Hz. As the engine has a natural forcing frequency of 192 Hz, i.e., $11,500 \div 60$, this reed is in some long-term danger of fracture from resonance-induced fatigue, for it passes through its resonant frequency each time the machine is revved to 11,500 rpm in each gear when employed as a racing motorcycle engine. Presumably, as the lifetime of a racing engine is rarely excessive, this design is acceptable. Indeed, it is even desirable in that it will vibrate readily at the forcing frequency of the engine. However, it should not be regarded universally as good design practice for an engine, such as a production outboard, where durability over some 2000 hours is needed to satisfy the market requirements.

Of some interest is the photograph in Plate 4.1, which is just such a 250 cc twin-cylinder racing motorcycle engine, i.e., each cylinder is 125 cc just like the design being discussed. On this engine is the reed valve induction system which can be observed in the upper right-hand corner of the photograph. The plastic molding attaches the carburetor to the reed block and locks it onto the crankcase. The interior of the molding is profiled to make a smooth area transition for the gas to flow from the round section of the carburetor to the basically rectangular section at the reed block entrance.

6.3.4 Concluding remarks on reed valve design

It is not uncommon to find flat plate reed valves being used in small outboard motors and in other low specific power output engines. A flat plate reed block is just that, a single plate holding the reed petals at right angles to the gas flow direction. The designer should use the procedure within Prog.6.4 of declaring a "dummy" reed block angle, ϕ_{rb}, of 40° and proceed with the design as usual. Ultimately the plan view of the working drawing will appear with the reed plate, per petal and port that is, exactly as the lower half of the sketch in Fig. 6.27.

With this inside knowledge of the reed design and behavior in advance of engine modeling or experimentation, the engineer is in a sound position to tackle the next stage of theoretical or experimental design and development. Should you assemble a computer program to simulate an engine and its reed valve motion, together with the theory presented in Chapters

2 and 5, then this section will permit a better initial selection of the data for use within such a model.

6.4 Empirical design of disc valves for two-stroke engines

This subject is one which I rarely discuss without a certain feeling of nostalgia, because it was the motorcycle racing engines from MZ in East Germany, with their disc valve inlet systems and expansion chamber exhaust pipes, which appeared on the world's racing circuits during my undergraduate student days. They set new standards of specific power performance and, up to the present day, relegated the then all-conquering four-stroke power units into second place. It was, literally, a triumph of intellect and professional engineering over lack of finance, resources, materials and facilities. For many years thereafter, the disc valve intake system was acknowledged to be the superior method of induction control for high specific output engines. However, it was not long before it was learned how to produce equal, if not superior, performance from piston control of the intake system. The current technical position is that the reed valve has supplanted the piston-controlled intake port for racing engines and is probably the most popular intake control method for all engines from the cheapest 2 hp brushcutter to the most expensive 300 hp V8 outboard motor. Nevertheless, the disc-valved engine, in the flat form made popular by MZ, and identical to the Rotax design shown in Plate 1.8, is still very successful on the racing circuits. Racing engine design has a fashion element, as well as engineering logic, at work; and fashion always comes full circle. Therefore, it is important that a design procedure for disc valve induction be among the designer's options and that engine modeling programs are available to predict the behavior of the disc-valved engine. This enables direct comparisons to be made of engine performance when fitted with all of the possible alternatives for the induction process.

As with other data required for engine modeling programs, the designer needs to be able to establish empirical factors so that a minimum of guesswork is required when faced with the too-numerous data bank of a major computer program or when rapid optimization is required for the design. The following discussion sets out a logic procedure which by now should be familiar, for it repeats the same basic empirical methodology used for the piston-controlled intake and the reed valve induction systems.

6.4.1 Specific time area analysis of disc valve systems

Figs. 1.7(a), Plate 1.8, and Figs. 6.28 and 6.29 show the mechanical positioning, porting control and timing of the disc valve intake device. The earlier explanations in this chapter regarding the influence of the specific time area on the bmep attainable are just as relevant here, as the disc valve merely gives an asymmetrical element to the inlet port area diagrams of Fig. 6.1 for piston-controlled ports. Consequently, Fig. 6.28 shows that it is possible to (i) open the intake port early, thereby taking full advantage of induction over the period when the crankcase experiences a sub-atmospheric pressure, and (ii) shut it early before the piston forces out any trapped air charge on the crankcase compression stroke. Actually, this is an important feature, for the intake pressure wave is weaker by virtue of taking place over a longer time period, and so the ramming wave is not so strong as to provide the vigorous

Chapter 6 - Empirical Assistance for the Designer

ramming effect observed for the piston-controlled induction system. In Fig. 5.3 and Fig. 6.28, the angle required to open the port into the engine is represented by ϕ_p in degrees of crankshaft rotation. The total cutaway angle on the disc is shown as ϕ_d. If the disc valve opens at $\theta_{dvo}°$btdc and closes at $\theta_{dvc}°$atdc, then the total opening period is θ_{max}, where:

$$\theta_{max} = \theta_{dvo} + \theta_{dvc} \tag{6.4.1}$$

It is also straightforward geometry to show that the relationship between θ_{max}, ϕ_p and ϕ_d is given by, as in Eq. 5.2.8:

$$\theta_{max} = \phi_p + \phi_d \tag{6.4.2}$$

The port opens and closes in $\phi_p°$ as shown in Fig. 6.28(b) and, either precisely for a port as sketched in Fig. 5.3, or approximately for any other shape such as a round hole, that port area rises linearly to a maximum value of A_{max}, the total inlet port area. Such geometry is easily programmed into an overall engine simulation model. Using the nomenclature of Figs. 5.3 and 6.28, the total value of the maximum port area, A_{max}, is calculated from its occupancy of a segment of a circular annulus, less the corner radii residues. It is already shown in Eq. 5.2.9, and repeated here.

$$A_{max} = \pi \frac{\phi_d}{360} \left(r_{max}^2 - r_{min}^2 \right) - r_p^2 (4 - \pi) \tag{6.4.3}$$

The time area with respect to crankshaft angle, $\int Ad\theta$, is directly calculable from the trapezoid in Fig. 6.28(b) as:

$$\int_{\theta=\theta_{dvc}}^{\theta=\theta_{dvo}} Ad\theta = A_{max} \left(\theta_{max} - \phi_p \right) \tag{6.4.4}$$

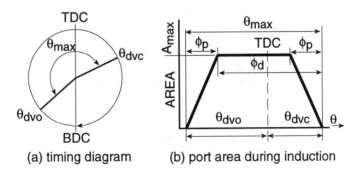

(a) timing diagram   (b) port area during induction

Fig. 6.28 Disc valve areas and timings during induction.

Design and Simulation of Two-Stroke Engines

From the previous discussion regarding the relevance of specific time area, the numerical level for the intake in question as a function of bmep is given by Eq. 6.1.9 and repeated here as:

$$A_{svi} = \frac{bmep + 1.528}{774} \quad s/m \tag{6.4.5}$$

When the target requirement of bmep for a given swept volume, V_{sv}, at an engine speed, rpm, is declared, there is available a direct solution for the design of the entire device. From Eq. 6.1.12:

$$\text{Specific time area, } A_{svi} = \frac{\sum_{\theta=\theta_{dvo}}^{\theta=\theta_{dvc}} A_\theta d\theta}{6V_{sv}rpm} \quad s/m \tag{6.4.6}$$

If you glance ahead at Fig. 6.29, the data listed in the upper left-hand side of that figure contain all of the relevant input data for the calculation, as exhibited in Eqs. 6.4.1 to 6.4.6. It is seen that the unknown data value which emerges from the calculation is the value of the outer port edge radius, r_{max}. This is the most convenient method of solution. Eqs. 6.4.1-6.4.6 are combined, estimates are made of all data but r_{max}, and the result is a direct solution for r_{max}. This is not as difficult as it might appear, for disc valve timings are not subject to great variations from the lowest to the highest specific power output level. For example, the timing events, θ_{dvo} and θ_{dvc}, would be at 150 and 65 for a small scooter engine and at 140 and 80, respectively, for a racing engine. For completeness, the actual solution route for the determination of r_{max} and the carbureter flow diameter (from Fig. 5.5), d_{tv}, is given below:

$$\int A d\theta = 6 A_{svi} V_{sv} rpm \tag{6.4.7}$$

$$A_{max} = \frac{\int A d\theta}{\theta_{max} - \phi_p} \tag{6.4.8}$$

$$r_{max} = \sqrt{\frac{360(C_c A_{max} + r_p^2(4 - \pi))}{\pi \phi_p} + r_{min}^2} \tag{6.4.9}$$

$$d_{tv} = \sqrt{\frac{4 C_c A_{max}}{\pi}} \tag{6.4.10}$$

Chapter 6 - Empirical Assistance for the Designer

The incorporation of the expansion coefficient, C_c, is very similar to that used for the reed valve case and is in the range of 1.35-1.45. Actually, it should be noted that in the reed valve design theory it is employed as the reciprocal number, at 0.70. In the computer solution presented in Prog.6.5, the actual value encoded is 1.4.

Note, as with previous solutions of similar equations, if the data for swept volume in Eq. 6.4.7 are inserted in the conventional units of cm^3, then the next line will fortuitously evaluate the maximum port area, A_{max}, as mm^2. If strict SI units are used, then V_{sv} is in m^3 in Eq. 6.4.7, and Eq. 6.4.8 will produce the output for A_{max} in the equally strict SI unit of m^2. However, I issue my familiar warning, stick to SI units throughout all design calculations and units will never become an arithmetic problem!

6.4.2 A computer solution for disc valve design, Prog.6.5

So that the designer may concentrate more on the design process and less on the arithmetic tedium of solution, the above equations are programmed into a computer program, Prog.6.5, DISC VALVE DESIGN, available from SAE. As with the majority of the programs referred to in this book, maximum advantage is taken of the screen graphics capability of the desktop computer. An example of the use of the calculation is given in Fig. 6.29, which shows the computer screen image of the program being used to design a disc valve option for induction to a 125 Grand Prix engine, one of the power units used throughout these chapters as a working example of a high specific output engine. The screen displays the correctly scaled drawing of the disc valve and the intake port, based on both the input data dimensions inserted by the designer and the output data as calculated by the program. Indeed, the computer screen picture is almost identical to the view given of the left-hand cylinder in Plate 1.8, or as sketched in Fig. 5.3.

The input data are listed in the upper half of the printed portion of the diagram. As discussed above, the performance target requirements based on cylinder swept volume and the

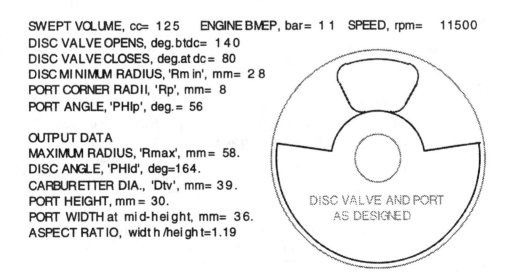

```
SWEPT VOLUME, cc= 125    ENGINE BMEP, bar= 11    SPEED, rpm=    11500
DISC VALVE OPENS, deg.btdc= 140
DISC VALVE CLOSES, deg.atdc= 80
DISC MINIMUM RADIUS, 'Rmin', mm= 28
PORT CORNER RADII, 'Rp', mm= 8
PORT ANGLE, 'PHIp', deg.= 56

OUTPUT DATA
MAXIMUM RADIUS, 'Rmax', mm= 58.
DISC ANGLE, 'PHId', deg=164.
CARBURETTER DIA., 'Dtv', mm= 39.
PORT HEIGHT, mm = 30.
PORT WIDTH at mid-height, mm= 36.
ASPECT RATIO, width/height=1.19
```

Fig. 6.29 Computer screen output from Prog.6.5, DISC VALVE DESIGN.

bmep to be attained at a peak power engine speed are both needed, and are followed by an estimate of the disc valve timings for the opening and closing positions. The latter are not difficult to estimate for the reasons already stated. The principal parameters requiring design experience are those for the minimum radius, r_{min}, and the port opening angle, ϕ_p.

The minimum radius, r_{min}, is determined by the geometry of the engine crankshaft. Refer to Fig. 1.7(a), where it can be seen that the value of this minimum radius is decided by the path of the gas flow past the outer diameter of the engine's main bearing and crankcase gas seal on that side of the crankshaft. Therefore, this element of the disc valve design must be carried out in conjunction with the design of the crankshaft and crankcase. It will be an iterative process, going to and fro between the two design procedures until a satisfactory compromise has been achieved for the disc valve and the crankshaft with its bearings and gas seals.

The output data listed on the screen, or subsequently line-printed, give the ensuing values for r_{max}, the carburetor flow diameter, d_{tv}, and the disc valve cutaway angle, ϕ_d. They also give the port aspect ratio at the mid-port height. The aspect ratio is defined as width to height, and the designer should attempt to retain it within the band of 1.0 to 1.25 by suitably adjusting the input parameter of ϕ_p. The gas flow from the (usually) round carburetor exit area to the port area needs a minimum expansion path to follow so that the coefficient of discharge of the junction is as high as possible. A simple guide to follow is to arrange the carburetor flow diameter to be approximately equal to the width of the intake port at mid-height. In the context of the output in Fig. 6.29, for a 125 racing engine, it can be seen that the predicted carburetor diameter is 39 mm, and is 1 mm larger than that found for the reed valve engine in Fig. 6.27 and for the actual engine in Chapter 5. It is common experience that disc valve racing engines need slightly larger carburetors than when reed valves are employed.

Note that the dimension, r_{max}, is not the outer diameter of the disc but the outer radius of the port. The actual disc needs an outer diameter that will seal the face of the intake port during the crankcase compression stroke. This will depend on several factors, such as the disc material, its thickness and stiffness, but a value between 2 and 4 mm of overlap in the radial direction would be regarded as adequate and conventional.

6.5 Concluding remarks

Having reached this juncture in the book, you will realize that the end of a design cycle has been achieved. The first chapter introduced the subjects in general, the second elaborated on unsteady gas flow, the third on scavenging, the fourth on combustion, the fifth on engine modeling, and this, the sixth, on the assembly of information so that the engine simulation can be performed in a logical manner. It is now up to you to make the next move.

Some will want to use the various computer programs as soon as possible, depending on their level of experience or need. Some will want to create computer software from the theory presented in Chapter 2, so that they can become more familiar with unsteady gas flow as a means of understanding engine behavior. Some will acquire engine modeling programs, of the type used to illustrate Chapter 5, to perform the design process for actual engines. The book is liberally sprinkled with the data for all such programs so that the initial runs can be effected by checking that a computer listing has been inserted accurately or that the results tally with those that illustrate this book.

The most important global view expressed here for you to consider is that the actual engine, like the engine model, uses the "separate" technologies of Chapters 1 to 4 to behave like the paper engines of Chapter 5. The best engine designer is one who can think in terms of a paper engine turning in his mind, each turn bringing images of the effect of changing the unsteady gas flow regime, the scavenging regime, and the combustion regime together with the mutual interaction of each of those effects. This implies a thorough understanding of all of those topics, and the more thorough that understanding, the more complex can be the interaction which the mind will handle. Should one of those topics be missing or poorly understood, then the quality of the mental debate will suffer. It is hoped that the visual imagery provided by the computer programs referenced will accelerate the understanding of all of these topics.

As pointed out frequently thus far, our comprehension of many phenomena is in its infancy and it is only thirty-five years on from the general availability of computers and also the black art of much of the 1950s' technology. It behooves the thinking designer to follow and to study the progress of technology through the technical papers which are published in this and related fields of endeavor.

The remainder of this book will concentrate on selected specialized topics, and particularly on the future of the design and development of two-stroke engines.

References for Chapter 6

6.1 J.L.M. Morrison, B. Crossland, <u>An Introduction to the Mechanics of Machines</u>, 2nd Ed., Longman, London, 1971.

6.2 F.S. Tse, I.E. Morse, R.T. Hinkle, <u>Mechanical Vibrations, Theory and Applications</u>, 2nd Ed., Allyn and Bacon, 1978.

6.3 C. Bossaglia, <u>Two-Stroke High Performance Engine Design and Tuning</u>, Chislehurst and Lodgemark Press, London, 1972.

6.4 H. Naitoh, M. Taguchi, "Some Development Aspects of Two-Stroke Cycle Motorcycle Engines," SAE Paper No. 660394, Society of Automotive Engineers, Warrendale, Pa., 1966.

6.5 H. Naitoh, K. Nomura, "Some New Development Aspects of Two-Stroke Cycle Motorcycle Engines," SAE Paper No. 710084, Society of Automotive Engineers, Warrendale, Pa., 1971.

6.6 J.F.T. MacClaren, S.V. Kerr, "Automatic Reed Valves in Hermetic Compressors," IIR Conference, Commission III, September 1969.

6.7 R.Fleck, "Expanding the Torque Curve of a Two-Stroke Motorcycle Race Engine by Water Injection," SAE Paper No. 931506, Society of Automotive Engineers, Warrendale, Pa., 1993.

Chapter 7

Reduction of Fuel Consumption and Exhaust Emissions

7.0 Introduction

Throughout the evolution of the internal-combustion engine, there have been phases of concentration on particular aspects of the development process. In the first major era, from the beginning of the 20th Century until the 1950s, attention was focused on the production of ever greater specific power output from the engines, be they two- or four-stroke cycle power units. To accomplish this, better quality fuels with superior octane ratings were prepared by the oil companies so that engines could run at higher compression ratios without risk of detonation. Further enhancements were made to the fund of knowledge on materials for engine components, ranging from aluminum alloys for pistons to steels for needle roller bearings, so that high piston speeds could be sustained for longer periods of engine life. This book thus far has concentrated on the vast expansion of the knowledge base on gas dynamics, thermodynamics and fluid mechanics which has permitted the design of engines to take advantage of the improvements in materials and tribology. Each of these developments has proceeded at an equable pace. For example, if a 1980s racing engine had been capable of being designed in 1920, it would have been a case of self-destruction within ten seconds of start-up due to the inadequacies of the fuel, lubricant, and materials from which it would have been assembled at that time. However, should it have lasted for any length of time, at that period in the 1920s, the world would have cared little that its fuel consumption rate was excessively high, or that its emission of unburned hydrocarbons or oxides of nitrogen was potentially harmful to the environment!

The current era is one where design, research and development is increasingly being focused on the fuel economy and exhaust emissions of the internal-combustion engine. The reasons for this are many and varied, but all of them are significant and important.

The world has a limited supply of fossil fuel of the traditional kind, i.e., that which emanates from prehistorical time and is available in the form of crude oil capable of being refined into the familiar gasoline or petrol, kerosene or paraffin, diesel oil and lubricants. These are the traditional fuels of the internal-combustion engine and it behooves the designer, and the industry which employs him or her, to develop more efficient engines to conserve that dwindling fossil fuel reserve. Apart from ethical considerations, many governments have enacted legislation setting limits on fuel consumption for various engine applications.

The population of the world has increased alarmingly, due in no small way to a more efficient agriculture which will feed these billions of humans. That agricultural system, and the transportation systems which back it up, are largely efficient due to the use of internal combustion engine-driven machinery of every conceivable type. This widespread use of internal-combustion engines has drawn attention to the exhaust emissions from its employment, and in particular to those emissions that are harmful to the environment and the human species. For example, carbon monoxide is toxic to humans and animals. The combination of unburned hydrocarbons and nitrogen oxides, particularly in sunlight, produces a visible smog which is harmful to the lungs and the eyes. The nitrogen oxides are blamed for the increased proportion of the rainfall containing acids which have a debilitating effect on trees and plant growth in rivers and lakes. Unburned hydrocarbons from marine engines are thought to concentrate on the beds of deep lakes, affecting in a negative way the natural development of marine life. The nitrogen oxides are said to contribute to the depletion of the ozone layer in the upper atmosphere, which potentially alters the absorption characteristics of ultraviolet light in the stratosphere and increases the radiation hazard on the earth's surface. There are legitimate concerns that the accumulation of carbon dioxide and hydrocarbon gases in the atmosphere increases the threat of a "greenhouse effect" changing the climate of the Earth.

One is tempted to ask why it is the important topic of today and not yesterday. The answer is that the engine population is increasing faster than people, and so too is the volume of their exhaust products. All power units are included in this critique, not just those employing reciprocating IC engines, and must also encompass gas turbine engines in aircraft and fossil fuel-burning, electricity-generating stations. Actually, the latter are the largest single source of exhaust gases into the atmosphere.

The discussion in this chapter will be in two main segments. The first concentrates on the reduction of fuel consumption and emissions from the simple, or conventional, two-stroke engine which is found in so many applications requiring an inexpensive but high specific output powerplant such as motorcycles, outboard motors and chainsaws. There will always be a need for such an engine and it behooves the designer to understand the methodology of acquiring the requisite performance without an excessive fuel consumption rate and pollutant exhaust emissions. The second part of this chapter will focus on the design of engines with fuel consumption and exhaust pollutant levels greatly improved over that available from the "simple" engine. Needless to add, this involves some further mechanical complexity or the use of expensive components, otherwise it would be employed on the "simple" engine. As remarked in Chapter 1, the two-stroke engine, either compression or spark ignition, has an inherently low level of exhaust emission of nitrogen oxides, and this makes it fundamentally attractive for future automobile engines provided that the extra complexity and expense involved does not make the two-stroke powerplant non-competitive with the four-stroke engine.

Before embarking on the discussion regarding engine design, it is necessary to expand on the information presented in Chapter 4 on combustion, particularly relating to the fundamental effects of air-fuel ratio on pollutant levels and to the basic differences inherent in homogeneous and stratified charging, and homogeneous and stratified combustion.

Chapter 7 - Reduction of Fuel Consumption and Exhaust Emissions

7.1 Some fundamentals of combustion and emissions

The fundamental material regarding combustion is covered in Chapter 4, but there remains some discussion which is specific to this chapter and the topics therein.

The first is to reiterate the origins of exhaust emission of unburned hydrocarbons and nitrogen oxides from the combustion process, first explained in Chapter 4. Recall the simple chemical relationship posed in Eq. 4.3.3 for the stoichiometric combustion of air and gasoline, and the discussion wherein it is stressed that the combustion of fuel and air occurs with vaporized fuel and air, but not liquid fuel and air. The stoichiometric combustion equation is repeated here and expanded to include the unburned HC and NO_x emissions.

$$CH_n + \lambda_m(O_2 + kN_2) = \\ x_1CO + x_2CO_2 + x_3H_2O + x_4O_2 \\ + x_5H_2 + x_6N_2 + x_7CH_b + x_8NO_x \quad (7.1.1)$$

It is shown in Chapter 4 that dissociation [4.1] will permit the formation of CO emission simply as a function of the presence of carbon and oxygen at high temperatures. This is also true of hydrocarbons, shown above as CH_b, or of oxides of nitrogen, shown as NO_x in the above equation. Nevertheless, the major contributor to CO and HC emission is from combustion of mixtures which are richer than stoichiometric, i.e., when there is not enough oxygen present to fully oxidize all of the fuel.

Hydrocarbons are formed by other mechanisms as well. The flame may quench in the remote corners and crevices of the combustion chamber, leaving the fuel there partially or totally unburned. Lubricating oil may be scraped into the combustion zone and this heavier and more complex hydrocarbon molecule burns slowly and incompletely, usually producing exhaust particulates, i.e., a visible smoke in the exhaust plume.

A further experimental fact is the association of nitrogen with oxygen to form nitrogen oxides, NO_x, and this undesirable behavior becomes more pronounced as the peak combustion temperature is increased at higher load levels or is focused around the stoichiometric air-to-fuel ratio, as shown clearly in Appendices A4.1 and A4.2.

It is quite clear from the foregoing that, should the air-fuel ratio be set correctly for the combustion process to the stoichiometric value, even an efficient combustion system will still have unburned hydrocarbons, carbon monoxide, and actually maximize the nitrogen oxides, in the exhaust gas from the engine. Should the air-fuel ratio be set incorrectly, either rich or lean of the stoichiometric value, then the exhaust pollutant levels will increase; except NO_x which will decrease! If the air-fuel mixture is very lean so that the flammability limit is reached and misfire takes place, then the unburned hydrocarbon and the carbon monoxide levels will be considerably raised. It is also clear that the worst case, in general, is at a rich air-fuel setting, because both the carbon monoxide and the unburned hydrocarbons are inherently present on theoretical grounds.

It is also known, and the literature is full of technical publications on the subject, that the recirculation of exhaust gas into the combustion process will lower the peak cycle temperature and act as a damper on the production of nitrogen oxides. This is a standard technique for production four-stroke automobile engines to allow them to meet legislative requirements for nitrogen oxide emissions. In this regard, the two-stroke engine is ideally suited for this application, for the retention of exhaust gas is inherent from the scavenging process. This natural scavenging effect, together with the lower peak cycle temperature due to a firing stroke on each cycle, allows the two-stroke engine to produce much reduced nitrogen oxide exhaust emissions at equal specific power output levels.

Any discussion on exhaust emissions usually includes a technical debate on catalytic after-treatment of the exhaust gases for their added purification. In this chapter, there is a greater concentration on the design methods to attain the lowest exhaust emission characteristics before any form of exhaust after-treatment is applied.

As a postscript to this section, there may be those who will look at the relatively tiny proportions of the exhaust pollutants in Eq. 7.1.1 and wonder what all the environmental, ecological or legislative fuss is about in the automotive world at large. Let them work that equation into yearly mass emission terms for each of the pollutants in question for the annual consumption of many millions of tons of fuel per annum. The environmental problem then becomes quite self-evident!

7.1.1 Homogeneous and stratified combustion and charging

The combustion process can be conducted in either a homogeneous or stratified manner, and an introduction to this subject is given in Sec. 4.1. The words "homogeneous" and "stratified" in this context define the nature of the mixing of the air and fuel in the combustion chamber at the period of the flame propagation through the chamber. A compression-ignition or diesel engine is a classic example of a stratified combustion process, for the flame commences to burn in the rich environment of the vaporizing fuel surrounding the droplets of liquid fuel sprayed into the combustion chamber. A carburetted, four-stroke cycle, spark-ignition engine is the classic example of a homogeneous combustion process, as the air and fuel at the onset of ignition are thoroughly mixed together, with the gasoline in a gaseous form.

Both of the above examples give rise to discussion regarding the charging of the cylinder. In the diesel case, the charging of the cylinder is conducted in a stratified manner, i.e., the air and the fuel enter the combustion chamber separately and any mixing of the fuel and air takes place in the combustion space. As the liquid fuel is sprayed in some 35° before tdc, it cannot achieve homogeneity before the onset of combustion. In the carburetted, spark-ignition, four-stroke cycle engine, the charging of the engine is conducted in a homogeneous fashion, i.e., all of the required air and fuel enter together through the same inlet valve and are considered to be homogeneous, even though much of the fuel is still in the liquid phase at that stage of the charging process.

It would be possible in the case of the carburetted, four-stroke cycle engine to have the fuel and air enter the cylinder of the engine in two separate streams, one rich and the other a lean air-fuel mixture, yet, by the onset of combustion, be thoroughly mixed together and burn as a homogeneous combustion process. In short, the charging process could be considered as stratified and the combustion process as homogeneous. On the other hand, that same engine could be designed, viz the Honda CVCC automobile power unit, so that the rich and lean air-fuel streams are retained as separate entities up to the point of ignition and the combustion process is also carried out in a stratified manner. The main point behind this discussion is to emphasize the following points:

(i) If a spark-ignition engine is charged with air and fuel in a homogeneous manner, the ensuing combustion process is almost inevitably a homogeneous combustion process.

(ii) If an engine is charged with air and fuel in a stratified manner, the ensuing combustion process is possibly, but not necessarily, a stratified combustion process.

In the analysis conducted in Sec. 4.3 for the combustion of gasoline, the air-fuel ratio is noted as the marker of the relationship of that combustion process to the stoichiometric, or ideal. You can interpret that as being the ratio of the air and fuel supply rates to the engine. This will be perfectly accurate for a homogeneous combustion process, but can be quite misleading for a design where stratified charging is taking place.

Much of the above discussion is best explained by the use of a simple example illustrated in Fig. 7.1. The "engine" in the example is one where the combustion space can contain, or be charged with, 15 kg of air. Consider the "engine" to be a spark-ignition device and the discussion is equally pertinent for both two-stroke and four-stroke cycle engines.

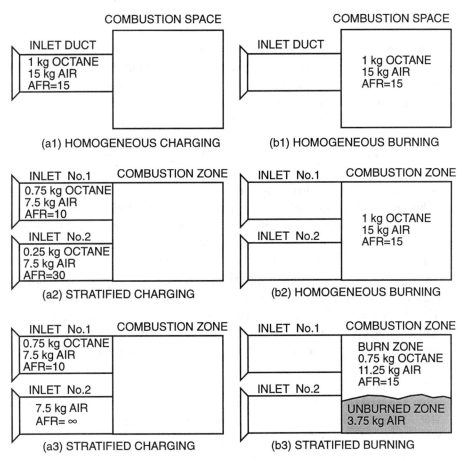

Fig. 7.1 Homogeneous and stratified charging and combustion.

At a stoichiometric air-fuel ratio for gasoline, this means that a homogeneously charged engine, followed by a homogeneous combustion process, would ingest 1 kg of octane with the air. This situation is illustrated in Fig. 7.1(a1) and (b1). The supplied air-fuel ratio and that in the combustion space are identical at 15.

If the engine had stratified charging, but the ensuing mixing process is complete followed by homogeneous combustion, then the situation is as illustrated in Fig. 7.1(a2) and (b2). Although one of the entering air-fuel streams has a rich air-fuel ratio of 10 and the other is lean at 30, the overall air-fuel ratio is 15, as is the air-fuel ratio in the combustion space during burning. The supplied air-fuel ratio and that in the combustion space are identical at 15. In effect, the overall behavior is much the same as for homogeneous charging and combustion.

If the engine has both stratified charging and combustion, then the situation portrayed in Fig. 7.1(a3) and (b3) reveals fundamental differences. At an equal "delivery ratio" to the previous examples, the combustion space will hold 15 kg of air. This enters in a stratified form with one stream rich at an air-fuel ratio of 10 and the second containing no fuel at all. Upon entering the combustion space, not all of the entering air in the second stream mixes with the rich air-fuel stream, but a sufficient amount does to create a "burn zone" with a stoichiometric mixture at an air-fuel ratio of 15. This leaves 3.75 kg of air unburned which exits with the exhaust gas. The implications of this are:

(i) The overall or supplied air-fuel ratio is 20, which gives no indication of the air-fuel ratio during the actual combustion process and is no longer an experimental measurement which can be used to help optimize the combustion process. For example, many current production (four-stroke) automobile engines have "engine management systems" which rely on the measurement of exhaust oxygen concentration as a means of electronically controlling the overall air-fuel ratio to precisely the stoichiometric value.

(ii) The combustion process would release 75% of the heat available in the homogeneous combustion example, and it could be expected that the bmep and power output would be similarly reduced. In the technical phrase used to describe this behavior, the "air-utilization" characteristics of stratified combustion are not as efficient as homogeneous combustion. The diesel engine is a classic example of this phenomenon, where the overall air-fuel ratio for maximum thermal efficiency is usually some 30% higher than the stoichiometric value.

(iii) The exhaust gas will contain a significant proportion of oxygen. Depending on the exhaust after-treatment methodology, this may or may not be welcome.

(iv) The brake specific fuel consumption will be increased, i.e., the thermal efficiency will be reduced, all other parameters being equal. The imep attainable is lower with the lesser fuel mass burned and, as the parasitic losses of friction and pumping are unaffected, the bsfc and the mechanical efficiency deteriorate.

(v) An undesirable combustion effect can appear at the interface between the burned and unburned zones. Tiny quantities of aldehydes and ketones are produced as the flame dies at the lean interface or in the end zones, and although they would barely register as pollutants on any instrumentation, the hypersensitive human nose records them as unpleasant odors [4.4]. Diesel engine combustion suffers from this complaint.

Chapter 7 - Reduction of Fuel Consumption and Exhaust Emissions

The above discussion may appear as one of praise for homogeneous combustion and derision of stratified combustion. Such is not the case, as the reality for the two-stroke engine is that stratified charging, and possibly also stratified combustion, will be postulated in this chapter as a viable design option for the reduction of fuel consumption and exhaust emissions. In that case, the goal of the designer becomes the maximization of air-utilization and the minimization of the potentially undesirable side-effects of stratified combustion. On the bonus side, the simple ported two-stroke engine has that which is lacking in the four-stroke cycle powerplant, namely an uncluttered cylinder head zone for the design and creation of an optimum combustion space.

7.2 The simple two-stroke engine

This engine has homogeneous charging and combustion and is spark-ignited, burning a volatile fuel such as gasoline, natural gas or kerosene. It is commonly found in a motorcycle, outboard motor or industrial engine and the fuel metering is conventionally via a carburetor. In general, the engine has fresh charge supplied via the crankcase pump. Indeed, the engine would be easily recognized by its inventor, Sir Dugald Clerk, as still embodying the modus operandum he envisaged; he would, it is suspected, be somewhat astonished at the level of specific power output that has been achieved from it at this juncture in the 20th Century!

The operation of this engine has been thoroughly analyzed in earlier chapters, and repetition here would be just that. However, to achieve the optimum in terms of fuel consumption or exhaust emissions, it is necessary to re-examine some of those operating characteristics. This is aided by Fig. 7.2.

The greatest single problem for the simple engine is the homogeneous charging, i.e., scavenging, of the cylinder with fuel and air. By definition, the scavenging process can never be "perfect" in such an engine because the exhaust port is open as fresh charge is entering the cylinder. At best, the designer is involved in a damage limitation exercise. This has been discussed at length in Chapter 3 and further elaborated on in Chapter 5 by the use of a computer model of the engine which incorporates a simulation of the scavenging process. It is proposed to debate this matter even further so that the designer is familiar with all of the available options for the improvement of those engine performance characteristics relating to fuel economy and exhaust emissions.

Linked to the scavenging problem is the necessity to tailor the delivery ratio curve of the engine to suit the application. Any excess delivery ratio over that required results in merely pumping air and fuel into the exhaust system.

One of the factors within design control is exhaust port timing and/or exhaust port area. Many engines evolve or are developed with the peak power performance requirement at the forefront of the process. Very often the result is an exhaust port timing that is excessively long even for that need. The end product is an engine with poor trapping characteristics at light load and low speed, which implies poor fuel economy and exhaust emissions. This subject will be debated further in this chapter, particularly as many of the legislative tests for engines are based on light load running as if the device were used as an automotive powerplant in an urban environment. Equally, the designer should never overlook the possibility of improving the trapping efficiency of any engine by suitable exhaust pressure wave tuning; this matter is fully covered in Chapters 2 and 5. Every 10% gain in trapping pressure is at least a 10%

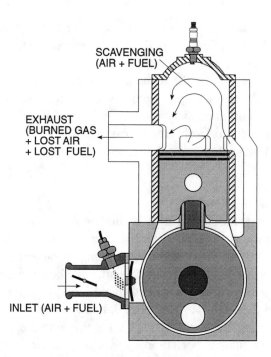

Fig. 7.2 The fuel consumption and emissions problem of the simple two-stroke engine.

reduction in bsfc and an even larger proportionate improvement in hydrocarbon exhaust emissions.

Together with scavenging and delivery ratio there is the obvious necessity of tailoring the air-fuel ratio of the supplied charge to be as close to the stoichiometric as possible. As a 10% rich mixture supplies more power at some minor expense in bsfc, the production engine is often marketed with a carburetor set at a rich mixture level, more for customer satisfaction than for necessity. Legislation on exhaust emissions will change that manufacturing attitude in the years ahead, but the designer is often presented by a cost-conscious management with the simplest and most uncontrollable of carburetors as part of a production package beyond designer influence. The air-fuel ratio control of some of these mass-produced cheap carburetors is very poor indeed. That this can be rectified, and not necessarily in an expensive fashion, is evident from the manufacturing experience of the automobile industry since the so-called "oil crisis" of 1973 [7.51].

Many of the simplest engines use lubricant mixed with the gasoline as the means of engine component oiling. In Great Britain this is often referred to as "petroil" lubrication. For many years, the traditional volumetric ratio of gasoline to oil was 25 or 30. Due to legislative pressure, particularly in the motorcycle and outboard field, this ratio is much leaner today, between 50 and 100. This is due to improvements in both lubricants and engine materials. For many applications, separate, albeit still total-loss lubrication methods, are employed with oil pumps supplying lubricant to selected parts of the engine. This allows gasoline-oil ratios to be varied from 200 at light loads to 100 at full load. This level of oiling would be closely aligned with that from equivalent-sized four-stroke cycle engines. It behooves the designer to continually search for new materials, lubricants and methods to further reduce the level of lubricant consumption, for it is this factor that influences the exhaust smoke output from a two-stroke engine at cold start-up and at light load. In this context, the papers by Fog *et al.* [7.22] and by Sugiura and Kagaya [7.25] should be studied as they contain much practical information.

One of the least understood design options is the bore-stroke ratio. Designers, like the rest of the human species, are prone to fads and fashions. The in-fashion of today is for oversquare engines for any application, and this approach is probably based on the success of oversquare engines in the racing field. Logically speaking, it does not automatically follow that it is the correct cylinder layout for an engine-driven, portable, electricity-generating set which will never exceed 3000 or 3600 rpm.

If the application of the engine calls for its extensive use at light loads and speeds, such as a motorcycle in urban traffic or trolling for fish with an outboard motor, then a vitally important factor is the maintenance of the engine in a "two-stroke" firing mode, as distinct from a "four-stroke" firing mode. This matter is introduced in Sec. 4.1.3. Should the engine skip-fire in the manner described, there is a very large increase in exhaust hydrocarbon emissions. This situation can be greatly improved by careful attention during the development phase to combustion chamber design, spark plug location and spark timing. Even further gains can be made by exhaust port timing and area control, perhaps leading to the incorporation of "active radical" combustion to solve this particular problem, and this option should never be neglected by the designer [4.34, 7.27].

In summary, the optimization of the simple two-stroke engine to meet performance and exhaust emission targets can be subdivided as follows:

(i)   Optimize the bore-stroke ratio.
(ii) Optimize the scavenging process.
(iii) Optimize the delivery ratio.
(iv) Optimize the port timings and areas.
(v)   Optimize the air-to-fuel ratio.
(vi) Optimize the combustion process.
(vii) Optimize unsteady gas-dynamic tuning.
(viii) Optimize the lubrication requirements.
(ix) Optimize the pumping and mechanical losses.

To satisfy many of the design needs outlined above, the use of a computer-based simulation of the engine is ideal. The engine models presented in Chapter 5 will be used in succeeding sections to illustrate many of the points made above and to provide an example for the designer that such models are not primarily, or solely, aimed at design for peak specific power performance.

7.2.1 Typical performance characteristics of simple engines

Before embarking on the improvement of the exhaust emission and fuel economy characteristics of the simple two-stroke engine, it is important to present and discuss some typical measured data for such engines.

In Chapter 4, on combustion, the discussion focuses on the origins of the emissions of carbon monoxide, nitric oxide, oxygen, carbon dioxide, and hydrogen, as created by that combustion process. Appendix A4.1 showed how hydrocarbon emissions due to scavenge losses may also be included with those emanating from combustion. Both types of gaseous emission creation are incorporated into the GPB simulation system. However, the use of simulation for design purposes is possible only if the computer model can be shown to provide similar trends to the measurements for the computed emissions, so calculations are presented to reinforce this point.

7.2.1.1 Measured performance data from a QUB 400 research engine

The first set of data to be presented is from the QUB 400 single-cylinder research engine [1.20]. This engine is water cooled with very good scavenging characteristics approaching that of the SCRE cylinder shown in Figs. 3.12. 3.13 and 3.18. The bore and stroke are 85 and 70 mm, respectively, and the exhaust, transfer and inlet ports are all piston-controlled with timings of 96° atdc, 118° and 60°, respectively.

The engine speed selected for discussion is 3000 rpm and the measured performance characteristics at that engine speed are given here as Figs. 7.3-7.8. Figs. 7.3-7.5 are at full throttle and Figs. 7.6-7.8 are at 10% throttle opening area ratio. In each set are data, as a function of air-fuel ratio, for bmep, bsfc, unburned hydrocarbon emissions as both ppm and bsHC values, and carbon monoxide and oxygen exhaust emission levels. It is worth noting that this engine employs a simple exhaust muffler and so has no exhaust pressure wave tuning to aid the trapping efficiency characteristic. Therefore, the performance characteristics attained are due solely to the design of the porting, scavenging, and combustion chamber. The engine does not have a high trapped compression ratio as the CR_t value is somewhat low at 6.7. Even without exhaust pressure wave tuning, that this is not a low specific power output engine is evident from the peak bmep level of 6.2 bar at 3000 rpm.

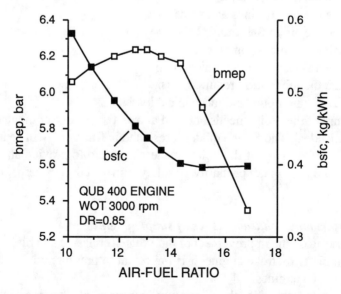

Fig. 7.3 Air-fuel ratio effect on bmep and bsfc at full throttle.

Chapter 7 - Reduction of Fuel Consumption and Exhaust Emissions

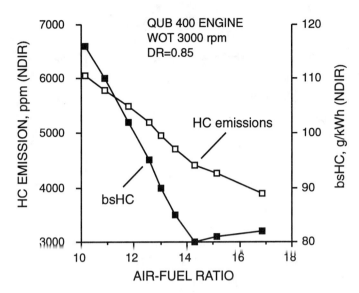

Fig. 7.4 Air-fuel ratio effect on HC emissions at full throttle.

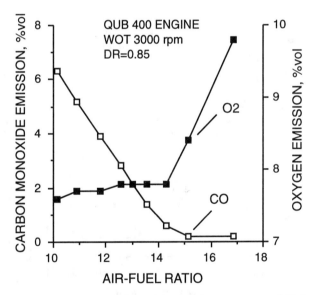

Fig. 7.5 Air-fuel ratio effect on CO and O_2 emission at full throttle.

Consider the measured data around the stoichiometric air-fuel ratio of 15. That this simple two-stroke engine must have good scavenging characteristics is seen from the hydrocarbon emission levels, which are at 80 g/kWh (NDIR) at full throttle and 17 g/kWh (NDIR) at a light load of 2.65 bar bmep. These probably translate into total HC emission figures, as would be measured by a FID system and as debated in Sec. 1.6.2, to 135 g/kWh and 29 g/kWh, respectively. The raw HC emission data (hexane) are 4200 ppm and 1250 ppm, respectively. The

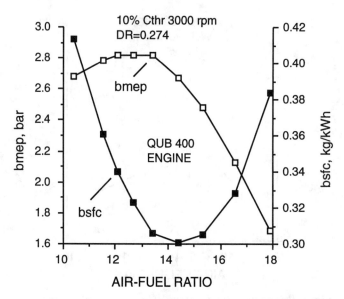

Fig. 7.6 Air-fuel ratio effect on bmep and bsfc at 10% throttle.

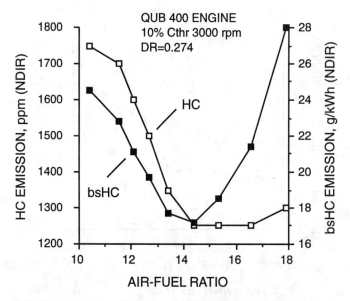

Fig. 7.7 Air-fuel ratio effect on HC emissions at 10% throttle.

bsfc is very low at 0.40 kg/kWh at 6.2 bar bmep and 0.30 kg/kWh at 2.65 bar bmep. The carbon monoxide level is also low at 0.2% by volume at full throttle and 0.1% at one-tenth throttle; while low CO values are to be expected close to the stoichiometric air-to-fuel ratio, these values are sufficiently low as to indicate very good combustion, and would be remarkably low for any four-stroke cycle engine. The oxygen emission is 7.5% at full throttle and 3% at one-tenth throttle; remember that the majority of the oxygen emission derives from the

Chapter 7 - Reduction of Fuel Consumption and Exhaust Emissions

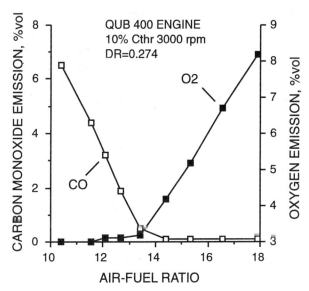

Fig. 7.8 Air-fuel ratio effect on CO and O_2 emissions at 10% throttle.

air lost during the scavenge process with about 1% coming from the inefficiency of the combustion process. These are probably the best fuel consumption and emission values recorded on a simple engine at QUB, particularly for an engine capable of 6.2 bar bmep without exhaust tuning. By comparing notes with colleagues in industrial circles, it is probable that these figures represent a "state-of-the-art" position for a simple, single-cylinder, two-stroke engine at this point in history. The levels of fuel economy and exhaust emissions at the lighter load point of 2.65 bar bmep would be regarded as particularly impressive, and there would be four-stroke cycle engines which could not improve on these numbers. It is noticeable that the carbon monoxide levels are at least as low as those from four-stroke cycle engines, but this is the one data value which truly emanates from the combustion process alone and is not confused by intervening scavenging losses.

Nevertheless, the values of hydrocarbon emission are, by automotive standards, very high. The raw value of HC emission caused by combustion inefficiency alone should not exceed 400 ppm, yet it is 4200 ppm at full load. This is a measure of the ineffectiveness of homogeneous charging, i.e., scavenging, of the simple two-stroke engine. It forever rules out the use of simple two-stroke engines in automotive applications against a background of legislated emissions levels.

To reinforce these comments, recall from Sec. 1.6.3 that exhaust oxygen concentration can be used to compute trapping efficiency, TE. The data sets shown in Figs. 7.5 and 7.8 would yield a TE value of 0.6 at full throttle and 0.86 at one-tenth throttle. This latter is a remarkably high figure and attests to the excellence of the scavenge design. What these data imply, however, is that if the fuel were not short-circuited with the air, and all other engine behavioral factors remained as they were, the exhaust HC level could possibly be about 350 ppm, but the full throttle bsfc would actually become 240 g/kWh and the one-tenth throttle bsfc would be at 260 g/kWh. These would be exceptional bsfc values by any standards and they illustrate the potential attractiveness of an optimized two-stroke engine to the automotive industry.

To return to the discussion regarding simple two-stroke engines, Figs. 7.3-7.8 should be examined carefully in light of the discussion in Sec. 7.1.1 regarding the influence of air-fuel ratio on exhaust pollutant levels. As carbon monoxide is the one exhaust gas emission not distorted in level by the scavenging process, it is interesting to note that the theoretical predictions provided by the equations in Sec. 4.3 for stoichiometric, rich, and lean air-fuel ratios are quite precise. In Fig. 7.5 the CO level falls linearly with increasing air-fuel ratio and it levels out at the stoichiometric value. At one-tenth throttle in Fig. 7.8, exactly the same trend occurs.

The theoretical postulations in terms of the shape of the oxygen curve are also observed to be borne out. In Fig. 7.5 the oxygen profile is flat until the stoichiometric air-fuel ratio, and increases linearly after that point. The same trend occurs at one-tenth throttle in Fig. 7.8, although the flat portion of the curve ends at an air-to-fuel ratio of 14 rather than at the stoichiometric level of 15.

The brake specific fuel consumption and the brake specific hydrocarbon emission are both minimized at, or very close to, the stoichiometric air-fuel ratio.

All of the theoretical predictions from the relatively simple chemistry described in Sec. 4.3 are shown to be relevant. In short, for the optimization of virtually any performance characteristic, the simple two-stroke engine should be operated as close to the stoichiometric air-fuel ratio as possible within the limits of the mechanical reliability of the components or of the onset of detonation. The only exception is maximum power or torque, where the optimum air-fuel ratio is observed to be at 13, which is about 13% rich of the stoichiometric level.

7.2.1.2 Typical performance maps for simple two-stroke engines

It is necessary to study the more complete performance characteristics for simple two-stroke engines so that you are aware of the typical characteristics of such engines over the complete load and speed range. Such performance maps are presented in Figs. 7.9-7.11 from the publication by Batoni [7.1] and in Fig. 7.12 from the paper by Sato and Nakayama [7.2].

The experimental data from Batoni [7.1]

In Figs. 7.9-7.11 the data are measured for a 200 cc motor scooter engine which has very little exhaust tuning to assist with its charge trapping behavior. The engine is carburetted and spark-ignited, and is that used in the familiar Vespa motor scooter. The units for brake mean effective pressure, bmep, are presented as kg/cm^2 where 1 kg/cm^2 is equivalent to 0.981 bar. The units of brake specific fuel consumption, bsfc, are presented as g/hp.hr where 1 g/hp.hr is equivalent to 0.746 g/kWh.

The bmep from this engine has a peak of 4.6 bar at 3500 rpm. Observe that the best bsfc occurs at 4000 rpm at about 50% of the peak torque and is a quite respectable 0.402 kg/kWh. Below the 1 bar bmep level the bsfc deteriorates to 0.67 kg/kWh. The map has that general profile which causes it to be referred to in the jargon as an "oyster" map.

The carbon monoxide emission map has a general level between 2 and 6%, which would lead one to the conclusion, based on the evidence in Figs. 7.5 and 7.8, that the air-fuel ratio used in these experimental tests was in the range of 12 to 13. By the standards of equivalent

Chapter 7 - Reduction of Fuel Consumption and Exhaust Emissions

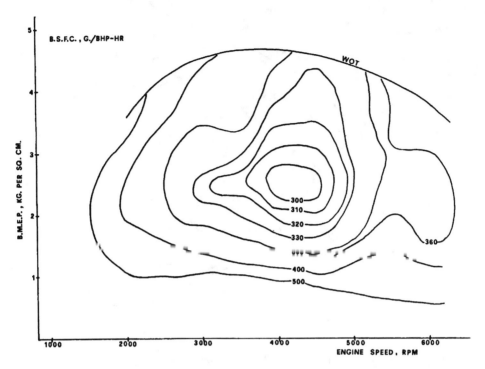

Fig. 7.9 Fuel consumption map for a 200 cc two-stroke engine (from Ref. [7.1]).

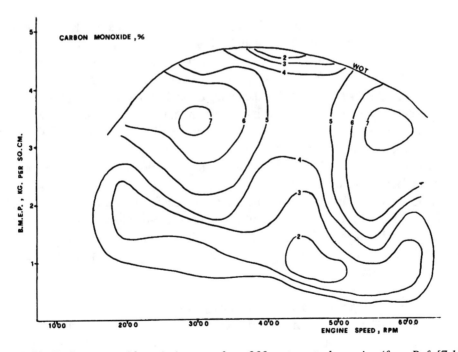

Fig. 7.10 Carbon monoxide emission map for a 200 cc two-stroke engine (from Ref. [7.1]).

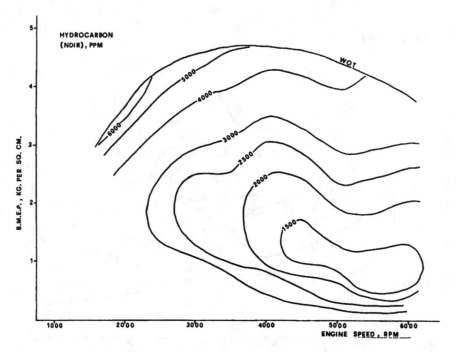

Fig. 7.11 Hydrocarbon map for a 200 cc two-stroke engine (from Ref. [7.1]).

four-stroke cycle engines, this level of CO emission would be quite normal or even slightly superior to it.

The hydrocarbon emission map, which has units in ppm from a NDIR measurement system, is directly comparable with Figs. 7.4 and 7.7, and exhibits values which are not dissimilar from those recorded for the QUB 400 engine. To be more specific, it would appear that the hydrocarbon emission levels from a simple two-stroke engine will vary from 5000 ppm at full load to 1500 ppm at light load; note that the levels quoted are those recorded by NDIR instrumentation. The recording of unburned hydrocarbons and other exhaust emission levels is discussed earlier in Sec. 1.6.2. As the combustion process is responsible for 300-400 ppm of those hydrocarbon emissions, it is clear that the scavenging process is responsible for the creation of between three and ten times the HC emission level which would be produced by an equivalent four-stroke cycle power unit.

In general, the best fuel economy and emissions occur at load levels considerably less than the peak value. This is directly attributable to the trapping characteristic for fresh charge in a homogeneously scavenged engine. The higher the scavenge ratio (whether the particular scavenging process is a good one or a bad one), the greater the load and power, but the lower the trapping efficiency.

The experimental data from Sato and Nakayama [7.2]

As the QUB 400 and the Batoni data do not contain information on nitrogen oxide emission, it is important that measurements are presented to provide you with the position occupied by the simple two-stroke engine in this regard. Such data have been provided by quite a few researchers, all of them indicating very low nitrogen oxide emission by a two-stroke powerplant. The data of Sato and Nakayama [7.2] are selected because they are very representative, are in the form of a performance engine speed map, and refer to a simple carburetted two-stroke engine with a cylinder capacity of 178 cc. The actual engine has two cylinders, each of that capacity, and is employed in a snowmobile. Although the air-to-fuel ratio is not stated, it is probably at an equivalence ratio between 0.85 and 0.9, which would be common practice for such engines.

The measured data are given in Fig. 7.12. As would be expected, the higher the load or bmep, the greater the peak cycle temperature and the level of the oxides of nitrogen. The values are shown as NO equivalent and measured as ppm on NDIR instrumentation. The highest value shown is at 820 ppm, the lowest is at 60 ppm, and the majority of the performance map is in the range from 100 to 200 ppm. This is much lower than that produced by the equivalent four-stroke engine, perhaps by as much as a factor of eight. It is this inherent characteristic, introduced earlier in Sec. 7.1.1, that has attracted several automobile manufacturers to indulge in research and development of two-stroke engines; this will be discussed further in later sections of this chapter, but it will not be a "simple" two-stroke engine which is developed for such a market requirement as its HC emissions are unacceptably high.

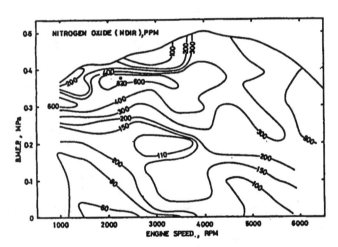

Fig. 7.12 Nitrogen oxide emission map for a 178 cc two-stroke engine (from Ref. [7.2]).

The theoretical simulation of a chainsaw and its exhaust emissions

In Chapter 4, in Sec. 5.5.1, and in Figs. 5.9-5.13, the computer simulation is shown to provide accurate correlation with the measured performance characteristics. The physical geometry of that engine is described in detail in that section. Here, the simulation is repeated at a single engine speed of 7200 rpm and the air-to-fuel ratio varied from 11.5 to 16.0.

The results of the chainsaw simulation at full throttle are given in Figs. 7.13-7.15 and are directly comparable with the measured data in Figs. 7.3-7.5 for the QUB 400 engine. In Fig. 7.13, the best torque is at an AFR of 12 and best bsfc at an AFR of 15.5; the trends are almost identical to the measured data in Fig. 7.3. In Fig. 7.14, the hydrocarbon emissions can be

Design and Simulation of Two-Stroke Engines

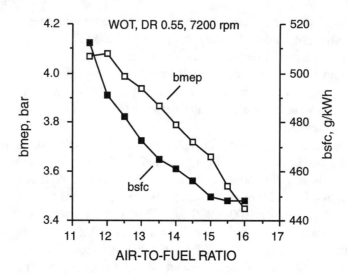

Fig. 7.13 Simulation of effect of AFR on chainsaw performance.

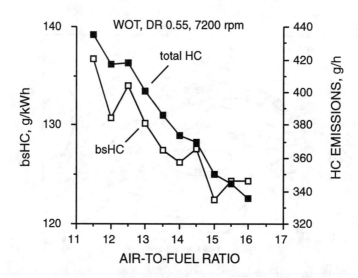

Fig. 7.14 Simulation of effect of AFR on chainsaw HC emissions.

compared with the profile of the measured data in Fig. 7.4; the rise in bsHC after the stoichiometric point is visible in both diagrams, whereas the total hydrocarbons continue to decline. In Fig. 7.15 the carbon monoxide emissions calculated are virtually identical to the profile of the measured values in Fig. 7.5 as the bsCO is seen to approach zero by the stoichiometric point; the oxygen content in the exhaust is near constant until just before the stoichiometric position and then increases thereafter. It is this constant oxygen content in the exhaust gas at rich air-to-fuel ratios, emanating from lost scavenge air flow, which is the basis for the estimation of trapping efficiency described in Sec. 1.6.3.

Chapter 7 - Reduction of Fuel Consumption and Exhaust Emissions

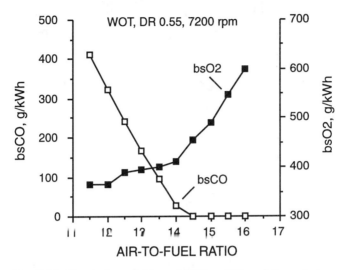

Fig. 7.15 Simulation of effect of AFR on CO and O_2 emissions.

The results of the chainsaw simulation at part throttle are given in Figs. 7.16-7.18 and are directly comparable with the measured data in Figs. 7.6-7.8 for the QUB 400 engine. The delivery ratio at a throttle area ratio, C_{thr}, of 0.15 has fallen to 0.30 from the full throttle value of 0.55. Actually, the "full throttle" condition has a throttle area ratio, C_{thr}, of 0.6, for this engine uses a diaphragm carburetor with significant obstruction from the throttle and choke butterfly plates at the venturi. The simulation value of delivery ratio of 0.30 is very comparable with the DR of 0.274 for the QUB 400 research engine. In Fig. 7.16, the best torque is at an AFR of 11.5 and best bsfc at an AFR of 15.5; thus the trends are very similar to the measured data in Fig. 7.6. In Fig. 7.17, the hydrocarbon emissions can be compared with

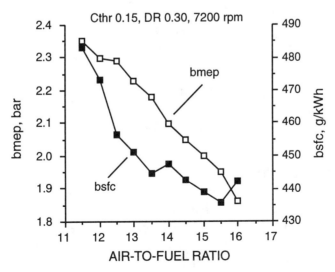

Fig. 7.16 Simulation of effect of AFR on part-throttle performance.

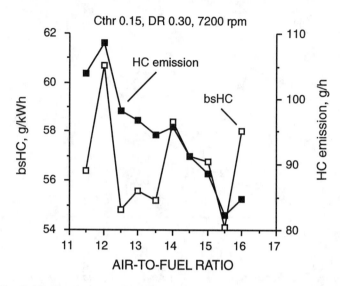

Fig. 7.17 Simulation of effect of AFR on part-throttle HC emissions.

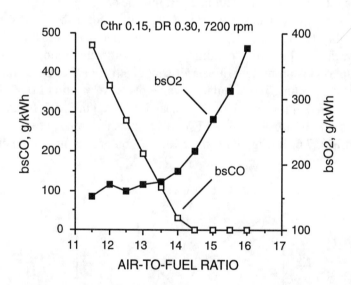

Fig. 7.18 Simulation of effect of AFR on part-throttle CO and O_2 emissions.

measured data in Fig. 7.7: the rise in bsHC after the stoichiometric point is visible in both diagrams, but that from the simulation shows greater variability, in part visually due to the scale chosen for that graph. The total hydrocarbons level off after the stoichiometric point and this effect is also seen from the simulated data. In Fig. 7.18, the carbon monoxide emissions calculated are identical to the profile of the measured values in Fig. 7.8. The calculated oxygen content in the exhaust is similar in profile to that measured in Fig. 7.8.

The results of the simulation, either in magnitude or in profile, are sufficiently close to those measured that the simulation can be employed in the design mode with some considerable degree of confidence that its predictions are suitably relevant and accurate.

The energy content in exhaust gas emissions

Exhaust gas which contains carbon monoxide and hydrocarbons is transmitting energy originally contained within the fuel into the exhaust system and the atmosphere. This energy content, $\dot{Q}_{ex}$, is determined from the power output, $\dot{W}$, the specific hydrocarbon emission rate, bsHC, the specific carbon monoxide emission rate, bsCO, the calorific value of the fuel, C_{flHC}, and the calorific value of carbon monoxide, C_{flCO}. They are related thus, in conventionally employed units, where the units of power output are kW, bsCO and bsHC are g/kWh, and the calorific values of carbon monoxide and the fuel are expressed as MJ/kg:

$$\dot{Q}_{ex} = \frac{\dot{W}}{3.6}(C_{flCO} \times bsCO + C_{flHC} bsHC) \quad W \qquad (7.2.1)$$

As a typical example of the simple two-stroke engine, consider an engine with a power output of 4 kW, specific carbon monoxide and hydrocarbon emission rates of 160 and 120 g/kWh, respectively, and with fuel and carbon monoxide calorific values of 43 and 10 MJ/kg, respectively. The exhaust energy content is found by:

$$\begin{aligned}\dot{Q}_{ex} &= \frac{\dot{W}}{3.6}(C_{flCO} \times bsCO + C_{flHC} bsHC) \\ &= \frac{4}{3.6}(10 \times 160 + 43 \times 120) = 7511 \text{ W}\end{aligned}$$

Observe that this amounts to 7.5 kW, or nearly twice the power output of the engine. The energy being "thrown away" in this fashion is insupportable in the environmental context.

Should this energy be realized in the exhaust system, either by a reactor or by a catalyst, the very considerable heat output would raise the exhaust gas temperature by many hundreds of degrees.

7.3 Optimizing fuel economy and emissions for the simple two-stroke engine

In Sec. 7.2, the problems inherent in the design of the simple two-stroke engine are introduced and typical performance characteristics are presented. Thus, you are now aware of the difficulty of the task which is faced, for even with the best technology the engine is not going to be competitive with a four-stroke engine in terms of hydrocarbon emission. In all other respects, be it specific power, specific bulk, specific weight, maneuverability, manufacturing cost, ease of maintenance, durability, fuel consumption, or CO and NO emissions, the simple two-stroke engine is equal, and in some respects superior, to its four-stroke competitor. There may be those who will be surprised to see fuel consumption in that list, but investigation shows that small-capacity four-stroke engines are not particularly thermally efficient. The reason is that the friction losses of the valve gear and oil pump begin to assume considerable proportions as the cylinder size is reduced, and this significantly deteriorates the mechanical efficiency of the engine.

In Sec. 7.2 there are options listed which are open to the designer, and the remainder of this section will be devoted to a closer examination of some of the options for the optimization of an engine. In particular, the computer simulation will be used to illustrate the relevance of some of those assertions. For others, experimental data will be introduced to emphasize the point being made. This will reinforce much of the earlier discussion in Chapter 5.

7.3.1 The effect of scavenging on performance and emissions

In Chapter 3, you are introduced to the experimental and theoretical methods for the improvement of the scavenging of an engine. For the enhancement of the performance characteristics of any engine, whatever its performance target, be they for a racing engine or one to meet legislated demands on exhaust emissions, better quality scavenging always translates into more air and fuel trapped with less of it lost to the exhaust system. To further comprehend this point, refer to the experimental data in the thesis by Kenny [3.18], or his associated papers listed in Chapter 3. To demonstrate the potential effect on emissions and performance by improving scavenging over a wider spectrum than that given experimentally by Kenny, a theoretical simulation is carried out. The simulation is for the chainsaw at full throttle as given in Sec. 5.4.1, at full throttle at 7200 rpm at an air-fuel ratio of 14.0 on premium unleaded gasoline. The scavenging is changed successively from the standard LOOPSAW characteristic, to UNIFLOW, SCRE, YAM12, and GPBDEF quality scavenging as defined in Table 3.16 and shown in Figs. 3.10-3.15. While it may be totally impractical to consider the incorporation of uniflow scavenging in a chainsaw, it is highly instructive to determine the effect on emissions and power of the employment of the (so-called) optimum in scavenging characteristics. It is equally useful to know the impact on these same performance-related parameters of the use of the worst scavenging, namely the YAM12 characteristics. Recall that the others are in between these extremes, with SCRE being a very good loop-scavenging system, while GPBDEF is an unconventional form of cross scavenging.

The results of the simulation are shown in Figs. 7.19 and 7.20. Fig. 7.19 gives the results for bmep, bsfc and bsHC. In Fig. 7.20 are the output data for delivery ratio, and the scavenging, trapping and charging efficiencies. It is not surprising that UNIFLOW scavenging gives the best performance characteristics. By comparison with YAM12, it is 16% better on bmep and power, 19% better on bsfc, 31% better on bsHC, 7% better on bsCO, 7.7% better on SE, and 11.4% better on trapping efficiency. The effect on hydrocarbon emissions is quite dramatic.

As comparison with uniflow scavenging is somewhat unrealistic, it is more informative for the designer to know what order of improvement is possible by the attainment of the ultimate in loop-scavenging characteristics, namely the SCRE scavenging quality. In making this comparison recall that the quality of the LOOPSAW scavenging is already in the "very good" category. Comparing the SCRE simulation results with those calculated using LOOPSAW scavenging, it is 5% better on bmep and power, 5.4% better on bsfc, 14% better on specific hydrocarbon emissions, 2% better on bsCO, 2.5% better on scavenging efficiency, and 3.4% better on trapping efficiency. The effect on bsHC is still considerable.

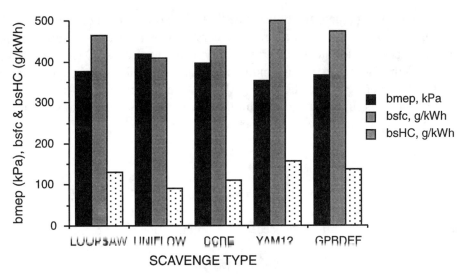

Fig. 7.19 Effect of scavenging on performance and emissions.

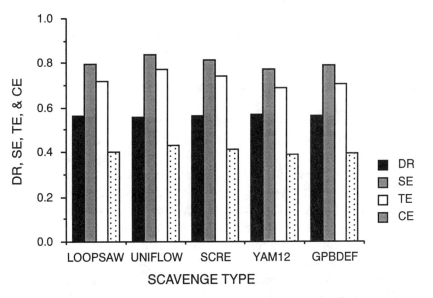

Fig. 7.20 Effect of scavenging system on charging of a chainsaw.

The order of magnitude of these changes of performance characteristics with respect to alterations in scavenging behavior are completely in line with those measured by Kenny [3.18] and at accuracy levels as previously seen in experiment/theory correlations I have carried out [3.35].

The contention that scavenging improves performance characteristics is clearly correct. However, the order of improvement may not be sufficient to permit the chainsaw to pass legislated levels for exhaust hydrocarbon emissions. The best performance seen is for

UNIFLOW scavenging where the brake specific hydrocarbon emission is still very high at 92.3 g/kWh. By comparison with a small industrial four-stroke cycle engine this remains inadequately excessive, for such an engine will typically have a bsHC emission of between 15 and 30 g/kWh. However, in some mitigation, many small four-stroke industrial engines have bsCO emission levels exceeding 200 g/kWh to help reduce the NO_x emission levels by running rich, whereas this chainsaw simulation shows its bsCO level to be possible at 25 g/kWh by operating close to the stoichiometric air-to-fuel ratio.

7.3.2 The effect of air-fuel ratio

It can be seen from the measured or calculated data in Figs. 7.3-7.18 that optimizing the air-to-fuel ratio means that it should be at one of two levels. If peak power and torque is the design aim then an equivalence ratio, λ, of 0.85 will provide that requirement. If the minimum emissions and fuel consumption are needed then optimization at, or close to, an equivalence ratio, λ, of unity is essential. For the simple two-stroke engine of conventional design that will almost certainly not be good enough to satisfy current or envisaged legislation.

The most important message to the designer is the vital importance of having the fuel metered to the engine in the correct proportions with the air at every speed and load. There are at least as large variations of bsfc and bmep with inaccurate fuel metering as there is in allowing the engine to be designed and manufactured with bad scavenging.

There is a tendency in the industry for management to insist that a cheap carburetor be installed on a simple two-stroke engine, simply because it is a cheap engine to manufacture. It is quite ironic that the same management will often take an opposite view for a four-stroke model within their product range, and for the reverse reason!

7.3.3 The effect of optimization at a reduced delivery ratio

It is clearly seen from Figs. 7.3-7.18 that a reduction of delivery ratio naturally reduces the power and torque output, but also very significantly reduces the fuel consumption and hydrocarbon emissions of the engine. The reason is obvious from Chapter 3—a reduction of scavenge ratio for any scavenging system raises the trapping efficiency.

Hence, at the design stage, serious consideration can be given to the option of using an engine with a larger swept volume and optimizing the entire porting and inlet system to operate with a lower delivery ratio to attain a more modest bmep at the design speed. The target power is then attained by employing a larger engine swept volume. In this manner, with an optimized scavenging and air-flow characteristic, the lowest fuel consumption and exhaust emissions will be attained at the design point.

The selection of the scavenging characteristic for such an approach is absolutely critical. The design aim is to approach a trapping efficiency of unity over the operational range of scavenge ratios. The candidate systems which could accomplish this are illustrated in Fig. 3.13. There are three scavenging systems which have a trapping efficiency of unity up to a scavenge ratio of 0.5. They are UNIFLOW, QUBCR and GPBDEF. The uniflow system can be rejected on the grounds that it is unlikely to be accommodated into a simple two-stroke engine. The remaining two are cross-scavenged engines, and the GPBDEF design is the better of these in that it has a trapping efficiency of unity up to a scavenge ratio (by volume) of 0.6. The physical arrangement of this porting is shown in Fig. 3.32(b).

Chapter 7 - Reduction of Fuel Consumption and Exhaust Emissions

To illustrate this rationale, the chainsaw engine geometry is redesigned to incorporate GPBDEF scavenging, and the exhaust, transfer and intake porting is reoptimized to flow air at 7200 rpm up to a maximum delivery ratio of 0.35 at full throttle. The porting, which resembles Fig. 3.32(b), has four exhaust and transfer ports each of 9 mm effective width and they are timed to open at 124° atdc and 132° atdc, respectively. The inlet port is now timed to open at 45° btdc and is a single port with the same width as before. The greatly reduced port timings and areas are obvious by comparison with the standard data given in Sec. 5.4.1. The standard data show the exhaust and transfer ports open at 108° atdc and 121° atdc, respectively, while the standard inlet port opens at 75° btdc. All other physical data are retained as given in Sec. 5.4.1. The computer simulation is run at 7200 rpm over a range of air-fuel ratios and presented in Figs. 7.21-7.25. Also incorporated on the same figures are the results of the simulation when the scavenging quality defined by GPBDEF is replaced by the original loop-scavenging quality of LOOPSAW. On the figures, the results of the two simulations are annotated as "loop" or "deflector."

The point is made above that selection of scavenging quality is critical in this design approach. This is shown very clearly in Figs. 7.21 to 7.25. The trapping efficiency in Fig. 7.25 shows that the GPBDEF scavenging reaches 95%, while the LOOPSAW scavenging gets to a very creditable, but critically lower, 90%. Both seem acceptably high, but the outcome in terms of hydrocarbon emissions is quite different. The deflector piston design achieves the target bsHC emission of 25 g/kWh, but the loop-scavanged design has double that value at 50 g/kWh. The specific oxygen emission figures in Fig. 7.13 reinforce the points made regarding the dissimilar trapping efficiency levels. The fuel consumption of the deflector piston engine is 10% better than the loop-scavenged design (see Fig. 7.22).

In Fig. 7.24 it is seen that the loop-scavenged engine breathes marginally more air than the deflector piston engine, but its higher trapping efficiency charges the engine, as observed from the charging efficiency values, some 10% higher at 0.34. Naturally, the better trapping efficiency of the deflector piston approach yielded a higher scavenging efficiency; this is seen in Fig. 7.25, where it is marginally better by 1% at 0.65. At a scavenging efficiency of 0.65,

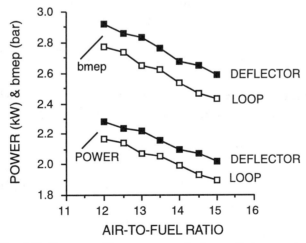

Fig. 7.21 Power and bmep output of a low emissions engine.

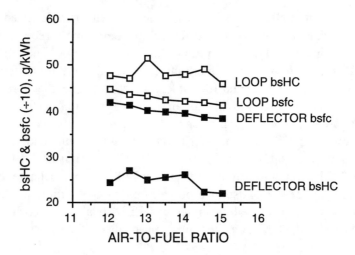

Fig. 7.22 Fuel consumption and bsHC output of a low emissions engine.

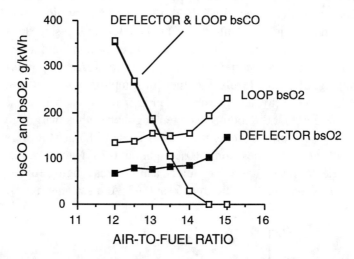

Fig. 7.23 Oxygen and carbon monoxide emission of a low emissions engine.

homogeneous combustion at a reasonable efficiency without misfire will still be possible. The bsfc level below 400 g/kWh seen in Fig. 7.22 is some 12% better than the full throttle case shown in Fig. 7.13, and is 10% better than the part-throttle chainsaw data in Fig. 7.16.

If you compare the performance characteristics for the chainsaw at part-throttle in Figs. 7.16 to 7.18 with those from the optimized low bmep engine in Figs. 7.21-7.25, you can see that more power and less HC emissions have been obtained at virtually the same delivery or scavenge ratios.

The design assumption is that the original chainsaw target regarding power performance is still required. It is 3.0 kW at 7200 rpm. The original chainsaw produced 3.8 bar bmep at an air-to-fuel ratio of 14 (see Fig. 7.13). The low bmep design approach, incorporating a deflec-

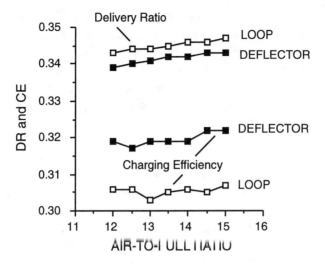

Fig. 7.24 Airflow and charging of a low emissions engine.

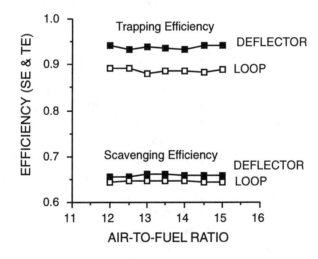

Fig. 7.25 Scavenging and trapping efficiencies of a low emissions engine.

tor piston and cross scavenging, yielded a bmep of 2.7 bar at the same fueling level (see Fig. 7.21). Thus, to achieve equality of power output at the same engine speed, the low bmep design would need to have a larger swept volume in direct proportion to these bmep levels, which would mean an engine displacement increase from 65 cm^3 to 91 cm^3. The specific hydrocarbons emitted would be reduced from 128 g/kWh to 25 g/kWh, which is over five times lower. In terms of energy emitted into the exhaust system, the total hydrocarbon emissions are reduced from 370 g/h to 72.3 g/h.

If the engine is operated at an air-to-fuel ratio of 14, the bsCO emission is 25 g/kWh. The energy content, $\dot{Q}_{ex}$, released into the exhaust system as calculated from Eq. 7.2.1, is 1.1 kW. The original design rejected 4.8 kW into the exhaust system at the same air-to-fuel ratio. The potential for the application of a catalyst to oxidize the bypassed fuel and carbon monoxide, without incurring an excessive rise in exhaust gas temperature, becomes a possibility for the optimized design but is a most unlikely prospect for the orginal concept.

The engine durability should also be improved by this methodology as the thermal loading on the piston will be reduced. There is a limit to the extent to which this design approach may be conventionally taken, as the engine will be operating ever closer to the misfire limit from a scavenging efficiency standpoint.

As a historical note, in the 1930s a motorcycle with a two-stroke engine was produced by the English company of Velocette [7.27]. The 250 cm^3 single-cylinder engine had a bore and stroke of 63 and 80 mm, respectively, with seperate oil pump lubrication and was cross scavenged by a deflector piston design very similar in shape to Fig. 3.32(a). The "part-spherical" combustion chamber was situated over the exhaust side of the piston and the port timings were not dissimilar to those discussed above for the optimized low bmep engine. It produced some 9 hp at 5000 rpm, i.e., a bmep of about 3 bar. It sold for the princely sum of £38 (about $52)! The road tester [7.27] noted that "slow running was excellent. The engine would idle at very low rpm without four-stroking—one of the bugbears of the two-stroke motor." Perhaps an optimized two-stroke engine design to meet fuel consumption and emissions requirements, not to speak of the incorporation of active radical combustion, is nothing new!

7.3.4 The optimization of combustion

The topic of homogeneous combustion is covered in Chapter 4. Since Chapter 1, and the presentation of Eq. 1.5.22, where it is shown that the maximum power and minimum fuel consumption will be attained at the highest compression ratio, you have doubtless been waiting for design guidance on the selection of the compression ratio for a given engine. However, as mentioned in Chapter 4, the selection of the optimum compression ratio is conditioned by the absolute necessity to minimize the potential of the engine to detonate. Further, as higher compression ratios lead to higher cylinder temperatures, and the emission of oxides of nitrogen are linked to such temperatures, it is self-evident that the selection of the compression ratio for an engine becomes a compromise between all of these factors, namely, power, fuel consumption, detonation and exhaust emissions. The subject is not one which is amenable to empiricism, other than the (ridiculously) simplistic statement that trapped compression ratios, CR_t, of less than 7, operating on a gasoline of better than 90 octane, rarely give rise to detonation.

The correct approach is one using computer simulation, and in Appendix A7.1 you will find a comprehensive discussion of the subject, using the "standard" chainsaw engine as the background input data to a computer simulation with a two-zone combustion and emissions model, as previously described in the Appendices to Chapter 4.

Active radical combustion

One aspect, active radical (AR) combustion, is described briefly in Sec. 4.1.3. It deserves further amplification as it will have great relevance for the optimization of the simple two-stroke engine to meet emissions legislation at light load and low engine speed, including the idle (no load) condition. The first paper on this topic is by Onishi [4.33] and the most recent is by Ishibashi [4.34]. The combustion process is provided by the retention of a large proportion of residual exhaust gas, i.e., at scavenging efficiencies of 65% or less. Onishi claims that the scavenge ducts should be throttled to give a low turbulence scavenge process to reduce mixing and so stratify the residual exhaust gas as to retain it at a high temperature. Ishibashi suggests that transfer duct throttling is not necessary but that exhaust throttling provides a similar outcome. He implies that raising the trapping pressure, i.e., increasing the apparent trapped compression ratio by throttling the exhaust system close to the exhaust port, is equally effective in providing the correct in-cylinder state and turbulence conditions to promote AR combustion.

The ensuing combustion process provides combustion on every cycle, which eliminates the "four-stroking" at similar load levels in homogeneous combustion. The AR combustion is smooth, detonation free, and has a dramatic effect on fuel consumption and HC emissions at low bmep levels. Ishibashi [4.34] describes a flat HC emissions profile at 20 g/kWh from 1 bar bmep to 3 bar bmep. In homogeneous combustion the HC emission ranges from equality with AR combustion at 3 bar bmep to three times worse at the 1 bar bmep level. The bsfc over this same bmep span is under 400 g/kWh, whereas with homogeneous combustion it rises to over 600 g/kWh.

It is clear that, where possible, AR combustion should be used in combination with the other methods described above to optimize engine-out emissions and reduce fuel consumption rates. Its incorporation requires high trapping pressures. The very low port timings of the optimized low bmep engine, as described above, may well satisfy this design criterion. However, where further exhaust throttling is needed to accomplish this aim, there are two approaches which can be adopted. There are two basic mechanical techniques to accomplish the design requirement for an exhaust port restriction to improve the light load behavior of the engine. The two methods are illustrated in Fig. 7.26 and discussed below.

The butterfly exhaust valve

Shown in Fig. 7.26(a) is a butterfly valve and the concept is much like that described by Tsuchiya *et al.* [7.3]. This is a relatively simple device to manufacture and install, and has a good record of reliability in service. The ability of such a device to reduce exhaust emissions of unburned hydrocarbons is presented by Tsuchiya [7.3], and Fig. 7.27 is from that paper. Fig. 7.27 shows the reduction of hydrocarbon emissions, either as mass emissions in the top half of the figure or as a volumetric concentration in the bottom half, from a Yamaha 400 cm^3 twin-cylinder road motorcycle at 2000 rpm at light load. The notation on the figure is for CR which is the exhaust port area restriction posed by the exhaust butterfly valve situated close to the exhaust port. The CR values range from 1, i.e., completely open, to 0.075, i.e., virtually closed. It is seen that the hydrocarbons are reduced by as much as 40% over a wide load variation at this low engine speed, emphasizing the theoretical indications discussed above with regard to its enhancement of trapping efficiency.

While Tsuchiya [7.3] reports that the engine behaved in a much more stable manner when the exhaust valve was employed at light load driving conditions in an urban situation, he makes no comment on active radical (AR) combustion and appears not to have experienced it during the experimental part of the research program.

Design and Simulation of Two-Stroke Engines

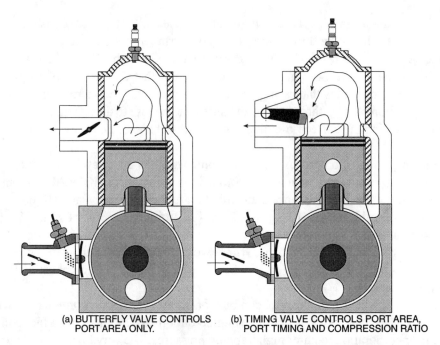

(a) BUTTERFLY VALVE CONTROLS PORT AREA ONLY.

(b) TIMING VALVE CONTROLS PORT AREA, PORT TIMING AND COMPRESSION RATIO

Fig. 7.26 Variable exhaust port area and port timing control devices.

The exhaust timing edge control valve

It is clear from Fig. 7.27(a) that the butterfly valve controls only the area at the exhaust port rather than the port opening and closing timing edges as well. On Fig. 7.27 is sketched the timing control valve originally shown in Fig. 5.2, which fits closely around the exhaust port and can simultaneously change both the port timing and the port area. The effectiveness of changing exhaust port timing is demonstrated in the simulations discussed in Secs. 5.2.2 and 7.3.4. There are many innovative designs of exhaust timing edge control valve ranging from the oscillating barrel type to the oscillating shutter shown in Fig. 7.27. The word "oscillating" may be somewhat confusing, so it needs to be explained that it is stationary at any one load or speed condition but it can be changed to another setting to optimize an alternative engine load or speed condition. While the net effect on engine performance of the butterfly valve and the timing edge control valve is somewhat similar, the timing edge control valve carries out the function more accurately and effectively. Of course, the butterfly valve is a device which is cheaper to manufacture and install than the timing edge control device.

Concluding remarks on AR combustion and port timing control

The fundamental message to the designer is that control over the exhaust port timing and area has a dramatic influence on the combustion characteristics, the power output, the fuel economy and the exhaust emissions at light load and low engine speeds.

7.3.5 Conclusions regarding the simple two-stroke engine

The main emphasis in the discussion above is that the simple two-stroke engine is capable of a considerable level of optimization by design attention to scavenging, combustion,

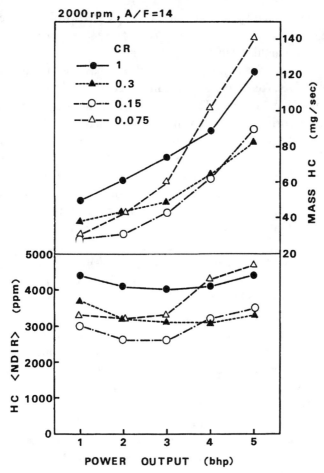

Fig. 7.27 Emissions reduction using a restriction at the exhaust port by a butterfly valve (from Ref. [7.3]).

carburetion, lubrication, and exhaust timing/area control. However, even the well-optimized design may still have an unacceptably high emission of unburned hydrocarbons even though the carbon monoxide and nitrogen oxide levels are acceptably low, perhaps even very low. This depends on the stringency of the legislation regarding the application of the engine. Legislated levels for exhaust pollutants from automobiles are extremely low. They are somewhat less stringent for handheld power tools such as weed trimmers and chainsaws.

The simple two-stroke engine, optimized at best, has a low CO and NO_x exhaust pollutant level, but a high HC and O_2 exhaust emission output. This leaves the engine with the possibility of utilizing an oxidation catalyst in the exhaust to remove the hydrocarbons and further lower the carbon monoxide levels.

An early paper on this subject by Uchiyama et al. [7.11] showed that a small Suzuki car engine could have the hydrocarbon emission reduced quite significantly by exhaust gas aftertreatment in this manner. They reported an 80% reduction of the hydrocarbon exhaust emission by an oxidizing catalyst.

More recently, and aimed specifically at the simple two-stroke engine used in mopeds, chainsaws and small motorcycles, Laimbock [7.21] presents experimental data on the effect of using the advice given in this chapter. He shows the results for a 125 cm^3 high-performance motorcycle engine when the scavenging and carburetion have been optimized and an exhaust timing edge control valve is used. For such small motorcycles there are emission control laws in Switzerland, Austria and Taiwan. The most severe of these is in Switzerland, where the machine must execute a driving cycle and emit no more than 8 g/km of CO, 3 g/km of HC and 0.1 g/km of NO_x. Laimbock shows that a production 125 cm^3 motorcycle engine, which has a peak bmep of 8 bar at 9000 rpm and is clearly a high specific output power unit, has emissions on this cycle of 21.7 g/km of CO, 16.9 g/km of HC and 0.01 g/km of NO_x. Clearly this motorcycle is unsuitable for sale within such regulations. By optimizing the scavenging and carburetion, the same machine will have emission characteristics on the same cycle of 1.7 g/km of CO, 10.4 g/km of HC and 0.03 g/km of NO_x. Thus, the optimization procedures already discussed in the chapter lowered the CO and HC significantly, but raised the NO_x levels. The HC level is still unacceptable from a Swiss legal standpoint. By introducing an oxidation catalyst into the tuned exhaust pipe of this engine in the manner shown in Fig. 7.28, Laimbock provides experimental evidence that the peak power performance of the motorcycle is barely affected, but the emissions are dramatically reduced. In this case, where the catalyst is of the oxidizing type, the test results on the Swiss driving cycle gave emission levels of 0.8 g/km of CO, 1.9 g/km of HC and 0.02 g/km of NO_x; such a machine is now well within the limits pending or proposed by many legislative bodies worldwide.

As far as fuel consumption is concerned, Laimbock [7.21] shows that the original 125 cc production motorcycle on the test driving cycle had a fuel consumption level of 20.8 km/liter, the model with improved scavenging and carburetion did 29.5 km/liter, while the final version with the exhaust catalysts fitted traveled 31.2 km/liter of gasoline.

There is no logical reason why a similar approach cannot be successful for any type of simple two-stroke cycle engine.

Finally, it is possible that the contribution of internal combustion engines to the atmospheric pollution by nitrous oxide, N_2O, may be a very important factor in the future [7.24]. There is every indication that a two-stroke engine produces this particular nitrogen oxide component in very small quantities by comparison with its four-stroke engine counterpart.

7.4 The more complex two-stroke engine

If the two-stroke engine is to have relevance in the wider automotive application, the raw level of unburned hydrocarbons in the exhaust system before catalytic after-treatment will have to be further reduced while the engine retains its high specific power output. Equally, levels of specific fuel consumption in the range of 250 to 300 g/kWh will be required over much of the speed and load range. In this case, it is essential that no fuel is ever lost to the exhaust system during scavenging, thereby deteriorating the thermal efficiency of the engine. It is clear from the earlier discussion in this chapter that the most miniscule quantity lost in this manner gives unacceptably high levels of HC emission. Clearly, an engine designed to accomplish these criteria is going to be much more mechanically or electronically complex.

The Achilles' heel of the simple two-stroke engine is the loss of fuel when it is supplied in conjunction with the scavenge air. Remove this problem, albeit with added complexity, and

Chapter 7 - Reduction of Fuel Consumption and Exhaust Emissions

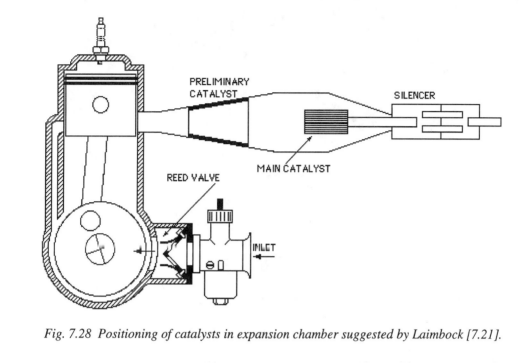

Fig. 7.28 Positioning of catalysts in expansion chamber suggested by Laimbock [7.21].

the fuel economy and hydrocarbon emissions of the engine are significantly improved. The fundamental requirement in design terms is shown in Fig. 7.29. Somewhere in the cylinder head or cylinder wall is placed a "device" which will supply fuel, or a mixture of fuel and air, into the cylinder in such a manner that none of the fuel is lost into the exhaust duct during the open cycle period. Although the sketch shows a two-stroke engine with crankcase scavenging, this is purely pictorial. The fundamental principle would apply equally well to an engine with a more conventional automotive type of crankshaft with pressure oil-fed plain bearings and the scavenge air supplied by a pump or a blower; in short, an engine as sketched in Fig. 1.6, described initially in Sec. 1.2.4, simulated in Sec. 5.4.3, or as designed by Thornhill [5.23].

The simplest idea which immediately comes to mind as a design solution is to use the diesel engine type of liquid injection system to spray the fuel into the cylinder after the exhaust port is closed. This straightforward approach is sketched in Fig. 7.30. Naturally, the fuel injection system is not limited to that generally employed for diesel engines and several other types have been designed and tested, such as those proposed by Beck [7.16], Schlunke [7.26], Stan [7.28] or Heimberg [7.54]. Not many analytical or experimental studies have been carried out on the direct injection of gasoline; however, the papers by Emerson [7.43], Ikeda [7.50] or Sinnamon [7.55] deserve study.

However, it is possible that this elegantly simple solution to the problem is not as obviously effective as it might seem. In Chapter 4, the combustion process by spark ignition of a fuel and air mixture is detailed as being between a homogeneous mixture of fuel vapor and air. It is conceivable that a liquid fuel injected in even the smallest droplets, such as between 10 and 15 μm, may still take too long to mix thoroughly with the air and evaporate completely

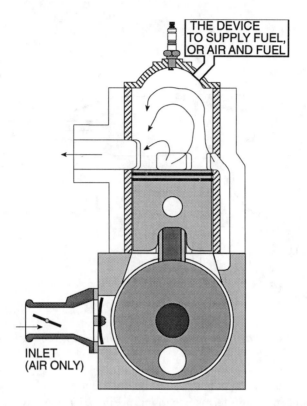

Fig. 7.29 The fundamental principle for the reduction of hydrocarbon on emissions and fuel consumption.

in the relatively short time period from exhaust port closure until the ignition point before the tdc position. Put in the simplest terms, if the end of the fuel injection of the gasoline is at 90° before the ignition point, and the engine is running at 6000 rpm, this implies a successful evaporation and mixing process taking place in 0.0025 second, or 2.5 ms. That is indeed a short time span for such an operation when one considers that the carburetted four-stroke cycle engine barely accomplishes that effect in a cycle composed of 180° of induction period, followed by 180° of compression process, i.e., four times as long. In the simple two-stroke cycle engine, the induction process into a "warm" crankcase for 180° of engine rotation helps considerably to evaporate the fuel before the commencement of the scavenge process, but even then does not complete the vaporization process, as is pointed out in the excellent presentation by Onishi *et al.* [7.8]. Further discussion on this is in Sec. 7.4.2 dealing with direct in-cylinder fuel injection.

Stratified charging

The potential difficulty regarding the adequate preparation of the fuel and air mixture prior to the combustion process opens up many solutions to this design problem. The patent

Chapter 7 - Reduction of Fuel Consumption and Exhaust Emissions

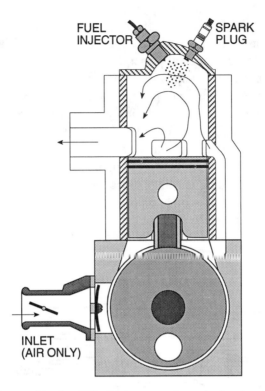

Fig. 7.30 Relative positioning of fuel injector and spark plug for direct in-cylinder injection of fuel in a spark-ignition engine.

literature is full of them. The basic approach is the stratified charging of the cylinder by the "device" of Fig. 7.29 with a premixed charge of vaporized fuel and air followed by mixing with the trapped cylinder charge and a homogeneous combustion process.

The fundamental advantages and disadvantages of stratified charging and stratified combustion processes have already been discussed in Sec. 7.1.2.

In the following section, some design approaches to stratified charging will be discussed in greater detail, presented together with the known experimental facts as they exist at this point in history. These are very early days in the development of this form of the two-stroke cycle engine.

7.4.1 Stratified charging with homogeneous combustion

Stratified charging has been tested in various forms and by several research groups and organizations. The most significant are discussed below in terms of their applicability for future use in production as power units which will rival the four-stroke engine in hydrocarbon emissions and fuel consumption levels, but which must retain the conventional advantages with respect to carbon monoxide and nitrogen oxide emissions.

Design and Simulation of Two-Stroke Engines

The QUB stratified charging engine

This early work is presented in technical papers published by Blair and Hill [7.9] and by Hill and Blair [7.10]. The fundamental principle of operation of the engine is illustrated in Fig. 7.31. The overriding requirement is to introduce a rich mixture of air and fuel into the cylinder during the scavenge process at a position as remote as possible from the exhaust port. Ideally, the remaining transfer ports would supply air only into the cylinder. The engine has two entry ports for air, a main entry for 80% of the required air into the crankcase, and a subsidiary one for the remaining air and for all of the necessary fuel into a long storage transfer port. That port and transfer duct would pump the stored contents of air and fuel into the cylinder during the succeeding scavenge process so that no fuel migrated to the crankcase. In the meantime, during the induction and pumping period, the fuel would have some residence time within the air and on the walls of the long rear transfer port so that some evaporation of the fuel would take place. In this manner the cylinder could be supplied with a premixed and partially evaporated fuel and air mixture in a stratified process. The resultant

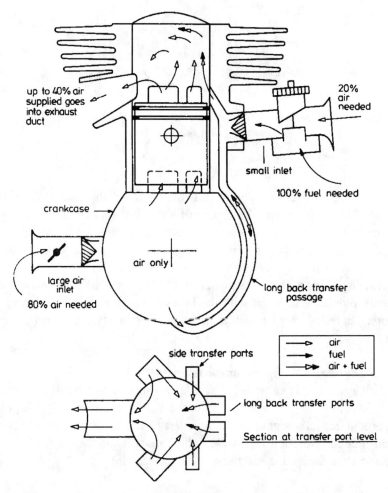

Fig. 7.31 Air and fuel flow paths in a QUB-type stratified charging system.

mixing with the trapped charge of cylinder air and retained exhaust gas would permit a homogeneous combustion process.

The test results for the engine, shown in Figs. 7.32 and 7.33 as fuel consumption and bmep levels at several throttle openings, reveal significantly low levels of fuel consumption. Most of the bmep range from 2 bar to 5.4 bar over a speed range of 1500 to 5500 rpm, but the bsfc levels are in the band from 360 to 260 g/kWh. These are particularly good fuel consumption characteristics, at least as good if not superior to an equivalent four-stroke cycle engine, and although the hydrocarbon emission levels are not recorded, they must be significantly low with such good trapping of the fuel within the cylinder. The power performance characteristics are unaffected by this stratified charging process, for the peak bmep of this engine at 5.4 bar is quite conventional for a single-cylinder engine operating without a tuned exhaust system.

The mechanical nature of the engine design is relatively straightforward, and it is one eminently suitable for the conversion of a simple two-stroke cycle engine. The disadvantages are the extra complication caused by the twin throttle linkages and the accurate carburetion of a very rich mixture by a carburetor. The use of a low-pressure fuel injection system to replace the carburetor [7.51] would simplify that element of the design at the further disadvantage of increasing the manufacturing costs.

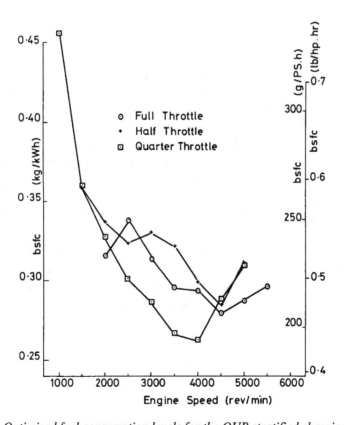

Fig. 7.32 Optimized fuel consumption levels for the QUB stratified charging engine.

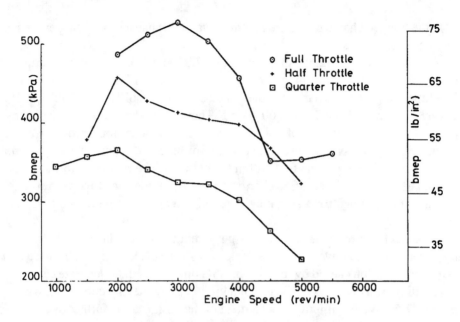

Fig. 7.33 BMEP levels at the optimized fuel consumption levels for the QUB stratified charging engine.

The Piaggio stratified charging engine

The fundamental principle of operation of this power unit is shown in Fig. 7.34 and is described in much greater detail in the paper by Batoni [7.1] of Piaggio. This engine takes the stratified charging approach to a logical conclusion by attaching two engines at the cylinder head level. The crankshafts of the two engines are coupled together in the Piaggio example by a toothed rubber belt. In Batoni's paper, one of the engines, the "upper" engine of the sketch in Fig. 7.34, has 50 cm^3 swept volume, and the "lower" engine has 200 cm^3 swept volume. The crankcase of both engines ingest air and the upper one inhales all of the required fuel for combustion of an appropriate air-fuel mixture in a homogeneous process. The crankcase of the upper engine supplies a rich mixture in a rotating, swirling scavenge process giving the fuel as little forward momentum as possible toward the exhaust port. The lower cylinder conducts a conventional loop-scavenge process with air only. Toward the end of compression the mixing of the rich air-fuel mixture and the remaining trapped cylinder charge takes place, leading to a homogeneous combustion process.

The results of the experimental testing of this 250 cm^3 Piaggio engine are to be found in the paper by Batoni [7.1], but are reproduced here as Figs. 7.35-7.37. A direct comparison can be made between this stratified charging engine and the performance characteristics of the 200 cm^3 engine that forms the base of this new power unit. Figs. 7.9-7.11, already discussed fully in Sec. 7.2.1.2, are for the 200 cm^3 base engine. Fig. 7.9 gives the fuel consumption behavior of the 200 cm^3 base engine, Fig. 7.10 the CO emission levels, and Fig. 7.11 the HC emission characteristics.

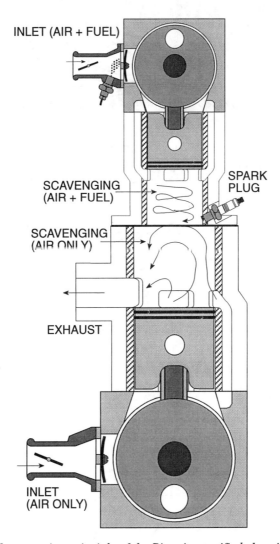

Fig. 7.34 The operating principle of the Piaggio stratified charging engine.

Fig. 7.35 shows the fuel consumption levels of the experimental engine (note that 1 g/kWh = 0.746 g/bhp.hr). The lowest contour in the center of the "oyster" map is 240 g/hp.hr or 322 g/kWh. The units of bmep on this graph are in kg/cm^2, which is almost exactly equal to a bar (1kg/cm^2 = 0.981 bar). These are quite good fuel consumption figures, especially when you consider that this engine is one of the first examples of stratified charging presented; the paper was published in 1978. The reduction of fuel consumption due to stratified charging is very clear when you compare Figs. 7.35 and 7.9. The minimum contour is lowered from 300 g/bhp.hr to 240 g/bhp.hr, a reduction of 33%. At light load, around 1 bar bmep and 1500 rpm, the fuel consumption is reduced from 500 to 400 g/bhp.hr, or 20%. This condition is particularly important for power units destined for automotive applications as so many of the test cycles for automobiles or motorcycles are formulated to simulate urban driving con-

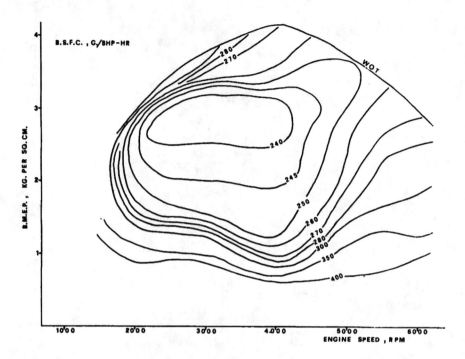

Fig. 7.35 Fuel consumption levels from the Piaggio stratified charging engine (from Ref. [7.1]).

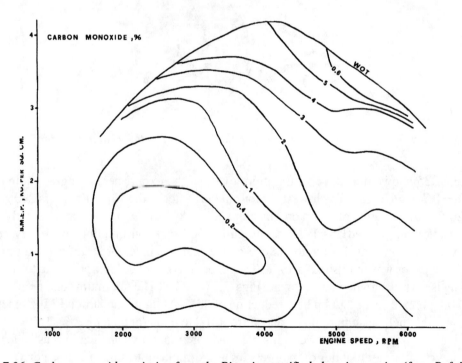

Fig. 7.36 Carbon monoxide emission from the Piaggio stratified charging engine (from Ref. [7.1]).

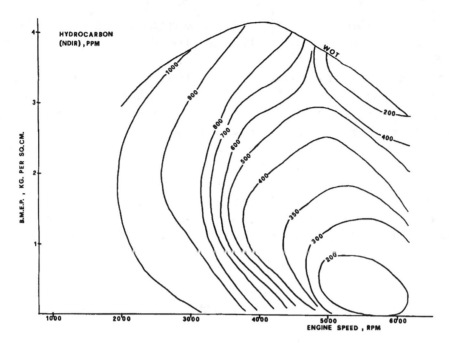

Fig. 7.37 Hydrocarbon emission levels from the Piaggio stratified charging engine (from Ref. [7.1]).

ditions where the machine is accelerated and driven in the 15-50 km/h zone. The proposed European ECE-R40 cycle is such a driving cycle [7.21].

The reduction of hydrocarbon emissions is particularly impressive, as can be seen from a direct comparison of the original engine in Fig. 7.11 with the stratified charging engine in Fig. 7.37. The standard engine, already discussed in Sec. 7.2.1, showed a minimum contour of 1500 ppm HC (C6, NDIR), at a light load but high speed point. In the center of the load-speed map in Fig. 7.11 the figures are in the 2500 ppm region, and at the light load point of 1 bar and 1500 rpm, the figure is somewhat problematic but 5000 ppm would be typical. For the stratified charging engine the minimum contour is reduced to 200 ppm HC, the center of the load-speed picture is about 500 ppm, and the all-important light load and speed level is somewhat in excess of 1000 ppm. This is a very significant reduction and is the level of diminution required for a successful automotive engine before the application of catalytic after-treatment.

A comparison of the carbon monoxide emission levels of the standard engine in Fig. 7.10 and of the stratified charging engine in Fig. 7.36 shows significant improvements in the two areas where it really matters, i.e., at light loads and speeds and at high loads and speeds. In both cases the CO emission is reduced from 2-3% to 0.2-0.3%, i.e., a factor of 10. The absolute value of the best CO emission at 0.2% is quite good, remembering that these experimental data were acquired in 1978.

Also note that the peak bmep of the engine is slightly reduced from 4.8 bar to 4.1 bar due to the stratified charging process, and there is some evidence that there may be some diminution in the air utilization rate of the engine. This is supplied by the high oxygen emission levels at full load published by Batoni [7.1, Fig. 8] where the value at 4 bar and 3000 rpm is shown as 7%. In other words, at that point it is almost certain that some stratified combustion is occurring.

This engine provides an excellent example of the benefits of stratified charging. It also provides a good example of the mechanical disadvantages which may accrue from its implementation. This design, shown in Fig. 7.34, is obviously somewhat bulky, indeed it would be much bulkier than an equivalent displacement four-stroke cycle engine. Hence, one of the basic advantages of the two-stroke engine is lost by this particular mechanical layout. An advantage of this mechanical configuration, particularly in a single-cylinder format, is the improved primary vibration balancing of the engine due to the opposed piston layout.

Nevertheless, a fundamental thermodynamic and gas-dynamic postulation is verified from these experimental data, namely that stratified charging of a two-stroke engine is a viable and sound approach to the elimination of much of the excessive fuel consumption and raw hydrocarbon emission from a two-stroke engine.

The Ishihara option for stratified charging

The fundamental principle of stratified charging has been described above, but other researchers have striven to emulate the process with either less physical bulk or less mechanical complication than that exhibited by the Piaggio device.

One such engine is the double piston device, an extension of the original split-single Puch engines of the 1950s. Such an engine has been investigated by Ishihara [7.7]. Most of these engines are designed in the same fashion as shown in Fig. 7.38. Instead of the cylinders being placed in opposition as in the Piaggio design, they are configured in parallel. This has the advantage of having the same bulk as a conventional twin-cylinder engine, but the disadvantage of having the same (probably worse!) vibration characteristics as a single-cylinder engine of the same total swept volume. The stratified charging is at least as effective as in the Piaggio design, but the combustion chamber being split over two cylinder bores lends itself more to stratified burning than homogeneous burning. This is not necessarily a criticism.

However, it is clear that it is essential to have the cylinders as close together as possible, and this introduces the weak point of all similar designs or devices. The thermal loading between the cylinder bores becomes somewhat excessive if a reasonably high specific power output is to be attained.

Another design worthy of mention and study, which has considerable applicability for such designs where the cost and complexity increase cannot be excessive due to marketing and packaging requirements, is that published in the technical paper by Kuntscher [7.23]. This design for a stratified charging system has the ability to reduce the raw hydrocarbon emission and fuel consumption from such engines as those fitted in chainsaws, mopeds, and small motorcycles.

The stratified charging engine from the Institut Français du Pétrole

This approach to stratified charging emanates from IFP and is probably the most significant yet proposed. The performance results are superior in most regards to four-stroke cycle engines, as is evident from the technical paper presented by Duret *et al.* [7.18]. The fundamental principle of operation is described in detail in that publication, a sketch of the engine

Chapter 7 - Reduction of Fuel Consumption and Exhaust Emissions

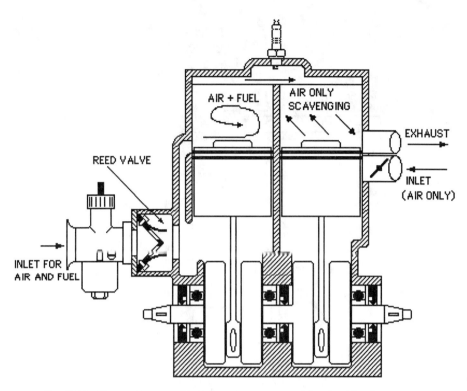

Fig. 7.38 Alternative stratified charging system of the double piston genre.

operating principle is given in Fig. 7.39, and a photograph of their engine is shown in Plate 7.1. The engine in the photograph is a multi-cylinder unit and, in a small light car operating on the EEC fuel consumption cycle at 90 and 120 km/h, had an average fuel consumption of 30.8 km/liter (86.8 miles/Imperial gallon or 73.2 miles/US gallon).

The crankcase of the engine fills a storage tank with compressed air through a reed valve. This stored air is blown into the cylinder through a poppet valve in the cylinder head. At an appropriate point in the cycle, a low-pressure fuel injector sprays gasoline onto the back of the poppet valve and the fuel has some residence time in that vicinity for evaporation before the poppet valve is opened. The quality of the air-fuel spray past the poppet valve is further enhanced by a venturi surrounding the valve seat. It is claimed that any remaining fuel droplets have sufficient time to evaporate and mix with the trapped charge before the onset of a homogeneous combustion process.

The performance characteristics for the single-cylinder test engine are of considerable significance, and are presented here as Figs. 7.40-7.43 for fuel consumption, hydrocarbons, and nitrogen oxides. The test engine is of 250 cm^3 swept volume and produces a peak power of 11 kW at 4500 rpm, which realizes a bmep of 5.9 bar. Thus, the engine has a reasonably high specific power output for automotive application, i.e., 44 kW/liter. In Fig. 7.40, the best bsfc contour is at 260 g/kWh, which is an excellent result and superior to most four-stroke cycle engines. More important, the bsfc value at 1.5 bar bmep at 1500 rpm, a light load and speed point, is at 400 g/kWh and this too is a significantly low value.

Design and Simulation of Two-Stroke Engines

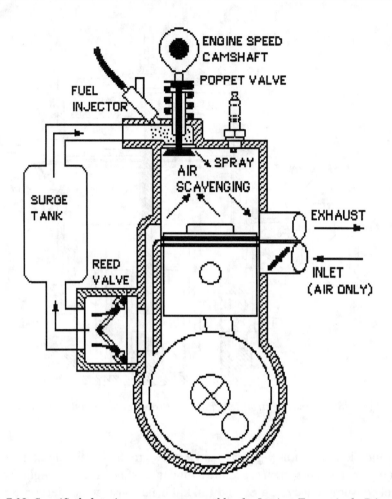

Fig. 7.39 Stratified charging system proposed by the Institut Français du Pétrole.

The unburned hydrocarbon emission levels are shown in Fig. 7.41, and they are also impressively low. Much of the important legislated driving cycle would be below 20 g/kWh. When an oxidation catalyst is applied to the exhaust system, considerable further reductions are recorded, and these data are presented in Fig. 7.43. The conversion rate exceeds 91% over the entire range of bmep at 2000 rpm, leaving the unburned hydrocarbon emission levels below 1.5 g/kWh in the worst situation.

Of the greatest importance are the nitrogen oxide emissions, and they remain conventionally low in this stratified charging engine. The test results are shown in Fig. 7.42. The highest level recorded is at 15 g/kWh, but they are less than 2 g/kWh in the legislated driving cycle zone.

The conclusions drawn by IFP are that an automobile engine designed and developed in this manner would satisfy the most stringent exhaust emissions legislation for cars. More important, the overall fuel economy of the vehicle would be enhanced considerably over an equivalent automobile fitted with the most sophisticated four-stroke cycle spark-ignition en-

Chapter 7 - Reduction of Fuel Consumption and Exhaust Emissions

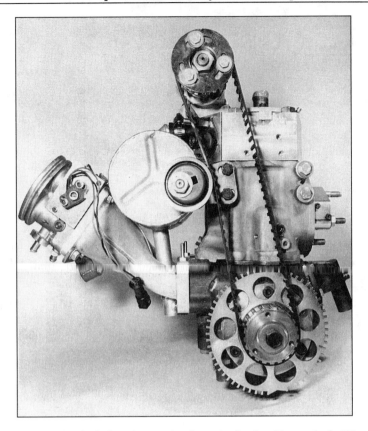

Plate 7.1 Stratified charging engine from the Institut Français du Pétrole (courtesy of Institut Français du Pétrole).

gine. The extension of the IAPAC design to motor scooters and outboards has been carried out [7.48, 7.52].

The bulk of the engine is increased somewhat over that of a conventional two-stroke engine, particularly in terms of engine height. The complexity and manufacturing cost is also greater, but no more so than that of today's four-stroke-engine-equipped cars, or even some of the larger capacity motorcycles or outboard motors.

Stratified charging by an airhead for the simple two-stroke engine

The stratified charging process can be carried out sequentially as well as in parallel. The basic principle is shown in Fig. 7.44. The engine uses crankcase compression and has a conventional supply of air and fuel into the crankcase. An ancillary inlet port, normally controlled by a reed valve, connects the atmosphere to the transfer ports close to their entry point to the cylinder. The ancillary inlet port contains a throttle to control the amount of air ingested. During the induction stroke the reed valve in the ancillary inlet port lifts and air is induced toward the crankcase, displacing air and fuel ahead of it. When crankcase compression begins the transfer duct is ideally filled with air only, and when scavenging commences it does so with an "airhead" in the van. The theory is that the scavenging is sequentially

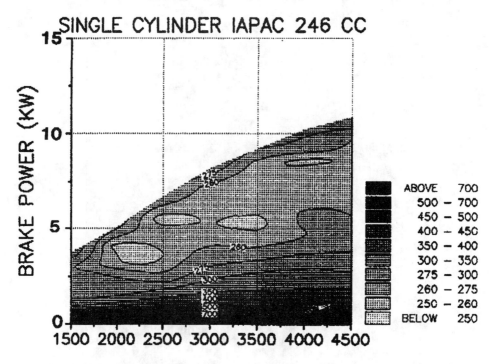

Fig. 7.40 Fuel consumption contours (g/kWh) of IFP stratified charging engine.

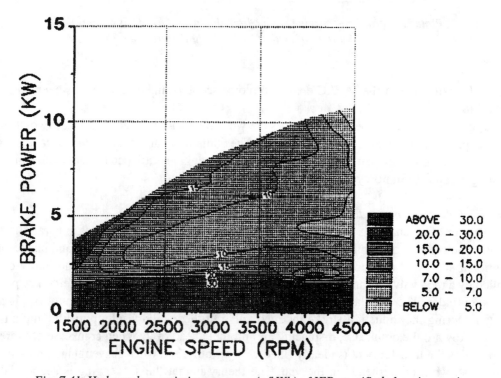

Fig. 7.41 Hydrocarbon emission contours (g/kWh) of IFP stratified charging engine.

Chapter 7 - Reduction of Fuel Consumption and Exhaust Emissions

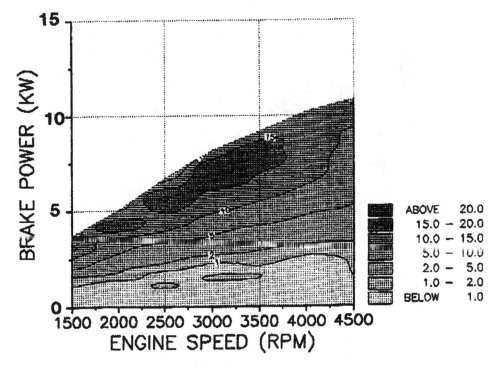

Fig. 7.42 Nitrogen oxide emission contours (g/kWh) of IFP stratified charging engine.

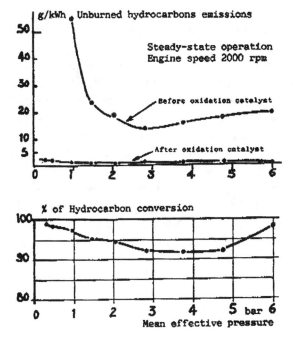

Fig. 7.43 Effect of oxidation catalyst on exhaust hydrocarbons in IFP engine.

Design and Simulation of Two-Stroke Engines

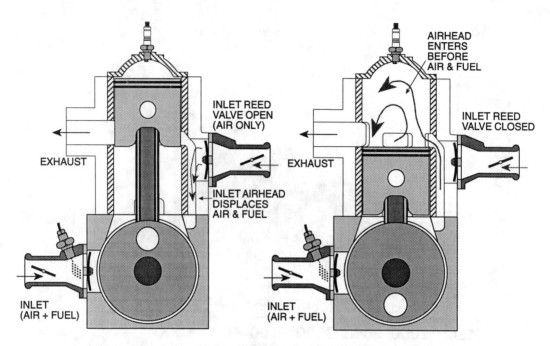

Fig. 7.44 Airhead stratified charging for the simple two-stroke engine.

stratified so that the final phase holding both air and fuel never reaches the exhaust ports. In the theoretical arena concerning two-stroke scavenging, my (somewhat bitter) experience is that the law proposed by Mr. Murphy is at least as influential as that by Mr. Bernoulli.

As it is preferable to apply the airhead to all of the scavenge ducts, the mechanical design is simplified if the transfer ports are located together. A cross-scavenged engine meets this criterion, as Fig. 3.32 illustrates. The application to a multi-cylinder, cross-scavenged engine is straightforward. The application to a single-cylinder, loop-scavenged engine is a more difficult design proposition and, from a scavenging standpoint, has a lower probability of success. The application to a multi-cylinder, loop-scavenged engine is a virtual impossibility.

One of the first publications to mention an airhead is that by none other than Lanchester [7.32] and there have been many papers and patents since on the topic. In recent times it is discussed by Saxena [7.31], and QUB has been active in this area with papers by Kee [7.29] and Magee [7.30].

At QUB [7.29] it was first applied to a cross-scavenged engine with a QUB-type piston, as shown in Fig. 1.13 or Plate 1.6. The scavenging quality is as QUBCR in Figs 3.12 and 3.13. The engine is of 270 cm^3 capacity with "square" dimensions of 70 mm bore and stroke. The test results for this engine have been fully reported by Kee [7.29], so a precis of them here will make the point that airhead stratification does lead to reductions in specific fuel consumption and in hydrocarbon emission, at no detriment to the performance characteristics of power and torque over the speed range.

The test results are shown in Figs. 7.45-47. Fig. 7.45 shows the attained bmep over the speed range. The main inlet throttle at the carburetor in the crankcase is fully open at all times during the test sequence shown in Figs. 7.45-7.47. The three curves illustrate little change in

Chapter 7 - Reduction of Fuel Consumption and Exhaust Emissions

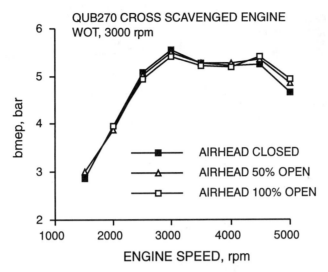

Fig. 7.45 Effect of airhead on engine torque.

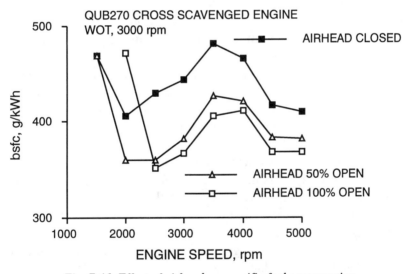

Fig. 7.46 Effect of airhead on specific fuel consumption.

overall air flow rate with the throttle valve in the airhead inlet either fully closed, or at one-half open, or fully open. The bmep is virtually unchanged in all three cases, although some extra air flow, leading to a minor torque increase, is visible at the highest engine speed of 5000 rpm. In Figs. 7.46 and 7.47, there is a noticeable improvement in bsfc and bsHC at engine speeds of 2000 rpm and above. The reduction in hydrocarbon emissions ranges from 55% at 2500 rpm to 20% at 5000 rpm. The reduction in brake specific fuel consumption ranges from 20% at 2500 rpm to 5% at 5000 rpm.

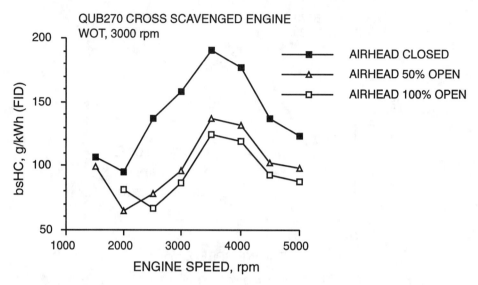

Fig. 7.47 Effect of airhead on hydrocarbon emissions.

While these are significant findings, the absolute levels of bsHC minimize at 60 g/kWh, which is considerably above the 25 g/kWh theoretically achieved by the low bmep optimization process described above for the simple two-stroke engine. It is not impossible to consider a combination of the two approaches.

One of the advantages of the airhead approach is that lost charge from the exhaust port tends to contain bypassed charge with a leaner air-to-fuel ratio than with homogeneous charging. This provides excess oxygen in the exhaust system and would improve the conversion efficiency of the pollutants by an oxidation catalyst placed in it.

7.4.2 Homogeneous charging with stratified combustion

As was pointed out earlier, direct in-cylinder fuel injection is one of the obvious methods of reducing, or even eliminating, the loss of fuel to the exhaust port during the scavenge process. Fig. 7.30 illustrates the positioning of such a fuel injector for the combustion of gasoline in a spark-ignition engine. Figs. 4.1(b) and 4.9 show the positioning of injectors for the injection of fuel into a compression-ignition engine, for it must not be forgotten that this engine falls directly, and more convincingly in the view of some, into this category.

Gasoline injection into spark-ignition engines

The potential difficulties of evaporating a fuel spray in time for a homogeneous combustion process to occur have been debated. The even more fundamental problem of attaining good flammability characteristics at light load and speed in a homogeneous combustion process has also been addressed. What, then, are the known experimental facts about in-cylinder fuel injection? Is the combustion process homogeneous, or does the desirable possibility of stratified burning at light load and speed exist? Does the fuel vaporize in time to burn in an efficient manner? The answers to these questions are contained in the technical papers published by several authors; answers to the question of cost and design approach for traditionally low-cost applications such as mopeds, scooters and outboards are also found in the literature [7.44, 7.45, 7.46, 7.48, 7.52].

Liquid gasoline injection

Among the first reports of experimental work in this field are the papers presented by Fuji Heavy Industries from Japan. The first of these was published as long ago as 1972 [7.17] and more recently by Sato and Nakayama [7.2] in 1987. Nuti [7.12] from Piaggio, and Plohberger *et al.* [7.19] from AVL, have also published experimental data on this subject and they too should be studied. The papers just mentioned all use high-pressure liquid-injection systems, not unlike that employed for diesel engines, but modified to attain smaller droplet sizes when using gasoline fuel. The droplet sizes required are of 10-15 μm mean diameter, usually measured by laser-based experimental techniques to acquire the Sauter Mean Diameter (SMD) of the fuel droplets [7.16].

The results measured by Sato [7.2] are reproduced here as Fig. 7.48. The peak power performance characteristics of the engine are unaffected by the use of fuel injection, as is their reported levels of NO_x emission. However, considerable reductions in the bsfc values are seen in the center of the "oyster" map, with the best contour being lowered to 300 g/kWh with fuel injection, from 380 g/kWh when the engine was carburetted. The equivalent picture is repeated for the hydrocarbon emission levels with the carburetted engine showing some 3000 ppm HC/NDIR values, whereas the fuel injection engine is reduced to 400 ppm. This effect is also reported by Nuti [7.12] and the fuel consumption levels in his engine would be even lower at the best possible condition, these being bsfc values of 270 g/kWh. These are significantly low levels of fuel consumption and hydrocarbon exhaust emissions.

However, Sato [7.2] also reports that the direct in-cylinder fuel injection did not improve the misfiring behavior (four-stroking) at light loads and speeds, and this can be seen in Fig. 7.48. In the lower left-hand corner of the "oyster" maps, at the light load and speed positions, the bsfc and hydrocarbon emission values are the same for the carburetted and fuel-injected engines. The values of bsfc are an unimpressive 500 to 600 g/kWh and the unburned hydrocarbons are at 3000-5000 ppm. In short, stratified burning at light load and low speed was not achieved.

Plohberger [7.19] shows very complete test data at light load over the speed range, and below 3000 rpm his results confirm that reported by Sato [7.2]. The direct in-cylinder injection of fuel has not solved the vital problem of light load and idle running, in these two instances.

Although Nuti [7.12] makes no comment on this situation, his HC emission is reportedly only slightly better than that given by Sato [7.2] and his CO levels are rising rapidly in that zone.

One might conclude from the experimental data presented that the direct fuel injection of liquid gasoline does not provide stratified combustion at light loads and speeds, and thereby does not improve the emissions and fuel economy of the two-stroke engine at the crucially important urban driving condition.

It would appear that there is sufficient time at higher speeds and loads to vaporize the gasoline and mix it with the air in time for a homogeneous combustion process to occur. The onset of injection required is remarkably early, and well before the trapping point at exhaust

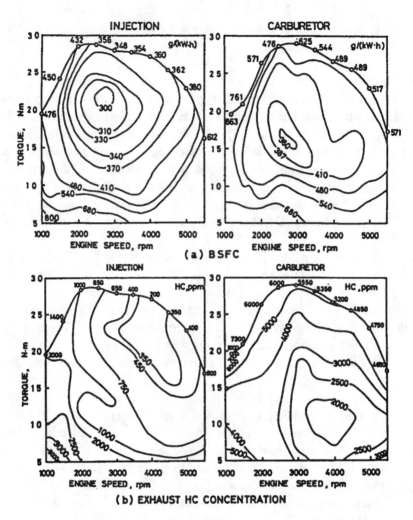

Fig. 7.48 Comparison of fuel consumption and hydrocarbon emissions between direct fuel injection and a carburetor (from Ref. [7.2]).

closure. Sato [7.2] describes the end of the dynamic injection process taking place at 10° abdc at 1000 rpm and 50° abdc at 5000 rpm. The exhaust port on this 356 cm³ twin-cylinder engine closed at 68° abdc.

However, in more recent times, with much learned about both stratified and homogeneous combustion since the publication of those earlier papers, liquid injection systems have greatly improved and consistent cyclic combustion in a stratified form down to the idle point is now possible [7.28, 7.39, 7.40, 5.23].

The ram-tuned liquid injection system

One such system, which departs from the conventional, comes from the TH Zwickau and is shown in Fig. 7.49. The basic principle of its operation is that it utilizes the water-hammer effect when a moving column of liquid is abruptly decelerated. Fig. 7.49 shows the circulation of fuel from the tank through the filter and pump to charge an accumulator at relatively low pressure, some 6 or 7 bar, flow rapidly through a smaller-diameter acceleration pipe, and the flow back to the tank if the solenoid valve is open. When the solenoid valve shuts, a high-pressure wave of some 25 bar is generated in the acceleration pipe and lifts the spring-backed needle off its seat to give fuel injection. A picture of the ensuing fuel spray is shown in Plate 7.2 and Carson [7.28] describes the spray characteristics as lying between 15 and 38 µm SMD, over the full fueling range. A pressure regulator controls the fuel line pressure to a set level and permits spillage back to the fuel tank. The damping valve is inserted into the circuit to diminish the amplitude of the reflected pressure waves prior to the next injection signal. One of the great advantages of this system is that it permits injection rates up to 6000 cycles per minute and that it can operate at even higher pressure levels of up to 300 bar at the nozzle, for use in compression-ignition engines [7.53].

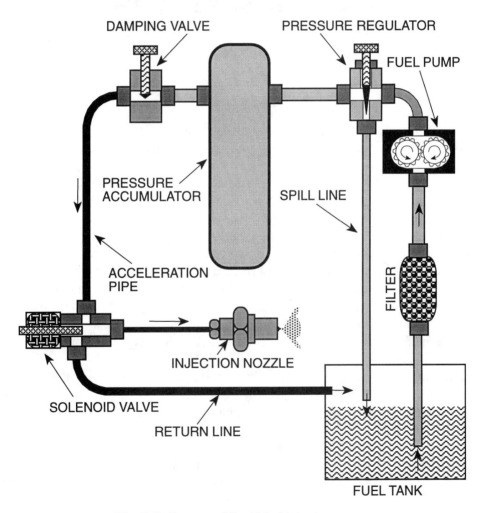

Fig. 7.49 Ram-tuned liquid fuel injection system.

Design and Simulation of Two-Stroke Engines

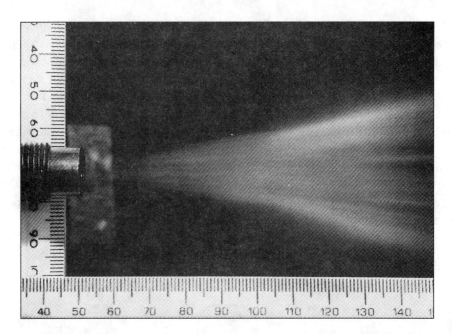

Plate 7.2 The spray pattern from the TH Zwickau ram-tuned liquid fuel injection system.

Its utilization in a QUB research engine is described by Carson *et al.* [7.28] and the test results of its behavior at light load and low speed are discussed below and shown in Figs. 7.60 to 7.63.

There is an alternative, and potentially simpler method of achieving the same hydraulic ram effect, proposed by Heimberg [7.54].

Air-blast injection of gasoline fuel into the cylinder

This is actually the oldest technique for injecting fuel into diesel engines, employed by Dr. Diesel no less [7.41], and was replaced (for compression-ignition engines) by the invention of the jerk pump giving liquid injection at the turn of this century. In a form suitable for the injection of gasoline into today's high-speed two-stroke engines, the system uses electromagnetic solenoid-actuated poppet valves, although that too would appear to have a precursor [7.42]. Landfahrer describes the inventive work of the research laboratory of AVL in this arena [7.44].

The method of operation of a modern air-blast fuel injector is sketched in Fig. 7.50. The injector is supplied with air at about 5 bar and liquid fuel at about 6 bar. The pressure difference between the air and the fuel is carefully controlled so that any known physical movement of the fuel needle would deliver a controlled quantity of fuel into the sac before final injection. In Fig. 7.50(a) the injector is ready for operation. Both of the electromagnetic solenoids are independently activated by a control system as part of the overall engine and vehicle management system. The fuel solenoid is electronically activated for a known period of lift and time, as seen in Fig. 7.50(b), the result being that a precise quantity of fuel is metered into the sac behind the main needle. The fuel needle is then closed by the solenoid, assisted by a spring, and the fuel awaiting final injection is heated within this space by conduction from the cylinder head; this period is shown in Fig. 7.50(c). At the appropriate juncture, the main (outwardly opening) poppet valve is activated by its solenoid and the high-pressure air supply can now act upon the stored fuel and spray it into the combustion chamber; this is shown in Fig. 7.50(d). The poppet is closed by its solenoid through electronic triggering, but it is further assisted by a spring to return to its seat, as in Fig. 7.50(a). This completes the cycle of operation. The spray which is created by such an "aerosol" method is particularly fine, and

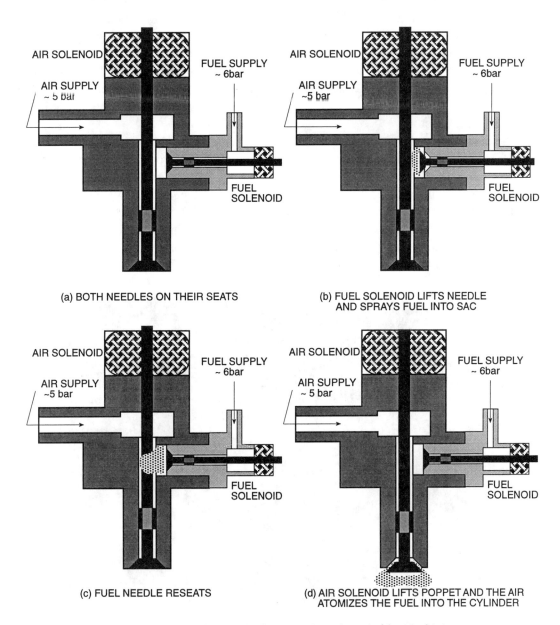

Fig. 7.50 Sequence of events in the operation of an air-blast fuel injector.

droplet sizes in the range of 3-20 μm SMD have been reported for these devices [7.34, 7.28]. Studies of the spray formation in such injectors have been provided by Ikeda [7.50] and by Emerson [7.43].

The spray pattern from an air-assisted injector, designed and manufactured at QUB for in-house R&D, is shown in Plate 7.3. It can be seen to have a wider hollow cone shape than the narrower cone emanating from liquid injection, shown in Plate 7.2. These are general comments, for it is possible to shape differently the spray patterns of most injector nozzles. The QUB air-assisted injector has a SMD value of between 3 and 12 μm over the full range of required delivery. An indication of the types of fuel sprays emanating from well-optimized fuel injectors are shown in Plates 7.4 and 7.5, for liquid and air-assisted fuel injectors, respectively. The "hollow-cone" nature of their spray patterns can be seen very clearly.

Actually, the device emanating from IFP [7.18], and discussed in Sec. 7.4.1, is a mechanical means of accomplishing the same ends, albeit with a greater mass of air present during the heating phase, but the final delivery to the cylinder is at a lower velocity than that produced by the air-blast injector.

The advantage of the air-assisted injector is that it potentially eliminates one of the liquid injector's annoying deficiencies, that of "dribbling" or leaving a droplet on the nozzle exit at low flow rates. This "dribble" produces misfire and excessive HC emission. A potential disadvantage is that the long-term reliability and accuracy of fuel delivery of the air-assisted injector is open to question. Solenoids exposed to high temperatures are not known for precision long-term retention of their electromagnetic behavior, and this would be unacceptable in vehicle service. That is why the injector is occasionally seen placed lower in the cylinder wall, or angled up from the transfer ports in some designs [7.17], i.e., away from, and not

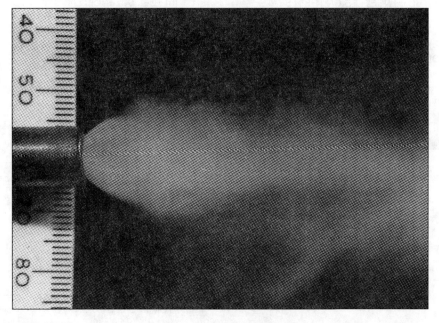

Plate 7.3 The spray pattern from a QUB-designed air-assisted fuel injection system.

Chapter 7 - Reduction of Fuel Consumption and Exhaust Emissions

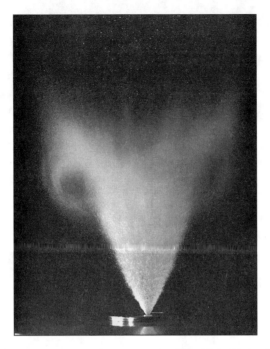

Plate 7.4 A spray pattern from an optimized liquid fuel injection system (courtesy of Mercury Marine).

Plate 7.5 A spray pattern from an optimized air-assisted fuel injection system (courtesy of Mercury Marine).

installed in direct thermal contact with, the hot cylinder head. However, this gives an undesirable subsidiary effect, as this leaves some residual hydrocarbons in the passage between the injector and the cylinder after the piston has passed by in the trapping process, and these are then carried into, and potentially out of, the cylinder on the next scavenge process or are exposed to the exhaust gas by the downward travel of the piston during the blowdown phase.

The Orbital Engine Corporation of Perth in Western Australia has been very active in the design and development of air-blast injectors and of vehicle engines fitted with such injectors [7.26, 7.38, 7.45]. Plate 7.6 shows one of their 1.2-liter, three-cylinder, naturally aspirated, spark-ignited engines with air-assisted fuel injection which has been employed in prototype automobiles [7.26, 7.38]. Plate 7.7 shows a moped-scooter two-wheeler fitted with the Orbital SEFIS system [7.45].

Direct fuel injection using liquid-only injection [Ref. 7.54] is illustrated in Plate 7.8. This system has been developed by Ficht in Germany in collaboration with Outboard Marine Corporation. The design originally stems from research work carried out at the Technische Hochschule of Zwickau [Ref. 7.23].

Plate 7.6 The Orbital three-cylinder 1.2-liter car engine with air-assisted direct fuel injection (courtesy of Orbital Engine Corporation Limited).

Chapter 7 - Reduction of Fuel Consumption and Exhaust Emissions

Plate 7.7 The Orbital SEFIS air-assisted direct fuel injection applied to a motor scooter (courtesy of Orbital Engine Corporation Limited).

Plate 7.8 A liquid fuel injection system for the direct injection of gasoline in an outboard motor (courtesy of Outboard Marine Corporation).

Tests on the direct-injected, spark-ignition QUB500rv engine

The tests are conducted using a specially constructed 500 cm^3 single-cylinder research engine, designated as the QUB500rv. It is shown in Fig. 7.51. This engine incorporates a reed valve inlet and a conventional crankcase pumping system. The cylinder scavenging system is of the QUB cross-scavenged type, described in detail in Secs. 1.2.2 and 3.5.4, and illustrated in Fig.3.34(b) and Plate 1.6. The fuel injector is mounted on the cylinder head and the spray is directed into the piston chamber formed by the deflector and the cylinder wall. A standard reach spark plug is positioned in the cylinder wall, as illustrated in Fig. 7.51. During the entire test program, the engine is operated with a box silencer exhaust system which does not include an oxidation catalyst. The engine specification is, briefly: bore and stroke 86 mm; CR_t 6.9; squish clearance 3 mm; and the exhaust and transfer ports open at 105° and 120° atdc, respectively.

For stratified charging, the air-assisted injector is designed and manufactured at QUB [7.28, 7.34, 7.47]. On this air blast injector (ABI), the automotive fuel injector and the solenoid are individually controlled to permit independent variation of timing and duration. The

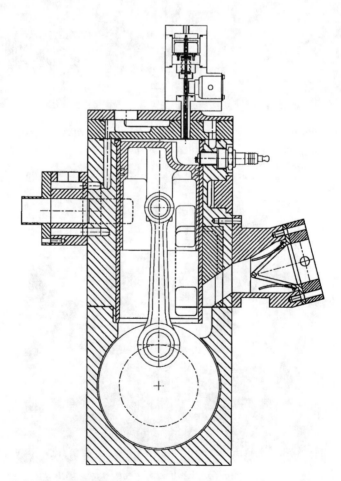

Fig. 7.51 The QUB500rv experimental engine with direct fuel injection.

Chapter 7 - Reduction of Fuel Consumption and Exhaust Emissions

start and duration of the metering of fuel into the injector body are referred to as "start of fuel" (SOF) and "fuel duration" (FD), respectively. Similarly, the start and duration of the air-assisted injection process are referred to as "start of air" (SOA) and "air duration" (AD), respectively.

The engine is also tested with the "perfect carburetor," i.e., an electronic fuel injector (EFI) mounted in the inlet tract so that the engine is homogeneously charged at all times. The word "perfect" is used in the sense that the operator can dial in any air-to-fuel ratio desired.

The engine is tested at full load at 3000 rpm, and at 1600 rpm at part load, with both stratified and homogeneous charging, i.e., ABI or EFI installed; the results of these tests are found in Figs. 7.52-7.55 and 7.56-7.59, respectively, over a very wide range of trapped air-fuel ratios, AFR_t. Ignition timing is always set at MBT (minimum advance for best torque). For the full load tests, stratified charging is recorded for three SOA timings, namely 110°, 125° and 140° atdc, and the air duration is 2 ms. For the part load tests, stratified charging is recorded for three SOA timings, namely 215°, 230° and 245° atdc, and the air duration is 5 ms.

In the entire test series, presented in Figs. 7.52-7.59, the profile of the relevant performance parameter for homogeneous charging can be compared with previous test results in Figs. 7.3-7.8 and theoretical results in Figs. 7.13-7.18. However, here in Figs. 7.55 and 7.59 is also shown the brake specific emissions of oxides of nitrogen with respect to air-to-fuel ratio, which can be compared for profile with theoretical computer simulations in Fig. A4.3.

During full load (wot) operation, the SOA timing preceded exhaust port closure by 115-145°. However, the SOA timings do not include any allowance for air solenoid delay. Tests conducted with a Hall effect sensor on the injector show that the poppet valve opening is delayed by approximately 2.5 ms. Therefore, the actual poppet valve opening occurred approximately 45° later than the nominal SOA timing. Nevertheless, it is clear that during this test a substantial proportion of the fuel is injected prior to exhaust port closure. The wot results, presented in Figs. 7.52-7.55, show that the air-assisted direct-injection system pro-

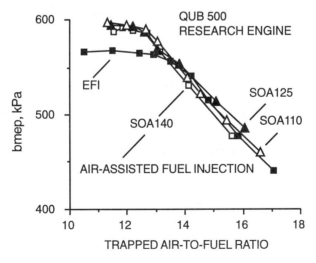

Fig. 7.52 Effect of fueling on bmep at full load, 3000 rpm.

Design and Simulation of Two-Stroke Engines

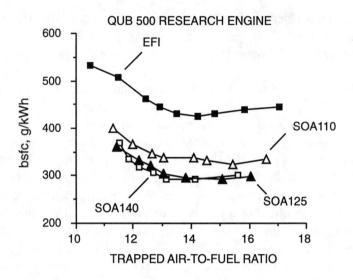

Fig. 7.53 *Effect of fueling on bsfc at full load, 3000 rpm.*

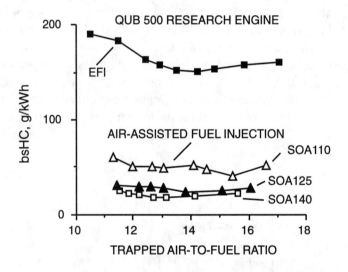

Fig. 7.54 *Effect of fueling on bsHC at full load, 3000 rpm.*

duced very significant improvements in fuel economy and hydrocarbon emissions over homogeneous charging (EFI). The minimum bsfc is reduced from 424 g/kWh to 293 g/kWh, and the minimum bsHC is reduced from 150 g/kWh to 19 g/kWh. These improvements are accompanied by a marginal increase in bmep, due to the additional air supply through the injector.

The results also illustrate the influence of SOA timing. Later injection gives an inadequate time for preparation of the mixture, and a consequent deterioration in combustion efficiency.

Chapter 7 - Reduction of Fuel Consumption and Exhaust Emissions

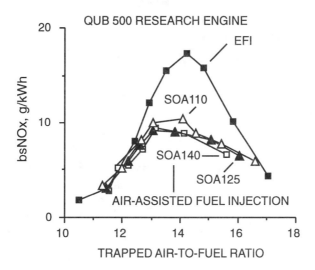

Fig. 7.55 Effect of fueling on bsNO$_x$ at full load, 3000 rpm.

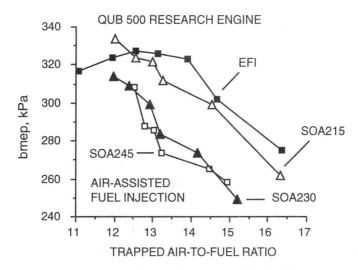

Fig. 7.56 Effect of fueling on bmep at part load, 1600 rpm.

The bsNO$_x$ characteristics, presented in Fig. 7.55, show that air-assisted injection gives no significant change during operation at rich or lean air-fuel ratios. Note that it peaks at the stoichiometric value. However, a reduction in NO$_x$ is noted during stratified charge operation, close to the stoichiometric air-fuel ratio. This feature may be due to the stratification of the charge and the reduced NO$_x$ formation in local combustion zones which are variably rich or lean.

The part load (throttle) results, presented in Figs. 7.56-7.59, show similar highly significant improvements in fuel economy and hydrocarbon emissions. The minimum bsfc is reduced from 374 g/kWh to 303 g/kWh, and the minimum bsHC is reduced from 61 g/kWh to

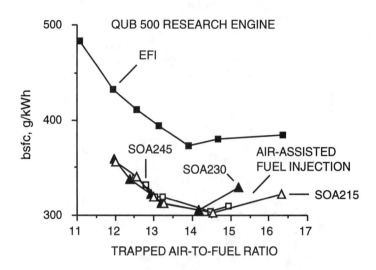

Fig. 7.57 Effect of fueling on bsfc at part load, 1600 rpm.

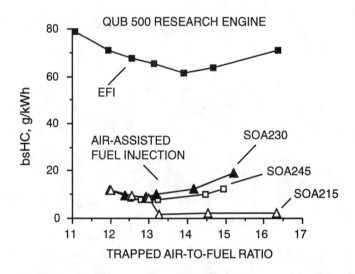

Fig. 7.58 Effect of fueling on bsHC at part load, 1600 rpm.

7.7 g/kWh. The NO_x results in Fig. 7.59 illustrate the inherently very low part load emissions of the two-stroke engine. Air-assisted injection did not significantly change these highly desirable characteristics.

At light load with ram-tuned liquid and air-assisted injection

The above test series on the QUB500rv engine is extended to include light load operation at 1600 rpm and to make comparisons between air-assisted injection and ram-tuned liquid injection. It is reported more extensively by Carson *et al.* [7.28]. The ram-tuned system employed is designed and developed for the test program at QUB by Prof. C. Stan of TH Zwickau.

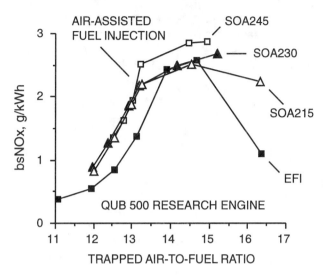

Fig. 7.59 Effect of fueling on bsNO$_x$ at part load, 1600 rpm.

Another direct-injection system using liquid injection [Ref. 7.54], originally emanating from research work at the same University but employing a somewhat different hydraulic process, is shown in Plate 7.8.

The ram-tuned liquid injection (RTL) system is operated at SOF settings of 50°, 70° and 80° btdc and the air blast injector (ABI) is operated at a SOA setting of 80° btdc. The fueling is varied over a wide range and the test results are shown in Figs. 7.60-7.63, with respect to bmep for bsfc, bsHC, bsNO$_x$ and bsCO, respectively. The hydrocarbon emissions are mainly

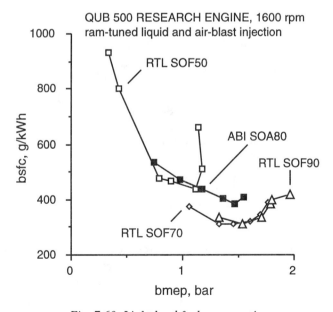

Fig. 7.60 Light load fuel consumption.

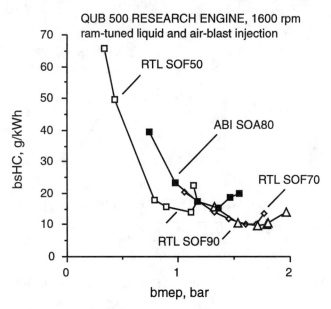

Fig. 7.61 Light load hydrocarbon emissions.

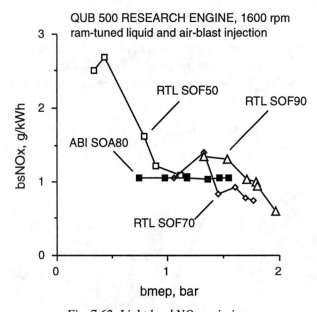

Fig. 7.62 Light load NO_x emissions.

at, or below, 10 g/kWh at this low load level, which is a very creditable result. The bsfc is at 300 g/kWh for some of the RTL settings; that too is an excellent result and would translate into very low on-road fuel consumption figures for an automobile. The $bsNO_x$ values mostly hover around 1.0 g/kWh, or less, which remains as remarkably low as always. In Fig. 7.63, the bsCO minimizes at 20 g/kWh and at 1.5 bar. That it is as high as 20 g/kWh at 1.5 bar bmep on

Chapter 7 - Reduction of Fuel Consumption and Exhaust Emissions

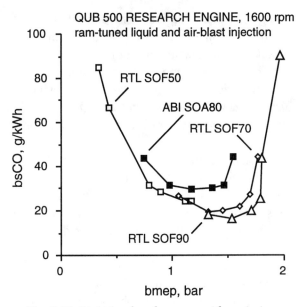

Fig. 7.63 Light load carbon monoxide emissions.

the "knee" of the curve suggests that combustion may not be complete and that the trapped air-to-fuel ratio is approximately stoichiometric; in Fig. 7.61, the bsHC figures of 9 g/kWh, at the same experimental point, would support this view.

The liquid injection system permits stratified combustion down to much leaner overall air-fuel ratios than the air-assisted injection system. The minimum value of 0.35 bar achieved at 1600 rpm is not far off an idle condition for a "real" engine, as this QUB 500rv test engine drives no ancillaries, such as an oil pump, an alternator, an ABI air pump, nor the RTL fuel pump, etc. The ability of the RTL system to be fueled down to 0.35 bar at the same air throttle setting implies an overall air-to-fuel ratio of some 50:1 while stratified burning locally at or about stoichiometric.

Stratified burning in general and in the QUB500rv in particular

Stratified combustion for a spark-ignition engine, as pointed out in Sec. 7.1.2, is a relatively difficult phenomenon to orchestrate successfully [7.37]. The sketch in Fig. 7.30 is deliberately drawn to show the basic principle. At light load the injection is timed late and the ignition early. The fuel spray is aimed so that it remains relatively cohesive and within the vicinity of the spark plug at ignition. As the piston is approaching tdc it is advantageous to provide a pocket in the piston crown into which the fuel is sprayed, which is located so as to move ever closer to the environs of the spark plug electrodes by ignition. The scavenge process is designed to not have localized high-speed air jets which would widely distribute the fuel spray over the combustion chamber.

The combustion chamber of the loop-scavenged engine, shown in Fig. 7.30, is specifically designed to accommodate these needs. It is a "total-offset" chamber of the type shown in Fig. 4.13 and designed by Prog.4.5. The jargon for the chamber shown in Fig. 7.30 is a "jockey cap," for obvious shape reasons. Most of the loop-scavenged engines have to employ extended electrode spark plugs to ensure that ignition reaches into the vaporizing fuel plume, and that the ignition period is reasonably extensive in time terms. For that reason the rapid rise, but short duration, spark from a capacitor discharge ignition system has to be replaced by the inherently longer lasting spark of an inductive ignition system. Actually, the "bowl-in-piston" chamber is even more effective than the "total-offset" design but, for crankcase compression engines, the piston becomes much too hot at high load; this design could operate well in an externally scavenged engine where it can receive the conventional under-piston coolant flow from an oil jet.

For the QUB deflector piston engine, such as the QUB500rv shown in Fig. 7.51, the deflector very conveniently provides the "pocket" into which fuel is to be injected. The proximity of the spark plug electrodes to the injected fuel plume is evident and the necessity of using extended electrode spark plugs is virtually eliminated; which means that an important item of concern regarding engine durability has been removed. It does not mean that further detailed experimental optimization of this chamber to enable good stratified burning is eliminated, but it is more readily optimized than the more open geometry of a loop-scavenged design. The test results given above for both the RTL and ABI injection were acquired without the extensive optimization of combustion chamber geometry which characterizes similar R&D efforts for loop-scavenged engines [7.38-7.40].

All of the above comments on design for stratified combustion doubtless seem instantly logical to you, but be assured that it was learned more slowly and with greater difficulty than that.

Therefore, from the light load test results at 1600 rpm, the ram-tuned, liquid-injection system performed in a superior manner to the QUB air-assisted injection system, but this does not necessarily imply that one is universally superior to the other for stratified combustion. It simply means that for the single geometry experiment conducted it was superior. In another physical configuration that might not prove to be the case.

Stratified combustion at idle conditions in the QUB500rv engine

During the conduct of experiments down to the idle condition, the ram-tuned liquid system alone accomplished this ultimate test of combustion stratification. A stable idle condition was achieved with the RTL system at 900 rpm with the following characteristics: DR is measured at 0.37; SOF is 50° btdc; ignition timing is 35° btdc; AFR_o is 42.7; exhaust gas oxygen content is 19.9% by volume; exhaust gas carbon monoxide content is 1.8% by volume; exhaust gas hydrocarbon content is 1714 ppm (FID); and the exhaust gas temperature is 88°C. This latter value is of some concern from an emissions standpoint as an oxidation catalyst would not "light-off" at this low temperature.

A multi-cylinder engine based on the QUB500rv

The test results presented above show that the QUB500rv engine design is well suited to the direct air-assisted injection or ram-tuned injection system. The excellent fuel economy and emissions, allied with the inherent manufacturing advantages of QUB cross-scavenging, make this engine design a serious contender for automotive two-stroke engines. At QUB, a three-cylinder engine has been constructed and tested based on the above knowledge. Due to confidentiality, the test results cannot be debated, except to say that in the light load and low speed regions, which characterize most urban emissions test cycles for automobiles, they are superior in every regard to those recorded for the single-cylinder unit.

7.5 Compression-ignition engines

Very little has been said above about diesel engines. It is only comparatively recently, and principally in Europe, that there has been any renewed interest in the topic. In one area at least, the diesel is immediately superior to the two-stroke gasoline engine, for it conducts stratified combustion at all speeds and loads with no difficulty and at high efficiency!

For an automobile, or for a truck, the externally scavenged engine should be re-examined from a design standpoint for its ability to take advantage of new knowledge, new materials, new supercharger or turbocharger designs, or new injection systems. This has already started, and the outcome is that this design and development area is being revisited. Early indications are that higher power density, better thermal efficiency and better exhaust emissions are possible for the automobile, the bus, the tractor or the truck. In Europe, there is already a high proportion of on-road vehicles fitted with turbocharged four-stroke cycle diesel engines, most of which have, to American eyes, astonishing performance characteristics.

It is quite conceivable that a 2-liter, two-stroke diesel car engine would have a flat bmep curve of 7.5 bar from 1000 to 4500 rpm and have a (best point) specific fuel consumption of 250 g/kWh, as an IDI design. If combustion can be attained as a DI engine, then that bmep is raised to 8.3 bar and the bsfc drops to 210 g/kWh. Both these designs have the inherently low HC emission common to all diesel engines, but the two-stroke engine has $bsNO_x$ levels below 1.6 g/kWh. The bmep information translates into peak power outputs of 112 kW and 125 kW for the IDI and the DI engine, respectively, which means that they would out-accelerate equivalent gasoline-engined vehicles due to the shape of the torque curve, may even have a higher terminal speed, and will certainly post considerably superior fuel consumption figures for both urban and highway driving.

Needless to add, as QUB has had a long history of R&D in diesel engines in general, and two-stroke diesel engines in particular, much activity is devoted to the theoretical and experimental advancement of these concepts.

7.6 Concluding comments

It is too early in the development history of two-stroke engines for precise conclusions to be drawn regarding their future role; this process has really been in progress only for some thirty years. Nevertheless, the general level of improvement in performance characteristics of the simple engine has been greatly enhanced by new research and development techniques for cycle calculation, scavenging and combustion design, fueling methods and exhaust tuning; much has been learned in the last half-decade [7.36]. Even more remarkable are the performance characteristics of some of the advanced two-stroke engines aimed at future automotive applications, some with behavioral characteristics at the first experimental attempt which rival or exceed those available from current four-stroke engine practice. Often, some of these new two-stroke engines are produced on research budgets and by engineering teams whose size is a mere fraction of that being devoted to the onward development of an existing four-stroke production engine. The end of this century, and the turn of the next, is going to be a very important period in the development of the two-stroke cycle engine and it is not impossible that an automobile fitted with such an engine will be in series production before the end of this millenium. In that context, the prophetic remark passed by Professor Dr.h.c. Alfred Jante, in an SAE paper [3.5] presented in May 1968, is particularly significant: "The stricter that the exhaust emissions standards become, the more the two-stroke engine will regain importance."

References for Chapter 7

7.1 G. Batoni, "An Investigation into the Future of Two-Stroke Motorcycle Engine," SAE Paper No. 780710, Society of Automotive Engineers, Warrendale, Pa., 1978.

7.2 T. Sato, M. Nakayama, "Gasoline Direct Injection for a Loop-Scavenged Two-Stroke Cycle Engine," SAE Paper No. 871690, Society of Automotive Engineers, Warrendale, Pa., 1987.

7.3 K. Tsuchiya, S. Hirano, M. Okamura, T. Gotoh, "Emission Control of Two-Stroke Motorcycle Engines by the Butterfly Exhaust Valve," SAE Paper No. 800973, Society of Automotive Engineers, Warrendale, Pa., 1980.

7.4 N. Hata, T. Iio, "Improvement of the Two-Stroke Engine Performance with the Yamaha Power Valve System," SAE Paper No. 810922, Society of Automotive Engineers, Warrendale, Pa., 1981.

7.5 S. Ohigashi, "Exhaust Pipe Contraction in Two-Stroke Engines," Japanese Patent No. Showa 38-19659, 1963.

7.6 H. Nishizaki, S. Koyama, "EK34 Type Engine and Idle Silence Valve for Subaru Rex," Internal Combustion Engines, Vol 11, No 128, Sept. 1972.

7.7 S. Ishihara, "An Experimental Development of a New U-Cylinder Uniflow Scavenged Engine," SAE Paper No. 850181, Society of Automotive Engineers, Warrendale, Pa., 1985.

7.8 S. Onishi, S.H. Jo, P.D. Jo, S. Kato, "Multi-Layer Stratified Scavenging (MULS)—A New Scavenging Method for Two-Stroke Engine," SAE Paper No. 840420, Society of Automotive Engineers, Warrendale, Pa., 1984.

7.9 G.P. Blair, B.W. Hill, A.J. Miller, S.P. Nickell, "Reduction of Fuel Consumption of a Spark-Ignition Two-Stroke Cycle Engine," SAE Paper No. 830093, Society of Automotive Engineers, Warrendale, Pa., 1983.

7.10 B.W. Hill, G.P. Blair, "Further Tests on Reducing Fuel Consumption with a Carburetted Two-Stroke Cycle Engine," SAE Paper No. 831303, Society of Automotive Engineers, Warrendale, Pa., 1983.

7.11 H. Uchiyama, T. Chiku, S. Sayo, "Emission Control of Two-Stroke Automobile Engine," SAE Paper No. 770766, Society of Automotive Engineers, Warrendale, Pa., 1977.

7.12 M. Nuti, "Direct Fuel Injection: An Opportunity for Two-Stroke SI Engines in Road Vehicle Use," SAE Paper No. 860170, Society of Automotive Engineers, Warrendale, Pa., 1986.

7.13 N.J. Beck, R.L. Barkhimer, M.A. Calkins, W.P. Johnson, W.E. Weseloh, "Direct Digital Control of Electronic Unit Injectors," SAE Paper No. 840273, Society of Automotive Engineers, Warrendale, Pa., 1084.

Chapter 7 - Reduction of Fuel Consumption and Exhaust Emissions

7.14 E. Vieilledent, "Low Pressure Electronic Fuel Injection System for Two-Stroke Engines," SAE Paper No. 780767, Society of Automotive Engineers, Warrendale, Pa., 1978.

7.15 R. Douglas, G.P. Blair, "Fuel Injection of a Two-Stroke Cycle Spark Ignition Engine," SAE Paper No. 820952, Society of Automotive Engineers, Warrendale, Pa., 1982.

7.16 N.J. Beck, W.P. Johnson, R.L. Barkhimer, S.H. Patterson, "Electronic Fuel Injection for Two-Stroke Cycle Gasoline Engines," SAE Paper No. 861242, Society of Automotive Engineers, Warrendale, Pa., 1986.

7.17 G. Yamagishi, T. Sato, H. Iwasa, "A Study of Two-Stroke Fuel Injection Engines for Exhaust Gas Purification," SAE Paper No. 720195, Society of Automotive Engineers, Warrendale, Pa., 1972.

7.18 P. Duret, A. Ecomard, M. Audinet, "A New Two-Stroke Engine with Compressor-Air Assisted Fuel Injection for High Efficiency Low Emissions Applications," SAE Paper No. 880176, Society of Automotive Engineers, Warrendale, Pa., 1988.

7.19 D. Plohberger, K. Landfahrer, L. Mikulic, "Development of Fuel Injected Two-Stroke Gasoline Engine," SAE Paper No. 880170, Society of Automotive Engineers, Warrendale, Pa., 1988.

7.20 S. Ishihara, "Experimental Development of Two New Types of Double Piston Engines," SAE Paper No. 860031, Society of Automotive Engineers, Warrendale, Pa., 1986.

7.21 F. Laimbock, "Der abgasame Hochleitungszweitaktmotor," Dritte Grazer Zweiradtagung, Technische Universitat, Graz, Austria, 13-14 April, 1989.

7.22 D. Fog, R.M. Brown, D.H. Garland, "Reduction of Smoke from Two-Stroke Engine Oils," Paper C372/040, Institution of Mechanical Engineers, London, 1989.

7.23 O.V. Kuntscher, A. Singer, "Mixture Injection Application for Avoiding Charge Exchange Losses in Two-Stroke Cycle Engines," Paper C372/025, Institution of Mechanical Engineers, London, 1989.

7.24 M. Prigent, G. De Soete, "Nitrous Oxide N_2O in Engine Exhaust Gases, A First Appraisal of Catalyst Impact," SAE Paper No. 890492, Society of Automotive Engineers, Warrendale, Pa., 1989.

7.25 K. Sugiura, M. Kagaya, "A Study of Visible Smoke Reduction from a Small Two-Stroke Engine using Various Engine Lubricants," SAE Paper No. 770623, Society of Automotive Engineers, Warrendale, Pa., 1977.

7.26 K. Schlunke, "Der Orbital Verbrennungsprozess des Zweitaktmotors," Tenth International Motor Symposium, Vienna, April 1989, pp63-78.

7.27 Anon, "The 249cc Model GTP Velocette," The Motor Cycle, London, 12 December 1934.

7.28 C.E. Carson, R.J. Kee, R.G. Kenny, C. Stan, K. Lehmann, S. Zwahr, "Ram-Tuned and Air-Assisted Direct Fuel Injection Systems Applied to a SI Two-Stroke Engine," SAE Paper No. 950269, Society of Automotive Engineers, Warrendale, Pa., 1995.

7.29 R.J. Kee, G.P. Blair, C.E. Carson, R.G. Kenny, "Exhaust Emissions of a Stratified Charge Two-Stroke Engine," Funfe Grazer Zweiradtagung, Technische Universitat, Graz, Austria, 12-23 April, 1993.

7.30 S.J. Magee, R. Douglas, G.P. Blair, C.E. Carson, J-P. Cressard, "Reduction of Fuel Consumption and Emissions of a Small Capacity Two-Stroke Cycle Engine," SAE Paper No. 931539, Society of Automotive Engineers, Warrendale, Pa., 1993.

7.31 M. Saxena, H.B. Mathur, S. Radzimirski, "A Stratified Charging Two-Stroke Engine for Reduction of Scavenged-Through Losses," SAE Paper No. 891805, Society of Automotive Engineers, Warrendale, Pa., 1989.

7.32 F.W. Lanchester, R.H. Pearsall, "An Investigation into Certain Aspects of the Two-Stroke Engine for Automobile Vehicles," The Automobile Engineer, London, February 1922.

7.33 T. Sato, M. Nakayama, "Gasoline Direct Injection for a Loop-Scavenged Two," SAE Paper No. 720195, Society of Automotive Engineers, Warrendale, Pa., 1972.

7.34 G.P. Blair, R.J. Kee, C.E. Carson, R.Douglas, "The Reduction of Emissions and Fuel Consumption by Direct Air-Assisted Fuel Injection into a Two-Stroke Engine," Vierte Grazer Zweiradtagung, Technische Universität, Graz, Austria, 8-9 April, 1991, pp114-141.

7.35 D. Campbell, G.P. Blair, "Anordnung von kleinvolumigen Zweitaktmotoren um gute Kraft und arme Emissionen zu erhalten," 4th International Motorcycle Conference, Verein Deutscher Ingenieure, Munich, 5-7 March 1991.

7.36 G. P. Blair, <u>The Basic Design of Two-Stroke Engines</u>, SAE R-104, Society of Automotive Engineers, Warrendale, Pa., February 1990, ISBN 1-56091-008-9.

7.37 Yamaha Hatsudoki Kabushiki Kaisha, Iwata, Japan, "Combustion Chamber for Injected Engine," US Patent 5,163,396, November 17, 1992.

7.38 B. Cumming, "Two-Stroke Engines—Opportunities and Challenges," Aachen Colloquium on Automobile and Engine Technology, Institut fur Kraftfahrwesen, Lehrstuhl fur Angewandte Thermodynamik, Aachen, 15-17 October 1991.

7.39 G.E. Hundleby, "Development of a Poppet-Valved Two-Stroke Engine-the Flagship Concept," SAE Paper No. 900802, Society of Automotive Engineers, Warrendale, Pa., 1990.

7.40 P. Conlon and R. Fleck, "An Experimental Investigation to Optimize the Performance of a Supercharged, Two-Stroke Engine," SAE Paper No. 930982, Society of Automotive Engineers, Warrendale, Pa., 1993.

7.41 R. Diesel, "Method of and Apparatus for Converting Heat into Work," German Patent No. 542846, 1895.

7.42 J.R. Pattison, "Electrically Controlled Gas Engine Fuel System," United States Patent No. 1,288,439, 1918.

7.43 J. Emerson, P.G. Felton and F.V. Bracco, "Structure of Sprays from Fuel Injectors Part 3: The Ford Air-assisted Fuel Injector," SAE Paper No. 890313, Society of Automotive Engineers, Warrendale, Pa., 1989.

7.44 K. Landfahrer, D. Plohberger, H. Alten and L. Mikulic, "Thermodynamic Analysis and Optimization of Two-Stroke Gasoline Engines," SAE Paper No. 890415, Society of Automotive Engineers, Warrendale, Pa., 1989.

7.45 S.R. Leighton, S.R. Ahern, "The Orbital Small Engine Fuel Injection System (SEFIS) for Directed Injected Two-Stroke Cycle Engines," Funfe Grazer Zweiradtagung, Technische Universitat, Graz, Austria, 12-23 April, 1993.

7.46 G.K. Fraidl, R. Knoll, H.P. Hazeu, "Direkte Gemischeinblasung am 2-Takt-Ottomotor," VDI Bericht, Dresden, 1993.

7.47 R.G. Kenny, R.J. Kee, C.E. Carson, G.P. Blair, "Application of Direct Air-Assisted Fuel Injection System to a SI Cross Scavenged Two-Stroke Engine," SAE Paper No. 932396, Society of Automotive Engineers, Warrendale, Pa., 1993.

7.48 G. Monnier, P. Duret, "IAPAC Compressed Air-Assisted Fuel Injection for High Efficiency, Low Emissions, Marine Outboard Two-Stroke Engines," SAE Paper No. 911849, Society of Automotive Engineers, Warrendale, Pa., 1991.

7.49 K-J. Yoon, W-T. Kim, H-S. Shim, G-W. Moon, "An Experimental Comparison between Air-Assisted Injection and High-Pressure Injection System at Two-Stroke Engine," SAE Paper No. 950270, Society of Automotive Engineers, Warrendale, Pa., 1995.

7.50 Y. Ikeda, T. Nakajima, N. Kurihara, "Spray Formation of Air-Assist Injection for Two-Stroke Engine," SAE Paper No. 950271, Society of Automotive Engineers, Warrendale, Pa., 1995.

7.51 M. Nuti, R. Pardini, "A New Simplified Carburetor for Small Engines," SAE Paper No. 950272, Society of Automotive Engineers, Warrendale, Pa., 1995.

7.52 P. Duret, G. Monnier, "Low Emissions IAPAC Fuel-Injected Two-Stroke Engines for Two-Wheelers," Funfe Grazer Zweiradtagung, Technische Universität, Graz, Austria, 12-23 April, 1993.

7.53 C. Stan, "Concepts for the Development of Two-Stroke Engines," SAE Paper No. 931477, Society of Automotive Engineers, Warrendale, Pa., 1993.

7.54 W. Heimberg, "Ficht Pressure Surge Injection System," ATA Paper No. 93A079, ATA-SAE Small Engine Technology Conference, Pisa, 1-3 December 1993, SAE Paper No. 931502.

7.55 J.F. Sinnamon, D.R. Lancaster, J.R. Steiner, "An Experimental and Analytical Study of Engine Fuel Spray Technologies," SAE Paper No. 800135, Society of Automotive Engineers, Warrendale, Pa., 1980.

7.56 Automotive Handbook, U. Adler (Ed.), Robert Bosch, Stuttgart, 18th Edition, 1989, ISBN 0-89883-510-0.

7.57 G.T. Kalghatgi, P. Snowdon, C.R. McDonald, "Studies of Knock in a Spark-Ignition Engine with 'CARS' Temperature Measurements and Using Different Fuels," SAE Paper No. 950690, Society of Automotive Engineers, Warrendale, Pa., 1995.

Appendix A7.1 The effect of compression ratio on performance characteristics and exhaust emissions

Appendices A4.1 and A4.2 describe a two-zone combustion model, with equilibrium thermodynamics giving dissociation for the burn zone constituents and a formation model for nitric oxide based on reaction kinetics. This model is sufficiently fundamentally based as to be able to calculate the local thermodynamic conditions during combustion and, when combined with a GPB simulation of the open cycle behavior of an engine and its ducting, is able to provide design guidance for the effect of compression ratio on the engine performance characteristics of power output, fuel consumption, the intake and exhaust noise emissions, and the gaseous exhaust emissions. In short, as mentioned at the beginning of Sec. 7.3.3, a full debate can be conducted on the design compromises which inevitably have to be made for any given engine to satisfy the market and legislative requirements for it.

The engine used as the design example is the standard chainsaw engine, used frequently throughout the book, but first introduced with geometrical input data in Sec. 5.5.1. The data are as presented there, but with the trapped compression ratio, CR_t, changed successively from 6.5 to 7.0 (the "standard" value), 7.5 and 8.0. The engine will assuredly give some detonation at a CR_t value of 8.0. For each of the simulations the engine speed employed is 9600 rpm, together with an air-to-fuel ratio of 13.0 on unleaded gasoline, which translates to a λ value of 0.9. The combustion model is exactly as described in Sec. 5.5.1, i.e., as it is shown in Fig. 4.7(d), with an ignition timing at 24° btdc used within each simulation.

The results of the modeling are shown in Figs. A7.1-A7.4. It should be noted that the delivery ratio and charging efficiency are ostensibly constant for all of these simulations, thus any changes in performance characteristics are almost entirely due to combustion variations.

The effect of compression ratio on power and fuel consumption

Eq. 1.5.22 predicts higher power and thermal efficiency for increasing compression ratio. Thermal efficiency is inversely related to brake specific fuel consumption, which can be seen

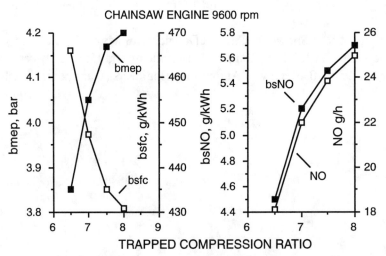

Fig. A7.1 Effect of compression ratio on torque, fuel consumption and NO_x emissions.

Eq. 1.5.22 predicts higher power and thermal efficiency for increasing compression ratio. Thermal efficiency is inversely related to brake specific fuel consumption, which can be seen from Eqs. 1.6.4 and 1.6.5, i.e., the higher the thermal efficiency, the lower the specific fuel consumption. That this holds true in practice is seen in Fig. A7.1. The bmep rises and the bsfc falls with increasing compression ratio, although the rate of change is beginning to decrease between a CR_t of 7.5 and 8.0. The power output has increased and the bsfc has decreased by about 8% over the CR_t simulation range from 6.5 to 8.0.

Eq. 1.5.22 predicts that the change of thermal efficiency for the same change of compression ratios, i.e., from 6.5 to 8.0, is 7.2%. Thus even the most fundamental of thermodynamics does give some basic guidance. However, it should be pointed out that Eq. 1.5.22 forecasts a thermal efficiency of 56.5% for a compression ratio of 8.0, and 52.7% for a CR_t of 6.5, when the actual values are less than half that!

The effect of compression ratio on nitric oxide (NO) emissions

Fig. A7.1 shows the brake specific nitric oxide emissions, bsNO, and the total nitric oxide emissions, NO, in g/kWh and g/h units, respectively. A somewhat dramatic increase with compression ratio is observed in both cases. The bsNO value has increased by 26.7% from 4.5 to 5.7 g/kWh. The total NO emission has gone from 18 to 25 g/h, which is an increase of 39%. The word dramatic is not too strong a word to employ, as those engineers who have to meet emissions legislation usually struggle with gains or losses between 1 and 2% during R&D experimentation, and will visibly blanch at a simulation indicating changes of the order shown above.

The origins of the change are to be found in Figs. A7.2 and A7.3. In Fig. A7.2 the peak temperatures in the burn zone are shown to rise with respect to CR_t from 2340°C to 2390°C. As commented on in Appendix A4.1, the formation of NO is exponentially connected to these temperatures, so that even an apparently modest rise of only 50°C above 2340°C has the

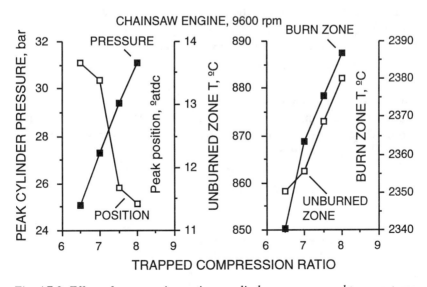

Fig. A7.2 Effect of compression ratio on cylinder pressures and temperatures.

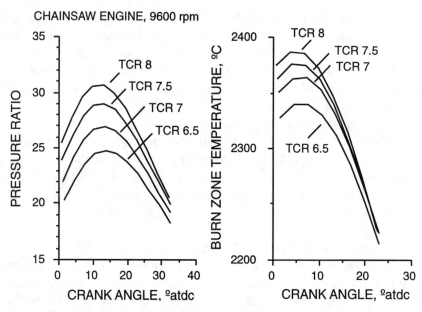

Fig. A7.3 Effect of compression ratio on combustion rates.

dramatic consequences described.

The effect of compression ratio on combustion

Fig. A7.2 summarizes the combustion information. With increasing compression ratio, the peak cylinder pressure is shown to rise from 25 bar to 31 bar and the location of that peak is shown to occur ever earlier from 13.7°atdc to 11.4°atdc. The burned and unburned zone peak temperatures are shown on the same figure. The profiles of these changes are shown in Fig. A7.3, and the rate of pressure rise in the cylinder, in terms of pressure change per unit time, is observed to increase. In actuality, the combustion pressure wave created will be sharper as the compression ratio rises and will compress the gas in the unburned zone more noticeably. This model, which does not include a combustion pressure wave, nevertheless shows that the unburned zone peak temperature increases from 858 to 882°C.

While considering this information, it is important to remember that the simulation employs the same Vibe function, and the same ignition timing and ignition delay, for each simulation at differing compression ratios. Clearly, the use of a retarded ignition timing will allay some of the thermal stress and loading seen at the higher compression ratios, albeit at the expense of power and thermal efficiency.

The effect of compression ratio on detonation

The profile with respect to time of the compression and, more important, the heating of the unburned zone is shown in Fig. A7.4. It can be seen that there is considerable similarity of temperature between compression ratios of 6.5 and 7, although the NO formation rate given in Fig. A7.1 is affected much more severely. At higher compression ratios the unburned zone temperatures are significantly higher.

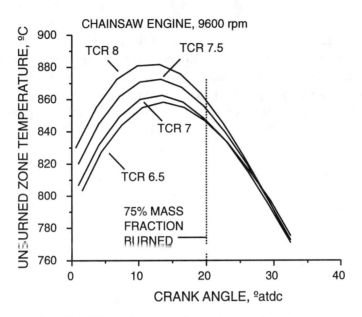

Fig. A7.4 Effect of compression ratio on unburned zone.

at the same instant in each simulation. It is observed that the end zone temperature at this instant, at a trapped compression ratio of 8.0, is above the peak value for a compression ratio of 6.5. The end zone contains air and vaporized fuel at these state conditions of temperature and pressure. If the temperature in this end zone, or in smaller localized pockets of this end zone, is above the auto-ignition temperature of any fuel in that pocket, then auto-ignition can occur and detonation, or "knocking," will be experienced.

It is reported by Bosch [7.56] that detonation can occur by auto-ignition in gasoline-fuel vapor mixtures at temperatures at and above 880°C. It is observed that this condition is met by the chainsaw simulation for the entire unburned zone at 12° atdc at a trapped compression ratio of 8.0, where it is 882°C. Detonation will assuredly occur at this point.

This opinion is reinforced by the data from Kalghatgi [7.57] who shows measured temperature data for the combustion of iso-octane in a four-stroke cycle engine at 11° atdc. At a temperature of 830°C, knocking or detonation was not present. At a temperature of 887°C at the same crankshaft position, knocking was present. The fine limits between the presence of detonation, or its absence, are clearly to be found within this publication.

The peak temperature in the unburned zone at a CR_t of 7.5 is 873°C. Detonation may not occur at this point, but my more cynical view, already mooted above, is that Murphy, whose Law applies always and everywhere, will have arranged already to have localized pockets of air and fuel meet the auto-ignition criterion at a trapped compression ratio of less than 8.0, but maybe not as low as 7.5.

In terms of the GPB simulation techniques used here, it can be reasonably safely assumed that detonation will not occur if the unburned zone temperatures are detected not to exceed 870°C, at any simulated speed or load within the operating range of the engine.

General

The onset of detonation, as can be seen from the above, and even with the extended information available from the above, is still not totally straightforward to predict. It is not simply a matter of using the highest possible detonation-free compression ratio so as to obtain the best power and fuel consumption, as the accompanying NO emission levels may well be above that required by legislation. Nevertheless, this simulation clearly provides the type of in-depth information upon which the design decision regarding the maximum, or the optimum, compression ratio for a given engine can be made.

Chapter 8

Reduction of Noise Emission from Two-Stroke Engines

8.0 Introduction

The subject of noise emission from an internal-combustion engine, and its reduction, is a specialized topic. Many textbooks and technical papers have been written on the subject, so it is not possible to completely cover the matter in a single chapter in this book. Instead, it is intended to orient you to noise emission pertaining to two-stroke engines and to the problems of silencer design for this particular type of power unit. Note that the American word for a "silencer" is a "muffler." Many texts discuss silencer design as if the only engine in existence is the spark-ignited, multi-cylinder, four-stroke automobile engine and merely describe the techniques applicable to that area. As that can create considerable misconceptions for the designer of silencers for two-stroke engines, this provides sufficient justification for the inclusion of this chapter within this book.

The opening section of this chapter gives some general background to the subject of noise, indeed it repeats in brief what may be found more completely in other books and papers which are suitably referenced for your wider education. This opening section is included so that the remainder of the chapter will be more immediately meaningful to the novice.

The next section deals with the fundamental nature of pressure-wave-created noise emission and to the theoretical methods available for its prediction. The latter part of this section debates the future for the technology of silencer design.

The succeeding sections will cover the more empirical approaches to silencer design and incorporate pragmatic advice on the design of such devices for two-stroke engines.

8.1 Noise

In many texts, you will find "noise" described as "unwanted sound." This is a rather loose description, as that which is wanted by some may be unwanted by others. A 250 cm^3 Yamaha V4 racing two-stroke, producing 72 unsilenced horsepower at 14,000 rpm, and wailing its way up the Mountain in the Isle of Man TT race of 1967, produced a noise which was music to the ears of a thousand racing motorcycle fans. A nearby farmer, the owner of a thousand chickens vainly trying to lay eggs, viewed the same sound from an alternate standpoint. This simple example illustrates the quite subjective nature of noise assessment. Nevertheless, between the limits of the threshold of human hearing and the threshold of damage to the human ear, it is possible to physically measure the pressure level caused by sound and to assign an experimental number to that value. This number will not detail whether the sound is "wanted" or "unwanted." As already pointed out, to some it will always be described as noise.

8.1.1 Transmission of sound

As discussed in Sec. 2.1.2, sound propagates in three dimensions from a source through the air (or a gas) as the medium of its transmission. The fundamental theory for this propagation is in Sec. 2.1.2. The speed of the propagation of a wave of acoustic amplitude is given by a_0, where:

$$a_0 = \sqrt{\gamma R T_0} = \sqrt{\frac{\gamma p_0}{\rho_0}} \qquad (8.1.1)$$

As shown in Sec. 2.1.6, the value for the ratio of specific heats, γ, is 1.4 for air and 1.375 for exhaust gas when both are at a temperature around 25°C, and for exhaust gas emanating from a stoichiometric combustion. At such room temperature conditions, the value of the gas constant, R, is 287 J/kgK for air and 291 J/kgK for the exhaust gas. Treating exhaust gas as air in calculations for sound wave attenuation in silencers produces errors of no real significance. For example, if the temperature is raised to 500 K, where γ and R for air and exhaust gas are taken from Sec. 2.1.6 and Table 2.1.3,

$$a_{air} = \sqrt{1.373 \times 287 \times 500} = 444 \quad m/s$$

$$a_{exhaust} = \sqrt{1.35 \times 290.8 \times 500} = 443 \quad m/s$$

As exhaust gas in a two-stroke engine contains a significant proportion of air which is short-circuited during the scavenge process, this reduces the already negligible error even further.

8.1.2 Intensity and loudness of sound

The propagation of pressure waves is already covered thoroughly in Sec. 2.1, so it is not necessary to repeat it here. Sound waves are but small pressure waves. However, the propagation of these small pressure pulses in air, following one after the other, varying in both spacing and amplitude, gives rise to the human perception of the pitch and of the amplitude of the sound. The frequency of the pressure pulsations produces the pitch and their amplitude denotes the loudness. The human ear can detect frequencies ranging from 20 Hz to 20 kHz, although as one becomes older that spectrum shortens to a maximum of about 12 kHz. For an alternative introductory view of this topic, consult the books by Annand and Roe [1.8] and by Taylor [8.11].

More particularly, the intensity, I, is used to denote the physical energy of a sound, and loudness, β, is <u>defined in this book</u> as the human perception of that intensity in terms of sound pressure level. I am well aware that the term "loudness" is often defined differently in other texts [1.8], but as this word is most meaningful to you, and to any listener, as a way to express perceived noise level as measured by pressure, I feel justified in using it in that context within this text.

The relationship between intensity and loudness is fixed for sounds which have a pure tone, or pitch, i.e., the sound is composed of sinusoidal pressure waves of a given frequency. For real sounds that relationship is more complex. The intensity of the sound, being an energy value, is denoted by units of W/m². Noise meters, being basically pressure transducers, record the "effective sound pressure level," which is the root-mean-square of the pressure fluctuation about the mean pressure caused by the sound pressure waves. This rms pressure fluctuation is denoted by dp, and in a medium with a density, ρ, and a reference speed of sound, a_0, the intensity is related to the square of the rms sound pressure level by:

$$I = \frac{dp^2}{\rho a_0} \qquad (8.1.2)$$

The pressure rise, dp, can be visually observed in Plates 2.1-2.3 as it propagates away from the end of an exhaust pipe.

The level of intensity that can be recorded by the human ear is considerable, ranging from 1 pW/m² to 1 W/m². The human eardrum, our personal pressure transducer, will oscillate from an imperceptible level at the minimum intensity level up to about 0.01 mm at the highest level when a sensation of pain is produced by the nervous system as a warning of impending damage. To simplify this wide variation in physical sensation, a logarithmic scale is used to denote loudness, and the scale is in units called a Bel, the symbol for which is B. Even this unit is too large for general use, so it is divided into ten subdivisions called decibels, the nomenclature for which is dB. The loudness of a sound is denoted by comparing its intensity level on this logarithmic scale to the "threshold of hearing," which is at an intensity, I_0, of 1.0 pW/m² or a rms pressure fluctuation, dp_0, of 0.00002 Pa, which is 0.0002 μbar. Thus, intensity level of a sound, I_{L1}, where the actual intensity is I_1, is given by:

$$\begin{aligned} I_{L1} &= \log_{10}\left(\frac{I_1}{I_0}\right) \text{ B} \\ &= 10 \log_{10}\left(\frac{I_1}{I_0}\right) \text{ dB} \end{aligned} \qquad (8.1.3)$$

In a corresponding fashion, a sound pressure level, β_1, where the actual rms pressure fluctuation is dp_1, is given by:

$$\beta_1 = \log_{10}\left(\frac{I_1}{I_0}\right) \text{ B}$$

or,
$$\beta_1 = 10\log_{10}\left(\frac{I_1}{I_0}\right) = 10\log_{10}\left(\frac{dp_1}{dp_0}\right)^2 = 20\log_{10}\left(\frac{dp_1}{dp_0}\right) \quad dB \qquad (8.1.4)$$

8.1.3 Loudness when there are several sources of sound

Imagine you are exposed to two sources of sound of intensities I_1 and I_2. These two sources would separately produce sound pressure levels of β_1 and β_2, in dB units. Consequently, from Eq. 8.1.4:

$$I_1 = I_0 \text{ antilog}_{10}\left(\frac{\beta_1}{10}\right) \qquad I_2 = I_0 \text{ antilog}_{10}\left(\frac{\beta_2}{10}\right)$$

The absolute intensity that you experience from both sources simultaneously is the superposition value, I_s, where:

$$I_s = I_1 + I_2$$

Hence, the total sound pressure level experienced from both sources is β_s, where:

$$\beta_s = 10\log_{10}\left(\frac{I_s}{I_0}\right) = 10\log_{10}\left\{\text{antilog}\left(\frac{\beta_1}{10}\right) + \text{antilog}\left(\frac{\beta_2}{10}\right)\right\} \qquad (8.1.5)$$

Consider two simple cases:
(i) You are exposed to two equal sources of sound which are at 100 dB.
(ii) You are exposed to two sources of sound, one at 90 dB and the other at 100 dB.
First, consider case (i), using Eq. 8.1.5:

$$\begin{aligned}\beta_s &= 10\log_{10}\left\{\text{antilog}\left(\frac{\beta_1}{10}\right) + \text{antilog}\left(\frac{\beta_2}{10}\right)\right\} \\ &= 10\log_{10}\left\{\text{antilog}\left(\frac{100}{10}\right) + \text{antilog}\left(\frac{100}{10}\right)\right\} \\ &= 103.01 \quad dB\end{aligned} \qquad (8.1.6)$$

Second, consider case (ii):

$$\begin{aligned}\beta_s &= 10\log_{10}\left\{\text{antilog}\left(\frac{\beta_1}{10}\right) + \text{antilog}\left(\frac{\beta_2}{10}\right)\right\} \\ &= 10\log_{10}\left\{\text{antilog}\left(\frac{100}{10}\right) + \text{antilog}\left(\frac{90}{10}\right)\right\} \\ &= 100.41 \quad dB\end{aligned} \qquad (8.1.7)$$

In case (i), it is clear that the addition of two sound sources, each equal to 100 dB, produces an overall sound pressure level of 103.01 dB, a rise of just 3.01 dB due to a logarithmic scale being used to attempt to simulate the response characteristics of the human ear. To physically support this mathematical contention, recall that the noise of an entire brass band does not appear to be so much in excess of one trumpet at full throttle.

In case (ii), the addition of the second weaker source at 90 dB to the noisier one at 100 dB produces a negligible increase in loudness level, just 0.41 dB above the larger source. The addition of one more trumpet to the aforementioned brass band does not raise significantly the total noise level as perceived by the listener.

There is a fundamental message to the designer of engine silencers within these simple examples: If an engine has several different sources of noise, the loudest will swamp all others in the overall sound pressure level. The identification and muffling of that major noise source becomes the first priority on the part of the engineer. Expanded discussion on these topics is found in the books by Harris [8.12] and Beranek [8.13] and in this chapter.

8.1.4 Measurement of noise and the noise-frequency spectrum

An instrument for the measurement of noise, referred to as a *noisemeter*, is basically a microphone connected to an amplifier so that the system is calibrated to read in the units of dB. Usually the device is internally programmed to read either the total sound pressure level known as the linear value, i.e., dBlin, or on an A-weighted or B-weighted scale to represent the response of the human ear to loudness as a function of frequency. The A-weighted scale is more common and the units are recorded appropriately as dBA. To put some numbers on this weighting effect, the A-weighted scale reduces the recorded sound below the dBlin level by 30 dB at 50 Hz, 19 dB at 100 Hz, 3 dB at 500 Hz, 0 at 1000 Hz, then increases it by about 1 dB between 2 and 4 kHz, before tailing off to reduce it by 10 dB at 20 kHz. The implications behind this weighting effect are that high frequencies between 1000 and 4000 Hz are very irritating to the human ear, to such an extent that a noise recording 100 dBlin at around 100 Hz only sounds as loud as 81dB at 1000 Hz, hence it is recorded as 81 dBA. To quote another example, the same overall sound pressure level at 3000 Hz appears to be as loud as 101 dB at 1000 Hz, and is noted as 101 dBA. An example of the effect of the A-weighting on noise over a range of frequencies is shown in Fig. 8.23 for the exhaust noise spectra from a chainsaw.

Equally common is for the noisemeter to be capable of a frequency analysis, i.e., to record the noise spectra over discrete bands of frequency. Usually these are carried out over one-octave bands or, more finely, over one-third octave bands. A typical one-octave filter set on a noisemeter would have switchable filters to record the noise about 31.25 Hz, 62.5 Hz, 125 Hz, 250 Hz, 500 Hz, 1000 Hz, 2000 Hz, 4000 Hz, 8000 Hz and 16,000 Hz. A one-third octave filter set carries out this function in narrower steps of frequency change. The latest advances in electronics and computer-assisted data capture allow this process of frequency analysis to be carried out in even finer detail.

The ability of a measurement system to record the noise-frequency spectrum is very important for the researcher who is attempting to silence, say, the exhaust system of a particular engine. Just as described in Sec. 8.1.3 regarding the addition of noise levels from several sources, the noisiest frequency band in the measured spectrum is that band which must be silenced first and foremost as it is contributing in the major part to the overall sound pressure level. The identification of the frequency band of that major noise component will be shown later as the first step toward its eradication as a noise source.

The measurement of noise is a tedious experimental technique in that a set procedure is not just desirable, but essential. Seemingly innocent parameters, such as the height of the microphone from the ground during a test, or the reflectivity of the surface of the ground in the vicinity of the testing, e.g., grass or tarmac, can have a major influence on the numerical value of the dB recorded from the identical engine or machine. This has given rise to a plethora of apparently unrelated test procedures, such as those in the SAE Standards [8.16]. In actual fact, the logic behind their formulation is quite impeccable and anyone embarking on a silencer design and development exercise will be wise to study them thoroughly and implement them during experimentation.

8.2 Noise sources in a simple two-stroke engine

The sources of noise emanating from a two-stroke engine are illustrated in Fig. 8.1. The obvious ones are the intake and the exhaust system, where the presence of gas pressure waves has been discussed at length in Chapter 2. As these propagate into the atmosphere they produce noise. The series of photographs in Chapter 2, Plates 2.1-2.4, illustrate the rapid nature of the pressure rise propagating into the atmosphere and toward the ear of the listener. The common belief is that the exhaust is the noisier of the two, and in general this is true. However, the most rudimentary of exhaust silencers will almost inevitably leave the intake system as the noisier of these two sources, so it also requires silencing to the same level and extent. If

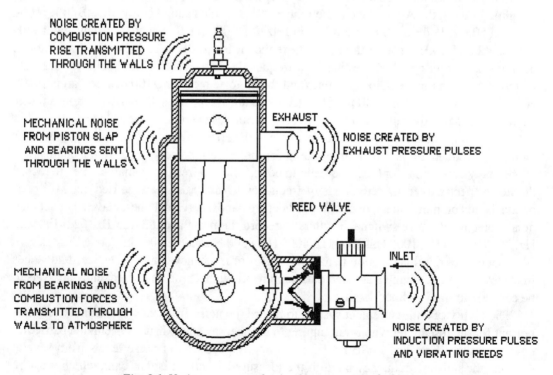

Fig. 8.1 Various sources of noise from a two-stroke engine.

the intake system contains a reed valve, it produces its own unique high-frequency noise components, much like the vibrating reed in an oboe or a clarinet which has a special quality described occasionally as "honking." This adds to the design complexity of the intake silencer for such engines.

Remember that any vibrating metal surface can act as a noise source, in the same manner as does the vibrating diaphragm of a loudspeaker, hence the pressure signal emanating from the combustion pressure rise is transmitted through the cylinder walls and can be propagated away from the outer surface of the cylinder and the cylinder head. If the engine is air-cooled, the cylinder and head finning are ideally suited to become metal diaphragms for this very purpose. The need to control this form of noise transmission is obvious; observe that air-cooled motorcycle engines have rubber damping inserted between the cooling fins for this very purpose. The problem is greatly eased by the use of liquid (water) cooling of a cylinder and cylinder head, as the intervening water layer acts as a damper on the noise transmission. That does not totally solve this problem, for the noise can be transmitted through the piston to the connecting rod, via the bearings to the crankcase walls, and ultimately to the atmosphere. The use of pressure-fed plain bearings on an engine crankshaft is superior to ball, roller, or needle roller bearings as a means of suppression of this type of noise transmission because the hydrodynamic oil film in the plain bearing does not as readily transmit the combustion vibrations. However, the employment of pressure-fed plain bearings does not lend itself to the use of the crankcase as the air pump of the simple two-stroke engine.

The rotating bearings of an engine produce noise characteristics of their own, emanating from the vibrations of the mechanical components. Another mechanical noise source is that from piston slap, as the piston rocks on the gudgeon pin within the cylinder walls around the tdc and bdc positions.

However, the discussion regarding noise suppression in this chapter will concentrate on the gas pressure-wave-generated noise from the exhaust and the inlet systems.

8.3 Silencing the exhaust and inlet system

In the matter of silencing, the two-stroke engine has some advantages and disadvantages by comparison with an equivalent four-stroke cycle engine. These are catalogued below in the first instance and discussed at greater length in succeeding sections as the need arises.

The disadvantages of the two-stroke engine
  (i)   The engine operates at double the frequency of creation of gas pressure waves and humans dislike exposure to higher-frequency noise.
  (ii) The ports in a two-stroke engine open faster than the poppet valves of the four-stroke engine and so the pressure wave fronts are steeper, thereby creating more high-frequency noise components within the sound spectrum.
  (iii) Many two-stroke engines are used in applications calling for an engine with low bulk and weight, thereby further reducing the space available for muffling and potentially giving the engine type an undeserved reputation for being naturally noisy.

(iv) Two-stroke engines use ball, roller, and needle roller element bearings and these tend to be noisy by comparison with pressure-fed hydrodynamic bearings.

The advantages of the two-stroke engine
(i) The engine with the tuned exhaust pipe produces a high specific output, but this is achieved by choking the final outlet diameter, thereby simplifying the design of an effective exhaust silencer.
(ii) The crankcase pump induces air by pumping with a low compression ratio and, as this reduces the maximum values of air intake particle velocity encountered in its time history, this lowers the higher-frequency content of the sound produced.
(iii) The peak combustion pressures are lower in the equivalent two-stroke cycle engine, so the noise spectrum induced by that lesser combustion pressure is reduced via all of the transmission components of the cylinder, cylinder head, piston and crankshaft.

8.4 Some fundamentals of silencer design

If you study the textbooks or technical papers on acoustics and on silencer design, such as many of those referenced below, you will find that the subject is full of empirical design equations for the many basic types of silencers used in the field of internal combustion engines. However useful these may be, you will get the feeling that a fundamental understanding of the subject is not being acquired, particularly as the acoustic theory being applied is one oriented to the propagation of acoustic waves, i.e., waves of infinitesimal amplitude, rather than finite amplitude waves, i.e., waves of the very considerable amplitude to be found in the inlet and exhaust systems of the internal-combustion engine. The subject matter to be found in the acoustic treatment of the theory is somewhat reminiscent of that for the topic of heat transfer, producing an almost infinite plethora of empirical equations for which the authors admit rather large error bands for their implementation in practice. This has always seemed to me as a most unsatisfactory state of affairs. Consequently, a research program was instigated at QUB some years ago to determine if it was possible to predict the noise spectrum emanating from the exhaust systems of internal-combustion engines using the approach of the calculation of the propagation of finite amplitude waves by the method of characteristics as described in Ref. [7.36]. This resulted in the technical publications by Blair, Spechko and Coates [8.1-8.3]. A much more complete exposition of the work of Coates is to be found in his doctoral thesis [8.17].

8.4.1 The theoretical work of Coates [8.3]

In those publications, a theoretical solution is produced [8.3] that shows that the sound pressure level at any point in space beyond the termination of an exhaust system into the atmosphere is, not empirically but directly, capable of being calculated. The amplitude of the nth frequency component of the sound pressure, p_n, is shown to be primarily a complex function of: (a) the instantaneous mass flow rate leaving the end of the pipe system, $\dot{m}$, and (b) the location of the measuring microphone in both distance and directivity from the pipe end, together with other parameters of some lesser significance.

Chapter 8 - Reduction of Noise Emission from Two-Stroke Engines

For any sinusoidal variation, the mean square sound pressure level, p_{rms}, of that nth frequency component is given by:

$$p_{rms\ n}^2 = \frac{p_n^2}{2} \qquad (8.4.1)$$

Coates shows that the mean square sound pressure emanating from a sinusoidal efflux from a pipe of diameter, d, is given by:

$$(p_{rms})^2 = \frac{1}{2}\left(\frac{\pi d^2}{4}\right)^2 \left(\frac{f_n}{r_m}\right)^2 |\rho c|^2 \left\{ \frac{2J_1\left(\frac{\pi f_n d \sin \theta}{a_0}\right)}{\frac{\pi f_n d \sin \theta}{a_0}} \right\}^2 \qquad (8.4.2)$$

where J_1 is the notation for the first-order Bessel function of the terms within. It can be seen from Eq. 8.4.2 that the combination of the second and fourth terms is the instantaneous mass flow rate at the aperture of the exhaust, or intake, system. The final term, in the curly brackets, indicates directivity of the sound, where θ is the angle between the receiving microphone from the centerline particle flow direction of the aperture. The variable, r_m, is the distance of the microphone from the pipe aperture to the atmosphere. From the theory in Sec. 2.2.3, and Eq. 2.2.11 in particular, the instantaneous mass flow rate at the aperture to the atmosphere is given by:

$$\dot{m} = G_5 a_0 \rho_0 \left(\frac{\pi}{4}d^2\right)(X_i + X_r - 1)^{G5}(X_i - X_r) \qquad (8.4.3)$$

where the X_i and X_r values are the time-related incident and reflected pressure amplitude ratios, respectively, of the pressure pulsations at the aperture to the atmosphere. An engine simulation of the type demonstrated in Chapter 5, using the theory of Chapter 2, inherently computes the instantaneous mass flow rate at every section of the engine and its ducting, including the inlet and exhaust apertures to the atmosphere. The instantaneous mass flow rate at the aperture to the atmosphere is collected at each time step in the computation and a Fourier analysis of the resulting periodic function is performed numerically to include all harmonics within the audible range, giving a series of the form:

$$\dot{m}_t = \varphi_0 + (\varphi_{a1} \sin \omega t + \varphi_{b1} \cos \omega t)\ldots\ldots(\varphi_{an} \sin n\omega t + \varphi_{bn} \cos n\omega t) \qquad (8.4.4)$$

The amplitude of each harmonic is then used in Eq. 8.4.2 to calculate the mean square sound pressure due to that particular component:

$$\beta_n = 10 \log_{10}\left(\frac{p_{rms\ n}}{dp_0}\right)^2 \text{ dB} \qquad (8.4.5)$$

Hence, the overall mean square sound pressure, β_s, can be obtained by simple addition of that of the harmonics, using the procedure for the addition of sound energy in Eq. 8.1.5.

How successful that can be may be judged from some of the results presented by Coates [8.3], although that work was carried out using the "homentropic method of characteristics" for the unsteady gas flow along the duct, and employed isentropic pipe end boundary conditions, both of which approaches are shown by Kirkpatrick [2.41, 5.20, 5.21] to be of a lesser accuracy than the method presented in Chapter 2. The implication of that statement is that Coates' ability to accurately calculate mass flow rates at the pipe exit, upon which accurate noise computation is predicated, is impaired. Consequently, this research work is ongoing at QUB using the theory given in Chapter 2, to investigate the order of improvement, if any, in accuracy of prediction of noise transmitted into space from the ducting of engines.

8.4.2 The experimental work of Coates [8.3]

The experimental rig used by Coates is described clearly in Ref. [8.3], but a summary here will aid the discussion of the experimental results and their correlation with the theoretical calculations. The exhaust system is simulated by a rotary valve that allows realistic exhaust pressure pulses of cold air to be blown down into a pipe system at any desired cyclic speed for those exhaust pressure pulsations. The various pipe systems attached to the exhaust simulator are shown in Fig. 8.2, and are defined as SYSTEMS 1-4. Briefly, they are as follows:

SYSTEM 1 is a plain, straight pipe of 28.6 mm diameter, 1.83 m long and completely unsilenced.

SYSTEM 2 has a 1.83 m plain pipe of 28.6 mm diameter culminating in what is termed a diffusing silencer which is 305 mm long and 76 mm diameter. The tail-pipe, of equal size to the entering pipe, is 152 mm long.

SYSTEM 3 is almost identical to SYSTEM 2 but has the entry and exit pipes re-entering into the diffusing silencer so that they are 102 mm apart within the chamber.

SYSTEM 4 has what is defined as a side-resonant silencer placed in the middle of the 1.83 m pipe, and the 28.6-mm-diameter through-pipe has 40 holes drilled into it of 3.18 mm diameter.

More formalized sketches of diffusing, side-resonant and absorption silencers are found in Figs. 8.7-8.9. Further discussion of their silencing effect, based on an acoustic analysis, is in Sec. 8.5. It is sufficient to remark at this juncture that:

(i) The intent of a diffusing silencer is to absorb all noise at frequencies other than those at which the box will resonate. Those frequencies which are not absorbed are called the pass-bands.

(ii) The intent of a side-resonant silencer is to completely absorb noise of a specific frequency, such as the fundamental exhaust pulse frequency of an engine.

(iii) The intent of an absorption silencer is to behave as a diffusing silencer, but to have the packing absorb the resonating noise at the pass-band frequencies.

The pressure-time histories within these various systems, and the one-third octave noise spectrograms emanating from these systems, were recorded. Of interest are the noise spectra and these are shown for SYSTEMS 1-4 in Figs. 8.3-8.6, respectively. The noise spectra are presented in the units of overall sound pressure level, dBlin, as a function of frequency. There

Chapter 8 - Reduction of Noise Emission from Two-Stroke Engines

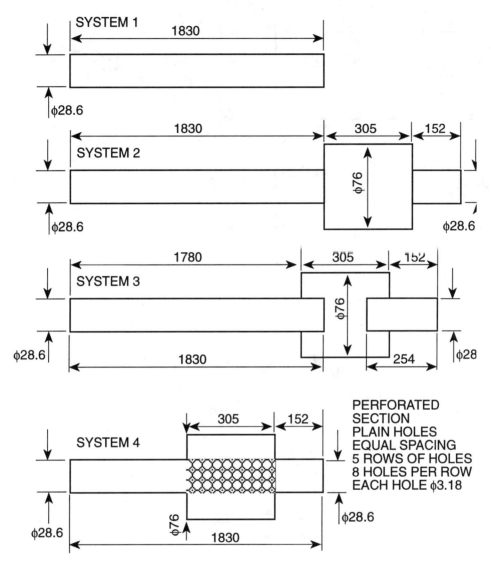

Fig. 8.2 Various exhaust systems and silencers used by Blair and Coates [8.3].

are four spectrograms on any given figure: the one at the top is where the measuring microphone is placed directly in line with the pipe end and the directivity angle is declared as zero, and the others are at angles of 30°, 60° and 90°. The solid line on any diagram is the measured noise spectra and the dashed line is that emanating from the theoretical solution outlined in Sec. 8.4.1. It can be seen that there is a very high degree of correlation between the calculated and measured noise spectra, particularly at frequency levels below 2000 Hz.

Perhaps the most important conclusion to be drawn from these results is that the designer of silencing systems, having the muffling of the noisiest frequency as a first priority, can use an unsteady gas-dynamic simulation program as a means of prediction of the mass flow rate at the pipe termination to atmosphere, be it the inlet or the exhaust system, and from that mass

Design and Simulation of Two-Stroke Engines

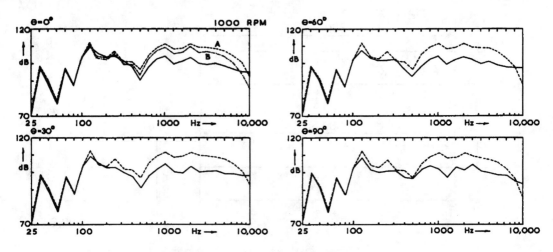

Fig. 8.3 *One-third noise spectrogram from SYSTEM 1.*

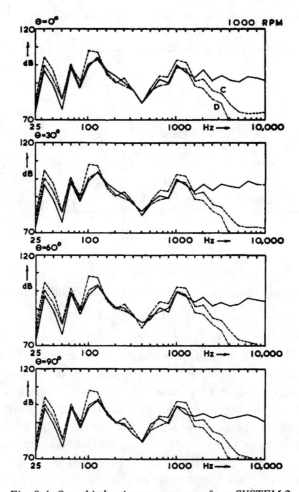

Fig. 8.4 *One-third noise spectrogram from SYSTEM 2.*

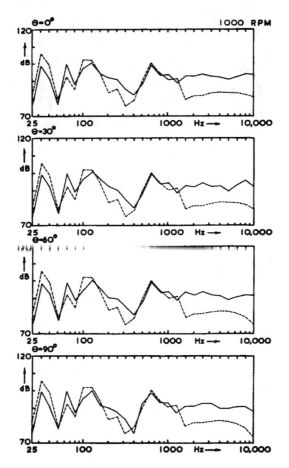

Fig. 8.5 One-third noise spectrogram from SYSTEM 3.

flow rate calculation over a complete cycle determine the noisiest frequency to be tackled by the silencer.

Of direct interest is the silencing effect of the various silencer elements attached to the exhaust pipe by Coates, SYSTEMS 2-4, the noise spectra for which are shown in Figs. 8.4-8.6. The SYSTEM 2, a simple diffusing silencer without re-entrant pipes, can be seen to reduce the noise level of the fundamental frequency at 133 Hz from 116 dB to 104 dB, an attenuation of 12 dB. From the discussion in Sec. 8.1.2, an attenuation of 12 dB is a considerable level of noise reduction. It will also be observed that a large "hole," or strong attenuation, has been created in the noise spectra of SYSTEM 2 at a frequency of 400 Hz. In Sec. 8.5.1 this will be a source of further comment when an empirical frequency analysis is attempted for this particular design.

The noise spectra for SYSTEM 3, a diffusing silencer with re-entrant pipes, is shown in Fig. 8.5. The noise level of the fundamental frequency of 133 Hz has been reduced further to 98 dB. The "hole" of high attenuation is now at 300 Hz and deeper than that recorded by SYSTEM 2.

Design and Simulation of Two-Stroke Engines

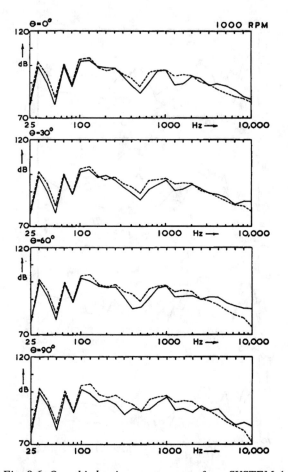

Fig. 8.6 One-third noise spectrogram from SYSTEM 4.

The noise spectra for SYSTEM 4 is shown in Fig. 8.6. This is a side-resonant silencer. The noise level of the fundamental frequency of 133 Hz is nearly as quiet as SYSTEM 2, but a new attenuation hole has appeared at a higher frequency, about 500 Hz. This, too, will be commented on in Sec. 8.5.2 when an empirical acoustic analysis is presented for this type of silencer. It can also be seen that the noise level at higher frequencies, i.e., above 1000 Hz, is reduced considerably from the unsilenced SYSTEM 1.

The most important conclusion from this work by Coates is that the noise propagation into space from a pipe system, with or without silencing elements, can be predicted by a theoretical calculation based on the motion of finite amplitude waves propagating within the pipe system to the pipe termination to the atmosphere. In other words, designers do not have to rely on empirically based acoustic equations for the design of silencers, be they for the intake or the exhaust system, for internal-combustion engines.

8.4.3 Future work for the prediction of silencer behavior

It has always seemed to me that this pioneering work of Coates [8.3, 8.17] has never received the recognition it deserves. Worse, it has tended to be ignored [8.25], due in part to those involved in silencer design not tackling their problems by the use of theory based on unsteady gas dynamics but persisting with theory based on acoustics, i.e., pressure waves with an infinitesimal amplitude [8.24]. No matter how mathematically sophisticated the acoustic theory used may be, it is fundamentally the wrong approach to solve for the noise created by the propagation of finite amplitude pressure waves into the atmosphere. A subsidiary reason, and Coates comments on this, is that silencer elements tend to be somewhat complicated in geometrical terms and have gas particle flow characteristics which are more three-dimensional in nature. As unsteady gas-dynamic calculations are normally one-dimensional, their potential for accurate prediction of the pressure wave reflection and transmission characteristics of real silencer elements is definitely reduced. Nevertheless, the fundamental accuracy of the theoretical premise by Coates is clearly demonstrated and should be the preferred route for the researchers of this subject to correctly deal with the real problems posed by the complex geometry of engine silencers. The theoretical means for so doing is now at hand.

In Sec. 3.4 it is shown how Computational Fluid Dynamics (CFD) is employed to solve the three-dimensional flow behavior inside engine cylinders, with an unsteady gas-dynamic calculation controlling the flow at the entry and exit boundaries of the system. There is no logical reason why this same theoretical approach should not be employed for the design of the most complex of silencers. The engine, together with its inlet and exhaust systems, would be modeled by the unsteady gas dynamics (UGD) method given in Chapter 2, and described further in Chapter 5, up to the boundaries of the various silencers, and that information fed to a CFD program for three-dimensional pursuit through the entire silencer geometry. The result of the CFD calculation would then be the accurate prediction of the mass flow rate spectrum entering the atmosphere, from which the ensuing noise spectra can be correctly assessed by the methods of Coates [8.3] already presented briefly in Sec. 8.4.2. This would be a realistic design process for the engine, together with its ducting and silencers, rather than the empirically based acoustics still in vogue today [8.18]. Acoustics theory could then be relegated to the useful role which befits all empiricism, namely providing more appropriate initial estimations of the geometry of silencer systems as input data, so that a CFD/UGD simulation would not be so wasteful of expensive human and computer time.

8.5 Acoustic theory for silencer attenuation characteristics

The behavior of silencers as treated by the science of acoustics is to be found in a plethora of texts and papers, of which the book by Annand and Roe [1.8] deserves most of your attention. Further useful papers and books are to be found in the References at the end of this chapter. This section will concentrate on the characteristics of the most common types of exhaust silencers, the diffusing, the side-resonant and absorption silencers as shown in Figs. 8.7-8.10, and the low-pass intake silencer sketched in Fig. 8.11.

8.5.1 The diffusing type of exhaust silencer

The sketch of a diffusing silencer element is shown in Fig. 8.7. It has entry and exit pipes of areas A_1 and A_2, respectively, which have diameters, d_1 and d_2, if they are of a circular cross-section. The pipes can be re-entrant into the box with lengths, L_1 and L_2, respectively. The final tail-pipe leaving the box has a length, L_t. The box has a length, L_b, and a cross-sectional area, A_b, or a diameter, d_b, should the box be of a circular cross-section. It is quite

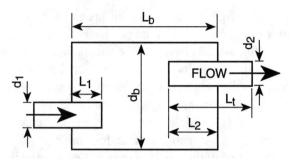

Fig. 8.7 Significant dimensions of a diffusing silencer element.

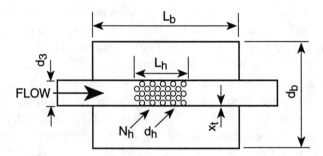

Fig. 8.8 Significant dimensions of a side-resonant silencer element.

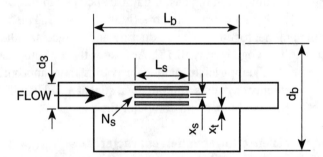

Fig. 8.9 A side-resonant silencer element with slits.

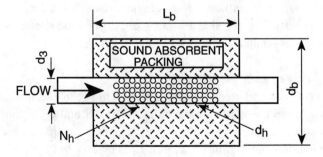

Fig. 8.10 Significant dimensions of an absorption silencer.

Chapter 8 - Reduction of Noise Emission from Two-Stroke Engines

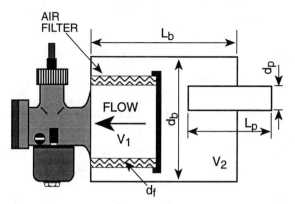

Fig. 8.11 Significant dimensions of an intake silencer.

common, for reasons of manufacturing simplicity, for silencers to have a circular cross-section for both the pipes and the box. If the pipes are of the re-entrant type, then any theory calling for a computation of the volume of the box, V_b, should take into account the box volume occupied by those pipes, thus:

$$V_b = A_b L_b - (A_1 L_1 + A_2 L_2) \tag{8.5.1}$$

The transmission loss, or attenuation, β_{tr}, of a diffusing silencer is basically a function of two parameters: the expansion ratio, A_r, i.e.,

$$A_{r1} = \frac{A_b}{A_1} \qquad A_{r2} = \frac{A_b}{A_2}$$

and the relationship between the wavelength of the sound, Λ, to the length of the box, L_b.

The wavelength of the sound is connected to the frequency of that sound, f, and the acoustic velocity, a_0. The acoustic velocity is calculated by Eq. 8.1.1. The wavelength, Λ, is found from:

$$\Lambda = \frac{a_0}{f} \tag{8.5.2}$$

As the frequency of the gas-borne noise arriving into the diffusing silencer varies, the box will resonate, much like an organ pipe, at various integer amounts of half the wavelength [1.8], in other words at

$$2L_b, \ \frac{2L_b}{2}, \ \frac{2L_b}{3}, \ \frac{2L_b}{4}, \ \frac{2L_b}{5}, \ \text{etc.}$$

From Eq. 8.5.2, this means at frequencies of

$$\frac{a_0}{2L_b}, \frac{2a_0}{2L_b}, \frac{3a_0}{2L_b}, \frac{4a_0}{2L_b}, \frac{5a_0}{2L_b}, \text{etc.}$$

At these frequencies the silencer will provide a transmission loss of zero, i.e., no silencing effect at all, and such frequencies are known as the "pass-bands." It is also possible for the silencer to resonate in the transverse direction through the diametral dimension, d_b, and provide further pass-band frequencies at what would normally be a rather high frequency level.

There are many empirical equations in existence for the transmission loss of such a silencer, but Kato and Ishikawa [8.18] state that the theoretical solution of Fukuda [8.9] is found to be useful. The relationship of Fukuda [8.9] is as follows for the transmission loss of a diffusing silencer, β_{tr}, in dB units:

$$\beta_{tr} = 10 \log_{10}(A_{r2}F(\kappa, L))^2 \quad \text{dB} \tag{8.5.3}$$

where

$$F(\kappa, L) = \frac{\sin(\kappa L_b) \times \sin(\kappa L_t)}{\cos(\kappa L_1) \times \cos(\kappa L_2)} \tag{8.5.4}$$

and

$$\kappa = \frac{2\pi f}{a_0} \tag{8.5.5}$$

Not surprisingly, in such empirical relationships there are many correcting factors offered for the modification of the basic relationship to cope with the effects of gas particle velocity, end correction effects for pipes which are re-entrant or flush with the box walls, boxes which are lined with absorbent material but not sufficiently dense as to be called an absorption silencer, etc. I leave it to you to pursue these myriad formulae, should you be so inclined, which can be found in the references.

To assist you with the use of the basic equations for design purposes, a simple computer program is included in the Appendix Listing of Computer Programs as Prog.8.1, DIFFUSING SILENCER. The attenuation equations programmed are those seen above from Fukuda, Eqs. 8.5.3-8.5.5.

To determine if such a design program is useful in a practical sense, an analysis of SYSTEM 2 and SYSTEM 3, emanating from Coates [8.3], is attempted and the results shown in Figs. 8.12 and 8.13. They show the computer screen output from Prog.8.1, which are plots of the attenuation in dB as a function of noise frequency up to a maximum of 4 kHz, beyond which frequency most experts agree the diffusing silencer will have little silencing effect. The theory would continue to predict some attenuation to the highest frequency levels, indeed beyond the upper threshold of hearing. Also displayed on the computer screen are the input data for the geometry of that diffusing silencer according to the symbolism presented in Fig.

8.7 and in this section. However, the input data are in the more conventional length dimensions in mm units and temperature in Celsius values; they are converted within the program to strict SI units before being entered into the programmed equations for the attenuation of a diffusing silencer according to Fukuda [8.9], Eqs. 8.5.3-8.5.5. You can check the input dimensions of SYSTEM 2 and SYSTEM 3 from Fig. 8.2.

The measured noise characteristics of SYTEM 2 are shown in Fig. 8.4. The attenuation of SYSTEM 2, as predicted by the theory of Fukuda and shown in Fig. 8.12, has a first major attenuation of 14 dB at 320 Hz and the first two pass-band frequencies are at 550 and 1100 Hz. If you examine the measured noise frequency spectrum in Fig. 8.4, you will find that an attenuation hole of 12 dB is created at a frequency of 400 Hz. Thus the correspondence with the theory of Fukuda, with regard to this primary criterion, is quite good and gives some confidence in its application for this particular function. There is also some evidence of the narrow pass-band at 550 Hz and there is no doubt about the considerable pass-band frequency at 1100 Hz in the measured spectra. There is no sign of the predicted attenuation, nor the pass-band holes, at frequencies above 1.5 kHz in the measured spectrum. There is also no evidence from the empirical solution of the reason for the attenuation in the measured spectrum of the fundamental pulsation frequency of 133 Hz, as commented on in Sec. 8.4.2. The general conclusion as far as SYSTEM 2 is concerned is that Prog.8.1 is a useful empirical design calculation method for a diffusing silencer up to a frequency of 1.5 kHz.

The measured noise characteristics of SYTEM 3 are shown in Fig. 8.5. The attenuation of SYSTEM 3, as predicted by the theory of Fukuda and shown in Fig. 8.13, has a first major

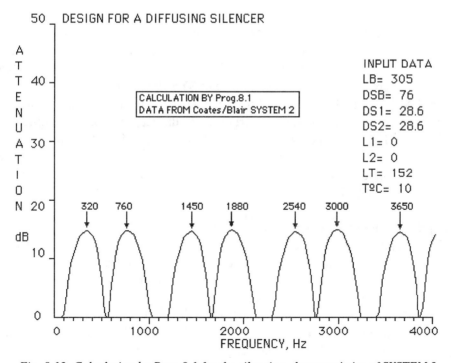

Fig. 8.12 Calculation by Prog.8.1 for the silencing characteristics of SYSTEM 2.

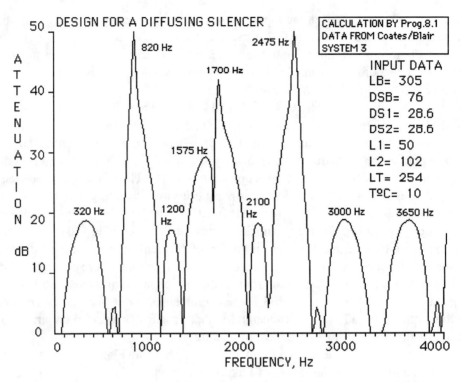

Fig. 8.13 Calculation by Prog.8.1 for the silencing characteristics of SYSTEM 3.

attenuation of 19 dB at 320 Hz, i.e., greater than that of SYSTEM 2. It is true that the measured silencing effect of SYSTEM 3 is greater than SYSTEM 2, but not so at the frequency of 400 Hz where the measured attenuations are both at a maximum and are virtually identical. The theory would predict that the pass-band hole of SYSTEM 2 at 1100 Hz would not be so marked for SYSTEM 3, and that can be observed in the measured noise spectra. The theoretically predicted pass-band at 600 Hz is clearly seen in the measured noise diagram. Again, there is little useful correlation between theory and experiment after 1500 Hz.

The general conclusion to be drawn, admittedly from this somewhat limited quantity of evidence, is that Prog.8.1 is a useful empirical design program for diffusing silencers up to a frequency of about 1500 Hz.

8.5.2 The side-resonant type of exhaust silencer

The fundamental behavior of this type of silencer is to absorb a relatively narrow band of sound frequency by the resonance of the side cavity at its natural frequency.

The side-resonant silencer with round holes in a central duct

The sketch in Fig. 8.8 shows a silencing chamber of length, L_b, and area, A_b, or diameter, d_b, if the cross-section is circular. The exhaust pipe is usually located centrally within the silencer body, with an area, A_3, or diameter, d_3, if it is a round pipe, and has a pipe wall thickness, x_t. The connection to the cavity, whose volume is V_b, is via holes, or a slit, or by a

Chapter 8 - Reduction of Noise Emission from Two-Stroke Engines

short pipe in some designs. The usual practice is to employ a number of holes, N_h, each of area, A_h, or diameter, d_h, if they are round holes for reasons of manufacturing simplicity. The volume of the resonant cavity is V_b where, if all cross-sections are circular:

$$V_b = A_b L_b - \frac{\pi L_b (d_3 + 2x_t)^2}{4} \tag{8.5.6}$$

According to Kato [8.18], the length, L_h, that is occupied by the holes should not exceed the pipe diameter, d_3, otherwise the system should be theoretically treated as a diffusing silencer. The natural frequency of the side-resonant system, f_{sr}, is given by Davis [8.4] as:

$$f_{sr} = \frac{a_0}{2\pi} \sqrt{\frac{K_h}{V_b}} \tag{8.5.7}$$

where K_h, the conductivity of the holes which are the opening, is calculated from:

$$K_h = \frac{N_h A_h}{x_t + 0.8 A_h} \tag{8.5.8}$$

The attenuation or transmission loss in dB of this type of silencer, β_{tr}, is given by Davis as:

$$\beta_{tr} = 10 \log_{10}(1 + Z^2) \tag{8.5.9}$$

where the term, Z, is found from:

$$Z = \frac{\frac{\sqrt{K_h V_b}}{2A_3}}{\frac{f}{f_{sr}} - \frac{f_{sr}}{f}} \tag{8.5.10}$$

When the applied noise frequency, f, is equal to the resonant frequency, f_{sr}, the value of Z becomes infinite as does the resultant noise attenuation in Eq. 8.5.9. Clearly this is an impractical result, but it does give credence to the view that such a silencer has a considerable attenuation level in the region of the natural frequency of the side-resonant cavity and connecting passage.

To help you use these acoustic equations for design purposes, a simple computer program is referenced in the Appendix Listing of Computer Programs as Prog.8.2, SIDE-RESONANT SILENCER. The attenuation equations programmed are those discussed above from Davis [8.4], Eqs. 8.5.6.-8.5.10.

To demonstrate the use of this computer design program, and to determine if the predictions emanating from it are of practical use to the designer of side-resonant elements within an exhaust muffler for a two-stroke engine, the geometrical and experimental data pertaining to SYSTEM 4 of Coates [8.3] are inserted as data and the calculation result is illustrated in Fig. 8.14. The information in that figure is the computer screen picture as seen by the user of Prog.8.2. The information displayed is the input data for, in this instance, SYSTEM 4 and you can check the methodology of data input from Fig. 8.2, and the output data which are the attenuation in dB over a frequency range up to 4 kHz. You may wonder about 10°C being the declared data value for exhaust temperature, but the simulation work of Coates was conducted by a rotor valve delivering very realistic exhaust pulses, but in very cold air! QUB is in Ulster and Ulster is not in the tropics.

The calculated sound attenuation of SYSTEM 4, as seen in Fig. 8.14, shows a large peak of 50 dB transmission loss at 600 Hz, with further attenuation stretching to 2.5 kHz. If you examine the measured noise spectrum of SYSTEM 4 in Fig. 8.6, and compare it to the unsilenced SYSTEM 1 in Fig. 8.3, you can see that there is considerable attenuation produced by this side-resonant silencer at 500 Hz, although it is not as profound as 50 dB. Of even greater interest to the designer is the visibly strong attenuation of SYSTEM 4 stretching up to 10 kHz, which is not present in either of the diffusing silencer designs discussed in Sec. 8.5.1. Also note that the calculated attenuation stretches down to 100 Hz, i.e., below the

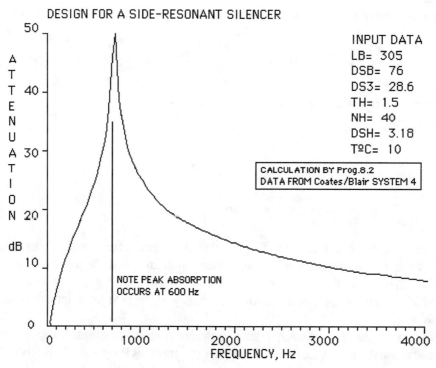

Fig. 8.14 Calculation by Prog.8.2 for the silencing characteristics of SYSTEM 4.

fundamental frequency of 133 Hz determined for Coates' experiments in Sec. 8.4.2, and that the measured diagram does show attenuation of that frequency at least equal to the level observed for the diffusing silencers.

The evidence is that the theoretical work of Davis [8.4], as programmed in Prog.8.2, is of direct use in determining the fundamental muffling frequency of a side-resonant silencer and that such a device is very useful in practical terms for attenuating noise at frequencies above 1 kHz.

The side-resonant silencer with slits in a central duct

The sketch in Fig. 8.9, which is very similar to Fig. 8.8, shows a silencing chamber of length, L_b, and area, A_b. The exhaust pipe is normally located centrally in the silencer, and is of area, A_3, or diameter, d_3, if it is a round pipe, and has a pipe wall thickness, x_t. The connection to the cavity, whose volume is V_b, is via a slit, or a number of slits, N_s. Each slit is of length, L_s, and of width, x_s.

The theoretical solution for the behavior of this type of side-resonant silencer is virtually identical to that given above and expressed as Eqs. 8.5.6-8.5.10. Only the relationship for the conductivity of the opening, K_s, is different and is given by Roe [1.8] as:

$$K_s = \frac{N_s x_s L_s}{x_t + 0.92 \kappa_s \sqrt{x_s L_s}} \tag{8.5.11}$$

where κ_s is directly related to the length-to-width ratio of the slot by:

$$\kappa_s = 1.0287 - \sqrt{1.0579 - 0.0088763\left(120.09 - \frac{L_s}{x_s}\right)} \tag{8.5.12}$$

The solution for the transmission loss, β_{tr}, of this variation of the side-resonant silencer is obtained by solving the same equations as before, Eqs. 8.5.6-8.5.10, with the conductivity for the round holes, K_h, replaced by that for the slits, K_s.

8.5.3 The absorption type of exhaust silencer

It is generally held that an absorption silencer, as illustrated in Fig. 8.10, acts as a diffusing silencer [8.18] and that the effect of the packing is to absorb noise in the pass-bands appropriate to a diffusing silencer. The geometry of the system is for a pipe of area, A_3, passing through a box of length, L_b, and area, A_b. If the cross-sections are circular, the diameters are d_3 and d_b, respectively. The holes through the pipe are normally round and are N_h in number, each of area, A_h, or diameter, d_h. The physical geometry is very similar to the side-resonant silencer, except the total cross-sectional area of all of the holes is probably in excess of five times the pipe area, A_3. This is the reason for the theoretical acoustic analysis being more akin to a diffusing silencer than a side-resonant silencer, particularly as the area ratio of holes to pipe in a side-resonant silencer is normally less than unity. Expressed numerically, the following are conventional empiricism:

$$\text{absorption silencer} \quad \frac{N_h A_h}{A_3} > 5$$

$$\text{side-resonant silencer} \quad \frac{N_h A_h}{A_3} < 1$$

The experimental evidence is that the absorption silencer is very effective at attenuating high frequencies, i.e., above 2 kHz, and is relatively ineffective below about 400 Hz. Most authorities agree that the theoretical design of an absorption silencer, either by the methodology of Coates [8.3] or by an acoustic procedure, is somewhat difficult because it depends significantly on the absorption capability of the packing material. The packing materials are usually a glass-reinforced fiber material or mineral wool.

You should attempt the design using Prog.8.1, but arrange for as many holes of an appropriate diameter in the central pipe as is pragmatic. If the hole size is too large, say in excess of 3.5 mm diameter, the packing material within the cavity will be blown or shaken into the exhaust stream. If the hole size is too small, say less than 2.0 mm diameter, the particulates in the exhaust gas of a two-stroke engine will ultimately seal them over, thereby rendering the silencing system ineffective. The normal hole size to be found in such silencers is between 2.0 and 3.5 mm diameter.

One detailed aspect of design which needs special comment is the configuration of the holes in the perforated section of the silencer. The conventional manufacturing method is to roll, then seam weld, the central pipe from a flat sheet of perforated mild steel. While this produces an acceptable design for the perforated pipe, a superior methodology is to produce the pipe by the same production technique, but from a mild steel sheet in which the holes have been somewhat coarsely "stabbed," rather than cleanly excised. Indeed, the stabbed holes can be readily manufactured in pre-formed round pipe by an internal expanding tool. The result is shown in sketch form in Fig. 8.15 for the finished pipe section. This has the effect of reducing the turbulent eddies produced by the gas flowing in either direction past the sharp edges of the clean-cut holes of conventional perforations. This type of turbulence has an irritating high-frequency content; it will be recalled that a whistle produces noise by this very edge effect.

Positioning an absorption silencer segment

The published literature [1.8] on the topic agrees on several facets of the design as commented on above, but the principal function is to remove the high-frequency end of the noise spectra. The high-frequency part of any noise spectrum emanates from two significant sources: the first is due to the sharp pressure rise at the front of the exhaust pulse, always faster in a two-stroke than a four-stroke engine (see Sec. 8.6.1), and the second is from the turbulence generated in the gas particle flow as it passes sharp edges, corners and protuberances into the gas stream. The considerable noise content inherent in turbulence is quite visible in Plate 2.4

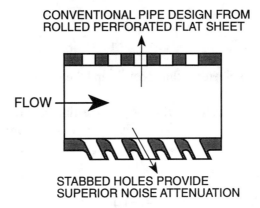

Fig. 8.15 Absorption silencers perform better with stabbed holes.

where the whirling smoke-ring of particles is seen to follow the spherical pressure wave front into the atmosphere. Thus, an absorption silencer element should always be the final segment of a series of elements making up a complete muffler. The logic for that statement is that there is little point in muffling the high-frequency part of the noise spectrum by an absorption silencer element which is then followed by a re-entrant pipe diffusing silencer, as this would produce further turbulence-generated noise from the eddies in the wake off the sharp, protruding pipe edges.

8.5.4 The laminar flow exhaust silencer

One of the most interesting and effective silencers, and which is very simple to design and manufacture, is that suggested by Roe [1.8, 8.27]. A variation on that theme, developed at QUB, is shown sketched in Fig. 8.16. I have shown it to be applicable to a motorcycle [8.23].

The basic principle is that the exhaust gas flows through a final exit section of the silencer which is a narrow annular gap of radial dimension, x_g. The word "laminar" has been used to describe this final phase of the outflow. For that to be true, the Reynolds number in the annular section would need to be less than 2000. The Reynolds number in the gap is given by:

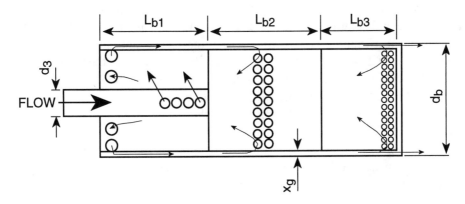

Fig. 8.16 A QUB design for a laminar flow silencer.

$$\mathbf{Re}_g = \frac{\rho_g c_g d_g}{\mu_g} \tag{8.5.13}$$

where d_g is the hydraulic diameter at this location and the subscript "g" denotes the local gas properties.

Let it be assumed that the area of flow, A_g, in the final annulus phase is equated to the entering duct flow area, A_3. This is a somewhat generous proportionality if very effective silencing is required, but can be used as a design criterion if the device is for muffling a high-performance engine.

$$A_3 = \frac{\pi}{4} d_3^2 = A_g = \frac{\pi}{4} \left(d_b^2 - \left(d_b - 2x_g \right)^2 \right) \tag{8.5.14}$$

In which case the gap dimension is given by:

$$x_g = \frac{d_b - \sqrt{d_b^2 - d_3^2}}{2} \tag{8.5.15}$$

The hydraulic diameter is correctly evaluated as:

$$d_g = \frac{4 \times \text{Area of gap}}{\text{wetted perimeter}} = \frac{d_b^2 - \left(d_b - 2x_g \right)^2}{2d_b - 2x_g} = 2x_g \tag{8.5.16}$$

To put some numbers on this contention, if the pipe diameter, d_3, is 36 mm and the inside silencer body diameter, d_b, is 90 mm, then the gap dimension, x_g, becomes 3.76 mm. Taking the argument further, using the theory of Sec. 2.3.1, assume that an exhaust pressure pulse of 1.2 atm appears in the entry duct at a reference temperature of 500°C. The theory in Sec. 2.3.1 shows that the Reynolds number in the entry duct is 37,900, the local particle velocity is 73.5 m/s, and the friction factor for the flow is 0.0057. If it is assumed that equality of particle velocity is maintained in the annular gap, to complement the assumption of equality of flow area, then the Reynolds number in the final annulus is 7900 and the friction factor is 0.008. The convection heat transfer coefficients, C_h, in the entry duct and the final annulus are predicted as 173 and 244 W/m²K. If the mean velocity of the flow in the final annulus section is reduced below that of the pressure pulsations in the entering duct, then it can be observed that the possiblity of laminar flow occurring is quite real.

Irrespective of the occurrence of laminar flow, the viscous effects and the area over which they apply, is significantly increased. This decreases turbulence levels and dampens other acoustic oscillations in a manner significantly greater than for conventional silencers. Heat transfer rate is increased, as is the area over which it is applied in the annulus, so that too will provide a sharper temperature profile along it, thus giving continous wave reflections to further damp the pressure oscillations (see Sec. 2.5).

There are several other factors that contribute to the overall effectiveness of this silencer. They are:

(i) The efflux of hot exhaust gas occurs through an annular ring into air which is at a much lower temperature. There is a large contact area between the two. This gives considerable damping of the turbulence in the exiting exhaust gas plume, thereby reducing the high-frequency noise inherent from that source. This approach to silencing turbulence noise is conventional practice for aircraft gas turbines.

(ii) The central body houses side-resonant cavities which can be used to tune out particular frequencies with a high noise content. The flow passing the entrances to these cavities is moving at right angles to those apertures, and at particle velocities closer to acoustic levels than in the conventional silencers shown in Figs. 8.7-8.9. Thus the assumptions inherent in acoustic theory for Helmholz resonators is approached more closely and the application of that theory can be applied with more confidence in the quality of the outcome.

(iii) The exhaust pressure pulsations entering the first chamber, a diffusing silencer section, have the normal and considerable amplitude associated with such waves. The first box reduces the magnitude of that pressure oscillation prior to it entering the annulus en route to the atmosphere, passing the several resonant cavities as it goes. However, the outside skin of the silencer does not experience the forces due to the full magnitude of the original oscillation in the first box, as is the case in the other silencers shown in Figs. 8.7-8.9, and so the outside skin vibrates less than those shown in the sketches. Later, in Sec. 8.6, there is discussion that double skinning of the outside of a silencer can be a vital issue to prevent that vibration from being a significant source of noise.

8.5.5 Silencing the intake system

Probably the simplest and most effective form of intake silencer is of the type sketched in Fig. 8.11. The geometry illustrated is for a single-cylinder engine but the same arrangement can also apply to a multi-cylinder design, and you will find such a discussion in the paper by Flaig and Broughton [1.12]. From that paper is reproduced the attenuation curve for a V8 two-stroke outboard motor, shown in Fig. 8.17. You can see that the peak attenuation required is mainly in the band from 400-600 Hz. Note that this corresponds to a forcing frequency from the fundamental induction pulses of this eight-cylinder engine at 3000-4500 rpm.

The acoustic design of the low-pass intake silencer

Fig. 8.11 supplies the basic mechanism of induction silencing, being a volume connected to the induction system, however many air intakes to the several cylinders there may be. On the atmospheric side of this box is a pipe of length, L_b, and area, A_b, or with diameter, d_b, if it has a circular cross-section. The air cleaner is placed within the box, helping to act as an absorption silencer of the various high-frequency components emanating from the edges of throttles or carburetor slides. Assuming that it is reasonably transparent to the air flow, this has no real effect on the silencing behavior of the box volume, V_b.

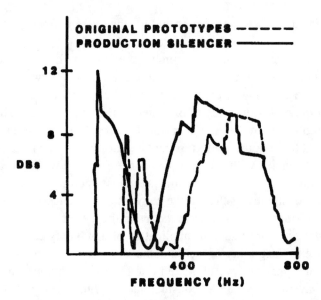

Fig. 8.17 Attenuation characteristics of an intake silencer for the OMC V8 outboard (from Ref. [1.12]).

This type of silencer is known as a low-pass device and has no silencing capability below its lowest resonating frequency. The sharp drop in noise attenuation behavior associated with this effect is clearly seen in Fig. 8.17 for the OMC V8 outboard motor, which has a lowest resonating frequency of about 200 Hz (8 cylinders at a minimum full throttle operating speed of 1500 rpm is equivalent to 200 Hz as a fundamental frequency).

The natural frequency of this type of silencer [1.8] is given by f_i, where:

$$f_i = \frac{a_0}{2\pi}\sqrt{\frac{A_p}{L_{eff}V_b}} \qquad (8.5.17)$$

where the effective length of the intake pipe, L_{eff}, is related to the actual length, L_p, and the diameter, d_p, by:

$$L_{eff} = L_p + \frac{\pi d_p}{4} \qquad (8.5.18)$$

The design of the silencer is accomplished by setting this natural frequency of the silencer, f_i, to correspond to the natural frequency of the induction pulses from the engine, f_e.

Returning to the attenuation behavior of the intake silencer of the OMC V8 engine outboard motor in Fig. 8.17, it can be seen that the peak transmission losses occur in two distinct bands, i.e., at 100 Hz and 500 Hz, a ratio of five; the lower absorption frequency at 100 Hz fundamentally corresponds to an engine speed of 750 rpm.

From this discussion, it is clear that the engine forcing frequency of direct design interest, f_e, is that corresponding to the engine speed of rotation and the number of cylinders,

Chapter 8 - Reduction of Noise Emission from Two-Stroke Engines

$$f_e = \frac{\text{number of cylinders} \times \text{rpm}}{60} \qquad (8.5.19)$$

and, as the intake silencer is of the low-pass type, the engine speed in question is at the lower end of the usable speed band. Annand and Roe [1.8] give some further advice on this particular matter.

There are several other criteria to be satisfied as well in this design process, such as ensuring that the total box volume and the intake pipe area are sufficiently large so as not to choke the engine induction process and reduce the delivery ratio. Such a calculation is normally accomplished using an unsteady gas-dynamic engine model, extended to include the intake and exhaust silencers [8.21]. However, an approximate guide to such parameters is given by the following relationship, where N_{cy} is the number of cylinders, V_{sv} is the swept volume of any one cylinder, and d_2 is the flow diameter of the intake duct or the carburetor, as shown in Fig. 5.5:

$$10 \times V_{sv}\sqrt{N_{cy}} < V_b < 20 \times V_{sv}\sqrt{N_{cy}} \qquad (8.5.20)$$

$$0.6 d_2 < d_p < 0.8 d_2 \qquad (8.5.21)$$

To put this advice in numerical form, revisit the 65 cm³ chainsaw engine employed as a design example in Chapters 5, 6 and 7. It is designed to run between 5400 and 10,800 rpm, cutting wood in a wide-open-throttle condition. The carburetor on the engine has a 22 mm flow diameter. The lowest engine speed of direct interest is 5400 rpm, where the engine has a fundamental induction frequency, f_e, of 90 Hz. Therefore, for design purposes it will be assumed that this 65 cm³ chainsaw engine will not be operated at full throttle below 5400 rpm, consequently the minimum value for f_i is 90 Hz. From Eq. 8.5.20 the minimum box volume, V_b, should be 650 cm³. No chainsaw can ever be so profligate with space, so 500 cm³ is probably the maximum which can be afforded for the box. The minimum pipe diameter, d_p, from Eq. 8.5.21, is approximately 15 mm. When the air temperature is at 20°C, so the acoustic velocity, a_0, from Eq. 8.1.1, is 343 m/s. Testing a real pipe length, L_p, of 125 mm, by using Eq. 8.5.18,

$$L_{eff} = L_p + \frac{\pi d_p}{4} = 125 + \frac{\pi \times 15}{4} = 136.8 \text{ mm}$$

From Eq. 8.5.17:

$$f_i = \frac{a_0}{2\pi}\sqrt{\frac{A_p}{L_{eff} V_b}} = \frac{343}{2\pi}\sqrt{\frac{\frac{\pi}{4}\left(\frac{15}{1000}\right)^2}{\frac{136.8}{1000} \times \frac{500}{10^6}}} = 87.7 \text{ Hz}$$

The conclusion from this calculation is that a silencing pipe 125 mm in length and 15 mm internal diameter inserted into a box of 500 cm^3 volume would provide an adequate intake silencing characteristic for this particular chainsaw engine. If the pipe is to be re-entrant into the box, shown as partly so in Fig. 8.11, you should remember that it is the effective internal box volume which is to be employed as input data.

It is interesting to note that these acoustic criteria are also satisfied by a pipe length of 160 mm and a diameter of 17 mm, and again by another of 200 mm length and 19 mm diameter, respectively. As the pipe diameters are larger, these silencers should be less restrictive and pass air more readily from the atmosphere into the silencer box. Their relative effectiveness is determined by a much more rigorous simulation method below.

It is important to emphasize that this is an acoustic, not an absolute, design calculation and the experienced designer will know that this prediction is normally a prelude to an intensive period of experimental development of the actual intake silencing system [1.12].

8.5.6 Engine simulation to include the noise characteristics

Noise characteristics may also be obtained by simulation, using the same software code employed earlier in Chapters 5-7 and incorporating the theory of Sec. 8.4.1. The example chosen is the chainsaw engine featured immediately above and frequently throughout Chapters 5, 6 and 7. The layout of the intake and exhaust silencers, and their connection to the engine cylinder, is shown in Fig. 8.18.

The intake system of the chainsaw

The silencer geometries are more completely seen in Fig. 8.18, which ties together the previous sketches shown as Figs. 5.5, 5.6 and 8.11. The standard intake system is, like many on industrial engines, somewhat rudimentary, whereas that shown in Fig. 8.11 or 8.18 is more sophisticated, as is the data acoustically acquired above for the intake silencer. The filter gives little obstruction to the flow as it has an effective flow diameter, d_f, of 30 mm. The filter

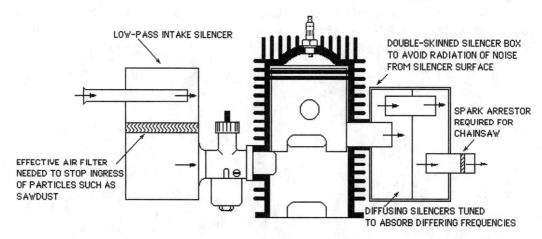

Fig. 8.18 Compact silencer designs needed for engines such as chainsaws where the bulk of the entire engine must be minimized.

splits the total box volume into two parts, labeled as V_1 and V_2, and the simulation models the entire system as two complete boxes separated by the filter element in precisely the same fashion that it does for the exhaust system. The geometry is detailed in Table. 8.1.

Table 8.1 Intake silencer geometry for the chainsaw

Name	d_f (mm)	d_2 (mm)	d_p (mm)	L_p (mm)	V_1 (cm³)	V_2 (cm³)	f_i (Hz)
S	30	22	22	50	200	100	240
E	30	22	17	160	200	300	88.9
F	30	22	15	125	200	300	88.3
G	30	22	19	200	200	300	80.1

It will be seen that those characterized in Table 8.1 above as E, F, and G are the various alternatives as designed by the acoustic criteria, whereas that called S is the original "standard" system.

The unsteady gas-dynamic simulation prepares an output file of the mass flow-time history at the intake from the atmosphere. For the four systems analyzed, this is shown in Fig. 8.19 as the results of simulation at 9600 rpm, at a slightly different, i.e., leaner, air-to-fuel ratio than that employed for the more complete data map in Fig. 5.9. The theory of Sec. 8.4.1 conducts a Fourier analysis of the data in Fig. 8.19 and produces the noise spectra shown in Fig. 8.20, at a distance, r_m, of 1.0 m from the intake and exhaust exit points, at an orientation angle, θ, of zero. Not surprisingly, the more open silencer, S, has the greater mass flow fluctuations and has a higher noise spectra profile than any of the more optimized silencers, E, F or G. They are all reasonably similar, but the silencer, F, has the smoothest mass flow-time history and comes out with the lowest peak noise level at 640 Hz, the fourth harmonic. Thus the design, F, one of those determined above by the acoustic criteria, emerges as being superior from a noise standpoint to the standard silencer, S. The overall noise levels, the summations of these spectra, are given in Table 8.2.

Table 8.2 Performance and intake silencing for the chainsaw

Name	Intake (dBlin)	Intake (dBA)	Exhaust (dBA)	Total (dBA)	bmep (bar)	DR
S	103.6	98.5	89.0	99.0	3.65	0.525
E	92.2	87.2	88.8	91.1	3.82	0.546
F	90.1	82.9	88.8	89.8	3.79	0.541
G	93.8	89.5	88.8	92.2	3.79	0.550

The intake geometry of F is overall quieter than E and G by 4 dB or more. That is very much quieter, for, recall from Eq. 8.1.6, two equal sound sources give an increase of approximately 3 dB. The acoustic criteria gave no guidance on this matter. The standard silencer, S, is actually noisier than the exhaust system.

Design and Simulation of Two-Stroke Engines

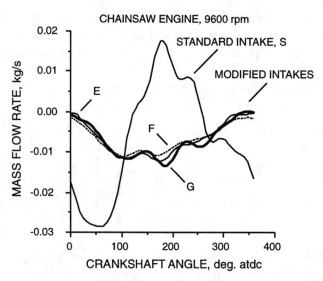

Fig. 8.19 Effect of modified intake system on the intake mass flow rate.

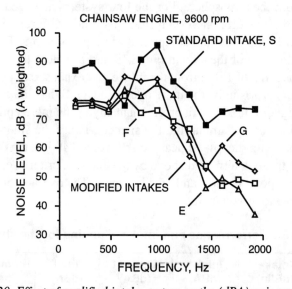

Fig. 8.20 Effect of modified intake system on the (dBA) noise spectra.

The power of the simulation technique is that it gives the designer the trade-off in terms of noise reduced by comparison with power or torque, lost or gained. Here, the standard and noisiest silencer, S, actually gave less torque, as bmep, by comparison with all of the modified silencers, simply because it breathed less air, as shown by the delivery ratio, DR, results. The best silencer, F, breathed in marginally less air than E or G, and lost a little torque as a consequence, but at a very considerable noise reduction. The overall noise level for F, however, is only 1.2 dB less than E, due to the exhaust silencer being the single biggest source of noise output. To silence the chainsaw engine fitted with intake system F any further, the exhaust system would need to be quietened, as extra noise reduction for the intake side can have little bearing on the final outcome.

When the engine has intake silencer, S, fitted, then the opposite is the case. Tackling the exhaust system for noise reduction in this case is pointless until the intake system noise level is reduced, at least to equality. Such arguments are reinforced in the concluding remarks to this chapter.

Silencing the untuned exhaust system of a chainsaw

The design of an untuned exhaust system for a two-stroke engine is discussed in Chapter 5, particularly in connection with the example of the chainsaw engine, where the small bulk required of the entire powerplant precludes the availability of adequate space for effective silencers, be they intake or exhaust. The problems in this regard are sketched in Fig. 8.18. There are other types of two-stroke engines with untuned or relatively untuned exhaust pipes, and these are to be found on agricultural and electricity-generating equipment where powerplant space may not be at the same premium as it is on a handheld power tool such as a chainsaw or a brushcutter. Another relatively straightforward example is the outboard motor, where the virtually unsilenced exhaust gas is directed underwater into the propeller wash which provides very effective exhaust silencing without any particular acoustic skills being employed by a designer. Whether this underwater racket is offensive to fish is not known.

The most difficult engine for which to design an adequate silencing characteristic is the handheld power tool, i.e., the chainsaw or similar device. It is self-evident that, as noise is a function of the square of the dp, or pressure, fluctuation as seen in Eq. 8.1.2, then the larger the silencer volume into which the exhaust pulses are "dumped," the lesser will be the dp value transmitted into the atmosphere. The designer of the chainsaw is continually looking for every available cubic centimeter of space to become part of the exhaust silencer. The space may be so minimal that the two-box design shown in Fig. 8.18 becomes an impossible luxury. The designer is left with the basic option of trimming out the fundamental firing frequency using a diffuser type or side-resonant type of silencer as described in Sec. 8.5.

As a rule of thumb, the designer of a silencer for the handheld power tool will know that there is going to be real silencing difficulties if there is not a total volume available for a silencer which is at least twelve to fifteen times greater than the cylinder swept volume. If that is not the case, then the only design methodology left open is to choke the exhaust system by a restrictive silencer, thereby reducing the delivery ratio, and accept the consequential loss of power. Further design complications arise from the legal necessity in many applications for the incorporation of a spark arrestor in the final tail-pipe (see Fig. 8.18) before entry to the atmosphere [8.22]. By definition, a spark arrestor is a form of area restriction, as it has to inhibit, and more importantly extinguish, any small glowing particles of carbon leaving the silencer. This need is obvious for engines such as chainsaws and brushcutters, where the device is being used in an environment with a known fire hazard.

In any event, of equal importance is the necessity to damp the vibrations of the skin of the silencer, for this compact device is exposed to the normal "impact" of the unattenuated exhaust pulses. Consequently, double-skinning of the silencer surface is almost essential.

The design example chosen is the same chainsaw used several times before in various chapters. The data for the chainsaw in Sec. 5.4.1 give the geometrical details of the exhaust silencer and its ducting, in terms of Fig. 5.6. The individual box volumes, V_A and V_B, are 300 and 260 cm^3, respectively, and it is clear that this is very small as it is not even ten cylinder volumes. The use of the acoustic program for a diffusing silencer, Prog.8.1, shows that there is absolutely no attenuation of sound below 800 Hz.

For this simple silencer, it has two diffusing silencer segments. The question which the designer asks is the extent to which the final outlet can be throttled before unacceptable power loss occurs. The simulation answers this type of query with precision. The "standard" outlet pipe diameter is 12 mm. As a design exercise, this outlet pipe is crimped at the very end to give equivalent end orifices of 11, 10 and 9 mm diameter, labeled as X, Y and Z, respectively. The GPB simulation models these effects, using the intake silencer, F, in each case.

The result of so doing is shown for the mass flow-time history at the exhaust exits in Fig. 8.21, the noise spectra in Fig. 8.22, and the overall effect on noise and performance characteristics in Table 8.3.

Table 8.3 Performance and exhaust silencing for the chainsaw

Name	Intake (dBA)	Exhaust (dBlin)	Exhaust (dBA)	Total (dBA)	bmep (bar)	DR
F	82.9	92.6	89.0	89.8	3.79	0.539
X	83.0	91.3	88.4	89.5	3.71	0.535
Y	83.1	89.8	87.4	88.7	3.61	0.524
Z	83.1	88.2	85.9	87.7	3.50	0.509

The expected results are achieved, namely choking the exhaust drops the delivery ratio and loses bmep. The amount is 0.29 bar, apparently a meager drop, but nearly 10% of the power output has actually been lost. The exhaust noise is lowered by some 3 dBA, and that

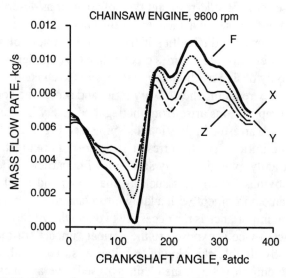

Fig. 8.21 Effect of modified exhaust system on the exit mass flow rate.

Chapter 8 - Reduction of Noise Emission from Two-Stroke Engines

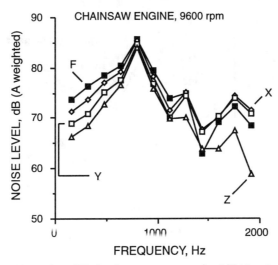

Fig. 8.22 Effect of modified exhaust system on the (dBA) noise spectra.

reduces the overall figure by 2 dBA, supporting the previous contention that reducing the exhaust noise would rapidly lower the total, which is still somewhat higher than the intake noise. Observe that the intake noise of Z is slightly higher, by some 0.2 dB, than that for system F. There is a slight penalty from the exhaust choking on the intake noise spectra.

From Fig. 8.21, the exhaust outlet choking is seen to dampen the mass flow oscillations, from whence comes the reduced noise levels observed in either Table 8.3, or in the (dBA) spectra in Fig. 8.22.

It was observed previously that an acoustic analysis of this exhaust silencer by Prog.8.1 showed noise attenuation only above 800 Hz. In Fig. 8.22 that is indeed seen very clearly, but there appears to be equally strong damping below 800 Hz; so is Prog.8.1 sufficiently adequate for design purposes, as this information apparently contradicts it? The answer to that query is given in Fig. 8.23, where both the total noise (as dBlin) and the A-weighted spectra are presented. The matter is now clarified. It is the A-weighting characteristic which is providing the reduced noise spectrum profile below 800 Hz. The total noise spectrum is relatively high and flat until 800 Hz when the diffusion behavior begins to take effect. In other words, Prog.8.1 does not include the effect of a noise weighting scale.

The relevance of exhaust silencer volume for a chainsaw

The space to accommodate an adequate volume of exhaust or intake silencer for a simple two-stroke engine is often a critical design element. Much comment on this issue has already been made above. To illustrate this point, and of the effectiveness of simulation in the design process, the modeling of the design, referred to above as F, is expanded to include slightly larger exhaust silencers with approximately 10 and 25% greater volumes. These designs are referred to as P and Q, and the changes to the exhaust box volumes are summarized in Table 8.4, with the nomenclature as in Fig. 5.6. The computations are run at the single engine speed of 9600 rpm.

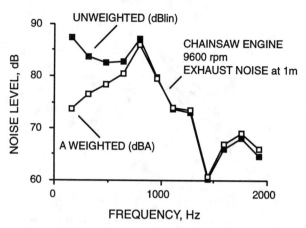

Fig. 8.23 *Effect of noise weighting on noise spectra.*

Table 8.4 Effect of exhaust box volume on silencing of a chainsaw

Name	V_A (cm³)	V_B (cm³)	β_{tr} (dBA)	Exhaust (dBA)	bmep (bar)	DR
F	300	260	−6.5	88.8	3.79	0.539
P	330	290	−8.0	87.0	3.93	0.548
Q	400	320	−10.0	84.7	3.95	0.557

The results of the simulations are shown in Table 8.4 and in Fig. 8.26. In Sec. 5.4.1, where the modeling of this same chainsaw engine is described in great detail as an example of all such industrial engines, the restrictive nature of the small volume exhaust silencer was discussed. It comes as no surprise, therefore, that even this modest increase of box volume lifts both the delivery ratio and the bmep, which are shown numerically in Table 8.4. The hydrocarbon emissions for design P is marginally superior to design F, but the difference is negligible. The exhaust noise is lessened significantly by over 4 dBA for the design Q. The reason for this is shown in Fig. 8.26, as the larger, i.e., longer with a bigger section area, lowers the frequency above which attenuation takes place. The use of Prog.8.1 for the relevant data shows that the zero attenuation frequency for boxes F, P and Q are approximately 600, 550 and 500 Hz, respectively. The associated transmission losses, β_{tr}, are predicted by Prog.8.1 and shown in Table 8.4 as 6.5, 8 and 10 dB, respectively. There is some correspondence with the transmission loss predicted by a simple acoustic analysis and that determined by the more complex GPB simulation, in that the difference in exhaust noise from F to Q is 4.1 dBA, and that for the acoustic transmission loss is 3.5 dB.

The natural exhaust pulsation frequency is 160 Hz at 9600 rpm. The fifth order at 800 Hz is dominant, which is observable from the exhaust pressure records in Fig. 5.22. From Fig. 8.26, the largest box Q is clearly the minimum volume which will make a significant impact on attenuation of the fundamental frequency.

Chapter 8 - Reduction of Noise Emission from Two-Stroke Engines

The messages for the designer are obvious; to silence an engine, the frequency at which maximum noise is being created is that which must be tackled as a first priority and, for the silencing of the exhaust of the small industrial engine every extra cubic centimeter of exhaust box volume is well worth the ingenuity spent in its acquisition.

8.5.7 Shaping the ports to reduce high-frequency noise

The profiling of the opening edge of a port changes the mass flow-time history at that port and has an influence on the noise that gas flow creates.

The intake ports

If the intake system of the engine is controlled by a port and piston skirt or is a disc valve intake engine as introduced initially in Secs. 1.0 and 1.3, the profile of the opening of that port affects the noise level. An expanded discussion on this matter is found below for exhaust ports, where the topic is of even greater importance, so it will be sufficient at this stage to draw your attention to the upper half of Fig. 8.24. In this diagram, the designer is presented with the option of shaping the opening or timing edge of the intake port so that the flow area-time characteristics of the port have a shallower profile. This provides an induction pulse pressure-time and mass flow-time behavior which is less steep, and this reduces the high-frequency content of the intake noise spectrum [8.15]. By definition, this implies that, for equality of intake port time-areas, a shallow profile designed on the timing edge can require a longer port timing duration, otherwise there may be a reduction of the delivery ratio. Nevertheless, this is a viable design option, particularly for such engine applications where there is little room available for a conventional low-pass intake silencer; as usual, the handheld power tool is the archetypical example of such an application.

The exhaust ports

The effect of profiling the exhaust port in a chainsaw, and its ensuing behavior in both noise and power terms, is discussed in some detail in a paper by Johnston [8.14]. It is impor-

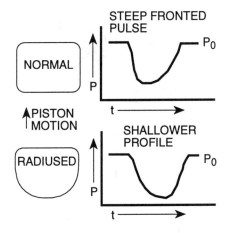

Fig. 8.24 Intake port profiling affects the pulse shape.

Design and Simulation of Two-Stroke Engines

tant to emphasize that slowing down the exit rate of exhaust gas flow during the blowdown phase inevitably leads to longer port timing durations and some reduction of trapping efficiency. This can be alleviated to some extent by making the exhaust port wider without unduly increasing the total timing duration, as the shallower area-time profile necessary for noise reduction also permits a wider port without endangering the mechanical trapping of the piston ring by the timing edge.

Even with a suitably profiled exhaust port timing edge, the pressure-time rise characteristics of an exhaust pulse in a two-stroke power unit is much faster than that in a four-stroke cycle engine where a poppet valve is employed to release exhaust gas from the cylinder. As already remarked, the noise spectrum from a rapidly rising exhaust pressure wave is "rich" in high-frequency content, irritating to the human ear.

In Fig. 8.25 it can be seen that the shaping of the exhaust port can be conducted in the same fashion as for the bottom of the inlet port. However, Johnston [8.14] shows more extreme profiles than the simple radiusing and proposed the "Q" port shape shown on the same figure, with a view to enhancing the effects already discussed.

Simulation of the effect by engine modeling

In the input data to the chainsaw engine simulation, both the exhaust and intake ports have a maximum width of 28 mm. The simulation result, described above as F, is compared with heavily radiused intake and exhaust ports, labeled as R. The original top and bottom radii for the intake and exhaust ports is 2 and 3 mm, respectively. For the simulation of R, this radius is increased to 10 mm in both cases. However, this reduces the port area and so it is necessary to open the exhaust port 1° earlier, at 107° atdc, and to open the inlet port 3° earlier, at 78° btdc, so as to provide for the same delivery ratio to pass through the engine. The result of the simulation of design R is shown in Table 8.5, where it is compared with design F.

The reduction of intake and exhaust noise as forecast by the theory is achieved, even though the modification of changing the port radii from 2 or 3 mm to 10 mm is a relatively modest change, by comparison with the "Q port" shown in Fig. 8.25. The equality of delivery

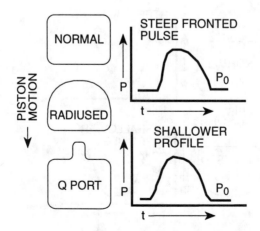

Fig. 8.25 Exhaust port profiling to reduce noise.

Table 8.5 Effect of radiused porting on performance and noise

Name	Intake (dBA)	Exhaust (dBA)	bmep (bar)	bsfc (g/kWh)	bsHC (g/kWh)	DR
F	82.9	89.0	3.75	480	127	0.539
R	82.5	88.3	3.61	488	134	0.530

ratio is virtually achieved and so the power performance is reasonably similar, yet is significantly different in a very important area. The trapping efficiency has been deteriorated so there is a rise in specific fuel consumption and a reasonably significant increase in hydrocarbon emissions. Thus the trade-off could be regarded as unacceptable by a manufacturer who must meet emission standards as a first priority, and may well feel that more effective engine silencing can be achieved by other means. Under pressure from emissions legislation, even the unthinkable may become acceptable, i.e., somehow make room on the device for larger volume intake and exhaust silencers. Before the millenium, as an exhaust catalyst may have to be installed to satisfy exhaust emissions legislation, that manufacturing bullet just may have to be bitten. The effective silencing of handheld power tools should have been carried out by their manufacturers many years ago and it is ironic that it is the latest U.S. and other exhaust emissions standards which are finally going to accomplish it.

8.6 Silencing the tuned exhaust system

You may find the following section useful as an exercise in data collection which could be the starting point for an engine simulation program to optimize any design, or the commencement of an experimental program for the same purpose. A design for a tuned pipe system is proposed and the data for it are collected using the acoustic theories given earlier.

The muffling of a tuned exhaust pipe is relatively straightforward and poses little difficulty in attaining the dual aims of good silencing while retaining the pressure wave tuning action necessary to attain a high bmep and high specific power output. In Sec. 7.3.4 it is shown how even the introduction of a catalyst can be accommodated in a tuned pipe without deterioration of the many design goals. Fig. 7.28 shows how that is accomplished, including the presence of a silencer after the tuned section. This explanation is expanded here by Fig. 8.26. The design of the tuned section of the pipe is obtained using a simulation as in Chapter 5, or the empiricism in Chapter 6, and the acoustic design of the silencing section can be carried out by the recommendations in this chapter. After the tuned section of the pipe, depending on the availability of space within the application for the power unit, there will be several chambers dedicated to the removal of specific acoustic frequencies or frequency bands. Typical of such chambers are those defined as diffusing silencers, side-resonant silencers, and absorption silencers. The detailed discussion of the noise suppression behavior of each of these types is found in Sec. 8.5.

Perhaps the most important point to be made at this introductory stage is that, having reduced the area of the exhaust pipe deliberately to attain a "plugging" reflection pulse for power performance reasons, the first silencing design action has already been taken to the benefit of the specific power output of the engine. To reinforce the point, the optimum tail-

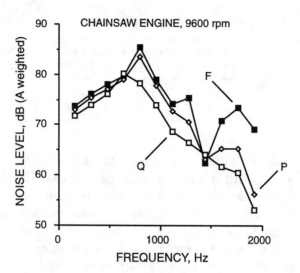

Fig. 8.26 Effect of box volume on the exhaust (dBA) noise spectra.

pipe diameter, d_6 or d_7 in Fig. 5.7, is always about half that of the initial exhaust pipe diameter, d_1. In other words, the area of the exhaust pipe which left the engine cylinder has been reduced by a factor of about four. This is quite unlike the situation for a four-stroke engine where any restriction of the exhaust pipe almost inevitably leads to a drop in delivery ratio and a consequential linear fall in power output. As a direct result, the silencing of a four-stroke engine is a more delicate design procedure. Equally, it implies that a high-performance street motorcycle with a two-stroke engine can be designed relatively easily to be legally quiet and the manufacturers of all production motorcycles ensure that this is the case. The somewhat noisy after-market, or even "diy," devices which cause irritation to the general public are the result of modifications by the owner, aimed more at machismo than machine efficiency!

The earlier design discussion on tuned expansion chambers did not dwell on the material to be used in its mechanical construction, but custom and practice show that this is typically fabricated from mild steel sheet metal which is about 1.1-1.3 mm thick. This may be acceptable for a racing engine where the mass of the exhaust system, as a ratio of the total machine mass, must be as low as possible. For a silenced system, as the exhaust pulses of a two-stroke engine are particularly steep-fronted, a thin metal outer skin of an expansion chamber and silencer will act as the diaphragm of a metal loudspeaker, causing considerable noise transmission to the atmosphere [8.26]. Therefore, the designer must inevitably introduce double-skinning of the most sensitive parts of the system, and often a layer of damping material is inserted between the double-skinned surface. Occasionally it is found to be adequate to line the internal surface of the silencer with a high-temperature plastic damping compound, and this is clearly an economic alternative in many cases.

8.6.1 A design for a silenced expansion chamber exhaust system

To reinforce the points made above, consider the design of a silenced and tuned expansion chamber system for an imaginary 125 cm^3 road-going motorcycle engine. The machine will be designed to produce peak power at 7500 rpm and be capable of pulling top gear at 100 km/h on a level road at 6000 rpm. The speed of 6000 rpm also coincides with the engine speed attained during acceleration noise testing for legislation purposes [8.16]. Not unnaturally, the manufacturer wishes to ensure that this is a well-silenced condition, and clearly the engine will be operating at a reasonably high bmep level at that point. The tuned pipe section of the pipe has been designed, it is assumed by the techniques already discussed in Chapters 5 and 6, and the optimum tail-pipe diameter has already been settled at 16 mm diameter with a box mid-section of 80 mm diameter. Thus, the types of pipe and silencer which would be applicable to this design are drawn in Fig. 8.27, showing a dimensioned sketch of the silencing section of the system and the ensuing transmission losses of the three boxes which make up the first portion of the silencer.

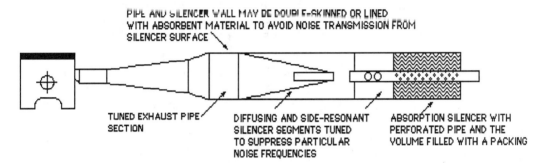

Fig. 8.27 Silencer design principles for a tuned pipe to retain high specific power output and to provide sufficient exhaust noise reduction.

The first item on the design agenda is a decision regarding the fundamental noisiest frequency to be silenced. From the discussion in Sec. 8.5.6, it is clear that the frequency in question is some five times that of the firing frequency, although the use of an engine simulation model for this imaginary 125 cm^3 motorcycle engine, with the program organized to predict the mass flow rate spectrum at the tail-pipe outflow, is a rather more accurate method of arriving at that conclusion for this, or any other, power unit. Assuming that the frequency multiplication factor is correct at five, the noisiest frequency will be at this fundamental value, i.e., 500 Hz.

The second item on the design agenda is the disposition of the silencing segments. From Sec. 8.5.3, the absorption silencer will come last, before the exit to the atmosphere. The first segment should be that aimed at suppressing the fundamental noise, i.e., a side-resonant silencer, followed by a diffusing silencer which has more broad-band muffling effectiveness. There might be a need for a third box to cope with any pass-bands that cannot be removed by the first two boxes.

The basic decisions having been made, the design of the first two boxes, A and B, is attempted. In Fig. 8.28 are the transmission losses of boxes A and B, together with the input data to Progs.8.1 and 8.2 for those boxes that correspond to the dimensioned sketch. The output from the individual calculations using Progs.8.1 and 8.2 are superimposed on top of each other.

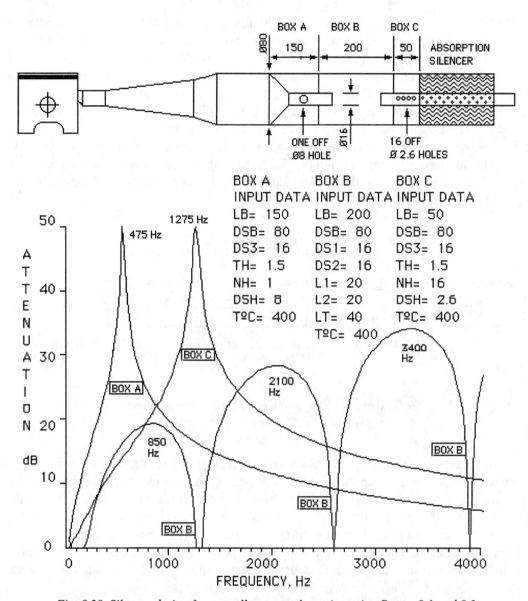

Fig. 8.28 Silencer design for a small motorcycle engine using Progs. 8.1 and 8.2.

The box A, a side-resonant silencer, has a single 8-mm-diameter hole in the 16 mm pipe with a 1.5-mm-thick wall. The exhaust temperature is assumed to be 400°C. Note that, if the real box has sloping walls as in the sketch, then the computed box volume may not ultimately agree with the actual box volume, in which case a modified length, L_b, should be inserted as data so that the calculation operates with a realistic box volume. You will understand the importance of that statement more fully if the relevant acoustic theory is re-examined in Sec. 8.5.2 and Eqs. 8.5.6-8.5.16. That comment aside, the computed acoustic transmission loss is seen to peak at 475 Hz, which is sufficiently close to the target frequency of 500 Hz for maximum suppression.

Chapter 8 - Reduction of Noise Emission from Two-Stroke Engines

The box B, a diffusing silencer, has 16-mm-diameter pipes entering and leaving, each pipe being re-entrant by 20 mm. The section length, L_b, is 200 mm. The transmission loss is quite broad-banded, but the most important frequency of suppression is at 850 Hz, this lesson being the principal one learned from the work of Coates [8.3] with his SYSTEM 2 and discussed in Sec. 8.5.1. The first pass-band is also of direct interest, and it is seen to exist at 1300 Hz.

If further design of this silencing system was left at this point, and the removal of the pass-band hole assigned to the absorption silencer segment, there would exist the strong possibility that this could turn out to be rather noisy at this frequency. With the first two boxes, A and B, there is a good transmission loss for the silencer up to 1300 Hz, and the evidence from Sec. 8.5.1 is that the pass-band hole at that frequency would exist. Thus, a third small box of the side-resonant type, box C, is added and is targeted to have high attenuation at that specific frequency. It can be seen that a short box, just 50 mm long, with 16 holes of 2.6 mm diameter in the 16-mm-diameter through-pipe with a 1.5-mm-thick wall, has a transmission loss which peaks at 1275 Hz. The addition of this third box, C, making three silencing segments before the absorption silencer portion, would provide strong attenuation of the noise up to about 3500 Hz, with no visible pass-band holes in that transmission loss. The absorption silencer can then be designed to remove the remainder of the high-frequency content, either in the conventional manner or by using the annular flow system recommended in Sec. 8.5.4.

You should recall the discussion regarding the sound radiation possibilities from the outside surface of silencer and pipe walls and, in a real design situation, ensure that the wall is double-skinned or suitably damped acoustically along the entire length of the system.

8.7 Concluding remarks on noise reduction

The principal concepts which the designer or developer has to retain at the forefront of the thought process when silencing the two-stroke engine, or indeed any reciprocating engine, are:

(i) Identify the principal source of noise from the engine, be it mechanical, induction or exhaust noise, and suppress that as a first priority. The other sources of noise could be completely eliminated and the total noise from the engine would be virtually unaffected if the single noisiest source remains unmuffled.

(ii) Within any given noise source, the technique for suppressing it is to determine the frequency spectrum of that noise and to tackle, as a first priority, the frequency band with the highest noise content. Should that remain unmuffled, then the other frequency bands could be completely silenced and the total noise content would virtually remain at the original level.

(iii) In the identification of noise sources, it is useful to remember that noise level, or loudness, is a function of the square of the distance from any given source [8.3] to the microphone recording that noise level. Thus placing the microphone close to the inlet tract end, or the exhaust pipe end, or a gearbox, in a manner which differentially distances the other sources, permits an initial and approximate identification of the relative contributions of the several sources of noise to the total noise output.

References for Chapter 8

8.1 G.P. Blair, J.A. Spechko, "Sound Pressure Levels Generated by Internal Combustion Engine Exhaust Systems," SAE Paper No. 720155, Society of Automotive Engineers, Warrendale, Pa., 1972.

8.2 G.P. Blair, S.W. Coates, "Noise Produced by Unsteady Exhaust Efflux from an Internal Combustion Engine," SAE Paper No. 730160, Society of Automotive Engineers, Warrendale, Pa., 1973.

8.3 S.W. Coates, G.P. Blair, "Further Studies of Noise Characteristics of Internal Combustion Engines," SAE Paper No. 740713, Society of Automotive Engineers, Warrendale, Pa., 1974.

8.4 D.D. Davis, G.M. Stokes, D. Moore, G.L. Stevens, "Theoretical and Experimental Investigation of Mufflers," NACA Report No.1192, 1954.

8.5 P.O.A.L. Davies, R.J. Alfredson, "The Radiation of Sound from an Engine Exhaust," *J.Sound Vib.*, Vol 13, p389, 1970.

8.6 P.O.A.L. Davies, "The Design of Silencers for Internal Combustion Engines," *J.Sound Vib.*, Vol 1, p185, 1964.

8.7 P.O.A.L. Davies, "The Design of Silencers for Internal Combustion Engine Exhaust Systems," Conference on Vibration and Noise in Motor Vehicles, *Proc.I.Mech.E.*, 1972.

8.8 G.H. Trengrouse, F.K. Bannister, "The Reduction of Noise Generated by Pressures Waves of Finite Amplitude Attenuation due to Rows of Holes in Pipes," University of Birmingham, Research Report No. 40, 1964.

8.9 M. Fukuda, H. Izumi, "Kuudougata Shouonki no Tokusei ni Kansuru Kenkyuu," *Trans.Japan Soc.Mech.E*, Vol 34, No 263, p1294, 1968.

8.10 P.M. Nelson, "Some Aspects of Motorcycle Noise and Annoyance," SAE Paper No. 850982, Society of Automotive Engineers, Warrendale, Pa., 1985.

8.11 R. Taylor, Noise, Penguin, London, 1975.

8.12 C.M. Harris, Handbook of Noise Control, McGraw-Hill, New York, 1957.

8.13 L.L. Beranek, Noise Reduction, McGraw-Hill, New York, 1960.

8.14 M.B. Johnston, "Exhaust Port Shapes for Sound and Power," SAE Paper No. 730815, Society of Automotive Engineers, Warrendale, Pa., 1973.

8.15 K. Groth, N. Kania, "Modifications on the Intake Ports with the Aim to Reduce the Noise of a Two-Stroke Crankcase-Scavenged Engine," SAE Paper No. 821069, Society of Automotive Engineers, Warrendale, Pa., 1982.

8.16 SAE Noise Standards, Society of Automotive Engineers, Warrendale, Pa.:
SAE J47—Maximum Sound Level Potential for Motorcycles
SAE J331a—Sound Levels for Motorcycles
SAE J192—Exterior Sound Level Measurement for Snowmobiles
SAE J986—Light Vehicle Noise Test Procedures
SAE J1174—Operator Ear Sound Level Measurement Procedure for Small Engine Powered Equipment
SAE J1046a—Exterior Sound Level Measurement Procedure for Small Engine Powered Equipment

SAE J1175—Bystander Sound Level Measurement Procedure for Small Engine Powered Equipment

SAE J1074—Engine Sound Level Measurement Procedure

SAE J1207—Measurement Procedure for Determination of Silencer Effectiveness in Reducing Engine Intake or Exhaust Sound

SAE J986—Light Vehicle Noise Test Procedures

SAE J1470—Light Vehicle Noise Test Procedures (ISO362 1981 equivalent)

8.17 S.W. Coates, "The Prediction of Exhaust Noise Characteristics of Internal Combustion Engines," Doctoral Thesis, The Queen's University of Belfast, April 1974.

8.18 E. Kato, R. Ishikawa, "Motorcycle Noise," Dritte Grazer Zweiradtagung, Technische Universitat, Graz, Austria, 13-14 April, 1989.

8.19 N.A. Hall, <u>Thermodynamics of Fluid Flow</u>, Longmans, London, 1957.

8.20 G.M. Jenkins, D.G. Watts, <u>Spectral Analysis and its Applications</u>, Holden-Day, San Francisco, 1968.

8.21 G.P. Blair, "Computer-Aided Design of Small Two-Stroke Engines for both Performance Characteristics and Noise Levels," I.Mech.E Automobile Division Conference on Small Internal Combustion Engines, Isle of Man, May 31-June 2, 1978, Paper No. C120/78.

8.22 SAE J335b, "Multiposition Small Engine Exhaust System Fire Ignition Suppression," Society of Automotive Engineers, Warrendale, Pa.

8.23 G.P. Blair, W.L. Cahoon, C.T. Yohpe, "Design of Exhaust Systems for V-Twin Motorcycle Engines to meet Silencing and Performance Criteria," SAE Paper No. 942514, Society of Automotive Engineers, Warrendale, Pa., 1994.

8.24 C.V. Beidl, K. Salzberger, A. Strebenz, M. Buchberger, "Theoretical and Empirical Approaches to Reduce the Noise Emission of Intake Silencers and Exhaust Systems with Catalysts," Funfe Grazer Zweiradtagung, Technische Universitat, Graz, Austria, 22-23 April, 1993.

8.25 R. Kamiya, "Prediction of Intake and Exhaust System Noise," Fourth Grazer Zweiradtagung, Technische Universitat, Graz, Austria, 8-9 April, 1991.

8.26 C. Poli, A. Pucci, "Vibro-Acoustical Analysis of a New Two-Stroke Engine Exhaust Muffler for Increasing its Sound Deadening," SAE Paper No. 931572, Society of Automotive Engineers, Warrendale, Pa., 1993, pp855-863.

8.27 G.E. Roe, "The Silencing of a High Performance Motorcycle," *J. Sound & Vibration*, 33, 1974.

Postscript

I can think of no better postscript to this book than the poem, er, excuse me, the doggerel verse which appeared on the flyleaf of the original book. The reason is that the sentiment and sentiments expressed are still as biting as they were then, and maybe more so. . . .

THE ORIGINAL MULLED TOAST

To Sir Dugald Clerk I raise my glass,
his vision places him first in class.
For Alfred Scott let's have your plaudits,
his Squirrels drove the four-strokes nuts.

To Motorradwerk Zschopau I doff my cap,
their Walter Kaaden deserves some clap.
Señores Bulto and Giro need a mention,
their flair for design got my attention.

Brian Stonebridge I allege,
initialled my thirst for two-stroke knowledge.
In academe I found inspiration
in Crossland and Timoney and Rowland Benson.

All of my students are accorded a bow,
their doctoral slavery stokes our know-how.
The craftsmen at Queen's are awarded a medal,
their precision wrought engines from paper to metal.

For support from industry I proffer thanks,
their research funds educate many Ulster cranks,
but the friendships forged I value more
as they span from Iwata to Winnebago's shore.

To the great road-racers I lift my hat,
they make the adrenalin pump pitter-pat,
for the Irish at that are always tough,
like Dunlop and Steenson and Ray McCullough.

In case you think, as you peruse this tome,
that a computer terminal is my mental home,
I've motorcycled at trials with the occasional crash
and relieved fellow-golfers of some of their cash.

Gordon Blair, May 1989

Appendix Listing of Computer Programs

Programs associated with Chapter 1:
  Prog.1.1, PISTON POSITION
  Prog.1.2, LOOP ENGINE DRAW
  Prog.1.3, QUB CROSS ENGINE DRAW
  Prog.1.4, EXHAUST GAS ANALYSIS

 Programs associated with Chapter 2:
   Prog.2.1, WAVE FLOW
   Prog.2.2, SUPERPOSITION
   Prog.2.3, FRICTION & HEAT TRANSFER
   Prog.2.4, INFLOW & OUTFLOW
   Prog.2.5, THREE-WAY BRANCH

Programs associated with Chapter 3:
   Prog.3.1, BENSON-BRANDHAM MODEL
   Prog.3.2, GPB SCAVENGING MODEL
   Prog.3.3a, GPB CROSS PORTS
   Prog.3.3b, QUB CROSS PORTS
   Prog.3.4, LOOP SCAVENGE DESIGN
   Prog.3.5, BLOWN PORTS

Programs associated with Chapter 4:
    Prog.4.1, SQUISH VELOCITY
    Prog.4.2, HEMI-SPHERE CHAMBER
    Prog.4.3, HEMI-FLAT CHAMBER
    Prog.4.4, BATHTUB CHAMBER
    Prog.4.5, TOTAL OFFSET CHAMBER
    Prog.4.6, BOWL IN PISTON
    Prog.4.7, QUB DEFLECTOR

Simulation Models associated with Chapter 5:
    Piston-ported, spark-ignition industrial engine
    Reed-valve, spark-ignition engine with expansion chamber exhaust

Programs associated with Chapter 6:
- Prog.6.1, TIMEAREA TARGETS
- Prog.6.2v2, EXPANSION CHAMBER
- Prog.6.3, TIME-AREAS
- Prog.6.4, REED VALVE DESIGN
- Prog.6.5, DISC VALVE DESIGN

Programs associated with Chapter 8:
- Prog.8.1, DIFFUSING SILENCER
- Prog.8.2, SIDE-RESONANT SILENCER

The Simulation Models and the set of computer Programs are available from SAE as executable files on diskettes for either IBM® PC (or compatible) or Macintosh® desktop computers.

Call SAE Customer Sales and Satisfaction at (412) 776-4970 for ordering information.

Index

Active radical (AR) combustion
  and engine performance/emissions optimization, 490-491
  and four-stroking, 285
  and trapping pressure, 491
Ahmadi-Befrui
  predictive CFD scavenging calculations, 250
Aircraft engines
  Junkers Jumo, 3
  Napier Nomad, 3
Air-fuel ratio
  general
    defined, 29-30
    equivalence ratio, 299-300, 304-305
    flammability limits of (SI engines), 284
    importance of, 465
    overall (in engine testing), 38
  and brake mean effective pressure
    low-emissions engine, 487
    QUB 400 engine, 472, 474, 476
  and combustion products
    molecular AFR, 297-298
    stoichiometric (ideal) AFR, 298, 465
    see also Exhaust emissions/exhaust gas analysis
Airhead charging. *See under* Stratified charging
Alternative fuel combustion
  turbulence, advantages of, 339
Annand, W.J.D.
  heat transfer analyses
    crankcase, 375-378
    racing motorcycle engine simulation, 397-399
    SI engines, 305-307
  see also Combustion processes (heat transfer analysis)
Applications, typical
  automobile racing, 2-3
  handheld power tools, 1-3, 14
  outboard motors, 2, 4
  port timings, typical, 19
  see also Compression ignition engines
Auto Union vehicles, 2

Bannister, F.K.
  unsteady gas flow analyses
    moving shock waves, 62, 201-204
    particle velocity, 55, 197-200
    simulation of engines, 143
Basic engine. *See* Two-stroke engine (general)
Batoni, G.
  Piaggio stratified charging engine, 500
Bearings (friction losses in)
  compression-ignition engines with plain bearings, 379
  SI engines with plain bearings, 379
  SI engines with rolling-element bearings, 378-379
  see also Friction/friction losses
Bel, defined, 543
Benson, R.S.
  "constant pressure" criterion, 98-99, 103, 104
  engine simulation (in unsteady gas flow), 142
Bingham, J.F.
  coefficient of discharge (in unsteady gas flow), 208
Blair, G.P.
  and QUB stratified charging engine, 498
  see also GPB engine simulation model
Blowdown (initial exhaust)
  basic two-stroke engine, 6, 7
Blown engines. *See* Turbocharged/supercharged engines
Bore-stroke ratio
  defined, 44
  of diesel engines, typical, 43
  effect on loop scavenging, 267
  optimum for uniflow scavenging, 253
  oversquare engines, load considerations for, 471
  square *vs.* oversquare (discussion), 45
  of various SI engine types (typical), 43
Brake mean effective pressure
  See under Performance measurement
Brake power output
  See under Performance measurement
Brake torque
  See under Performance measurement

Burmeister & Wain
  marine diesel engines, 3-4
Butler, Edward, 1
Butterfly exhaust valves
  effect on exhaust emissions/engine performance, 491-492, 493

CAD/CAM techniques
  in scavenging port design, 274, 275
Carbon dioxide
  in closed cycle, two-zone combustion model, 353, 354-355
  in QUB SP apparatus (discontinuous exhaust), 188-189
  see also Exhaust emissions/exhaust gas analysis
Carbon monoxide. See under Exhaust emissions/exhaust gas analysis
Carson, C.E.
  QUB 500rv engine tests, 516, 527-529
Cartwright, A.
  mass fraction burned data (SI engine), 312-313
Catalysis, exhaust
  oxygen catalysis, effect of (IFP stratified charging engine), 506, 509
  in simple two-stroke engine, 494, 495
Cathedral engine (Harland & Wolff), 3, 5
CFD (Computational Fluid Dynamics)
  in scavenging analysis
    introduction, 244
    Ahmadi-Befrui predictive calculations, 250
    charge purity plots (Yamaha test cylinders), 246-248
    in future design/development (discussion), 274, 275-276
    grid structure for scavenging calculations, 245
    PHOENICS CFD program, flow assumptions in, 245-246
    as a port design tool (loop scavenging), 264-265
    SR-TE-SR plots (CFD vs. experimental values), 248-250
  see also Scavenging
Chainsaws
  exhaust ducting (typical untuned systems), 371, 372
  Homelite (typical), 3
  mass fraction burned characteristics, 312
  port timing for, typical, 19
  power output, typical, 45
  trapping efficiency (simulated engine), 479-481
  two-zone closed cycle combustion model
    CO formation, 354
    CO_2 formation, 353, 354-355
    hydrogen formation, 355
    NO formation, 349-352
    oxygen (effect on NO formation), 352-354
    oxygen mass ratio (in burn zone), 352-354
  see also Computer modeling (chainsaw engine simulation); Specific time area (A_{sv})
Charging efficiency (CE)
  defined
    as function of TE and SR, 29
    in ideal scavenge model, 213
    in terms of mass ratio, 29
  spilled charge, effect of
    in chainsaw engine simulation, 387-388
    in racing motorcycle engine simulation, 400
  tuned exhaust system, effect of
    in racing motorcycle engine simulation, 399-400
  see also Homogeneous charging (various); Stratified charging (various)
Chen, C.
  simulation of engines (in unsteady gas flow), 142
Clearance volume
  design for (discussion), 334-336
Clerk, Sir Dugald
  cross scavenging deflector piston design, 9, 10
  invents two-stroke engine, 1
  use of crankshaft as air pump, 12
Coates, S.W.
  experimental silencer configurations, 550-554
  sound pressure analysis, 548-550
Coefficient of discharge (unsteady gas flow)
  introduction, 205
  Bingham analytical approach, 208
  caveats regarding experimental technique, 206-207
  determination of C_d for engine simulation, 207-208
  expression for (as mass flow rate ratio), 206, 207

initial conditions, assumptions regarding, 205
mass flow rate, theoretical, 205-206
measurement at exhaust port (typical values), 208-210
measurement of (experimental set-up), 205-206
in QUB SP single-pulse experimental apparatus, 172
Combustion chambers (general)
  bathtub chamber
    typical configuration, 336
  bowl in piston chamber
    in loop-scavenging design, 271
    typical configuration, 336
  central chamber
    squish flow area, calculation of, 329
  clearance volume, design for, 334-336
  computer-designed configurations
    introduction, 335
    alternative chamber types, 336
    example: hemispheric chamber, 335, 337
    GPB CROSS PORTS program, 259-260, 261
    HEMI-SPHERE program, 335, 337
    QUB CROSS ENGINE DRAW, 25-26
    QUB CROSS PORTS program, 260, 261-263
  deflector chamber
    squish flow area, calculation of, 329, 330
  detonation/knocking, influence on, 286, 287
  heat release analysis
    See Heat release
  hemi-flat chamber
    typical configuration, 336
  hemisphere chamber
    design example (HEMI-SPHERE program), 335, 337
    typical configuration, 336
  for homogeneous charge combustion, 338-339
  IDI diesel engine, geometry of, 317
  Mexican hat chamber, 283, 314
  offset chamber
    squish flow area, calculation of, 329
    see also Combustion chambers, total offset
  QUB deflector chamber
    QUB CROSS ENGINE DRAW, 25, 26
    QUB CROSS PORTS program, 260, 261-263
    typical configuration, 336
  and squish action
    introduction and discussion, 331
    central squish system, 338-339
    combustion chamber design for, 334-337
    design for clearance volume, 334-336
    diesel engines, design considerations for, 334, 335
    squish flow area, 326, 328-330
    squish velocity, 331, 332
    squished kinetic energy, 331-332, 334-335
    see also Squish action
  for stratified charge combustion, 337
  total offset chamber
    computer design program for, 335, 336
    squish flow area, calculation of, 329
    see also Combustion chambers, offset
  wall wetting, 314
  see also Combustion processes; Ports (various); Scavenging
Combustion processes
  introduction, 281-282
  active radical (AR) combustion
    and engine performance/emissions optimization, 490-491
    and four-stroking, 285
  air-fuel ratio (general)
    flammability limits of (SI engines), 284
    importance of, 465
    see also Air-fuel ratio; Exhaust emissions/ exhaust gas analysis
  alternative fuel combustion
    turbulence, advantages of, 339
  closed cycle model, single-zone
    introduction, 318
    simple model, 318-319
    complex model, 319-321
    gas purity throughout, 321
    limitations of, 322
  closed cycle model, two-zone (chainsaw engine)
    introduction and theoretical model, 347
    CO formation, 354
    CO_2 formation, 353, 354-355
    cylinder temperature *vs.* crankshaft angle, 349-350
    equations of state, thermodynamic, 348
    heat transfer between zones, 348-349
    hydrogen formation, 354, 355
    masses in burned/unburned zones, 347

Combustion processes *(continued)*
  closed cycle model, two-zone (chainsaw engine) *(continued)*
    nitric oxide (NO) formation, 349-352, 353
    oxygen (its effect on NO formation), 352-354
    oxygen mass ratio, 352-354
    peak temperature *vs.* AFR, 349, 350
    purities in burned/unburned zones, 347-348
  combustion efficiency
    QUB LS400 SI engine, 294, 296
    vs. scavenging efficiency, 304
    vs. trapped charge purity, 304
  combustion optimization
    introduction, 490
    active radical (AR) combustion, 490-491
    butterfly exhaust valve, 491-492, 493
    exhaust edge timing control valve, 492
  compression ignition
    combustion chamber geometry (Ricardo Comet IDI engine), 317
    diffusion burning, 314
    exhaust smoke from, 288
    fast fuel *vs.* fast air approach, 314
    high-speed limit (IDI engine), 314
    lean mixture combustion, 300-301
    low-speed limit (DI engine), 314
    physical processes in, 288-289
    premixed combustion, 314, 316
    rich limit for, 288
    and squish action, 334
    typical combustion chamber/fuel injector, 283, 288
    see also heat release (compression-ignition engines) *(below)*
  dissociation, effects of, 297, 301-302
  equivalence ratio
    defined, 299
    and compression-ignition engines, 305
    in lean mixture combustion, 300
    in rich mixture combustion, 299
    and SI engines, 304-305
  exhaust gases/combustion products. *See* Exhaust emissions/exhaust gas analysis
  flame velocity
    flame front propagation velocity, 283-284
    one-dimensional model (discussion), 322-323
    and squish action (QUB loop-scavenged), 333

  fuels
    alternative fuel (turbulence in combustion), 339
    consumption of. *See* Fuel consumption/fuel economy
    properties of, 297
  heat availability
    and combustion efficiency (discussion), 303-304
    see also Heat release
  heat transfer analysis (Annand model)
    closed cycle model (general), 305-307
    crankcase analysis, 375-378
    heat transfer coefficients, 307-308
    racing motorcycle engine, 397-399
    see also Heat release; Heat transfer
  homogeneous combustion
    defined, 286
    alternative fuels, use of, 339
    detonation, susceptibility to, 338
    functional diagram, 467
    QUB deflector chamber, constraints on, 338
    and squish action, 338
    and stratified charging, 467-468
    vs. stratified combustion (discussion), 466-468, 469
    see also stratified combustion *(below)*; Stratified charging (with homogeneous combustion)
  lean mixture combustion, 300-301
  mass fraction burned
    Cartwright experimental data for (SI engine), 312-313
    from loop-scavenged QUB LS400 SI engine, 294, 295
    Reid experimental data for (SI engine), 310-312
    Vibe analytical approach, 295, 309-310
  rich mixture combustion, 288, 299-300
  spark-ignition
    air-fuel mixtures, flammability limits of, 284
    combustion model, three-dimensional, 323-324
    flame front propagation velocity, 283-284
    gasoline, free-surface flammability of, 282
    ignition process, physics of, 282-284
    SE effect on flammability, 285
    SI systems effectiveness, 285

Index

in two-stroke engine (schematic diagram), 283
and squish action
  burning characteristics, 333-334
  central squish system, design of, 338-339
  compression ignition, 334
  detonation reduction (and squish velocity), 333-334
  flame velocity (loop-scavenged, QUB), 333
  in homogeneous charge combustion, 338-339
  in stratified charge combustion, 337
stoichiometric combustion equation, 297-298, 465
stratified combustion
  air-blast gasoline injection (into cylinder), 516-521
  engine tests, light load (QUB 500rv), 526-529
  engine tests: QUB 500rv experimental engine, 521-526
  functional diagram, 467
  general comments (QUB 500rv engine), 529-531
  in Honda CVCC engine, 466-467
  liquid gasoline injection, 513-514
  ram-tuned liquid injection system (TH Zwickau), 514-516
  with stratified charging (discussion), 468
  see also homogeneous combustion *(above)*; Homogeneous charging (with stratified combustion)
see also Exhaust emissions/exhaust gas analysis
Compression-ignition
  See Combustion processes; Compression-ignition engines
Compression-ignition engines
  aircraft engines (Napier, Junkers), 3
  applications, typical, 3-5, 14
  bore-stroke ratios, typical, 43
  combustion chambers in
    IDI diesel engine, geometry of, 317
    Ricardo Comet chamber geometry (IDI), 317
    squish action, design considerations for, 334, 335
    see also Squish action
  combustion processes in
    See under Combustion processes
  Detroit Diesel Allison Series 92 engine, 14

exhaust emissions of
  exhaust gas composition (general), 14
  see also Exhaust emissions/exhaust gas analysis
Harland & Wolff "Cathedral" (marine), 3, 5
low-speed limit (DI engine), 314
performance of (discussion), 531
power output
  truck diesel, 45
scavenging in
  loop scavenging port design (blown 4-cylinder DI), 272-273
  uniflow scavenging, 11-12
turbocharged, 531
unpegged piston rings in, 131
Compression ratio
  crankcase, defined, 22
  effect on performance/emissions
    bmep, 536-537
    bsfc, 536-537
    bsNO emissions, 537-538
    combustion rate, 538
    cylinder temperature, 537-538
    detonation, 538-540
    peak cylinder pressure, 537-538
  effect on squish behavior, 34, 335
  geometric, defined, 22
  trapped, defined, 22
Computer modeling
  related terms:
    Computer modeling (basic engine)
    Computer modeling (chainsaw engine simulation)
    Computer modeling (multi-cylinder engine simulation)
    Computer modeling (racing motorcycle engine simulation)
    Computer programs
    GPB engine simulation model
    Specific time area (A_{sv})
Computer modeling (basic engine)
  introduction, 357-358
  computer model, key elements of, 358-359
  crankcase heat transfer analysis
    Annand model, 375-378
    Nusselt, Reynolds numbers in, 376
  cylinder porting *vs.* piston motion
    introduction, 359-360

595

Computer modeling (basic engine) *(continued)*
  cylinder porting *vs.* piston motion *(continued)*
    complex port layout with piston crown control, 362
    simple port layout with piston crown control, 360-362
    simple port layout with piston skirt control, 361, 362-363

Computer modeling (chainsaw engine simulation)
  bsfc and bsHC *vs.* rpm, 382
  charge purity *vs.* crankshaft angle, 386
  correlation of simulation with measurements, 380-383
  cylinder pressure *vs.* crankshaft angle, 382-383, 385-386
  cylinder temperature, pressure *vs.* crankshaft angle, 392, 393
  engine parameters, description of, 380
  exhaust system behavior (pressure, temperature), 389-390
  friction and pumping losses
    introduction and discussion, 380
    friction mep (defined), 378-379
    friction mep *vs.* rpm, 383-384
    mechanical efficiency *vs.* rpm, 384
    pumping mep *vs.* rpm, 383-384
    SI engines with plain bearings, 379
    SI engines with rolling-element bearings, 378-379
  gas flow through cylinder (discussion), 385-386
  intake system
    intake ducting (dimensions and discussion), 370-371
    intake ducting throttle area ratio, 371
    system behavior (pressure, temperature), 390-392
  internal pressures *vs.* crankshaft angle, 385-386
  internal temperatures *vs.* crankshaft angle, 386
  loopsaw scavenging, 380, 388
  mep and mechanical efficiency *vs.* rpm, 384-385
  power and BMEP *vs.* rpm, 380, 381
  scavenge gas temperatures *vs.* crankshaft angle, 388-389
  scavenging characteristics (measured *vs.* computed), 380-382
  scavenging pressure *vs.* crankshaft angle, 387-388
  SE, SR, TE, CE *vs.* crankshaft angle, 387-388
  tuned exhaust systems (high-performance, multi-cylinder)
    four-cylinder automotive engine, 374, 375
    three-cylinder automotive engine, 373, 374
    three-cylinder outboard engine, 373, 374
    V-8 outboard motor (OMC), 373, 375, 568
  tuned exhaust systems (high-performance, single-cylinder)
    dimensions of (typical), 372
    discussion, 371-372
    QUB 500 68bhp motorcycle engine, 373
  untuned exhaust systems
    compact industrial (discussion), 371
    dimensions of (chainsaw), 372
  valves
    control valves (for exhaust port edge timing), 363-365
    disc valves (intake), 365-367
    poppet valves, design guide for, 412-414
    poppet valves, engine simulation using, 363
    reed valves (intake), 367-370
  final note: further simulations involving this engine, 392-394
  see also Specific time area (A_{sv})

Computer modeling (multi-cylinder engine simulation)
  blower scavenging (four-cylinder supercharged engine)
    introduction and discussion, 407
    four cylinders, advantages of, 409
    open port period, three cylinder *vs.* four cylinder, 409
    open-cycle pressures and charging (cylinders 1-2), 407-409
  blower scavenging (three-cylinder supercharged engine)
    exhaust tuning, 405-407
    open-cycle pressures and charging (cylinders 1-2), 404-405
    temperature and purity in scavenging ports, 405, 406

Computer modeling (racing motorcycle engine simulation)
  cylinder pressure *vs.* crankshaft angle, 392-393, 398-399
  engine parameters, description of, 394-395
  exhaust system
    exhaust pressure diagrams, 397, 398

exhaust timing control valve, effect of, 396-397, 398
resonance effects (on CE), 400
heat transfer analysis (closed cycle Annand model), 397-399
high charging efficiency, origins of
  discussion, 401
  exhaust system resonance, effects of, 400
  plugging and suction pulses, 401
  pressure oscillations, effects of, 399-400
reed valve induction system, behavior of, 401-402
scavenging model assumptions, 395
final note: further simulations involving this engine, 402
Computer programs
  described, 20-21
  BENSON-BRANDHAM MODEL, 216-219
  BLOWN PORTS, 270-273
  DIFFUSING SILENCER, 558-560, 582
  DISC VALVE DESIGN, 459-460
  EXHAUST GAS ANALYSIS, 40
  EXPANSION CHAMBER, 444-445
  GPB CROSS PORTS, 259-260, 261
  HEMI-SPHERE, 335, 337
  LOOP ENGINE DRAW, 23, 24, 25
  LOOP SCAVENGE DESIGN 4, 268-269
  PHOENICS CFD, 245-246
  PISTON POSITION, 23, 24
  QUB CROSS ENGINE DRAW, 25-26
  QUB CROSS PORTS, 260, 261-263
  REED VALVE DESIGN, 452, 454-455
  SIDE RESONANT SILENCER, 558-560, 582
  SQUISH VELOCITY, 330-331
  TIMEAREA TARGETS, 423-424
  listing of (Appendix), 589-590
Control valves
  and combustion optimization, 492
  discussion and geometry (chainsaw engine simulation), 363-365
  effect of (racing motorcycle engine simulation), 396-397, 398
  simulation model, criteria for, 365
  trapped compression ratio (modified), 365
Cross-scavenged engines. *See under* Scavenging

Day, Joseph, 1
Delivery ratio
  defined, 27
  engine air flow (DR) *vs.* A_{svx}, 427-430
  in GPB engine simulation model, 168
Detonation
  combustion chamber influence of, 286, 287
  compression ratio, effect of, 538-540
  in cross scavenging, 10
  description of, 286
  detonation reduction (and squish velocity), 333-334
  in homogeneous charge combustion, 338
  in stratified charging, 288
  vs. knocking, 286
  see also Preignition
Detroit Diesel
  Allison Series 92 engine, 14
Diesel engines. *See* Compression ignition engines
Disc valves
  characteristics and origin of, 16
  configuration, typical, 16
  empirical design of
    introduction, 456
    disc *vs.* reed valves (discussion), 446-447
    see also specific time area (A_{sv}) analysis *(below)*
  in MZ racing motorcycles, 16-17, 456
  Rotax disc-valved engine, 17, 456
  specific time area (A_{sv}) analysis
    introduction, 456-457
    carburetor flow diameter, 458
    computer solution for (DISC VALVE DESIGN), 459-460
    conventional *vs.* SI units, 459
    design calculations for (intake system modeling), 365-367
    design example (DISC VALVE DESIGN program), 459-460
    disc valve timing *vs.* induction port area, 457
    maximum port area, 457, 458
    outer port edge radius, 458
    total opening period, 457
Dissociation
  basic reactions ("water-gas" reaction), 297, 346
  effects of, 301-302
  see also Exhaust emissions/exhaust gas analysis

DKW machines
  early supercharged models, 1
Duret, P.
  Institut Français du Pétrole stratified charging engine, 504
Dynamometer testing. *See* Performance measurement

Efficiency
  mechanical (defined), 37-38
  see also Friction/friction losses
Eight-stroking
  from inadequate scavenging, 218
Empirical design assistance
  introduction, 415-416
  disc valve design. *See under* Disc valves
  engine porting
    chainsaw performance, effect of porting changes on, 426-431
    determination of A_{sv} (measured), 424-426
    engine performance values *vs.* A_{sv} (QUB experience), 420-423
    mass flow rates, 417-419
    port area *vs.* crankshaft angle, 417
    specific time area (A_{sv}), derivation of, 419-420
    TIMEAREA TARGETS program, derived A_{sv} values from, 423-424
    see also practical considerations *(below)*; Specific time area (A_{sv})
  exhaust systems
    tuned (high-performance engine), 437-445
    untuned, 435-437
    see also Exhaust systems
  practical considerations
    introduction, 431
    basic engine dimensions, acquisition of, 431-432
    data selection, remarks on, 445-446
    empiricism in general (comment), 434-435
    exhaust ports, width criteria for, 432
    factors driving a new design, 431
    inlet ports, width criteria for, 433
    port timing criteria, 434
    scavenge port layout (piston ported engine), 433
    transfer ports, width criteria for, 433-434
    unpegged rings, 433-434
  reed valve design. *See under* Reed valves
Engine testing. *See* Performance measurement
Equivalence ratio
  defined, 299
  in compression-ignition engines, 305
  in lean mixture combustion, 300
  in rich mixture combustion, 299
  in SI engines, 304-305
  see also Air-fuel ratio
Exhaust closure. *See* Port design; Port timing; Trapping
Exhaust emissions/exhaust gas analysis
  air-fuel ratio (general)
    importance of (discussion), 465
    molecular, 297-298
    stoichiometric, 298-299
  air-fuel ratio, effect of (low emissions engine)
    bsCO emissions, 488
    bsHC emissions, 487-488
    bsO_2 emissions, 487-488
  air-fuel ratio, effect of (QUB 400 research engine)
    bsHC emissions, 473-476
    CO emissions, 473-475
    O_2 emissions, 473, 475
  air-fuel ratio, effect of (two-zone chainsaw engine simulation)
    brake specific hydrocarbon emissions, 479-480, 482
    on bsCO emissions, 480-482
    CO_2 mass ratio, 353, 354-355
    on CO mass ratio, 352, 354
    cylinder temperature (*vs.* crankshaft angle), 349-350
    hydrogen mass ratio, 354, 355
    NO growth rate, 349-352
    peak temperatures (burn/unburned zones), 349, 350
  Batoni performance maps (Vespa motor scooter engine)
    CO emissions, 476-478
    HC emissions, 478
  butterfly exhaust valve, effect of (HC emissions), 491-492, 493
  carbon dioxide (CO_2)
    two-zone closed cycle combustion model (chainsaw), 353, 354-355

Index

Exhaust emissions/exhaust gas analysis *(continued)*
  carbon monoxide (CO)
    AFR, effect of (two-zone chainsaw engine simulation), 480-482
    Batoni performance maps (Vespa motor scooter engine), 476-478
    bsCO, AFR effect on (low emissions engine), 488
    bsCO, AFR effect on (QUB 400 engine), 473-475
    bsCO emissions, combustion-derived equation for, 343
    in closed cycle, two-zone combustion model (chainsaw), 352, 354
    in exhaust emissions (introduction), 303
    in Piaggio stratified charging engine, 502-503
  combustion process (theoretical)
    brake specific pollutant rate, 343
    bsCO emissions, combustion-derived equation for, 343
    equilibrium reactions in, 346
    gas pollutant mass, 343
    HC emissions, combustion-derived, 343-344
    HC emissions, scavenge-derived, 344
    HC emissions, total, 344
    mass ratios (of exhaust components), 343
    nitrogen oxides, 344-345
  combustion processes (homogeneous) (SI engine), 467-468
    see also Stratified charging (with homogeneous combustion)
  complex two-stroke engine
    introduction, 494-495
    fuel and scavenge air, separation of (principle), 495-496
    fuel injector, relative positioning of, 495, 497
    fuel vaporization/mixing, the problem defined, 495-496
    stratified charging and fuel vaporization/mixing (discussion), 496-497
    see also Stratified charging
  compression ignition engines
    exhaust emissions, constraints on, 4
    turbocharged/supercharged (example), 14
  dissociation
    basic reactions of, 297
    effects of, 301-302
  exhaust gas composition (general)
    aldehydes, 468
    discussion of (general), 67-68
    of fuel-injected diesel engines, 14
    hydrocarbons, 302-303
    introduction to, 302-303
    ketones, 468
    lean mixture combustion, 300-301
    in lean mixture combustion, 301
    nitrogen oxides, 303
    of petroil-lubricated engines, 13, 470-471
    of pressure-lubricated engines, 13
    properties of (at low, high temperatures), 68
    in rich mixture combustion, 298, 299-300
    of turbocharged/supercharged engines, 13-14
  hydrocarbons
    Batoni performance maps for (Vespa motor scooter engine), 478
    bsHC at various fuelings (QUB 500rv engine, full load), 524
    bsHC at various fuelings (QUB 500rv engine, light load), 526
    bsHC *vs.* AFR (low emissions engine), 487-488
    bsHC *vs.* AFR (QUB 400 research engine), 473, 476
    bsHC *vs.* AFR (two-zone chainsaw engine simulation), 479-480, 482
    bsHC *vs.* rpm (chainsaw engine simulation), 382
    butterfly exhaust valve, effect on, 491-492, 493
    combustion-derived, 343-344
    in exhaust gas composition (general), 302-303
    Institut Français du Pétrole stratified charging engine, 506, 508, 509
    in liquid gasoline injected engines, 513-514
    Piaggio stratified charging engine, 503
    QUB270 cross-scavenged airhead-stratified engine, 511-512
    radiused porting, effect on bsHC (chainsaw engine), 579
    scavenge-derived, 344
    from skip-firing two-stroke engine (four stroking), 471

Exhaust emissions/exhaust gas analysis *(continued)*
  hydrocarbons *(continued)*
    total, 344
    vs. specific time areas (A_{svx}), 427-430
  laboratory testing for
    brake specific pollutant gas flow, 39
    bsCO emission rate, 39
    CO mass flow rate, 39
    EXHAUST GAS ANALYSIS (computer program), 40
    mass flow basis, importance of, 38-39
    NDIR (non-dispersive infrared) analysis, 40-41, 493
    O_2 concentration, 38
  nitrogen oxides
    AFR, effect of (two-zone chainsaw engine simulation), 349-352
    formation (as function of temperature), 303
    formation in combustion process (theoretical), 344-345
    Institut Français du Pétrole stratified charging engine, 506, 509
    oxygen, effect on NO formation (two-zone chainsaw engine simulation), 352-354
    Sako/Nakayama experimental data (178 cc snowmobile engine), 478-479
    various fuelings, effect of (QUB 500rv engine, full load), 525
    various fuelings, effect of (QUB 500rv engine, light load), 526, 527
  oxygen (O_2)
    AFR *vs.* bsO_2 emissions (low emissions engine), 487-488
    AFR *vs.* oxygen emissions (QUB 400 research engine), 473, 475
    O_2 concentration, laboratory testing for, 38
    O_2 mass ratio in burn zone (chainsaw model), 350, 353
  radiused porting, effect on bsHC (chainsaw engine), 579
  Sako/Nakayama experimental data (178 cc snowmobile engine)
    NO emissions, 478-479
  scavenging, effect of, 484-486
  simple two-stroke engine
    emissions evaluation of, 492-494
    exhaust catalysis in, 494, 495
    performance/emissions requirements, criteria for, 471
  skip firing (four-stroking), hydrocarbon emissions from, 471
  specific time area (A_{sv})
    HC emissions *vs.* A_{svx}, 427-430
  "water-gas" reaction, 297, 346
Exhaust systems
  catalysis in (simple engine), 494, 495
  tuned systems (general)
    dimensions of, typical, 372
    effect on CE (racing motorcycle engine), 399-401
    pressure oscillations in, 399-400
    resonance in (high-performance engines), 400
  tuned systems (high-performance, empirical design of)
    introduction, 437
    cylinder and exhaust-pipe pressure *vs.* engine speed, 438-440
    data selection for, 441, 445-446
    empirical design process, criteria for, 441
    exhaust temperature, significance of, 442
    Grand Prix pipe design (EXPANSION CHAMBER program), 444-445
    horn coefficient, 444
    multi-stage diffuser, use of, 441
    pressure wave phasing, description of, 437-438
    pressure wave phasing, optimum (graphical representation), 437
    racing tailpipe diameters, calculation of, 441-442
    reflection time (of plugging pulse), 442-443, 579
    typical design calculations, 443-444
    variable tuned length (L_T) *vs.* engine speed, 440
    water injection, tuning via, 440
  tuned systems (high-performance, multi-cylinder)
    four-cylinder automotive engine, 374, 375
    three-cylinder automotive engine, 373, 374
    three-cylinder outboard engine, 373, 374
    V-8 outboard motor (OMC), 373, 375, 568
  tuned systems (high-performance, single-cylinder)
    dimensions and geometry of, 372
    discussion, 371-372
    QUB 500 68bhp motorcycle engine, 373

Exhaust systems *(continued)*
  untuned systems (compact industrial)
    discussion, 371
    dimensions of (chainsaw), 372
  untuned systems (empirical design of), 435
    introduction, 435
    downpipe diameter *vs.* exhaust port area, 435, 436
    downpipe length, determination of, 436-437
    silencer box, placement of, 435
    silencer box volume *vs.* swept volume, 436-437
  see also Intake systems; Noise reduction/noise emission; Silencers/silencing
Exhaust timing control valves. *See* Control valves

FID (flame ionization detector) analysis
  of exhaust emissions, 40-41
First Law of Thermodynamics
  First Law, defined, 93
  and expansion wave inflow (at a bellmouth), 93-94
  in GPB engine simulation model
    heat transfer along duct (during time step), 156-158, 160
    heat transfer in cylinders and plenums (during time step), 163-165
    mass/energy transport after time stop, 156-158, 160
    open cycle flow through a cylinder, 163-164
  and moving shock waves (in unsteady gas flow), 202
  in pressure wave reflections
    at branches in a pipe, 120-121
    at contractions in pipes, 106, 107
    expansion wave reflection at plain end, 95
    at expansions in pipe area, 102, 104
    at inflow from a cylinder, 138, 139
    at outflow from cylinder, 130-131, 132
    at restrictions between differing pipe area, 110-112
  see also Heat transfer
Flame velocity
  flame front propagation velocity, 283-284
  one-dimensional model (discussion), 322-323
  and squish action (loop-scavenged, QUB), 331, 333

Flying Squirrel machines, 1
Four-stroking
  and active radical (AR) combustion, 285
  from inadequate scavenging, 218, 285
  and liquid gasoline fuel injection, 513
  skip-firing, hydrocarbon emissions from, 471
  and stratified-charge combustion, 288
Friction/friction losses
  bearings
    compression-ignition engines with plain bearings, 379
    SI engines with plain bearings, 379
    SI engines with rolling-element bearings, 378-379
  chainsaw engine simulation
    introduction and discussion, 380
    friction mep (defined), 378-379
    friction mep *vs.* rpm, 383-384
    mechanical efficiency *vs.* rpm, 384
    pumping mep *vs.* rpm, 383-384
  compression-ignition engines
    with plain bearings, 379
  friction and pumping losses (defined)
    friction mep, 38, 378-379
    pumping loss, 37-38
  in GPB engine simulation model, 151
  in pressure wave propagation
    frictional heat release, 82-83
    see also Pressure waves in pipes (friction loss during)
Fuel consumption/fuel economy
  introduction and discussion, 463-464
  brake specific fuel consumption
    defined, 37
    AFR, effect of (chainsaw engine), 479-481
    AFR, effect of (low emissions engine), 487-488
    AFR, effect of (QUB 400 engine), 472, 474, 476
    Batoni map for (Vespa motor scooter engine), 476-477
    effect of A_{svx} on, 427-430
    indicated specific, defined, 35
    Institut Français du Pétrole stratified charging engine, 505, 508
    in liquid gasoline injected engines, 513-514
    in Piaggio stratified charging engine, 501-502

Fuel consumption/fuel economy *(continued)*
  brake specific fuel consumption *(continued)*
    QUB 270 cross-scavenged airhead-stratified engine, 511
    QUB stratified charging engine, 499
    radiused porting, effect of (chainsaw engine), 579
    scavenging type, influence of, 484-486
    in stratified charging/combustion, 468
    various fueling, effect of (QUB 500rv engine, full load), 524
    various fueling, effect of (QUB 500rv engine, light load), 527
    Yamaha DT 250 engine, 227-228
  combustion processes
    air-fuel ratio, importance of, 465
    dissociation, effects of, 297, 301-302
    homogeneous charging/combustion (four-stroke SI engine), 466-467
    Honda CVCC engine, 466-467
    stoichiometric combustion equation, 297-298, 465
    stratified charging/combustion (diesel engine), 466
    stratified charging/combustion (SI engine), 467-469
    stratified charging/homogeneous combustion (SI engine), 467-468
    stratified *vs.* homogeneous processes (general discussion), 466-467, 469
    see also Combustion processes; Stratified charging (homogeneous combustion)
  simple two-stroke engine
    air-fuel ratio *vs.* performance (QUB 400 engine), 472-475
    bore-stroke ratio (oversquare engines), 471
    fuel consumption/exhaust emissions problems with (diagram), 470
    homogeneous charging, problems caused by, 469
    measured performance data (QUB 400 engine), 472-476
    performance/emissions requirements, criteria for, 471
    trapping efficiency, importance of, 469-470
    see also Exhaust emissions/exhaust gas analysis
Fuel injection. *See* Homogeneous charging *(various)*

Fuels
  alternative, turbulence in combustion of, 339
  properties of. *See* Combustion processes; Exhaust emissions/exhaust gas analysis
Fukuda, M.
  attenuation equations, 558

Gas flow
  related terms:
    Coefficient of discharge (unsteady gas flow)
    Gas flow (general)
    Gases, properties of
    GPB engine simulation model
    Particle velocity, determination of
    Pressure wave propagation, friction loss during
    Pressure wave propagation, heat transfer during
    Pressure wave reflection in pipes
    Pressure wave superposition in pipes
    Propagation/particle velocity (acoustic waves)
    Propagation/particle velocity (finite amplitude waves in free air)
    Propagation/particle velocity (finite amplitude waves in pipes)
    QUB SP single-pulse experimental apparatus
    Shock waves, moving (in unsteady gas flow)
    Simulation of engines (using unsteady gas flow)
Gas flow (general)
  introduction
    discussion, 49, 52
    exhaust pulse propagation (schlieren photographs), 50-51
    nomenclature for (compression, expansion), 52-54
    toroidal vortex (smoke ring) in exhaust pulse, 49, 51
Gases, properties of
  basic properties
    air, properties of (at 293K), 66-67
    air, properties of (at 500K and 1000K), 67
    common engine gases, properties of (table), 66
    exhaust gases, properties of (discussion), 67-68

gas constant/universal gas constant, 64
mass ratios (oxygen, nitrogen), 65
molal enthalpy, internal energy, 65
molal specific heat (at constant pressure, volume), 65
molal specific heats (oxygen, nitrogen), 66
molecular weight, average, 65
specific heat (at constant pressure, temperature), 64
specific heats ratio, 64

Gaussian Elimination method
for wave reflection at contractions in pipes, 107
for wave reflection at outflow from a cylinder, 132

General Motors
diesel road engines, 3

Geometric compression ratio, defined, 8

GPB engine simulation model
introduction
unsteady gas flow computation, discussion of, 142-144
basis of GPB model, 144
computational model, criteria for, 143-144
duct meshing for pressure wave propagation, 144-145
four-stroke engines, applicability to, 143
GPB model computational requirements, 145
physical parameters (e.g., pressure, density, etc.), 145
pressure wave propagation within mesh J *(illus.)*, 146
Reimann variables, use of, 142
superposition pressure amplitude ratio, 145
air flow into an engine
B parameter, assessment of, 167
delivery ratio, 168
scavenge ratio, 168
total mass air flow, 167
computation time, comments on, 191-192
concluding remarks about, 192
correlation with QUB SP experimental apparatus. *See under* QUB (Queen's University of Belfast)
friction and heat, changes due to, 151
inter-mesh boundaries (reflections after time stop)
introduction, 151-152
adjacent meshes in differing discontinuities (diagram), 152
parallel ducting, 152-153
tapered pipes, 153-154
interpolation procedure (wave transmission through a mesh)
gas dynamic parameters, values of, 149-150
groupings of variables ("known" terms), 148
pressure amplitude ratios, determination of, 148-149
propagation velocities, 147
singularities during interpolation procedure, 150
values of lengths x_p, x_q, 147
mass/energy transport along duct (after time stop)
introduction, 156
First Law of Thermodynamics, application of, 156-158, 160
gas dynamic parameters, values of (four cases), 158-159
"hand" of the flow, 157
mesh J, energy flow diagram for, 157
new reference conditions, determination of, 161-162
purity in mesh space J, 161
system state, change of, 160
transport at mesh J (four cases), 156-157
thermodynamics of cylinders and plenums (during time step)
introduction, 162
boundary conditions, application of, 163-164
First Law of Thermodynamics, application of, 163-165
heat transfer coefficient, determination of (discussion), 164
heat transfer from/to plenum, expression for, 164
new gas properties, purity, 166
new reference conditions (for next time step), 166
open cycle flow through cylinder (thermodynamic diagram), 163
sign conventions, importance of, 163
system state, change of, 164
system temperature, solution for, 165
time interval, selection of, 146

GPB engine simulation model *(continued)*
   wave reflections at end of pipe (after time stop)
      cylinder, atmosphere or plenum, 155-156
      discussion, 154-155
      "hand," determination of, 154-155
      restricted pipe, 155
   wave transmission (during time increment dt), 147

Harland & Wolff
   cathedral engine (diesel), 3, 5
Heat losses
   See Heat transfer
Heat release
   compression-ignition engines
      combustion chamber geometry, 316-317
      direct-ignition (DI) engine, 314-316
      fast fuel *vs.* fast air approach, 314
      heat loss by fuel vaporization, 308-309
      indirect-injection (IDI) engine, 316-318
      Woschni heat transfer equation, 305
   frictional (in pressure wave propagation), 82-83
   SI engines
      combustion chamber geometry, 289, 290
      heat loss by fuel vaporization, 308
      heat release/heat loss calculations, 291-293
      incremental heat loss (where Q_R is zero), 292
      incremental heat release (First-Law expression for), 291, 293
      incremental heat release (Rassweiler and Winthrow expression for), 293
      from loop-scavenged QUB LS400 SI engine, 294, 295
      Nusselt number, 305-306, 376
      polytropic exponents, determination of, 292, 293
      prediction from cylinder pressure diagram, 289-294
      Reynolds number, 306
      in stratified charging/combustion, 468
      thermodynamic equilibrium analysis, 291-293
      total heat released, 31, 309-310
   see also Combustion processes; Heat transfer
Heat transfer
   along duct during time step (GPB model)
      introduction, 156

      First Law of Thermodynamics, application of, 156-158, 160
      gas dynamic parameters, values of (for four cases), 158-159
      "hand" of the flow, 160
      mass/energy transport diagram (mesh J, four cases), 156-157
      mesh J, energy flow diagram for, 156-157
      new reference conditions, determination of, 161-162
      purity in mesh space J, 161
      system state, change of, 160
   in cylinders and plenums during time step (GPB model)
      introduction, 162
      boundary conditions, application of, 163-164
      First Law of Thermodynamics, application of, 163-165
      heat transfer coefficient, determination of (discussion), 164
      heat transfer from/to plenum, expression for, 164
      new gas properties, purity, 166
      new reference conditions (for next time step), 166
      open cycle flow through cylinder (thermodynamic diagram), 163
      sign conventions, importance of, 163
      system state, change of, 164
      system temperature, solution for, 165
   heat losses in GPB engine simulation model, 151
   heat transfer analysis (Annand model)
      closed cycle model (general), 305-307
      crankcase analysis, 375-378
      heat transfer coefficients, 307-308
      racing motorcycle engine, 397-399
   Nusselt number in
      defined, 84-85
      in convection heat transfer analysis, 84-85
      in crankcase heat transfer analysis, 376
   in open cycle flow through a cylinder
      First Law of Thermodynamics, application of, 163-164
   during pressure wave propagation, 84-85
   Reynolds number in
      in convective heat transfer analysis, 84-85

in crankcase heat transfer analysis, 376
Woschni heat transfer equation
(compression-ignition engines), 305
between zones (closed cycle, two-zone chainsaw engine), 348-349
see also Combustion processes; First Law of Thermodynamics; Heat release
Heimberg, W.
Ficht pressure surge injection system, 516, 535
HEMI-SPHERE program
typical combustion chamber design, 335, 337
Hill, B.W.
QUB stratified charging engine, 498
Hinds, E.T.
reed valves as pressure loaded cantilevered beam, 368
Homelite chainsaw, 3
Homogeneous charge combustion
See Combustion processes; Homogeneous charging (with stratified combustion); Stratified charging (with homogeneous combustion)
Homogeneous charging (with stratified combustion)
air-blast fuel injection (gasoline)
advantages/disadvantages of, 518
description, 516-517
injector placement, problems with, 518, 520
SEFIS motorscooter engine (Orbital Engine Corp.), 521
sequence of events *(illus.)*, 517
spray patterns, 517-519
three-cylinder car engine (Orbital Engine Corp.), 520
gasoline injection, technical papers on, 512-513
liquid gasoline injection
bsfc, bsHC emissions (fuel injected *vs.* carburetted), 513-514
dribble problem with, 518
ram-tuned liquid injection
description, 514-515
functional schematic diagram, 515
Heimberg alternative method, 516
QUB 500rv engine at light load, 516, 526-529
spray pattern from, 516
Honda CVCC engine, 466-467
Hopkinson, B.
perfect displacement scavenging, 213-214
perfect mixing scavenging, 214-215
seminal paper on scavenging flow, 211, 276
Hydrocarbon concentration
in exhaust emissions (introduction), 303
see also Exhaust emissions/exhaust gas analysis

IFP (Institut Français du Pétrole)
stratified charging engine
brake specific fuel consumption, 505, 508
description and functional diagram, 504-506
hydrocarbon emissions, 506, 508, 509
NO emissions, 506, 509
oxygen catalysis, effect of, 506, 509
photograph of, 507
Ignition (spark, compression). *See under* Combustion processes
Indicated mean effective pressure
discussion of, 34
see also Mean effective pressure
Indicated power output
defined, 34-35
see also Power output
Indicated specific fuel consumption
defined, 35
see also Fuel consumption/fuel economy
Indicated torque
defined, 35
see also Torque
Induction systems
disc valve
port area diagrams, 457
see also Disc valves
reed valve
empirical design of (discussion), 447-450
in racing motorcycle engine simulation, 401-402
see also Reed valves
see also Specific time area (A_{sv})
Intake systems
computer modeling of (chainsaw engine simulation)
intake disc valves, 365-367
intake ducting (dimensions and discussion), 370-371
intake ducting throttle area ratio, 371
intake reed valves, 367-370

Intake systems *(continued)*
  computer modeling of (chainsaw engine simulation) *(continued)*
    intake system temperature/pressure *vs.* crankshaft angle, 390-392
  intake disc valves
    specific time area (A_{sv}) analysis, 365-367
  intake manifolds
    bellmouth inflow, expansion wave reflection at, 93-95
    plain end inflow. expansion wave reflection at, 91, 95-97
    reflection possibilities in, 89
  see also Exhaust systems; Silencers/silencing
Ishibashi, Y.
  active radical (AR) combustion and engine optimization, 490-491
Ishihara, S.
  double-piston stratified charging engine, 504, 505

Jante method (scavenge flow assessment)
  advantages/disadvantages of, 223
  description of apparatus, 219-220, 221
  QUB experience with, 222-223
  typical velocity contours, 221-222
Jones, A.
  simulation of engines (using unsteady gas flow), 143
Junkers Jumo (aircraft engine), 3

Kinetic energy
  squished
    combustion chamber, influence of (discussion), 331
    compression ratio, effect of (diesel engines), 334, 335
    vs. squish clearance (various combustion chambers), 332
  turbulence kinetic energy (incremental, total), 330
Kirkpatrick, S.J.
  development of QUB SP experimental apparatus, 172
  simulation of engines (using unsteady gas flow), 142

Knocking
  description of, 286
  see also Detonation

Laimbock, F.
  catalysis in two-stroke engines, 494, 495
Lanchester, F.W.
  airhead stratified charging, 510
Lax, P.D.
  Lax-Wendroff computation time, 191-192
  simulation of engines (using unsteady gas flow), 142
Lean mixture combustion, 300-301
Losses, friction and pumping
  See Friction/friction losses
Lubrication
  petroil lubrication, 12-13, 14, 470
  pressure-lubricated engines, exhaust emissions of, 13, 470

Mackay, D.O.
  development of QUB SP experimental apparatus, 172
Marine engines
  diesel
    Burmeister & Wain, 3-4
    cathedral engine (Harland & Wolff), 3, 5
    Sulzer (Winterthur), 4
  tuned exhaust systems (three-cylinder outboard engine), 373, 374
  V-8 outboard motor (OMC), 373, 375, 568
McGinnity, F.A.
  pressure wave reflections in branched pipes, 117-119
McMullan, R.K.
  development of QUB SP experimental apparatus, 172
Mean effective pressure
  concept of (discussion), 34
  brake mep, defined, 37
  friction mep, determination of (in engine testing), 38
  pumping mep, determination of (in engine testing), 38
  see also Performance measurement

Mercury Marine
  air-assisted fuel injection system spray pattern, 519
  liquid fuel injection system spray pattern, 519
Mufflers. *See* Noise Reduction/noise emission; Silencers/silencing
MZ racing motorcycles
  disc valves in, 16-17, 456

Nakayama, M.
  liquid gasoline injection studies, 513
Napier Nomad (aircraft engine), 3
NDIR (non-dispersive infrared) analysis
  of exhaust emissions, 40-41, 493
Newton-Raphson method
  expansion wave reflected pressure, 96-97
  wave reflection at contractions, 107
  wave reflection at outflow from a cylinder, 132
Nitrogen oxide
  in exhaust emissions (introduction), 302-303
  in two-zone closed cycle model (chainsaw engine), 349-354
  see also Exhaust emissions/exhaust gas analysis
Noise reduction/noise emission
  introduction, 541
  noise sources, engine
    introduction, 546-547
    bearings, plain *vs.* rolling element, 547
    reed valve "honking," 547
    two-stroke engines, advantages/disadvantages of, 547-548
    water cooling, advantage of, 547
  sound, introduction to
    Bel, defined, 543
    human hearing, frequency response of, 542, 543
    human hearing, logarithmic response of, 543
    loudness (multiple sound sources), 544-545
    masking (of soft by loud sounds), 545
    noise, subjective nature of, 541-542
    noise measurement (A-, B-weighted scales), 545
    noisemeters, switchable filters in, 545
    sound intensity, perceived, 543
    sound pressure level, calculation of, 543-544
    sound propagation, speed of, 542
    test procedures, sensitivity of, 545-546

  see also Exhaust systems; Silencers/silencing
Nusselt number
  in crankcase heat transfer analysis, 376
  in heat release analysis (SI engines), 305-306
  in pressure wave propagation heat transfer, 84-85
Nuti, M.
  liquid gasoline injection HC emissions studies, 513

Onishi, S.
  active radical (AR) combustion and engine optimization, 285, 490-491
Operation, basic engine. *See* Two-stroke engine (general)
Orbital Engine Corp. (Perth)
  air-assisted fuel injection engines
    SEFIS motorscooter engine, 5214
    three-cylinder car engine, 520
Otto cycle. *See* Thermodynamic (Otto) cycle
Outboard Marine Corporation (OMC)
  liquid fuel injection outboard motor, 520, 521
  V-8 outboard motor (OMC), 373, 375
Oxygen concentration
  in exhaust emissions, 38
  see also Exhaust emissions/exhaust gas analysis; Purity
Oyster maps
  of bsfc, bsHC emissions (fuel injected *vs.* carburetted), 513-514

Particle velocity, determination of
  in finite amplitude waves (in pipes)
    compression wave particle velocity (c_e), 59
    expansion wave particle velocity (c_i), 61
    particle velocity (c) (value in air), 58
    shock wave particle velocity (c_{sh}), 62, 64
    wave reflection at cylinder outflow (Gaussian Elimination method), 132-133
  in finite amplitude waves (unconfined)
    Bannister, F.K., 55
    Earnshaw equation for, 55
  in pressure wave reflection (in pipes)
    at contractions in pipe area, 107-108
    at expansions in pipe area, 104-105
    inflow from a cylinder, 139-140

Particle velocity, determination of *(continued)*
  in pressure wave reflection (in pipes) *(continued)*
    outflow from a cylinder, 132-133
    at restrictions between differing pipe areas, 112-113
  in pressure wave superposition (in pipes)
    oppositely moving waves, 71
    supersonic particle velocity (oppositely moving waves), 74-77
  in unsteady gas flow
    Bannister derivation of, 197-200
  see also Gas flow *(related terms)*
Performance measurement
  brake mean effective pressure
    AFR, effect of (low-emissions engine), 487
    AFR, effect of (QUB 400 engine), 472, 474, 476
    defined, 37
    exhaust silencing, effect of (chainsaw engine), 574, 576
    influence of engine type on, 45
    intake silencing, effect of (chainsaw engine), 571
    and power output, 37
    QUB 270 cross-scavenged airhead-stratified engine, 510-511
    QUB stratified charging engine, 499, 500
    radiused porting, effect of (chainsaw engine), 579
    scavenging type, influence of, 484-486
    of typical two-stroke engines, 43
    various fueling, effect of (QUB 500rv engine, full load), 523
    various fueling, effect of (QUB 500rv engine, light load), 529
    vs. rpm (chainsaw engine simulation), 380, 381
    see also mean effective pressure *(below)*
  brake power output
    See power output *(below)*
  brake specific fuel consumption
    See under Fuel consumption/fuel economy
  brake thermal efficiency (defined), 37
  brake torque (defined), 36
  dynamometer testing
    introduction and discussion, 35-36
    dynamometer test stand (functional flow diagram), 36, 37

  performance parameters, 36-38
  principles of, 36
  friction mean effective pressure
    determination of (in engine testing), 38
  indicated mean effective pressure
    defined, 38
  mean effective pressure (mep)
    concept of (discussion), 34
  piston speed
    and brake power output, 44-45
  power output
    and brake mean effective pressure, 37
    brake power output (defined), 36
    brake power output *vs.* piston speed/swept volume, 44-45
    and brake specific fuel consumption, 37
    of chainsaws (typical), 45
    engine type, influence of, 45-46
    four-stroke engine (GPB model), 169
    indicated (defined), 34-35
    silencer design for high power output (motorcycle engines), 581
    two-stroke engine (GPB model), 169
    two-stroke engines, potential power output of, 43
    vs. rpm (chainsaw engine simulation), 380, 381
  pumping mean effective pressure
    determination of (in engine testing), 38
  swept volume
    and brake power output, 44-45
Petroil lubrication, 12-13, 14, 470
  see also Lubrication
Piaggio stratified charging engine
  introduction and functional description, 500-501
  benefits and advantages of, 504
  CO emissions, 502-503
  fuel consumption levels, 501-503
  hydrocarbon emissions, 503
Piston speed
  and brake power output, 44-45
  and speed of rotation, 44-45
Pistons
  cross-scavenged deflector (typical), 10, 11
  deflector designs. *See* Port design, scavenging
  for loop-scavenged engine, 10
  piston motion, importance of (in scavenging flow), 223-224

Index

PISTON POSITION (computer program), 23, 24
piston speed, influence on speed of rotation, 44-45
piston speed/swept volume and brake power output, 44-45
for QUB cross scavenging, 10, 11, 12
Plohberger, D.
  liquid gasoline injection load studies, 513
Pollution
  general discussion of, 4, 465-466
  see also Combustion processes; Exhaust emissions/exhaust gas analysis
Poppet valves
  effective seat area, 413
  engine simulation using, 363
  geometry of, 412
  maximum lift, calculation of, 414
  timing delays (QUB 500rv research engine), 523
  valve curtain area, 413-414
Port design
  cross scavenging, conventional
    basic layout, 254, 255
    deflection ratio, 256-257
    deflector height, importance of, 256
    deflector height ratio, 256
    deflector radius, determination of, 254, 256
    design advantages, 253-254
    ease of optimization, 253
    maximum scavenge flow area, 254
    multiple ports, 254
  cross scavenging, QUB
    introduction, 261
    basic layout, 12
    deflection ratio, importance of, 262
    experience, importance of, 263
    QUB CROSS PORTS (computer program), 260, 261-263
    squish area ratio, importance of, 262
    transfer port geometry, selection of, 262-263
  cross scavenging, unconventional (GPB)
    basic layout, 255
    deflection ratio, exhaust side, 258
    deflector area, exhaust side, 258
    deflector design characteristics, 257-258
    design procedure, 258
    evaluation of, 258-259
    GPB CROSS PORTS (computer program), 259-260, 261
  loop scavenging, external (blown engines)
    BLOWN PORTS (computer program), 270-273
    designs for (discussion), 269-270
    four-cylinder DI diesel, design for, 272-273
    typical layout, 270, 271
  loop scavenging, internal
    bore-stroke ratio, effect of, 267
    CFD as a future design tool, 264, 265
    cylinder size, effect of, 267
    design difficulties of, 263
    deviation angles (laser doppler velocimetry), 264, 265
    LOOP SCAVENGE DESIGN (computer program), 268-269
    main port orientation, importance of, 263-264
    main transfer port, design analysis of, 265-266
    rear and radial side ports (layout and design features), 266, 267
    side ports (layout and design features), 266, 267
    transfer port inner walls, 264, 266-267
    transfer port layout (after J.G. Smyth), 264
  loop scavenging (general)
    port plan layouts (typical), 10
  port area diagrams
    disc valves (induction), 457
  scavenging design and development (discussion)
    CAD/CAM techniques, 274, 275
    CFD in, 274, 275-276
    experimental experience, summary of, 273-274
    rapid prototyping (stereo lithography), 274, 276
  uniflow scavenging
    piston pegging, 251
    piston rings, 251-252
    port plan layout, geometry of, 250-252
    port plan layout, mechanical considerations in, 251
    port width ratio, defined, 252
    port-width-to-bore ratio, 252-253
    scavenge belt, use of splitters in, 252
    suitability to long-stroke engines, 253

see also specific valves; Computer modeling; Port timing; Scavenging
Port timing
  disc-valved engine, 18, 19, 20
  piston ported engine, 18-19
  reed valved engine, 18, 20
  timing diagrams, typical, 18
  timing events, symmetry/asymmetry of, 15-16
  see also specific valves; Port design; Scavenging
Power output
  brake power output
    defined, 36
    brake mean effective pressure, 36-38
    and piston speed, 44-45
    vs. piston speed/swept volume, 44-45
  of chainsaws
    typical values, 45
    vs. rpm (chainsaw engine simulation), 380, 381
  and engine type
    engine type, influence of, 45-46
    four-stroke engine (GPB model), 169
    racing motor, 45
    truck diesel, 45
    two-stroke engine (GPB model), 169
  indicated (defined), 34-35
  silencer design for high power output (motorcycle engines), 581
  see also Performance measurement
Preignition
  in cross scavenging, 10
  see also Detonation
Pressure oscillations
  in tuned exhaust systems (racing motorcycle engine), 399-400
Pressure wave propagation
  related terms:
    Coefficient of discharge (unsteady gas flow)
    Gas flow (general)
    Gases, properties of
    GPB engine simulation model
    Particle velocity, determination of
    Pressure wave propagation, friction loss during
    Pressure wave propagation, heat transfer during
    Pressure wave reflection in pipes
    Pressure wave superposition in pipes
    Propagation/particle velocity (acoustic waves)
    Propagation/particle velocity (finite amplitude waves in free air)
    Propagation/particle velocity (finite amplitude waves in pipes)
    QUB SP single-pulse experimental apparatus
    Shock waves, moving (in unsteady gas flow)
Pressure wave propagation, friction loss during
  friction factor (bends in pipes)
    introduction, 83
    pressure loss coefficient (C_b), 83-84
  friction factor (straight pipes)
    introduction, 81
    Reynolds number (**Re**), 82
    thermal conductivity (C_k), 81
    viscosity (μ), 82
    work and heat generated, 83
    discussion of results, 83
  in straight pipes
    introduction, 77
    compression *vs.* expansion waves, friction effect on, 81
    discussion of results, 81
    energy flow diagram, 77-78
    particle movement (dx), 81
    pressure amplitude ratios, 80
    shear stress (τ) at wall, 78
    single wave *vs.* train of waves (discussion), 81
    superposition pressures, 79
    superposition time interval, 80
Pressure wave propagation, heat transfer during
  introduction, 84
  heat transfer coefficient, convective (C_h), 85
  Nusselt number, 84-85
  pressure loss coefficient (C_b), 84
  total heat transfer, 85
Pressure wave reflection (in pipes)
  at branches in a pipe
    introduction, 114
    accuracy of theories (numerical examples), 122-124
    Benson superposition pressure postulate, 114-115
    complete solution (general discussion), 117-118

Pressure wave reflection (in pipes) *(continued)*
  at branches in a pipe *(continued)*
    complete solutions: one and two supplier pipes, 118-122
    First Law of Thermodynamics, application of, 120-121
    McGinnity non-isentropic solution for, 117-119
    net mass flow rate (at the junction), 115
    pressure amplitude ratios, general solution for, 116
    pressure loss equations (one/two supplier pipes), 121-122
    pressure/pressure amplitude ratios (one/two supplier pipes), 116-117
    stagnation enthalpies (one/two supplier pipes), 121
    temperature-entropy curves (one/two supplier pipes), 120
    unsteady flow at three-way branch (diagram), 115
  at contractions in pipe area
    introduction, 105
    Benson "constant pressure" criterion, 106
    First law of Thermodynamics, application of, 106, 107
    gas properties (functions of), 105
    isentropic flow in, 105
    mass flow continuity equation, 106, 107
    Newton-Raphson and Gaussian Elimination methods, 107
    numerical examples, 113-114
    reference state conditions, 106
    sonic particle velocity, solution for, 107-108
  at duct boundaries
    introduction, 88-90
    compression wave at closed end, 91-92
    compression wave at open end, 92-93
    engine manifolds, reflection possibilities in (diagram), 89
    expansion wave at plain open end, 95-97
    expansion wave inflow at bellmouth open end, 93-95
    Newton-Raphson method (for expansion wave reflected pressure), 96-97
    notation for reflection/transmission, 90-91
    wave reflection criteria (typical pipes), 91
  at expansions in pipe area
    Benson "constant pressure" criterion, 103, 104
    continuity equation (for mass flow), 102, 103
    First Law of Thermodynamics, application of, 102, 104
    flow momentum equation, 103, 104
    numerical examples, 113-114
    particle flow diagram (simple expansion/contraction), 102
    sonic particle velocity, solution for, 104-105
    temperature-entropy curves (isentropic, non-isentropic), 101
    turbulent vortices and particle flow separation, 101
  at gas discontinuities
    introduction, 85-86
    complex case: variable gas composition, 87-88
    conservation of mass and momentum, 86-87
    energy flow diagram, 86
    simple case: common gas composition, 87
  inflow from a cylinder
    introduction and discussion, 135-136
    First Law of Thermodynamics, application of, 138, 139
    flow diagram, 136
    gas properties (functions of), 136
    mass flow continuity equation, 138, 139
    numerical examples, 140-142
    pressure amplitude ratios, 138
    reference state conditions, 137
    sonic particle velocity, solution for, 139-140
    temperature-entropy diagrams, 136, 137
  outflow from a cylinder
    introduction and discussion, 127-129
    First Law of Thermodynamics, application of, 130-131, 132
    flow diagram, 128
    flow momentum equation, 131, 132
    gas properties (functions of), 130
    mass flow continuity equation, 130
    numerical examples, 133-135
    pressure amplitude ratios, 131
    reference state conditions, 130
    sonic particle velocity, solution for, 132-133
    stratified scavenging, significance of, 129, 163

Pressure wave reflection (in pipes) *(continued)*
  outflow from a cylinder *(continued)*
    temperature-entropy diagrams, 128-129
  at restrictions between differing pipe areas
    introduction, 108
    First Law of Thermodynamics, application of, 110-112
    flow momentum equation, 111, 112
    gas properties (functions of), 109
    mass flow continuity equation, 110, 111
    numerical examples, 113-114
    particle flow regimes (diagram of), 109
    reference state conditions, 109
    sonic particle velocity, solution for, 112-113
    temperature-entropy curves for, 108-109
  at sudden area changes
    introduction, 97
    Benson "constant pressure" criterion, 98-99
    energy flow diagram, 98
    examples: enlargements and contractions, 100-101
    nomenclature, consistency of, 98
  in tapered pipes
    introduction, 124-126
    dimensions and flow diagram, 125
    gas particle Mach number, importance of, 127
    separation of flow (from walls), 126-127
Pressure wave superposition (oppositely moving, in pipes)
  introduction, 69
  mass flow rate
    directional conventions for, 73-74
    numerical values of, 74
  supersonic particle velocity
    Mach number (defined), 74-75
    numerical values for, 77
    Rankine-Hugoniot equations (combined shock/reflection), 76-77
    superposition Mach number (determination of), 75-76
    weak shock concept (for modeling unsteady gas flow), 75, 77
  wave propagation
    acoustic, propagation velocities (numerical values for), 73
    acoustic velocity, sign conventions for, 73
    propagation velocities, sign conventions for, 73
  "wave interference during superposition" effect, 73
  wave superposition
    acoustic velocities, local, 69
    particle velocity, absolute, 70
    particle/propagation velocities, sign conventions for, 69, 71
    particle/propagation velocities (individual wavetop), 69
    pressure-time data, experimental (interpretation of), 71-72
    simplified pressure diagram, 70
    superposition particle velocity (analytical), 70-71
    superposition particle velocity (experimental), 71-72
    superposition pressure ratio, 71
Propagation/particle velocity (acoustic waves)
  pressure ratio, 54
  specific heats ratio (for air), 54
  velocity in air (after Earnshaw), 54
Propagation/particle velocity (finite amplitude waves in free air)
  particle velocity
    absolute pressure (p), 55
    gas constant (for air), 55
    gas particle velocity (for air), 55, 57
    pressure amplitude ratio, 55
    pressure ratio, 55
    specific heat (constant pressure/volume), 55
    specific heat ratio, functions of (for air), 56
    specific heats ratio, 55
  propagation velocity
    absolute propagation velocity, 57-58
    acoustic velocity, 57
    density, 57
    isentropic change of state, 57
Propagation/particle velocity (finite amplitude waves in pipes)
  basic parameters (values in air)
    particle velocity, 58
    pressure amplitude ratio, 58
    propagation velocity, 58
    reference acoustic velocity, 58, 59
    reference density, 59
  the compression wave
    absolute pressure, 59
    density, 60

local acoustic velocity, 60
local Mach number, 60
mass flow rate, 60
particle velocity, 59
pressure amplitude ratio, 59
propagation velocity, 59
the expansion wave
  absolute pressure, 60
  density, 60
  local acoustic velocity, 60
  local Mach number, 61
  mass flow rate, 61-62
  particle velocity, 61
  pressure amplitude ratio, 61
  propagation velocity, 61
shock formation (wave profile distortion)
  introduction, 62
  compression waves (discussion), 62, 64
  expansion waves (discussion), 63-64
  flow diagram of, 63
  particle velocity (shock wave), 62, 64
  propagation velocity (shock wave), 62, 63
  steep-fronting, 62
Pumping mean effective pressure
  determination of (in engine testing), 37-38
  see also Friction/friction losses
Purity
  charge purity
    CFD plots (Yamaha DT250 cylinders), 246-248
    vs. crankshaft angle (chainsaw engine simulation), 386
    in closed cycle combustion model (single-zone), 321
  exhaust port purity
    correlated theoretical model (Yamaha DT250 cylinders), 239-240
    theoretical calculation of, 238-239
    theoretical curves (eight test cylinders), 239-240
  idealized incoming scavenge flow, defined, 212
  importance of (in Benson-Brandham model), 217-218
  in mesh space J (during time step in GPB model), 161
  scavenging purity, defined, 28

QUB (Queen's University of Belfast)
  air-assisted fuel injection system spray pattern, 519
  Jante method, experience with, 222-223
  laminar flow type silencer design, 565
  QUB cross scavenging
    deflection ratio, importance of, 262
    and homogeneous charge combustion, 338
    piston design for (typical), 10, 11, 12
    QUB CROSS ENGINE DRAW ((computer program)), 25-26
    QUB CROSS PORTS ((computer program)), 260, 261-263
    scavenging efficiency of, 11
    squish action in, 262, 325, 338
    SR *vs.* SE (QUBCR cylinder), 230
    SR *vs.* TE (QUBCR cylinder), 231
  QUB single-cycle scavenging test apparatus
    functional description, 224-227
    see also Scavenging
  QUB SP single-pulse experimental apparatus
    introduction, 170
    coefficient of discharge for, 172
    convergent exhaust taper, 183-185
    design criteria of, 171
    divergent exhaust taper (long megaphone), 185-187
    divergent exhaust taper (short), 181-183
    effect of friction on outflow (straight pipe), 176
    exhaust pipe with discontinuity, 188-191
    functional description, 171-172
    reference gas properties (CO_2 and air, discontinuous exhaust), 188-189
    straight pipe (inflow process), 175-177
    straight pipe (outflow process), 173-175
    sudden exhaust expansion, 177-179
  QUB stratified charging engine
    air-fuel paths in (diagram), 498
    bmep (at optimized fuel consumption levels), 499, 500
    fuel consumption, optimized, 499
    functional description, 498-499
  QUB 250-cc racing-model engine, 1, 2
  QUB 270 cross-scavenged airhead-stratified engine
    advantages of, 512
    bmep *vs.* rpm, 510-511

QUB (Queen's University of Belfast) *(continued)*
  QUB 270 cross-scavenged airhead-stratified engine *(continued)*
    bsfc *vs.* rpm, 511
    bsHC *vs.* rpm, 511-512
  QUB 400 research engine
    AFR *vs.* oxygen emissions, 473, 475
    bmep, AFR effect on, 472, 474, 476
    bsCO, AFR effect on, 473-475
    bsHC *vs.* air-fuel ratio, 473, 476
    measured performance *vs.* AFR, 472-473
  QUB LS400 SI engine
    combustion efficiency, 294, 296
    heat release analysis, 294, 295
    mass fraction burned, 294, 295
  QUB 500rv research engine (various fueling)
    description and configuration, 522
    multi-cylinder engine, basis for, 530
    poppet valve timing, 523
    test parameters (described), 523
    stratified combustion in, 529-530
    bmep *vs.* AFR (full load), 523
    bmep *vs.* AFR (light load), 529
    bsfc *vs.* AFR (full load), 524
    bsfc *vs.* AFR (light load), 526
    bsHC *vs.* AFR (full load), 524
    bsHC *vs.* AFR (light load), 526
    bsNO$_x$ *vs.* AFR (full load), 525
    bsNO$_x$ *vs.* AFR (light load), 526, 527

Rankine-Hugoniot equations
  shock waves in unsteady gas flow, 75-77
Rassweiler, G.M.
  expression for incremental heat release (SI engines), 293
Reed valves
  introduction and discussion, 17
  charging alternatives for, 17
  design of, empirical
    carburetor flow diameter. determination of, 453
    dummy reed block angle, use of, 455
    empirical design, introduction to, 446-447
    empirical design process, criteria for, 451
    glass-fiber reed petals, durability of, 455
    as pressure-loaded cantilevered beam, 453-454
    reed block, rubber-coated (with steel reeds and stop plate), 447
    reed flow area (A_{rd}), determination of, 452
    reed petal materials, 446, 451, 455
    reed port area (A_{rp}), effective, 452
    reed tip lift behavior, 448-450, 453
    REED VALVE DESIGN computer program, 452, 454-455
    stop-plate radius, determination of, 454
    vibration and amplitude criteria, 453
    see also Specific time area (A_{sv})
  design of, general
    design dimensions (reed petal, reed block), 367
    as pressure-loaded cantilevered beam, 368-369
    in racing motorcycle engine simulation, 394, 401-402
    reed block, flow restrictions from, 369
    reed block, placement and function of, 368
    reed flow area (A_{rd}), determination of, 369
    reed port area (A_{rp}), effective, 369
    reed tip lift ratio, 370
  in Grand Prix motorcycle racing engine, 17, 19
  reed *vs.* disc valves (discussion), 446-447
  typical configurations, 16, 367
Reid, M.G.O.
  mass fraction burned experimental data (SI engine), 310-312
Reimann variables
  in simulation of engines (using unsteady gas flow), 142, 143
research funding, comment on, 192
Reynolds number
  in crankcase heat transfer analysis, 376
  and heat release analysis (SI engines), 305-306
  in heat release analysis (SI engines), 305-306
  in pressure wave propagation friction factor (straight pipes), 82
  in scavenging flow (experimental assessment of), 223
Ricardo Comet (IDI diesel engine)
  combustion chamber geometry, 317
Rich limit (for diesel combustion), 288
Rich mixture combustion, 288, 299-300
Rootes-Tilling-Stevens
  diesel road engines, 3

Roots blower
  in turbocharged/supercharged fuel-injected engine, 13-14, 15

Saab vehicles
  and Monte Carlo Rally, 2
SAE (Society of Automotive Engineers)
  Standard J604D, 27
Sako/Nakayama experimental data (178 cc snowmobile engine)
  NO emissions, 478-479
Sato, T.
  liquid gasoline injection studies, 513-514
Scavenge ratio (SR)
  defined, 27, 212
Scavenging
  definitions
    purity (idealized incoming scavenge flow), 212
    scavenge ratio, 27, 212
    scavenging efficiency (basic), 28, 212-213
    scavenging efficiency (perfect displacement), 213-214
    scavenging efficiency (perfect mixing), 214-215
    scavenging purity, 28
  fundamental theory of, 211-213
    see also isothermal scavenge model *(below)*
  Benson-Brandham model
    comparison with QUB test results, 233-236
    loop/cross/uniflow scavenging, relevance of model to, 218-219
    predictive value, inadequacy of, 233
    purity, importance of (and SE curve), 217-218
    trapping characteristics model, advantages of, 216-218
    two part (mixing/displacement) model, 215-217
    Yamaha DT250 cylinders: QUB test results *vs.* Benson-Brandham models, 234-235
    see also Computer modeling *(various)*
  blower scavenging (three-cylinder supercharged engine)
    exhaust tuning, 405-407
    open-cycle pressures and charging (cylinders 1-2), 404-405
    temperature and purity in scavenging ports, 405, 406
  blower scavenging (four-cylinder supercharged engine)
    introduction and discussion, 407
    four cylinders, advantages of, 409
    open port period, three cylinder *vs.* four cylinder, 409
    open-cycle pressures and charging (cylinders 1-2), 407-409
  CFD (Computational Fluid Dynamics) in
    introduction, 244
    Ahmadi-Befrui predictive calculations, 250
    charge purity plots (Yamaha test cylinders), 246-248
    in future design and development (discussion), 274, 275-276
    grid structure for scavenging calculations, 244-245
    PHOENICS CFD port design code, flow assumptions in, 245-246
    as a port design tool (loop scavenging), 264-265
    SE-TE-SR plots (CFD *vs.* experimental values), 248-250
  cross scavenging
    Clerk, Sir Dugald (deflector piston), 9, 10
    combustion chamber design for, 339
    design advantages of, 253-254
    detonation/preignition potential of, 10
    ease of optimization, 253
    influence on SE-SR and TE-SR characteristics, 218-219
    manufacturing advantages of, 10-11
    piston for (conventional), 10, 11
    piston for (QUB type), 10, 11, 12
    port design for. *See* Port design, scavenging
    scavenging efficiency of, 10, 11
  design and development techniques (discussion)
    CAD/CAM techniques, 274, 275
    CFD, use of, 274, 275-276
    experimental experience, summary of, 273-274
    rapid prototyping (stereo lithography), 274, 276
    see also Computer modeling *(various)*
  effect on exhaust emissions, 484-486

Scavenging *(continued)*
  exit properties, determining by mass
    introduction, 242
      exhaust gas temperature, 242-243
      exit charge purity (by mass), 243
      incorporation into engine simulation, 244
      instantaneous SE, SR (conversion from mass to volumetric value), 243
  four-stroking (from inadequate scavenging), 218, 285
  Hopkinson, B.
    perfect displacement/perfect mixing scavenging, 213-215
    seminal paper on scavenging flow, 211, 276
  isothermal (ideal) scavenge model
    charging efficiency, 213
    mixing zone/displacement zone, 212
    physical representation of, 212
    scavenging efficiency, 212-213
    trapping efficiency, 213
  Jante test method
    introduction, 219
    advantages/disadvantages of, 223
    pitot tube comb (motorcycle engine test), 221
    QUB employment of, 222-223
    test configuration, 219-220
    "tongue" velocity patterns, 221-222
    velocity contours, typical, 220-222
  loop scavenging
    influence on SE-SR and TE-SR characteristics, 218-219
    invention of, 8-9
    litigation about, 10
    manufacturing disadvantages of, 10
    piston for (typical), 10
    port plan design for. *See* Port design
    scavenging efficiency of, 10
    SR *vs.* SE (QUB loop-scavenged test, Yamaha DT250 cylinders), 229-230, 232-233
  loopsaw scavenging (in chainsaw engine simulation), 380, 388
  perfect displacement, 213-214
  perfect mixing, 214-215
  perfect displacement/perfect mixing (combined), 215-216
  port designs (all types)
    See Port design

  positive-displacement scavenging
    engine configuration, typical, 15
    functional description, 13-14
  scavenge ratio (SR)
    defined, 27, 212
    chainsaw engine simulation, 386-389
    in GPB engine simulation model, 168
    instantaneous SR (conversion from mass to volumetric value), 243
    in isothermal scavenging (tested), 212
    in perfect mixing scavenging, 214-215
    in perfect displacement scavenging, 213-214
    in mixing/displacement scavenging combined, 215
    from QUB single-cycle test apparatus, 225-226
    QUB single-cycle test *vs.* Benson-Brandham model, 233-236
    with short-circuited air flow, 216
    SR plots (CFD *vs.* experimental values), 248-250
    vs. SE and TE (Benson-Brandham model), 216-219
    vs. SE (QUB loop-scavenged test, Yamaha DT250 cylinders), 229-230, 232-233
    vs. SE (QUB single-cycle test, Yamaha DT250 cylinders), 227, 228-229
  scavenging efficiency (SE)
    chainsaw engine simulation, 386-389
    defined, 28, 212-213
    effect on flammability (SI engines), 285
    evaluation of test results (discussion), 232-233
    instantaneous SE (conversion from mass to volumetric value), 243
    isothermal scavenging characteristics (tested, Yamaha DT250 cylinders), 227, 228-229
    in perfect displacement scavenging, 213-214
    in perfect mixing scavenging, 214-215
    in mixing/displacement scavenging combined, 215
    in QUB single-cycle gas scavenging tests, 226, 233-236
    racing motorcycle engine simulation, 395-396
    SE plots (CFD *vs.* experimental values), 248-250
    with short-circuited air flow, 216

Scavenging *(continued)*
  scavenging efficiency (SE) *(continued)*
    volumetric scavenging efficiency (tested), 226
    vs. AFR (low emissions engine), 487-489
    vs. SR and TE (Benson-Brandham model), 216-219
    vs. SR (QUB loop-scavenged test, Yamaha DT250 cylinders), 229-230, 232-233
    vs. SR (QUB single-cycle test, Yamaha DT250 cylinders), 227, 228-229
    see also scavenging flow, experimental assessment of *(below)*
  scavenging flow, experimental assessment of
    introduction, 219
    chainsaw engine simulation, 386-389
    comparison with wind-tunnel testing, 223
    dynamic similarity, importance of, 224-225, 226-227
    laminar *vs.* turbulent flow, accuracy of, 223-224
    liquid-filled single-cycle apparatus, 224
    loop, cross, uniflow scavenging compared (QUB apparatus), 227, 229-233
    piston motion, importance of, 223-224
    QUB apparatus and Benson-Brandham models compared, 233-236
    QUB single-cycle gas scavenging apparatus, 224-227
    Sammons' proposal for single-cycle apparatus, 224
    scavenging coefficients, experimental values for, 236
    visualization of (wet-dry methods), 219
    see also Jante test method *(above);* Computer modeling (engine)
  theoretical model with experimental correlation
    introduction and discussion, 237
    chainsaw engine simulation, 386-389
    description and equations of flow, 237-238
    exhaust port purity, calculation of, 238-239
    exhaust port purity, typical curves (eight test cylinders), 239-240
    Sher interpretation of profile linearity, 239
    final cautionary note, 240-241
  uniflow scavenging
    introduction, 11
    advantages *vs.* complexity of, 12
    bore-stroke ratio, optimum, 253
    in diesel engines, 11-12
    engine configurations for, typical, 13
    influence on SE-SR and TE-SR characteristics, 218-219
    port design for. *See* Port design, scavenging
    suitability to long-stroke engines, 253
    tendency to vortex formation, 253
  volumetric scavenging model (in engine simulation)
    introduction, 241
    exit charge temperature, determination of, 241-242
    temperature, trapped air/trapped exhaust gas, 242-243
    temperature differential factor, 241-242
  Yamaha DT250 cylinders scavenging test results
    compared with Benson-Brandham models, 234-235
    exhaust port purity (correlated theoretical model), 239-240
    full-throttle QUB tests, 227, 228-229
    loop-scavenged QUB tests, 229-230, 232, 233
    see also specific valves; Port design; Port timing; Trapping
Scott, Alfred
  and deflector piston design, 10
  Flying Squirrel machines, 1
Sher, E.
  interpretation of scavenging profile linearity, 239
Shock waves, moving (in unsteady gas flow)
  introduction, 201
  First Law of Thermodynamics, application of, 202
  flow diagram, 201
  gas particle velocity, 202
  momentum equation, 201-202
  pressure/density relationships, 201, 204
  temperature/density relationships, 203
  see also Gases, properties of
Short circuiting
  basic two-stroke engine, 8
Silencers/silencing
  silencer design, fundamentals of
    introduction, 548

Silencers/silencing *(continued)*
  silencer design, fundamentals of *(continued)*
    CFD, applicability of to silencer design, 555
    mean square sound pressure, calculation of, 549-550
    unsteady gas dynamics (UGD), importance of, 554-555
    see also Coates, S.W. *(below)*
  absorption type silencers
    configuration, 556, 563
    discussion of, 564
    holes-to-pipe area ratio, 564
    positioning, 564-565
    stabbed *vs.* perforated holes in, 564, 565
  chainsaw engine silencing
    exhaust port profiling, 577-578
    exhaust silencer volume, space constraints on, 573
    exhaust silencer volume *vs.* silencing effectiveness, 575-577
    exhaust system dimensions, general criteria for, 435
    intake port profiling, 577
    intake silencing geometry (various), 557, 570-571
    performance and silencing *vs.* geometry (in take silencer), 571-573
    performance and silencing *vs.* geometry (untuned exhaust silencer), 573-576
    placement of, 435
    radiused porting, effect on noise, 578-579
    silencing by exhaust choking, 573
    spark arrestors, need for, 573
    untuned exhaust system, silencing, 573-575
  Coates, S.W.
    experimental silencer designs, 550-551
    noise spectra (calculated *vs.* measured), 552-554
    theoretical work of, 548-550
  design philosophy, remarks on, 583
  diffusing type silencers
    design examples (from DIFFUSING SILENCER program), 558-560
    Fukuda attenuation equations, 558
    intake silencer, significant dimensions of, 557
    significant dimensions, 555-557
    transmission loss *vs.* frequency, calculation of, 557-558
  laminar flow type silencers
    advantages of (fluid dynamics), 566
    configuration of (QUB), 564, 565
    hydraulic diameter, calculation of, 566
    numerical values for, typical, 566
    overall effectiveness, factors contributing to, 567
  low-pass intake system silencers
    introduction and configuration, 557, 567
    attenuation characteristics, OMC V8 out board, 568
    box volume and pipe diameter, determination of, 569
    design analysis, typical (chainsaw), 569-570
    engine forcing frequency, 568-569
    lowest resonant frequency, significance of, 568
    natural frequency, determination of, 568
  motorcycle engine silencing
    introduction, 579-580
    box volume, effect of, 580
    design example (DIFFUSING SILENCER and SIDE-RESONANT SILENCER programs), 582
    design principles for high specific power output, 581
    materials and fabrication, influence of, 580
    pass-band holes, importance of, 583
  side-resonant type silencers
    attenuation (transmission loss) of, 561
    configuration, 556, 560-561
    design example (from SIDE RESONANT SILENCER computer program), 561-563
    holes-to-pipe area ratio, 564
    natural frequency of, 561
    with slitted central duct, 556, 563
  tuned exhaust system silencing
    discussion, 579-580
    design example: motorcycle engine simulation, 580-583
  untuned exhaust system silencing
    discussion, 573
    design example: chainsaw engine simulation, 573-575
    silencer box volume *vs.* swept volume, 436-437
  see also Exhaust systems; Noise reduction/noise emission

Index

Simulation, engine
See Computer modeling *(various)*; GPB engine simulation model
Smoke, exhaust
  from compression ignition, 288
  general comments on, 13, 288, 465
Spark-ignition systems
  effectiveness of, 285
  in two-stroke engines, 282-284
  see also Combustion processes
Specific time area (A_{sv})
  introduction to
    A_{sv} for exhaust blowdown (measured), 417, 420, 422-423
    A_{sv} for exhaust ports (measured), 420, 422-423
    A_{sv} for inlet ports (measured), 420-421, 423
    A_{sv} for transfer ports (measured), 420-421, 423
    basic geometry of, 417
    derivation of A_{sv}, 419-420, 424-425
    determination of measured A_{sv} values, 424-426
    mass flow rates, equations for, 417-419
    port areas *vs.* crankshaft angle (piston control), 417
    TIMEAREA TARGETS computer program, 423-424
  in chainsaw engine empirical design
    A_{sv} values for (from TIMEAREA TARGETS program), 424
    A_{sv} values for (measured), 426
    cylinder diagram (computer-generated), 418
    engine air flow (DR) *vs.* A_{svx}, 427-430
    engine torque *vs.* A_{svx}, 427, 429
    exhaust port timing, effect on A_{sv} of, 426-427
    exhaust port timing, effect on performance characteristics, 426-428
    hydrocarbon emissions *vs.* A_{svx}, 427-430
    specific fuel consumption *vs.* A_{svx}, 427-430
    transfer port timing, effect on A_{sv} of, 427, 429
    transfer port timing, effect on performance characteristics, 429-430
    typical A_{sv} (computed), 425
  in disc valve empirical design
    introduction, 456-457
    carburetor flow diameter, 458
    conventional *vs.* SI units, 459
    disc valve timing *vs.* induction port area, 457
    maximum port area, 457, 458
    outer port edge radius, 458
    total opening period, 457
  in racing motorcycle engine empirical design
    A_{sv} values for (from TIMEAREA TARGETS program), 424
    A_{sv} values for (measured), 426
    cylinder diagram (computer-generated), 425
  in reed valve empirical design
    introduction, 450-451
    A_{svi}, calculation of, 451
    carburetor flow diameter, estimation of, 453
    design criteria for, 451
    effective reed port area, A_{rp}, 369, 462
    REED VALVE DESIGN computer program, 452, 454-455
    required reed flow, A_{rd}, 369, 452
    vibration/amplitude criteria, 453-454, 455
    concluding remarks, 455-456
Speed of rotation
  piston speed, influence of, 44-45
Squish action
  introduction and discussion
    CFD analysis, discussion of, 325-326
    compression behavior (squish and bowl volumes), 327
    compression process in (ideal, isentropic), 327-328
    example: QUB cross-scavenged engine, 325
    gas mass flow, 328
    squish area *vs.* combustion chamber design, 329-330
    squish kinetic energy, 330, 331-332
    squish pressure ratio, derivation of, 327
    squish pressure *vs.* cylinder and bowl pressures, 327
    squish velocity, derivation of, 328
    state conditions, equalization of, 327
  combustion chamber designs for
    introduction, 331
    central squish system, 338-339
    design example: HEMI-SPHERE CHAMBER program, 337
    in DI diesel engines, 334, 335
    kinetic energy *vs.* chamber type, 331-332
    squish velocity *vs.* chamber type, 331-332

619

Squish action *(continued)*
  combustion chamber designs for *(continued)*
    squish velocity vs. flame speed, 333
  and combustion processes
    burning characteristics, effect on, 333-334
    detonation reduction (and squish velocity), 333-334
    flame velocity (QUB loop-scavenged), 331, 333
    homogeneous charge combustion, 338-339
    stratified charge combustion, 337
  computer programs for
    applicable combustion chamber types, 335-336
    BATHTUB CHAMBER program, 335
    design example: HEMI-SPHERE CHAMBER program, 337
    SQUISH VELOCITY program, 330-331
Steep-fronting
  shock formation in pipes, 62
  see also Propagation/particle velocity (finite amplitude waves in pipes)
Stratified charging (general)
  defined, 286, 288
  introduction and discussion, 337, 496-497
  detonation in, 288
  in diesel engines, 466
  and four-stroking, 288
  in SI engines, 467-469
  and specific fuel consumption, 468
  and squish action, 338-339
  stratified *vs.* homogeneous processes (general discussion), 466-469
  see also Combustion processes; Homogeneous charging (with stratified combustion)
Stratified charging (with homogeneous combustion)
  introduction, 467-468, 497
  airhead charging
    introduction and functional description, 507, 509-510
    bmep (QUB270 engine), 510-511
    bsfc (QUB270 engine), 511
    hydrocarbon emissions (QUB270 engine), 511-512
    QUB270 cross-scavenged engine (description), 510

IFP (Institut Français du Pétrole) engine
  brake specific fuel consumption, 505, 508
  description and functional diagram, 504-506
  hydrocarbon emissions, 506, 508, 509
  NO emissions, 506, 509
  oxygen catalysis, effect of, 506, 509
  photograph of, 507
Ishihara double-piston engine, 504, 505
Piaggio stratified charging engine
  introduction and functional description, 500-501
  benefits and advantages of, 504
  CO emissions, 502-503
  fuel consumption levels, 501-503
  hydrocarbon emissions, 503
QUB stratified charging engine
  air-fuel paths in (diagram), 498
  bmep (at optimized fuel consumption levels), 499, 500
  fuel consumption, optimized, 499
  functional description, 498-499
  stratified *vs.* homogeneous processes (general discussion), 466-469
  see also Combustion processes; Homogeneous charging (with stratified combustion)
Strouhal number, 223
Sulzer (Winterthur)
  marine diesel engines, 4
Supercharged engines. *See* Turbocharged/supercharged engines
Suzuki vehicles, 2
Swept volume
  and brake power output, 44-45
Symmetry/asymmetry
  of port timing events, 15-16
  see also Port timing

Temperature
  AFR *vs.* peak temperatures (burn/unburned zones), 349, 350
  determining exit properties by mass
    introduction, 242
    exit purity (at equality temperature), 243
    exit temperature (as function of trapped air/ exhaust gas enthalpy), 243-244
    SE as function of trapped air/exhaust gas temperature, 243

temperature *vs.* crankshaft angle (chainsaw engine simulation)
  cylinder (closed cycle two-zone chainsaw model), 349-350
  cylinder temperature, pressure, 392, 393
  exhaust system, 390
  intake system temperature, pressure, 390-392
  internal, 386
  scavenge model gases, 388-389
  volumetric scavenging model (in engine simulation)
    introduction, 241
    exit charge temperature, determination of, 241-242
    temperature differential factor, 241-242
Testing, engine. *See* Exhaust emissions/exhaust gas analysis; Performance measurement
Thermal efficiency
  brake thermal efficiency (defined), 37
  of Otto cycle, 32
Thermodynamic (Otto) cycle
  measured *vs.* theoretical values, 31-33
  thermal efficiency of, 32
  work per cycle, 33
Thermodynamic terms, defined
  air-fuel ratio, 29-30
  charging efficiency, 29, 213
  delivery ratio (DR), 27
  heat release (combustion, Q_R), 31, 309
  scavenge ratio (SR), 27
  scavenging efficiency, 28, 212-213
  scavenging purity, 28
  trapped charge mass, 30-31
  trapped fuel quantity, 30-31
  trapping efficiency, 28-29, 213
Thermodynamics of cylinders/plenums. *See under* GPB engine simulation model
Three port engine
  Day's original design, 1
Throttle area ratio, 371
Torque
  Brake torque (defined), 36
  indicated torque (defined), 35
Trapping
  definitions
    charging efficiency (CE) as function of TE and SR, 213
    trapped charge mass, 30-31

trapped compression ratio, 8
trapped fuel mass (at trapping point), 31
trapped mass (total, at trapping point), 31
trapping efficiency (basic), 28-29, 213
exhaust closure
  basic two-stroke engine, 6, 8
trapped compression ratio
  vs. squished kinetic energy (diesel engines), 334, 335
trapping efficiency
  defined (basic), 28-29, 213
  from exhaust gas analysis, 41-42
  measured performance (QUB 400 research engine), 472-473
  in perfect displacement scavenging, 213-214
  in perfect mixing scavenging, 215
  QUB single-cycle gas scavenging apparatus, 224-226
  of simulated chainsaw engine, 479-481
  TE plots (CFD *vs.* experimental values), 248-250
  vs. AFR (low emissions engine), 487-489
  vs. scavenge ratio (Benson-Brandham model), 216-219
trapping point
  in basic two-stroke engine, 8
trapping pressure
  and active-radical combustion, 491
see also Exhaust emissions/exhaust gas analysis; Scavenging
Turbocharged/supercharged engines
  blower scavenging, effect of (three-cylinder engine)
    introduction, 400
    exhaust tuning, 405-407
    open-cycle pressures and charging (cylinders 1-2), 404-405
    temperature and purity in scavenging ports, 405, 406
  compression ignition
    bmep and emissions of (discussion), 531
    Detroit Diesel Allison Series 92 diesel engine, 14
    in European four-stroke on-road engines, 531
    four-cylinder DI diesel, loop scavenging design for, 272-273
  configuration of, typical, 15
  exhaust emissions of, 14

Turbocharged/supercharged engines *(continued)*
  functional description, basic, 13-14
  loop scavenging port design, external (blown engines)
    BLOWN PORTS (computer program), 270-273
    designs for (discussion), 269-270
    four-cylinder DI diesel, design for, 272-273
    typical layout, 270, 271
  Roots blower in (fuel-injected engine), 13-14, 15
Turbulence
  advantages of (in alternative fuel combustion), 339
Two-stroke engine (general)
  advantages of, 5
  applications, typical
    automobile racing, 2-3
    handheld power tools, 1-2, 3
    outboard motors, 2, 4
    see also Compression ignition engines
  engine geometry, elements of
    compression ratio, 22
    computer programs used (introduction), 20-21
    LOOP ENGINE DRAW (computer program), 23, 24, 25
    PISTON POSITION (computer program), 23, 24
    piston position *vs.* crankshaft angle, 21, 22-23
    QUB CROSS ENGINE DRAW, 25-26
    swept volume/trapped swept volume, 21
    units used, 20
  fundamental operation (basic engine)
    charge transfer/scavenging, 6, 7-8
    exhaust closure (trapping), 6, 8
    fuel induction, alternatives for, 7
    functional diagram, 6
    geometric compression ratio (defined), 8
    initial exhaust (blowdown), 6, 7
    power stroke/induction, 6-7
    short circuiting, 8
    trapped compression ratio (defined), 8
    trapping point, 8
  opinions regarding, 4-5
  port timing, elements of
    introduction, 18
    disc-valved engine, 18, 19, 20
    piston-ported engine, 18-19
    reed-valved engine, 18, 20
    timing diagrams, typical, 18
  see also specific valves; Combustion processes; Port design; Scavenging

Valves/valving
  introduction to
    disc valves, 16, 17
    function of, 15-16
    poppet valves, 16
    port timing characteristics, typical (reed and disc valves), 18
    reed valves, 17
    typical configuration (disc valve), 16
    typical configuration (reed valve), 16, 19
  see also specific valves; Computer modeling; Port design; Port timing
Velocette motorcycle engine, 490
Vespa motor scooter engine
  Batoni performance maps (CO emissions), 476-478
Vibe, I.I.
  mass fraction burned analysis (SI engine), 295, 309-310
Vortex formation
  in uniflow scavenging, 253

Wallace, F.J.
  simulation of engines (using unsteady gas flow), 143
"Water-gas" reaction, 297, 346
  see also Combustion processes; Exhaust emissions/exhaust gas analysis
Wendroff, B.
  Lax-Wendroff computation time, 191-192
  simulation of engines (using unsteady gas flow), 142
Winthrow, L.
  expression for incremental heat release (SI engines), 293
Work
  pressure wave propagation (straight pipes)
    friction work and heat generated, 83
  work done during engine cycle (GPB model), 168-169

Index

work per thermodynamic (Otto) cycle, 33
see also Power output
Woschni, G.
  coefficient of heat transfer (determination of), 164
  heat transfer equation (compression-ignition engines), 305

Yamaha
  DT250 cylinders, scavenging test results for
    brake mean effective pressure, 227-228
    brake specific fuel consumption, 227-228
    comparison with Benson-Brandham models, 234-235
    full-throttle QUB scavenging flow test, 227-229
    isothermal scavenging characteristics (tested), 227-229
    loop-scavenged QUB test results for, 229-230, 232-233
  modified Yamaha cylinders
    CFD charge purity plots/analysis for, 246-248
    SE *vs.* SR plots for, 248-250

Zchopau (MZ motorcycles)
  valve design, 16-17, 456
Zwickau Technische Hochschule
  ram-tuned liquid injection research
    description and functional block diagram, 514-515
    light load testing at QUB, 526-529
    in Outboard Marine (OMC) engine, 520, 521
    spray pattern, 516